Cryobiology

Cryobiology

Edited by Michael Cedric

SYRAWOOD PUBLISHING HOUSE
New York

Published by Syrawood Publishing House,
750 Third Avenue, 9th Floor,
New York, NY 10017, USA
www.syrawoodpublishinghouse.com

Cryobiology
Edited by Michael Cedric

International Standard Book Number: 978-1-68286-519-4 (Hardback)

Cataloging-in-Publication Data

Cryobiology / edited by Michael Cedric.
p. cm.
Includes bibliographical references and index.
ISBN 978-1-68286-519-4
1. Cryobiology. 2. Biology. I. Cedric, Michael.
QH324.9.C7 C79 2018
571.464 5--dc23

Printed in China.

TABLE OF CONTENTS

PREFACE

Every book is initially just a concept; it takes months of research and hard work to give it the final shape in which the readers receive it. In its early stages, this book also went through rigorous reviewing. The notable contributions made by experts from across the globe were first molded into patterned chapters and then arranged in a sensibly sequential manner to bring out the best results.

Cryobiology is an upcoming field of science that has undergone rapid development over the past few decades. The effect that low temperature has on any biological material can be studied under the subject of cryobiology. The techniques used in cryobiology are tissue preservation and transplantation, cryosurgery, embryo and gamete preservation, etc. Cryopreservation primarily aims at preserving living cells. This book elucidates the concepts and innovative model around prospective developments with respect to cryobiology and cryopreservation. Scientists and students actively engaged in this field will find this book full of crucial and unexplored concepts.

It has been my immense pleasure to be a part of this project and to contribute my years of learning in such a meaningful form. I would like to take this opportunity to thank all the people who have been associated with the completion of this book at any step.

Editor

1

Effects of the Insemination of Hydrogen Peroxide-Treated Epididymal Mouse Spermatozoa on γH2AX Repair and Embryo Development

Jianfeng Xiao[1], Yanmei Liu[2], Zhiling Li[1]*, Yongcui Zhou[1], Hong Lin[1], Xiaoyan Wu[1], Man Chen[1], Wanfen Xiao[1]

1 Reproductive Center, The First Affiliated Hospital of Shantou University Medical College, Shantou University, Shantou, Guangdong, People's Republic of China,
2 Reproductive Center, Women and Children Healthcare Hospital, Qinzhou, Guangxi, People's Republic of China

Abstract

Background: Cryopreservation of human semen for assisted reproduction is complicated by cryodamage to spermatozoa caused by excessive reactive oxygen species (ROS) generation.

Methods and Findings: We used exogenous ROS (H_2O_2) to simulate cryopreservation and examined DNA damage repair in embryos fertilized with sperm with H_2O_2-induced DNA damage. Sperm samples were collected from epididymis of adult male KM mice and treated with capacitation medium (containing 0, 0.1, 0.5 and 1 mM H_2O_2) or cryopreservation. The model of DNA-damaged sperm was based on sperm motility, viability and the expression of γH2AX, the DNA damage-repair marker. We examined fertility rate, development, cell cleavage, and γH2AX level in embryos fertilized with DNA-damaged sperm. Cryopreservation and 1-mM H_2O_2 treatment produced similar DNA damage. Most of the one- and two-cell embryos fertilized with DNA-damaged sperm showed a delay in cleavage before the blastocyst stage. Immunocytochemistry revealed γH2AX in the one- and four-cell embryos.

Conclusions: γH2AX may be involved in repair of preimplantation embryos fertilized with oxygen-stressed spermatozoa.

Editor: John R. Battista, Louisiana State University and A & M College, United States of America

Funding: This work was supported by the National Natural Science Foundation of China (No. 81070542 and No. 30872771), the Natural Science Foundation of Guangdong Province of China (No. 10151503102000020 and No. 81515031102000010), the Medical Scientific Research Foundation of Guangdong Province, China (B2010214), and the Foundation for Distinguished Young Talents in Higher Education of Guangdong, China (LYM10076). The funders had no role in study design, data collection and analysis, decision to publish, or preparation of the manuscript.

Competing Interests: The authors have declared that no competing interests exist.

* E-mail: stlizhiling@126.com

These authors contributed equally to this work.

Introduction

Assisted reproduction centers, andrology laboratories and sperm banks cryopreserve human semen before assisted-reproduction treatment. Such cryopreservation provides patients with "fertility insurance" for intrauterine insemination, especially with cryopreserved donor semen, conventional *in vitro* fertilization or intracytoplasmic sperm injection [1]. However, during cryopreservation, spermatozoa are exposed to physical and chemical stress that results in adverse changes in membrane lipid composition, motility, viability and acrosome status of sperm [2,3,4]. These changes reduce the fertilization ability of human spermatozoa after cryopreservation [5,6].

Cold shock during sperm cryopreservation is associated with oxidative stress and reactive oxygen species (ROS) generation [7]. ROS-induced damage to spermatozoa is mediated by oxidative attack of the bis-allylic methylene groups of sperm phospholipid-bound polyunsaturated fatty acids, thus leading to lipid peroxidation [8,9]. The effects of lipid peroxidation include irreversible loss of motility, damage to sperm DNA or deficiencies in oocyte penetration and sperm–oocyte fusion [1,10].

The marker of DNA damage is the formation of γH2AX focus, which is phosphorylated by ataxia telangiectasia mutated kinase. H2AX is one of the H2A variants with a conserved Ser-Gln-Glu motif at the C terminus; its phosphorylation plays a key role in DNA damage [11,12]. γH2AX may promote recombination and conformational changes of chromatins, which may prevent the premature separation of broken ends, a function that would safeguard against harmful chromosome rearrangements [12].

Although numerous reports have described DNA damage to sperm caused by exogenous ROS (H_2O_2) or cryopreservation, only sperm motility variables and intracellular ROS and nitric oxide levels were examined [13,14,15]. The DNA damage to sperm needs to be repaired after fertilization to ensure correct genetic message transfer to progeny. Previous study showed that exogenous H_2O_2 increases intracellular ROS levels in spermatozoa [16].

Double-strand breaks are the most deleterious DNA lesions, which, if left unrepaired, may have severe consequences for cell survival, because they lead to chromosome aberrations, genomic instability, or cell death. The formation of DSBs activates many factors, including phosphorylation of the histone variant H2AX, producing γH2AX. Analysis of γH2AX expression can be used to detect the genotoxic effect of substances [11], for a repair marker of DNA damage. We aimed to examine embryos fertilized with sperm showing cryopreservation-similar DNA damage and DNA repair in terms of the organization and biological character of early embryos. We used an external source of ROS (H_2O_2) to simulate cryopreservation and observe the repair of DNA damage by the DNA-damage marker γH2AX in embryos fertilized with sperm with oxygen-induced DNA damage.

Results

DNA-damaged Spermatozoa Model Induced by H_2O_2

Sperm was exposed to capacitation medium containing 0, 0.1, 0.5 or 1 mM H_2O_2. The percentage motility and viability of sperm was lower for cryopreserved than fresh sperm ($P<0.05$; Fig. 1). The percentage motility and viability was lower with cryopreservation than with 0.1, 0.5 and 1 mM H_2O_2 ($P<0.05$) and decreased with increasing H_2O_2.

Induced DNA Damage in Sperm

We examined the presence of the DNA-damage repair marker γH2AX in spermatozoa by immunocytochemistry. γH2AX expression was absent in fresh sperm but was present in cryopreserved sperm, and the level increased with increasing concentration of H_2O_2 (Fig. 2). The DNA damage, as determined by γH2AX level, was close with cryopreservation and 1-mM H_2O_2 treatment, so we chose 1 mM H_2O_2 as the DNA damage model induced by oxidative stress in the following experiments.

Effect of H_2O_2 on Fertility and Embryo Development of Mice

Spermatozoa fertility and embryo development was similar with fresh spermatozoa and sperm treated with 1-mM H_2O_2 ($P>0.05$, Table 1).

Effect of H_2O_2 on Cleavage of Embryos

No embryos had cleaved 12 h after insemination with fresh spermatozoa and sperm treated with 1 mM H_2O_2. At 16 h, the percentage of cleaved embryos was 69% and 58% for fresh spermatozoa and sperm treated with 1 mM H_2O_2, respectively, and was 87% and 73% at 36 h. The cleavage of one- to two-celled embryos did not change between 36 and 48 h for fresh spermatozoa but peaked at 40 h, at 82%, for sperm treated with 1 mM H_2O_2.

At 32 h after insemination, 4% of four-celled embryos had cleaved with fresh spermatozoa. However, no embryos had cleaved with sperm treated with 1 mM H_2O_2. At 36 h, the percentage cleaved embryos was 11% and 7% for fresh spermatozoa and sperm treated with 1 mM H_2O_2, respectively, and at 48 hr, the percentage cleaved peaked at 84% and 80%, respectively. Thus, cleavage of one- and two-celled embryos was delayed in part in sperm treated with 1 mM H_2O_2 as compared with fresh spermatozoa after insemination (Fig. 3).

Expression of γH2AX in Embryos Fertilized with H_2O_2-treated Sperm

We examined the presence of γH2AX in fertilized mouse embryos. In embryos fertilized with fresh spermatozoa, no embryos showed γH2AX expression. However, γH2AX expression was detected in one- and four-celled embryos fertilized with sperm treated with 1-mM H_2O_2 (Fig. 4). Therefore, the DNA repair mechanism was functional in one- and four-celled embryos fertilized with oxygen-stressed sperm.

Discussion

Cryopreservation of human semen for assisted reproduction is complicated by cryodamage to spermatozoa caused by excess ROS generation. We used exogenous ROS (H_2O_2) to simulate cryopreservation and examined DNA damage repair in mouse embryos fertilized with mouse sperm with H_2O_2-induced DNA damage. Most of the one- and two-celled embryos fertilized with sperm treated with H_2O_2 showed a delay in cleavage before the blastocyst stage. Immunocytochemistry revealed the expresison of the DNA-repair marker γH2AX in the one- and four-celled

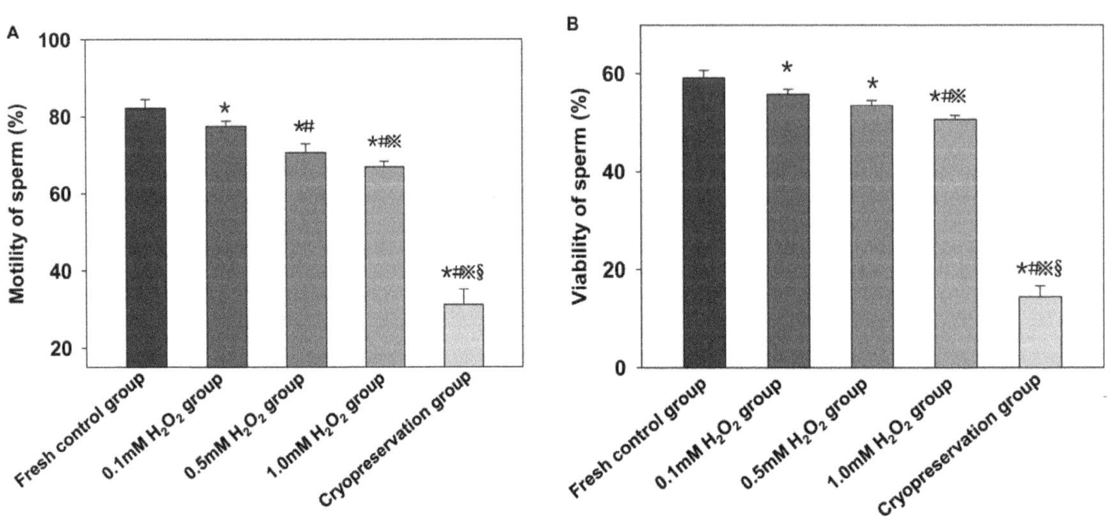

Figure 1. Motility of mouse spermatozoa (A) and viability of spermatozoa with different treatments (B). Data are mean±SD, *$P<0.05$ compared with fresh control group, # $P<0.05$ compared with 0.1 mM H_2O_2, ? $P<0.05$ compared with 0.5 mM H_2O_2, §$P<0.05$ compared with 1 mM H_2O_2.

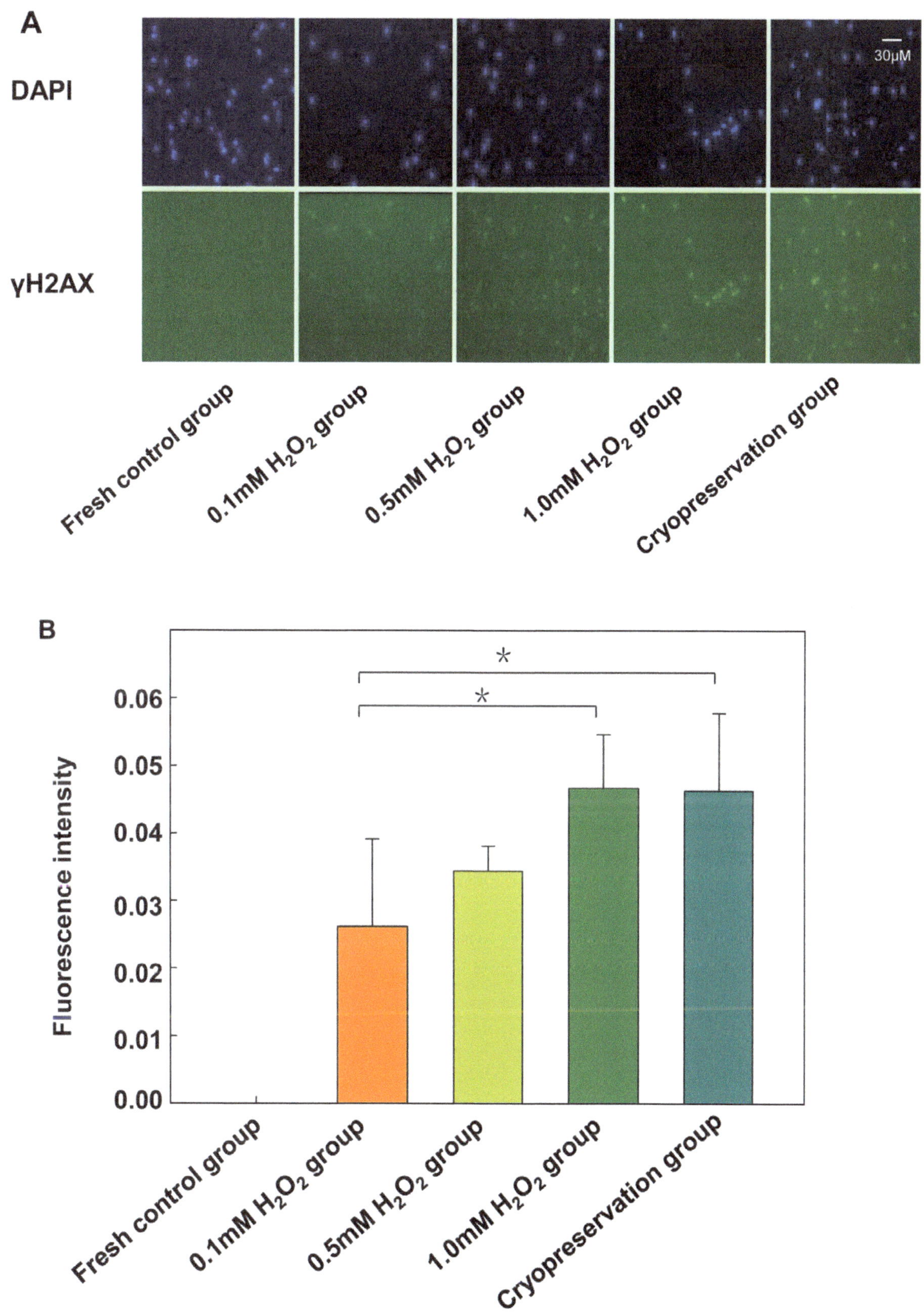

Figure 2. Fluorescence intensity of γH2AX in nuclei among fresh control, H_2O_2 and cryopreserved sperm (A) and quantification (B). Data are mean±SD, $*P<0.05$ compared with 0.1 mM H_2O_2. DAPI = staining of nuclei.

Table 1. Fertilization *in vitro* and development of mouse embryos with fresh spermatozoa and sperm treated with 1-mM H_2O_2.

Treatment	Total eggs	Fertilized eggs (%)	Two-celled embryos (%)	Four-celled embryos (%)
Fresh spermatozoa	97	52 (53.6)	45 (86.5)	38 (84.4)
1-mM H_2O_2	106	51 (48.1)	42 (82.4)	34 (81.0)

embryos. γH2AX may be involved in DNA repair of preimplantation embryos fertilized with oxygen-stressed spermatozoa.

The cryopreservation of human semen plays an important role in the clinical management of male infertility. However, cryopreservation has been found to exacerbate damage to sperm, with reduced sperm motility and viability [1]. Vitality, chromatin normality, morphologic features, and membrane integrity of spermatozoa were severely altered by a freeze–thaw technique [17]. During the freeze–thaw process, excessive ROS was generated. Oxidative stress has long been implicated as major etiological factor in DNA damage in sperm.

ROS play a key role in the modulation of sperm function in the physiologic and pathologic environment. Physiologic concentrations of ROS are necessary to maintain normal sperm function. However, high levels can cause sperm dysfunction [18]. We found the percentage motility and viability of spermatozoa decreased significantly with cryopreservation, with no difference between fresh sperm and treatment with 0.1, 0.5 and 1 mM H_2O_2. However, DNA damage was close between cryopreservation and 1 mM H_2O_2 treatment, so we chose 1 mM H_2O_2 as a DNA-damage model induced by oxidative stress in the experiments.

Semen ROS are generated by spermatozoa (especially defective or immature sperm) and semen leukocytes [18,19,20,21,22]. In contrast to somatic cells, sperm are vulnerable to oxidative stress because of their unique membrane structures, as well as limited antioxidants or protective enzymes [23]. Spermatozoa and seminal plasma possess an antioxidant system of taurine, reduced glutathione (GSH), glutathione peroxidase (GSH-Px), catalase and superoxide dismutase to prevent oxidative damage [24]. However, semen selected by gradient centrifugation is necessary in assisted reproductive technology. Seminal plasma, sperm fragmentation and pathogenic microorganisms are cleared through assisted reproductive technology. Seminal plasma is the source of antioxidant systems, so sperm is in a superoxidative state after semen is selected. [25,26].

We found no difference in fertilized egg rate and two- or four-celled embryos fertilized with fresh spermatozoa or 1-mM H_2O_2-treated sperm. In our previous study, the fertilized egg rate with cryopreserved sperm was lower than with fresh spermatozoa ($P<0.05$, unpublished data). The reasons for the loss of fertility are various, including factors affecting the proportion of survivors (e.g., cold shock susceptibility, cooling rate, diluent composition and osmotic stress) and factors influencing the functional status of survivors (e.g., membrane stability, oxidative damage, membrane receptor integrity, nuclear structure) [27]. The sperm motility rate and viability was lower with cryopreservation than with fresh sperm, with no difference between fresh sperm and 1-mM H_2O_2-treated sperm. However, the DNA damage was the same between cryopreserved and 1-mM H_2O_2–treated sperm on immunocytochemistry. DNA damage in sperm is negatively related with natural conception. The possibility of natural conception is zero with sperm DNA damge >30% [28]. The effects of sperm DNA damage on fertilization and embryonic development are still controversial. Sperm DNA damage increases the failure probability of *in vitro* fertilisation [29]. As well, DNA double-strand break is negatively related with fertilization rate [30]. Sperm DNA fragmentation seems to affect embryo post-implantation development in intracytoplasmic sperm injection: high sperm DNA fragmentation can compromise embryo viability, thus resulting in pregnancy loss [31].

We found no difference in fertilization rate or percentage of fertilized two- and four-celled embryos with fresh sperm and 1-mM H_2O_2-treated sperm. However, the time of cleavage to the two- and four-cell stage was delayed with 1-mM H_2O_2 treatment as compared with fresh sperm from 12 to 48 h after insemination. The cell checkpoint mechanism may play a role in embryo development. After sperm had been treated with 1 mM H_2O_2, phosphorylation of γH2AX was detected in one- and four-celled embryos, which suggests that the G2/M checkpoint and DNA repair mechanisms function in these embryos. The presence of γH2AX suggests that DNA repair mechanisms are functional in one- and four-celled embryos. γH2AX is required to maintain repair proteins at sites of DNA damage [32]. We did not observe phosphorylation of H2AX at one-, two-, four- and eight-celled embryos fertilized with fresh sperm.

Mature sperm cannot repair DNA damage. So the DNA damage must be repaired after fertilization to ensure the integrity of the DNA transferred to offspring [33]. The zygote mainly generates non-translated transcripts [34,35]. In the mouse, transcription becomes fully functional during zygotic gene activation (ZGA) in the G2-phase of two-celled embryos [36,37]. Hence, the mRNA and protein stored in the maternal gamete, the oocyte, must provide the zygote with the necessary molecules to perform all basic cellular functions until ZGA [38]. We observed this process indirectly. The γH2AX foci appeared in one-celled embryos and disappeared in two-celled embryos, then reappeared in four-celled embryos. The G2/M checkpoint mechanism seems to function insufficiently in one-celled embryos, but two-celled embryos are the stage of the maternal genome transforming into a zygotic genome. Perhaps the DNA damage caused by 1 mM H_2O_2 was repaired in part before ZGA. However, other DNA damage was not repaired, so the DNA damage was not enough to begin embryonic DNA damage repair. The unrepaired DNA damage signal was amplifed during embryonic development and activated the repair pathway after ZGA. The cleavage delay of the one-celled embryos to the two-celled stage and the two-celled embryos to the four-celled stage suggested that the G2/M checkpoint mechanism is operative.

Conclusions

The repair of DNA damage that originates from oxidative-stressed sperm after fertilization depends on the mRNA and protein stored in the maternal gamete. The DNA damage caused by 1-mM H_2O_2 in sperm was repaired in part before ZGA. Other DNA damage is repaired during the early pre-zygotic stage of life when DNA repair pathways are activated because we found γH2AX expression in four-celled embryos fertilized with oxygen-stressed sperm.

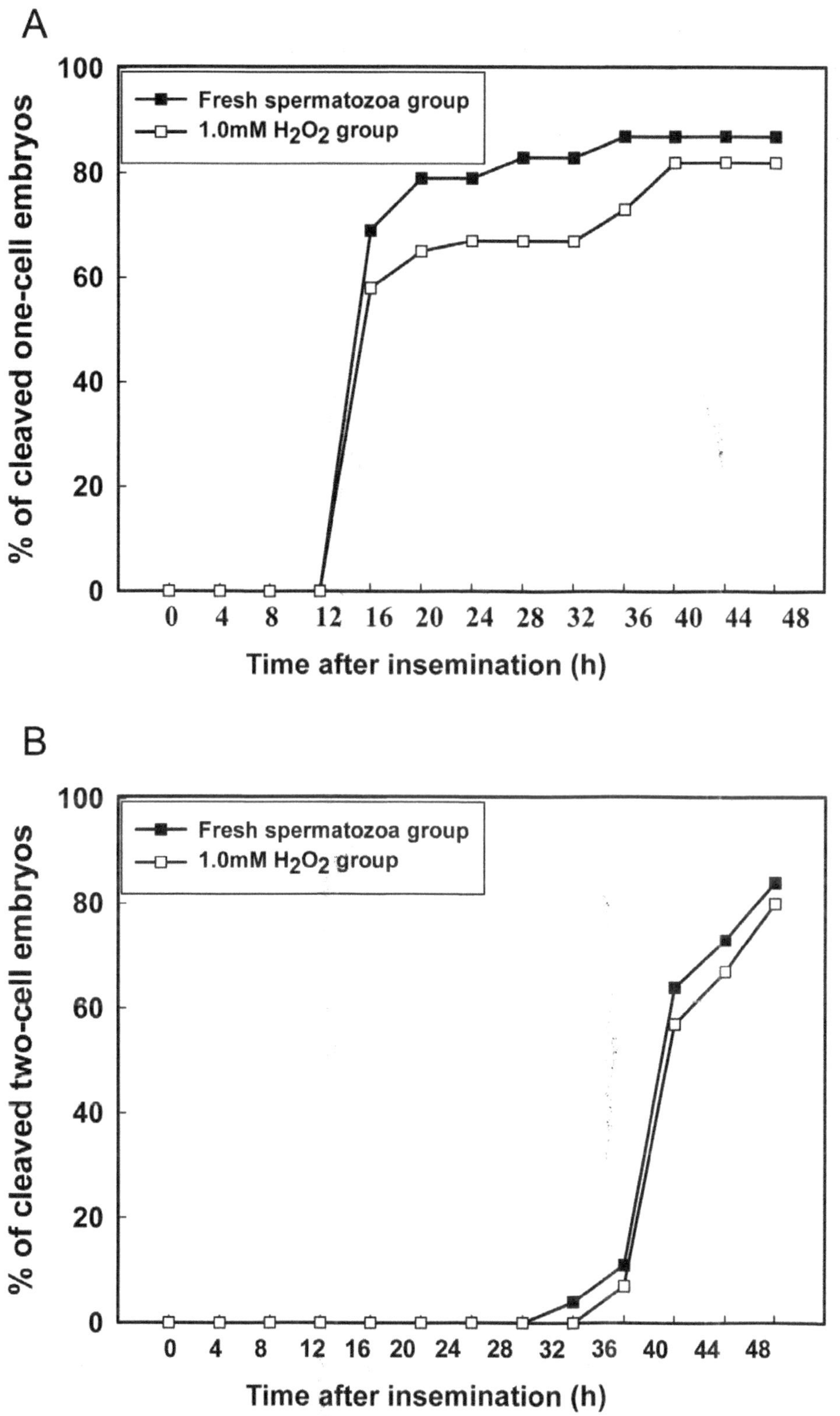

Figure 3. Delay in cleavage of one- and two-celled embryos after fertilization. Sperm was treated with 1 mM H_2O_2 and the time of embryo cleavage was examined. A: Cleavage rate of one- to two-celled embryos and B: two- to four-celled embryos.

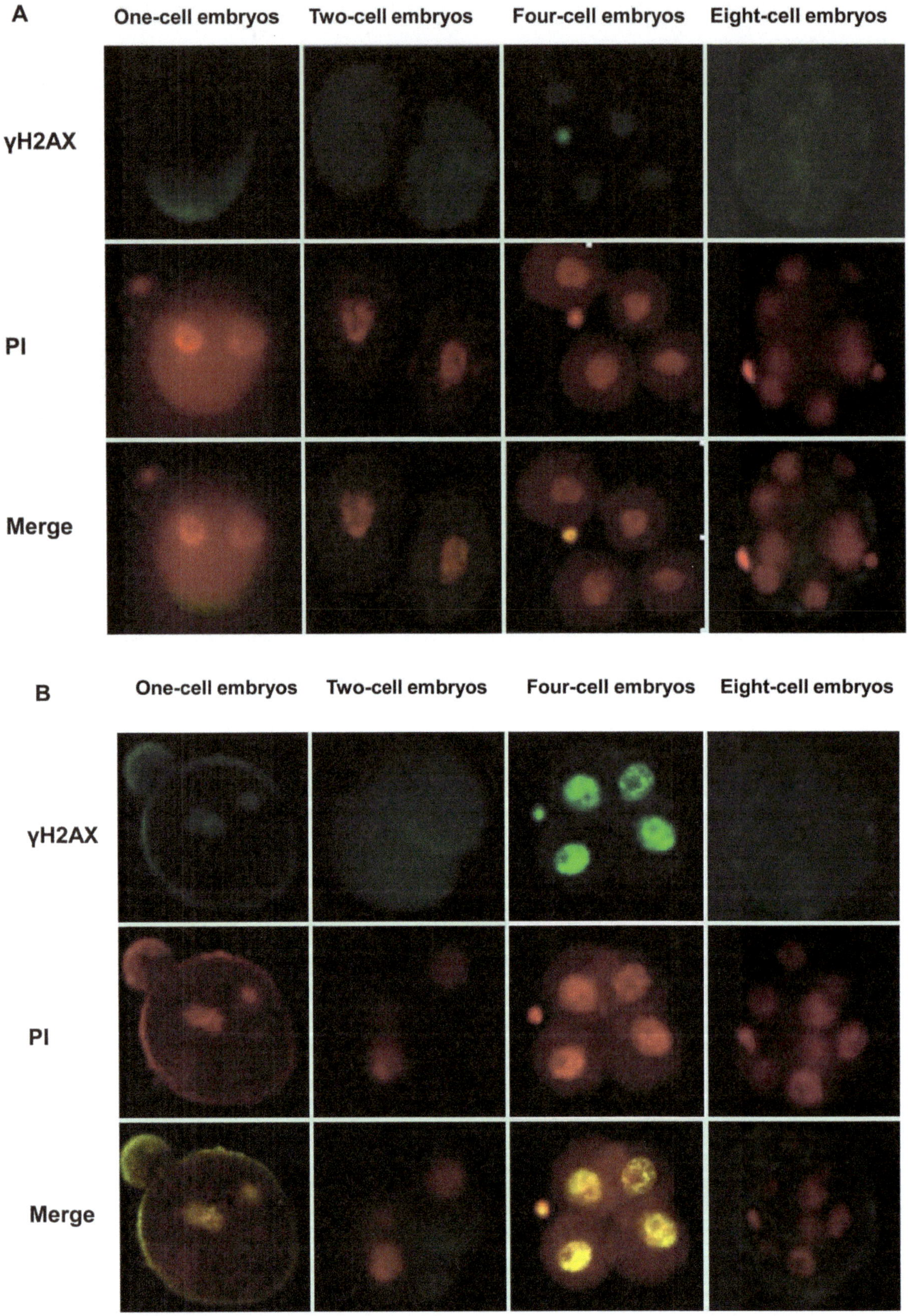

Figure 4. Immunohistochemistry of γH2AX expression in embryos. Changes in γ-H2AX expression in embryos fertilized by A: fresh sperm and B: sperm treated with H_2O_2. PI = propidium iodide staining.

Materials and Methods

Material and Reagents

All chemicals were of analytical grade. The mouse antibody for anti-phospho-histone H2AX (γH2AX; Ser 139) was from Millipore (USA) and goat anti-mouse Alexa Fluor 488 secondary antibody was from Invitrogen (Carlsbad, CA, USA). Human tubal fluid-HEPES (HTF-HEPES) was from SAGE (USA). Pancreatin was from Sigma (St. Louis, MO, USA). Phosphate buffered saline (PBS) was prepared in milli-Q water: 136.9 mM NaCl, 2.7 mM KCl, 0.9 mM $CaCl_2$, 0.5 mM $MgCl_2$, 7 mM Na_2HPO_4, 1.25 mM NaH_2PO_4, 1.5 mM KH_2PO_4, and 1 g/l PVA (290 mOsm/kg; pH 7.2). R18S3 was prepared with 1.8 g skim milk powder soluted into milli-Q water for 6% solution, then centrifuged at 4°C 15,000 g for 1 h. For 20 ml supernatant, 12.5 ml milli-Q water and 7.2 g raffinose was added, then milli-Q water was added to 40 ml; the mixture was agitated to solute raffinose, filtered, subpackaged and stored at −20°C. Pancreatin solution was prepared by the addition of 0.1% pancreatin to PBS, adjusting pH to 3.0 with HCl, filtering, subpackaging and storage at 4°C. Sperm capacitation liquid was prepared by adding 1.5% bovine serum albumin (BSA) to HTF solution. For fertilization liquid, 0.4% BSA was added to HTF solution. For embryo culture medium, 0.4% HTF and 10% fetal bovine serum was added to HTF solution. H_2O_2 sperm capacitation liquid was prepared by adding H_2O_2 to a final concentration of 0.1, 0.5 and 1 mM. T-PBS involved adding 0.05% Tween20 to PBS. Stationary liquid involved adding 4% paraformaldehyde to PBS. Membrane liquid was prepared by adding 0.2% Triton X-100 to PBS. Blocking buffer was prepared by adding 0.05% Tween20, 3% BSA and 10% goat serum to PBS. Sperm capacitation liquid, fertilization liquid and embryo culture medium were placed at 37°C in a 5% CO_2 incubator for 4 h.

Epididymal Sperm

Adult male KM mice (4–6 weeks old) **were** obtained from the animal center of Shantou University Medical College and treated in compliance with The Guide for the Care of Use of Laboratory Animals by the US National Institutes of Health (NIH Publication No. 85–23, revised 1996) and the rules of the National Animal Protection of China. All experimental protocols were approved by the Laboratory Animal Ethics Committee of our institution (SUMC2011-107). The study was approved by the Institutional Animal Care and Use Committee of Shantou University Medical College. Sperm was obtained from the caudae epididymis of mice and exposed to capacitation medium (containing 0, 0.1, 0.5 and 1 mM H_2O_2) at 37? in a 5% CO_2 incubator for 1.5 h.

Semen Cryopreservation

The cryoprotectant was thawed, mixed, and added to 35-mm culture dishes and warmed at 37°C in a 5% CO_2 incubator. Cauda epididymis tissue was put in 1-ml warming cryoprotectant and extruded slightly to ensure no sperm. Spermatozoa were collected from the cauda epididymis and cultured at 37°C in a 5% CO_2 incubator for 10 to 15 min. Sperm variables (density, motility rate and vitality) were recorded by light microscopy (×400 magnification). Spermatozoa were mixed with cryoprotectant and 100 μl was packed into 1.8-ml cryovials (Nalge Co., Rochester, New York) stored in a refrigerator at 4°C for 30 min. The samples were then frozen as follows: aliquots were suspended in liquid nitrogen vapor (8–10 cm above the level of liquid nitrogen at −80°C) for 10 min, plunged into liquid nitrogen (−196°C) completely, and stored in liquid nitrogen until use [1].

Semen Analysis

Sperm motility was assessed by phase-contrast microscopy (100× magnification) with a warm stage maintained at 37°C. A 5-μl drop of semen was wet-mounted on a microscope slide and covered with a coverslip. Sperm motility was estimated in 3 different microscopy fields for each semen sample. The mean of the 3 successive estimations was recorded as the final motility score [24].

Collection and Culture of Oocytes and Embryos

Fully grown germinal vesicle oocytes were collected from the ovaries of mice 13 to 15 h after injecting 10 IU human chorionic gonadotropin and 10 IU of the mate's serum gonadotropin ahead 48 h. Cumulus oocytes were collected in PBS, then moved to fertilization liquid with use of a pipette and washed. The oocytes were imseminated with spermatozoa that had been preincubated for 1.5 h in sperm capacitation liquid. At 4 h after insemination in HTF medium, the embryos were washed 3 to 5 times, then incubated.

Immunocytochemistry

A 5-μl sperm suspension was placed on slides, covered with polylysine and fixed with 4% paraformaldehyde for 15 min, washed with T-PBS 3 times, permeabilized with 0.2% Triton X-100 at room temperature for 30 min, washed with T-PBS 3 times, covered with goat serum working solution at room temperature for 2 h, incubated with primary antibody anti-γH2AX (1:1000 dilution) for at 4°C for 12 to 15 h, then with secondary FITC-conjugated goat anti-mouse IgG antibody (1:500; Santa Cruz Biotechnology, Santa Cruz, CA, USA) at 37°C for 1 h. The immunostained sperm were washed with T-PBS, counterstained with 4′,6-diamidino-2-phenylindole (DAPI), and observed under an Olympus FluoView FV1000 confocal microscope (Olympus Inc., Japan).

Embryos were washed with PBS and the zona pellucida was removed by digestion with pancreatin (0.1%) for 30 s, then embryos were fixed with 4% paraformaldehyde in T-PBS for 30 min, permeabilized for 30 min at room temperature with T-PBS containing 0.1 Triton X-100 and 3 mg/ml BSA, stained with anti-phospho-γH2AX antibody for 1 h at 37°C, then secondary FITC-conjugated antibody. The immunostained embryos were washed with T-PBS containing 0.1% Triton X, counterstained with propidium iodide, and observed under an Olympus FluoView FV1000 confocal microscope.

Stastistical Analysis

Results are expressed as mean±SD. Data analysis involved use of SPSS 13.0 (SPSS Inc, Chicago, IL). Comparisons between 2 groups involved chi-square test and more than 2 groups one-way ANOVA. $P<0.05$ was considered statistically significant.

Acknowledgments

We are grateful for the support from Center for Neuroscience, Shantou University Medical College for the utilization of the laser confocal microscopy.

Author Contributions

Conceived and designed the experiments: JX ZL YZ. Performed the experiments: YL MC WX. Analyzed the data: HL XW. Contributed reagents/materials/analysis tools: YL MC. Wrote the paper: JX YL ZL.

References

1. Li Z, Lin Q, Liu R, Xiao W, Liu W (2010) Protective effects of ascorbate and catalase on human spermatozoa during cryopreservation. J Androl 31: 437–444.
2. O'Connell M, McClure N, Lewis SE (2002) The effects of cryopreservation on sperm morphology, motility and mitochondrial function. Hum Reprod 17: 704–709.
3. Schiller J, Arnhold J, Glander HJ, Arnold K (2000) Lipid analysis of human spermatozoa and seminal plasma by MALDI-TOF mass spectrometry and NMR spectroscopy - effects of freezing and thawing. Chem Phys Lipids 106: 145–156.
4. Alvarez JG, Storey BT (1993) Evidence that membrane stress contributes more than lipid peroxidation to sublethal cryodamage in cryopreserved human sperm: glycerol and other polyols as sole cryoprotectant. J Androl 14: 199–209.
5. Mack SR, Zaneveld LJ (1987) Acrosomal enzymes and ultrastructure of unfrozen and cryotreated human spermatozoa. Gamete Res 18: 375–383.
6. Martinez-Soto JC, de DiosHourcade J, Gutierrez-Adan A, Landeras JL, Gadea J (2010) Effect of genistein supplementation of thawing medium on characteristics of frozen human spermatozoa. Asian J Androl 12: 431–441.
7. Chatterjee S, Gagnon C (2001) Production of reactive oxygen species by spermatozoa undergoing cooling, freezing, and thawing. Mol Reprod Dev 59: 451–458.
8. Aitken RJ (1989) The role of free oxygen radicals and sperm function. Int J Androl 12: 95–97.
9. Alvarez JG, Storey BT (1995) Differential incorporation of fatty acids into and peroxidative loss of fatty acids from phospholipids of human spermatozoa. Mol Reprod Dev 42: 334–346.
10. Aitken RJ (1995) Free radicals, lipid peroxidation and sperm function. Reprod Fertil Dev 7: 659–668.
11. Podhorecka M, Skladanowski A, Bozko P (2010) H2AX Phosphorylation: Its Role in DNA Damage Response and Cancer Therapy. J Nucleic Acids 2010.
12. Adiga SK, Toyoshima M, Shimura T, Takeda J, Uematsu N, et al. (2007) Delayed and stage specific phosphorylation of H2AX during preimplantation development of gamma-irradiated mouse embryos. Reproduction 133: 415–422.
13. du Plessis SS, McAllister DA, Luu A, Savia J, Agarwal A, et al. (2010) Effects of H(2)O(2) exposure on human sperm motility parameters, reactive oxygen species levels and nitric oxide levels. Andrologia 42: 206–210.
14. Thomson LK, Fleming SD, Aitken RJ, De Iuliis GN, Zieschang JA, et al. (2009) Cryopreservation-induced human sperm DNA damage is predominantly mediated by oxidative stress rather than apoptosis. Hum Reprod 24: 2061–2070.
15. Garg A, Kumaresan A, Ansari MR (2009) Effects of hydrogen peroxide (H2O2) on fresh and cryopreserved buffalo sperm functions during incubation at 37 degrees C in vitro. Reprod Domest Anim 44: 907–912.
16. Mahfouz R, Aziz N, Sharma R, Bykova M, Sabanegh E, et al. (2008) Assessment of intracelular human sperm reactive oxygen species after hydrogen peroxide exposure using four different probes. Fertility and sterility 90: S320–S321.
17. Hammadeh ME, Szarvasy D, Zeginiadou T, Rosenbaum P, Georg T, et al. (2001) Evaluation of cryoinjury of spermatozoa after slow (programmed biological freezer) or rapid (liquid nitrogen vapour) freeze-thawing techniques. J Assist Reprod Genet 18: 364–370.
18. Aitken J, Fisher H (1994) Reactive oxygen species generation and human spermatozoa: the balance of benefit and risk. Bioessays 16: 259–267.
19. Barroso G, Morshedi M, Oehninger S (2000) Analysis of DNA fragmentation, plasma membrane translocation of phosphatidylserine and oxidative stress in human spermatozoa. Human Reproduction 15: 1338–1344.
20. Sati L, Ovari L, Bennett D, Simon SD, Demir R, et al. (2008) Double probing of human spermatozoa for persistent histones, surplus cytoplasm, apoptosis and DNA fragmentation. Reprod Biomed Online 16: 570–579.
21. Muratori M, Piomboni P, Baldi E, Filimberti E, Pecchioli P, et al. (2000) Functional and ultrastructural features of DNA-fragmented human sperm. J Androl 21: 903–912.
22. Gomez E, Buckingham DW, Brindle J, Lanzafame F, Irvine DS, et al. (1996) Development of an image analysis system to monitor the retention of residual cytoplasm by human spermatozoa: correlation with biochemical markers of the cytoplasmic space, oxidative stress, and sperm function. J Androl 17: 276–287.
23. Lewis SE, Simon L (2010) Clinical implications of sperm DNA damage. Hum Fertil (Camb) 13: 201–207.
24. Bucak MN, Tuncer PB, Sariozkan S, Baspinar N, Taspinar M, et al. (2010) Effects of antioxidants on post-thawed bovine sperm and oxidative stress parameters: antioxidants protect DNA integrity against cryodamage. Cryobiology 61: 248–253.
25. Buffone MG, Calamera JC, Brugo-Olmedo S, De Vincentiis S, Calamera MM, et al. (2012) Superoxide dismutase content in sperm correlates with motility recovery after thawing of cryopreserved human spermatozoa. Fertility and sterility 97: 293–298.
26. Calamera JC, Buffone MG, Doncel GF, Brugo-Olmedo S, de Vincentiis S, et al. (2010) Effect of thawing temperature on the motility recovery of cryopreserved human spermatozoa. Fertility and sterility 93: 789–794.
27. Watson PF (2000) The causes of reduced fertility with cryopreserved semen. Anim Reprod Sci 60–61: 481–492.
28. Shen H, Ong C (2000) Detection of oxidative DNA damage in human sperm and its association with sperm function and male infertility. Free radical biology & medicine 28: 529–536.
29. Morris ID, Ilott S, Dixon L, Brison DR (2002) The spectrum of DNA damage in human sperm assessed by single cell gel electrophoresis (Comet assay) and its relationship to fertilization and embryo development. Human Reproduction 17: 990–998.
30. Harrouk W, Codrington A, Vinson R, Robaire B, Hales BF (2000) Paternal exposure to cyclophosphamide induces DNA damage and alters the expression of DNA repair genes in the rat preimplantation embryo. Mutation Research/ DNA Repair 461: 229–241.
31. Borini A, Tarozzi N, Bizzaro D, Bonu MA, Fava L, et al. (2006) Sperm DNA fragmentation: paternal effect on early post-implantation embryo development in ART. Human Reproduction 21: 2876–2881.
32. Fernandez-Capetillo O, Lee A, Nussenzweig M, Nussenzweig A (2004) H2AX: the histone guardian of the genome. DNA Repair (Amst) 3: 959–967.
33. Zini A, Libman J (2006) Sperm DNA damage: clinical significance in the era of assisted reproduction. Canadian Medical Association Journal 175: 495–500.
34. Matsumoto K, Anzai M, Nakagata N, Takahashi A, Takahashi Y, et al. (1994) Onset of paternal gene activation in early mouse embryos fertilized with transgenic mouse sperm. Mol Reprod Dev 39: 136–140.
35. Nothias JY, Miranda M, DePamphilis ML (1996) Uncoupling of transcription and translation during zygotic gene activation in the mouse. EMBO J 15: 5715–5725.
36. Schultz RM (2002) The molecular foundations of the maternal to zygotic transition in the preimplantation embryo. Human Reproduction Update 8: 323–331.
37. Hamatani T, Carter MG, Sharov AA, Ko MS (2004) Dynamics of global gene expression changes during mouse preimplantation development. Dev Cell 6: 117–131.
38. Derijck AA, van der Heijden GW, Giele M, Philippens ME, van Bavel CC, et al. (2006) gammaH2AX signalling during sperm chromatin remodelling in the mouse zygote. DNA Repair (Amst) 5: 959–971.

2

Circulating Angiogenic Cells can be Derived from Cryopreserved Peripheral Blood Mononuclear Cells

Tanja Sofrenovic[1,2], Kimberly McEwan[1,3], Suzanne Crowe[1], Jenelle Marier[1,2], Robbie Davies[4], Erik J. Suuronen[1,2]*, Drew Kuraitis[1,2]*

1 Division of Cardiac Surgery, University of Ottawa Heart Institute, Ottawa, Canada, **2** Department of Cellular and Molecular Medicine, University of Ottawa, Ottawa, Canada, **3** Department of Mechanical Engineering, University of Ottawa, Ottawa, Canada, **4** Department of Statistics, University of Ottawa Heart Institute, Ottawa, Canada

Abstract

Background: Cell transplantation for regenerative medicine has become an appealing therapeutic method; however, stem and progenitor cells are not always freshly available. Cryopreservation offers a way to freeze cells as they are generated, for storage and transport until required for therapy. This study was performed to assess the feasibility of cryopreserving peripheral blood mononuclear cells (PBMCs) for the subsequent *in vitro* generation of their derived therapeutic population, circulating angiogenic cells (CACs).

Methods: PBMCs were isolated from healthy human donors. Freshly isolated cells were either analyzed immediately or cryopreserved in media containing 6% plasma serum and 5% dimethyl sulfoxide. PBMCs were thawed after being frozen for 1 (early thaw) or 28 (late thaw) days and analyzed, or cultured for 4 days to generate CACs. Analysis of the cells consisted of flow cytometry for viability and phenotype, as well as functional assays for their adhesion and migration potential, cytokine secretion, and *in vivo* angiogenic potential.

Results: The viability of PBMCs and CACs as well as their adhesion and migration properties did not differ greatly after cryopreservation. Phenotypic changes did occur in PBMCs and to a lesser extent in CACs after freezing; however the potent $CD34^{+}VEGFR2^{+}CD133^{+}$ population remained unaffected. The derived CACs, while exhibiting changes in inflammatory cytokine secretion, showed no changes in the secretion of important regenerative and chemotactic cytokines, nor in their ability to restore perfusion in ischemic muscle.

Conclusion: Overall, it appears that changes do occur in cryopreserved PBMCs and their generated CACs; however, the $CD34^{+}VEGFR2^{+}CD133^{+}$ progenitor population, the secretion of pro-vasculogenic factors, and the *in vivo* angiogenic potential of CACs remain unaffected by cryopreservation.

Editor: Zoran Ivanovic, French Blood Institute, France

Funding: This work was supported by Grant-in-Aid T6793 from the Heart and Stroke Foundation of Ontario (to ES). TS was supported by Master's studentship award from the Heart and Stroke Foundation of Ontario, DK was supported by a Canadian Graduate Scholarship from the Canadian Institutes of Health Research. The funders had no role in study design, data collection and analysis, decision to publish, or preparation of the manuscript.

Competing Interests: The authors have declared that no competing interests exist.

* E-mail: dkuraitis@ottawaheart.ca (DK); esuuronen@ottawaheart.ca (ES)

Introduction

Although a unifying definition regarding their characterization does not exist [1,2], endothelial progenitor cells (EPCs) often identified as $CD34^{+}VEGFR2^{+}CD133^{+}$ cells have the ability to augment postnatal vasculogenesis [3–6]. The therapeutic revascularization that results from EPC transplantation is believed to be mediated by two main mechanisms: differentiation into new blood vessels [7,8] and paracrine signaling to augment endogenous vessel growth via the production of pro-vasculogenic cytokines [9,10]. In humans, patients with acute myocardial infarction who received an intracoronary infusion of bone marrow derived progenitors (sorted for markers CD34/CD45 and CD133) or peripheral blood-derived progenitors (plated for 3 days and positive for endothelial markers such as CD31, KDR (VEGFR2), von Willebrand factor and CD105 as well as uptake of low density lipoprotein and lectin binding) saw a beneficial effect in post-infarction remodeling processes, such as a global increase in ejection fraction and a decrease in infarct size [11,12]. The number of EPCs in the blood has been shown to be a predictor of cardiovascular health: low levels of circulating EPCs have been associated with increased risk of major cardiovascular events and vascular function [13].

EPCs can be generated from the culture of peripheral blood mononuclear cells (PBMCs) isolated from blood by density gradient centrifugation. PBMCs are cultured for 4–7 days in endothelial-promoting media on fibronectin and the subsequently generated therapeutic population is referred to as 'circulating angiogenic cells' (CACs), or early EPCs [2,14].

As cardiovascular disease is the number one leading cause of death in the Western world [15], there is a potential for CAC therapy to improve the quality of life for patients of this disease by aiding in the restoration of blood flow to the heart. However, EPCs and CACs are not available off-the-shelf and their frequency in circulating PBMCs is rather low, at about 0.0001% to 0.01% for EPCs [16] and 2% for CACs [17]. Furthermore, diabetes and cardiovascular disease decrease EPC numbers and function [18,19], making it difficult to obtain therapeutically-relevant and potent cells for application in therapy. Cryopreservation offers a method to maintain cells as they are generated, until they are required for therapy. More importantly, cryopreservation may allow a patient to store his or her own autologous cells until needed, thereby avoiding the risks and potential of graft-versus-host disease [20].

Cryopreservation has been applied for some time in the medical field, ranging from freezing of blood and bone marrow cells for transplantation, to embryo preservation for *in vitro* fertilization and long term gamete storage for cancer patients. This process preserves cells by dramatically reducing biological metabolism at low temperatures; however, cryopreservation also causes damage to some cell types, as well as potentially changing their function [21,22]. One study demonstrated that cryopreservation of T-cell subsets caused an increase in the expression of CXCR4 and CD69, while expression of L-selectin (CD62L) was decreased [23]. The consequence of cryopreservation on CACs, and their generation from PBMCs, remains to be thoroughly investigated.

The aims of this study were to investigate the outcome that cryopreservation has on the phenotype and function of: 1) freshly-isolated PBMCs; and 2) *in vitro* culture-generated CACs derived from fresh and cryopreserved PBMCs. In our study, we focused on the CACs (sometimes referred to elsewhere as early EPCs), which represent a highly heterogeneous population, thought to mostly exert their therapeutic effects through paracrine mechanisms.

Results

A summary of the cell populations and experiments is presented in Figure 1.

Viability of PBMCs but not of Generated CACs is Reduced after Cryopreservation

Fresh and cryopreserved cells were incubated with 7AAD$^-$ exclusion stain and the number of viable cells was quantified via flow cytometry. Fresh samples of PBMCs and CACs showed about 99.7±0.1% and 95.3±0.1% viability, respectively. Following cryopreservation, PBMCs sustained a non-significant reduction in viability on day 1 (93.1±1.5%) with a significant loss observed on day 28 (viability of 85.0±4.3%; $p = 0.0078$; Figure 2 A). However, the viability of CACs remained relatively stable over time post-cryopreservation at 88.7±1.4% on day 1 and 94.3±3.8% on day 28 ($p_B = 1$; Figure 2 A). Morphology of the thawed cells was preserved compared to their fresh counterpart sample as observed under a light microscope at 10× magnification (Figure 2 B).

Cryopreservation Affects PBMC Phenotype

The phenotype of the fresh and frozen cells was analyzed by staining the cells for surface markers: CD31, CD34, KDR (VEGFR2), CD133 and L-selectin and their appropriate isotype matched IgGs. The IgGs were used qualitatively [24] as there were no significant differences observed between the different sample time points (Figure 3 A–C). Surface markers CD34, KDR and CD133 were selected as they are most commonly used to describe a potent subset of CACs sometimes referred to as EPCs. Other surface markers investigated were CD31, an endothelial cell marker and L-selectin, an important adhesion protein for PBMCs and CACs. Expression of the endothelial marker CD31 remained stable at day 1 but by day 28 it was significantly increased compared to the earlier time points (Table 1). CD34 and VEGFR2 expressing cells followed a similar trend with a significant rise of these populations after cryopreservation compared to their fresh counterparts. Furthermore, the number of CD133 expression cells increased after cryopreservation while the number of L-selectin positive cells was reduced in the cryopreserved PBMC samples when compared to the fresh PBMC

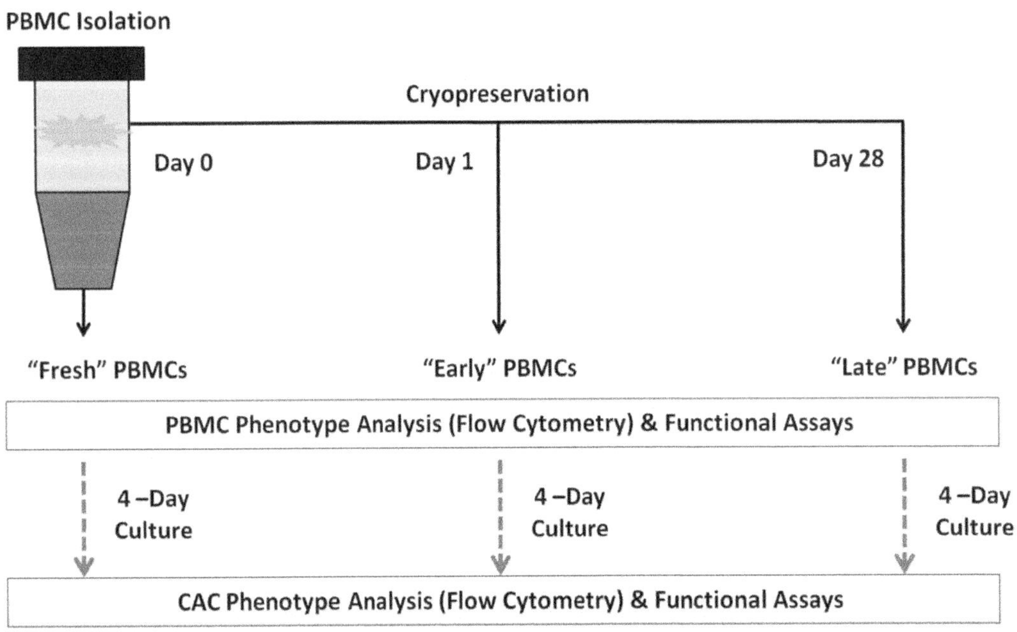

Figure 1. Summary of the methods. In brief, PBMCs were isolated and either analyzed (fresh), plated to generate CACs, or cryopreserved for 1 or 28 days after which the cells were thawed and analyzed or plated to generate CACs from cryopreserved PBMCs.

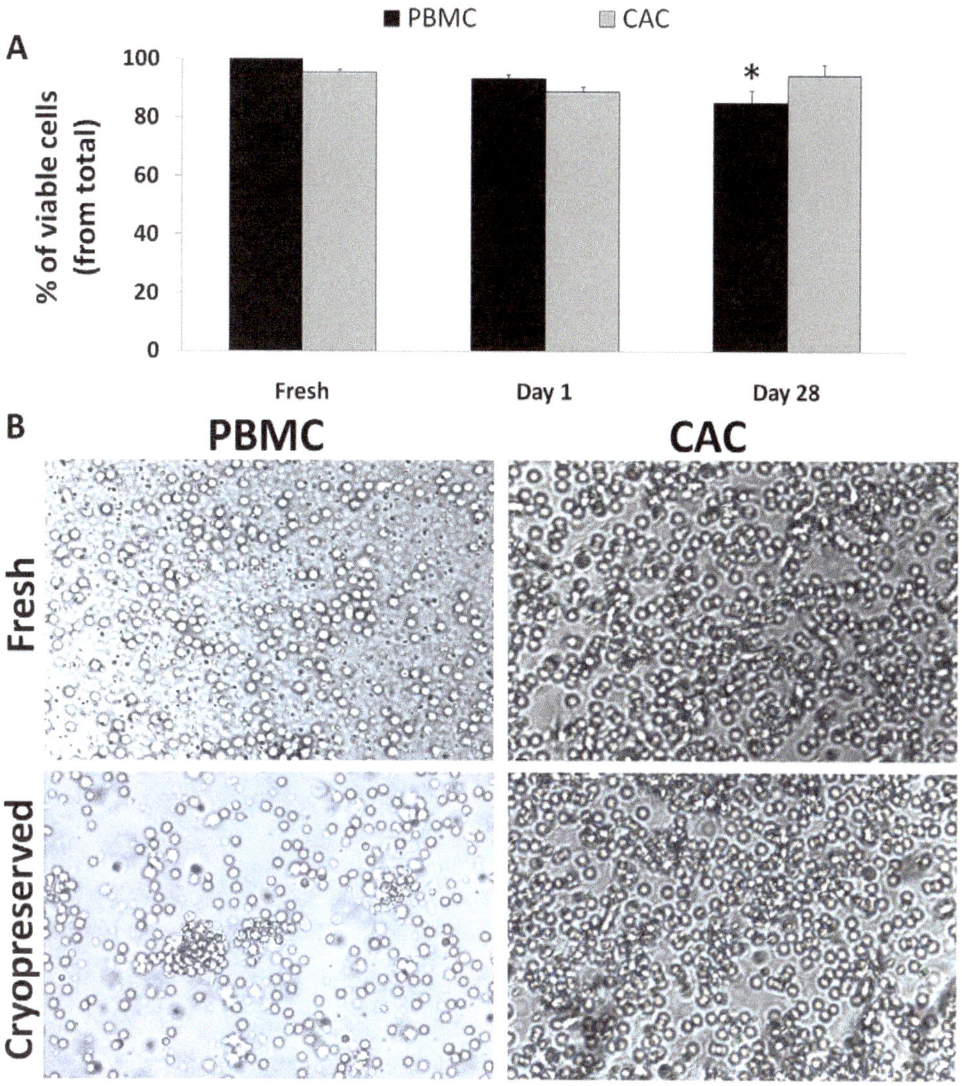

Figure 2. Viability of PBMCs, but not of generated CACs, is reduced after cryopreservation. 7AAD^{-} exclusion staining was used to enumerate the number of viable cells for fresh, early and late cryopreserved PBMCs and CACs. Prior to flow cytometry, the cells were incubated with 7AAD for 5 minutes. The proportion of viable cells is represented as a percentage ± SE (n = 7) (**A**). Representative images of fresh and cryopreserved PBMCs and CACs were taken by light microscopy at 10× magnification; scale bar = 50 μm. (**B**). *$p<0.05$ compared to the fresh sample.

samples (Table 1). Double, triple and even quadruple staining of cells positive for the markers described above was also investigated to look at certain subpopulations of PBMCs. Their summary is shown in Table 1.

Cryopreservation Affects CAC Phenotype

The number of CACs generated was not significantly different between fresh and cryopreserved samples. CAC analysis followed the same procedures as the PBMCs and investigated the number of cells expressing the five markers: CD31, CD34, CD133, VEGFR2 and L-selectin. Overall, EPC identifying markers (CD34, VEGFR2 and CD133) did not show any significant differences after cryopreservation compared to the fresh samples (Table 2). The number of CD31 positive cells was significantly decreased after 1 day of cryopreservation; however, a significant decrease was not observed for the 28-day frozen cells. L-selectin was significantly up-regulated in cells after 28 days of cryopreservation compared to the fresh samples. The double, triple and quadruple staining of cells for the selected markers described was also investigated to look at various subpopulations of CACs. Their summary is shown in Table 2.

Cryopreservation does Affect Lectin Binding and LDL Uptake of CACs

Lectin binding and LDL uptake are characteristic functions of EPCs and other circulating cells, such as leukocytes. We decided to investigate whether these functions are altered in the cells by the cryopreservation process. There were no significant differences in the uptake of LDL and binding of lectin when day 1 and day 28 cryopreserved PBMCs were compared to the fresh PBMCs (Figure 4 A, B). However, a significant increase in LDL uptake (by 2.4-fold) was observed in day 28 CACs generated after cryopreservation, compared to fresh CACs (Figure 4 C, $p = 0.004$). There was also a significant increase in lectin binding for CACs

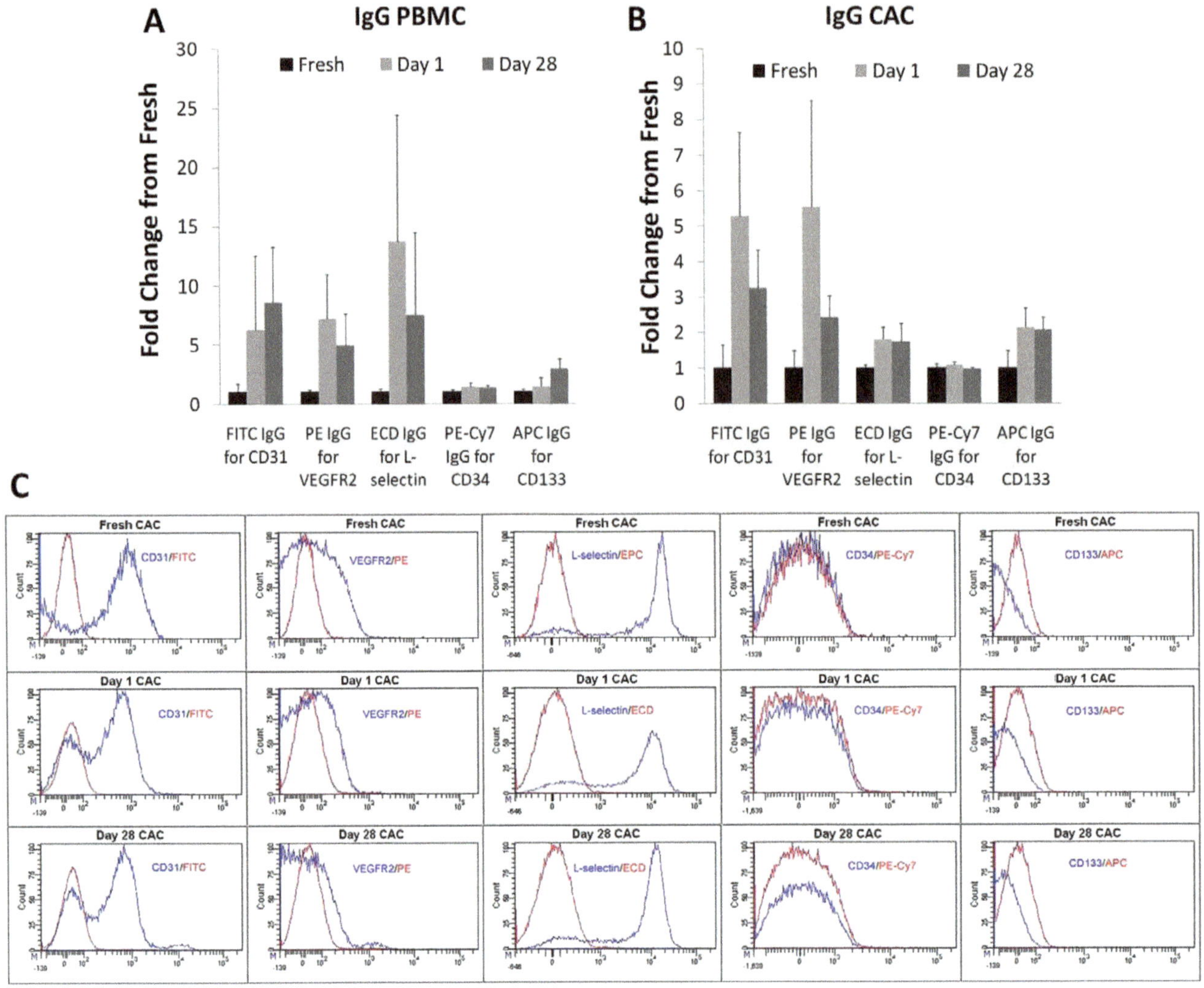

Figure 3. Isotype matched IgGs for each surface marker. Controls were conducted for background staining of fluorescent conjugates for markers CD31, VEGFR2, L-selectin, CD133 and CD34 on PBMCs (**A**) and CACs (**B**). No significant difference was observed in positively stained cells during fresh, day 1 and day 28 time points. The average IgG background staining is presented as fold-change ± SE compared to fresh samples (n = 7). Representative histograms of the various markers (blue) and corresponding IgG (red) analysis from one donor's fresh, day 1 and day 28 CACs (**C**).

after cryopreservation with an increase of ~1.1-fold compared to the fresh samples (Figure 4 D, $p = 0.005$).

Functional Capacities of PBMCs and CACs after Cryopreservation

Adhesion and migration are important functional properties of most cells. Since changes in some of the proteins involved in these processes (VEGFR2, L-selectin) were observed after cryopreservation, assays measuring the cellŝ adhesive and migratory abilities were conducted. Overall, cryopreserved PBMCs and CACs demonstrated no difference in adhesive capabilities on fibronectin coated plates compared to the fresh samples (Figure 5 A, C). Cryopreserved PBMCs were also unaffected in their ability to migrate using a VEGF chemokine (Figure 5 B), whereas CACs after 1 day of cryopreservation exhibited reduced migration ($p = 0.04$), compared to fresh CACs ($p = 0.001$). Migration of day 28 cryopreserved CACs was not significantly affected ($p = 0.12$; Figure 5 D).

Cytokine Secretion of CACs after Cryopreservation

The supernatant from CAC cultures was collected and a cytokine array was performed to evaluate the secretion of select cytokines by the cells. The secretion of pro-inflammatory cytokines interleukin-1α (IL-1α), IL-1β and tumor necrosis factor (TNF-α) and anti-inflammatory cytokine IL-10 was observed to be higher in frozen cells compared to their fresh counterparts. In most cases, day 1 cryopreserved cells seemed to have the highest secretion of these inflammatory cytokines compared to both fresh and day 28 samples (Figure 6 A–D). Secretion of angiopoietin-1, a protein that plays a role in vascular development, was significantly reduced in day 28 cells (Figure 6 E). Granulocyte colony-stimulating factor (GCSF), which stimulates the BM to produce cells, was increased in day 1 cryopreserved cells; however, this increase was not seen in the day 28 cells (Figure 6 F). Furthermore, intracellular adhesion molecule 1 (ICAM-1), important in cell-to-cell adhesion, was increased in cryopreserved cells, whereas tissue inhibitor of metalloproteinases 2 (TIMP-2), which plays a role in extracellular matrix degradation and suppression of EC proliferation [25], was

Table 1. PBMC phenotype is affected after cryopreservation.

PBMC	Fresh (%)	Day 1 (%)	Day 28 (%)
CD31⁺	67.38±2.17	70.42±4.12	76.88±2.10[a,b]
CD31⁻CD34⁺	10.12±0.47	7.69±0.62[a]	8.11±0.80[a,b]
CD31⁻CD34⁺VEGFR2⁺	0.61±0.14	1.55±0.17	2.16±0.15
CD31⁻VEGFR2⁺	1.64±0.39	3.88±0.42[a]	4.91±0.34[a]
CD31⁺CD34⁻VEGFR2⁺	3.36±0.57	8.04±0.73[a]	5.14±2.28[a]
CD31⁺CD34⁻L-Selectin⁺VEGFR2⁺	3.50±0.24	5.12±1.10	6.31±0.70[a]
CD31⁺CD34⁺	29.47±0.48	40.13±6.13	38.8±2.42
CD31⁺CD34⁺VEGFR2⁺	3.71±0.30	15.42±5.82	13.30±1.64
CD31⁺L-Selectin⁺	60.89±3.11	60.76±1.04	62.54±1.73
CD31⁺VEGFR2⁺	7.30±0.42	22.02±4.67	22.02±1.0
CD34⁺	35.89±1.36	48.14±2.72[a]	43.93±2.59[a]
CD34⁺L-Selectin⁻	4.72±0.82	5.42±0.89	4.72±1.02
CD34⁺L-Selectin⁻VEGFR2⁺	0.61±0.22	3.66±0.61[a]	2.58±0.76[a]
CD34⁺L-Selectin⁺	27.44±1.58	24.91±0.61	23.39±1.92[a]
CD34⁺L-Selectin⁺VEGFR2⁺	2.24±0.58	6.94±0.57[a]	4.63±2.15
CD34⁺CD133⁺	0.14±0.053	0.53±.077[a]	0.44±0.17
CD34⁺CD133⁺VEGFR2⁺	0.015±0.0082	0.15±0.043[a]	0.12±0.049
CD34⁺VEGFR2⁺	2.33±0.56	11.66±2.99[a]	7.63±2.81
L-Selectin⁺	84.37±1.52	74.79±2.43[a]	70.76±2.82[a]
L-Selectin⁻VEGFR2⁺	0.75±0.33	4.06±0.25[a]	3.57±0.61[a]
L-Selectin⁺VEGFR2⁺	3.66±0.56	6.40±0.79	5.08±1.61
CD133⁺	0.27±0.067	2.65±0.97[a]	1.97±0.23[a]
CD133⁺VEGFR2⁺	0.0027±0.0027	0.072±0.013[a]	0.084±0.031[a]
VEGFR2⁺	12.47±2.85	27.57±6.19[a]	27.07±5.44[a]

[a] $p<0.05$ vs. 'fresh' time point.
[b] $p<0.05$ vs. 'day 1' time point.
The values represent the percentage of cells within the total cell population (sample of 200,000 cells) that express the given phenotype.

reduced in the cryopreserved cells (Figure 6 G–H). The secretion of pro-vasculogenic cytokine basic fibroblast growth factor (bFGF), pro-myogenic cytokine insulin-like growth factor 1 (IGF-1) and chemotactic cytokines stromal cell derived factor 1 (SDF-1) and VEGF were all observed to be unaffected by cryopreservation (Figure 6 I–L).

Cryopreservation does not Affect *in vivo* Angiogenic Potential of CACs

Fresh or cryopreserved-derived human CACs were injected into the skeletal muscle of immunocompromised mice with hindlimb ischemia induced by femoral artery ligation. Laser Doppler perfusion analysis was conducted to measure blood flow over a 2-week post-treatment period, after which the animals were sacrificed. Animals receiving cryopreserved cell treatments displayed restoration in perfusion comparable to those receiving transplantation of fresh cells. No significant differences were observed in perfusion between treatments at days 4, 7, 10 and 14 post-op (Figure 7 A). Furthermore, there were no differences in the number of arterioles (assessed by smooth muscle actin (SMA) staining) between hindlimbs treated with cryopreserved cells or fresh cells (Figure 7 B). Staining the tissue with anti-human mitochondria antibodies yielded no detection of the transplanted human cells in the hindlimbs at the time of sacrifice, suggesting that transplanted cells did not persist in the tissue 2 weeks after delivery.

Discussion

Cryopreservation is an appealing method for storing an individual's own autologous stem and progenitor cells until needed for therapy; however, the effects of freezing cells for storage have not been fully characterized. In particular, there is limited data regarding cryopreserved PBMC-generated CACs, which have been a major focus of regenerative research. Our data demonstrates that cryopreservation does not adversely affect the viability of PBMCs and CACs but it does cause phenotypic changes in PBMCs and to a lesser degree in some subpopulations of CACs. However, the potent $CD34^+VEGFR2^+CD133^+$ population remained unaffected in the CACs. Also the CACs demonstrated no changes in the secretion of important regenerative and chemotactic cytokines thought to be involved in the CAC's regenerative ability via paracrine mechanisms. Furthermore, the angiogenic potential of CACs was examined *in vivo*, and no differences were seen between the 3 groups in their ability to restore perfusion to ischemic hindlimbs; demonstrating that cryopreservation did not affect the therapeutic potency of the cells. Nevertheless, some differences between the fresh and the cryopreserved CACs were observed, such as changes in inflammatory and other cytokine

Table 2. CAC phenotype is affected after cryopreservation.

CAC	Fresh (%)	Day 1 (%)	Day 28 (%)
CD31$^+$	72.81±2.94	61.97±1.09[a]	58.10±5.12
CD31$^-$CD34$^+$	8.38±1.25	12.51±0.58[a]	12.98±0.78[a]
CD31$^-$CD34$^+$VEGFR2$^+$	0.79±0.23	2.70±0.32[a]	1.76±0.53
CD31$^-$VEGFR2$^+$	1.80±0.50	7.35±0.96[a]	4.62±1.20
CD31$^+$CD34$^-$VEGFR2$^+$	3.83±0.66	8.83±0.32[a]	6.80±0.83[a]
CD31$^+$CD34$^-$L-Selectin$^+$VEGFR2$^+$	3.05±0.51	6.88±0.26[a]	4.65±0.72
CD31$^+$CD34$^+$	31.45±1.68	29.20±1.20	19.39±3.24[a]
CD31$^+$CD34$^+$VEGFR2$^+$	3.81±0.63	8.42±0.89[a]	5.27±0.82[b]
CD31$^+$L-Selectin$^+$	64.26±2.69	58.45±1.36	45.79±6.69[a]
CD31$^+$VEGFR2$^+$	6.75±1.08	16.04±1.14[a]	11.19±1.63
CD34$^+$	42.53±1.89	42.73±2.17	38.51±3.26
CD34$^+$L-Selectin$^-$	4.53±0.82	2.69±0.37	2.18±0.61
CD34$^+$L-Selectin$^-$VEGFR2$^+$	0.90±0.19	1.95±0.29[a]	0.78±0.24[b]
CD34$^+$L-Selectin$^+$	31.29±1.61	28.03±1.06	23.78±3.27
CD34$^+$L-Selectin$^+$VEGFR2$^+$	3.44±0.54	9.32±1.16[a]	6.31±1.29
CD34$^+$CD133$^+$	0.24±0.11	0.71±0.25	4.35±2.43
CD34$^+$CD133$^+$VEGFR2$^+$	0.070±0.042	0.23±0.064	0.44±0.22
CD34$^+$VEGFR2$^+$	3.57±0.63	10.07±1.28	5.88±0.99
L-Selectin$^+$	82.99±2.65	87.26±1.47	92.04±1.98[a]
L-Selectin$^-$VEGFR2$^+$	1.00±0.31	2.88±0.36[a]	2.83±0.72[a]
L-Selectin$^+$VEGFR2$^+$	3.97±0.67	10.34±0.67[a]	6.32±1.30[b]
CD133$^+$	2.54±1.15	4.93±1.02	3.60±1.89
CD133$^+$VEGFR2$^+$	0.050±0.030	0.12±0.022	2.39±1.40
VEGFR2$^+$	14.67±3.15	26.26±2.19	28.44±6.47

[a]$p<0.05$ vs. 'fresh' time point.
[b]$p<0.05$ vs. 'day 1' time point.
The values represent the percentage of cells within the total cell population (sample of 200,000 cells) that express the given phenotype.

secretion and the numbers of cells expressing certain non-progenitor surface markers (L-selectin and CD31).

The maintenance of PBMC viability following cryopreservation observed in the present work is in accordance with other studies that also did not find considerable cell death after cryopreservation of PBMCs [26,27]; however, it is believed that the cell viability achieved after cryopreservation is largely dependent on the expertise/technique of the laboratory that performs the cryopreservation [26,27]. In particular, one report showed cell viability post-cryopreservation ranging from 1% to more than 90% [27].

The effect of cryopreservation on CAC viability has not been studied in great detail [28,29], especially when considering PBMC-derived CACs. One investigation using 7-AAD, examined the viability of CACs from cryopreserved umbilical cord blood and found a significant level of apoptosis [29]. In contrast, our results illustrated no significant decline in CAC viability after 1 day and 28 days time points of cryopreservation. It is possible that this is attributable to differences between the cells under study or that after being cultured for 4 days the CAC viability differences are negated. The importance of cell viability after cryopreservation is imperative as CACs make up a small fraction of the total PBMCs. The ability to generate a large number of viable cells would allow for a better stem cell therapy outcome.

The cells' phenotype was analyzed by staining for cell surface markers CD31, L-selectin, CD34, CD133 and VEGFR2 using flow cytometry. In our investigation, CACs generated from cryopreserved PBMCs showed no difference in expression of surface markers CD34, VEGFR2 and CD133; however a difference in CD31 and L-selectin expression was observed. CD34, CD133 and VEGFR2 are markers used to characterize CACs and the potent subpopulation of EPCs, although they are not exclusive to these populations and can be expressed by other hematopoietic lineages [30]. As previously mentioned, CD34$^+$VEGFR2$^+$CD133$^+$ cells have been shown to be a potent subpopulation of cells involved in vascular regeneration [3–6]. The fact that this population and their vasculogenic derivatives (ex. CD34$^+$VEGFR2$^+$ cells) are unaffected by the cryopreservation in CACs, and are even increased in cryopreserved PBMCs, indicates that perhaps this population is able to better tolerate the harsh process of freezing without significant losses, compared to other populations.

CD31 is a marker of endothelial cells and L-selectin is an adhesion molecule found on CACs and other PBMCs. The expression of CD31 by day 28 cryopreserved PBMCs was increased; whereas the frequency of L-selectin$^+$ PBMCs decreased, from about 85% to 70%, after 28 days of cryopreservation. This result is in accordance with previous studies that have also found down-regulation of L-selectin in PBMCs following cryopreservation [31], which may be a cellular response to stress [32]. Regarding progenitor markers CD34, CD133 and VEGFR2, we

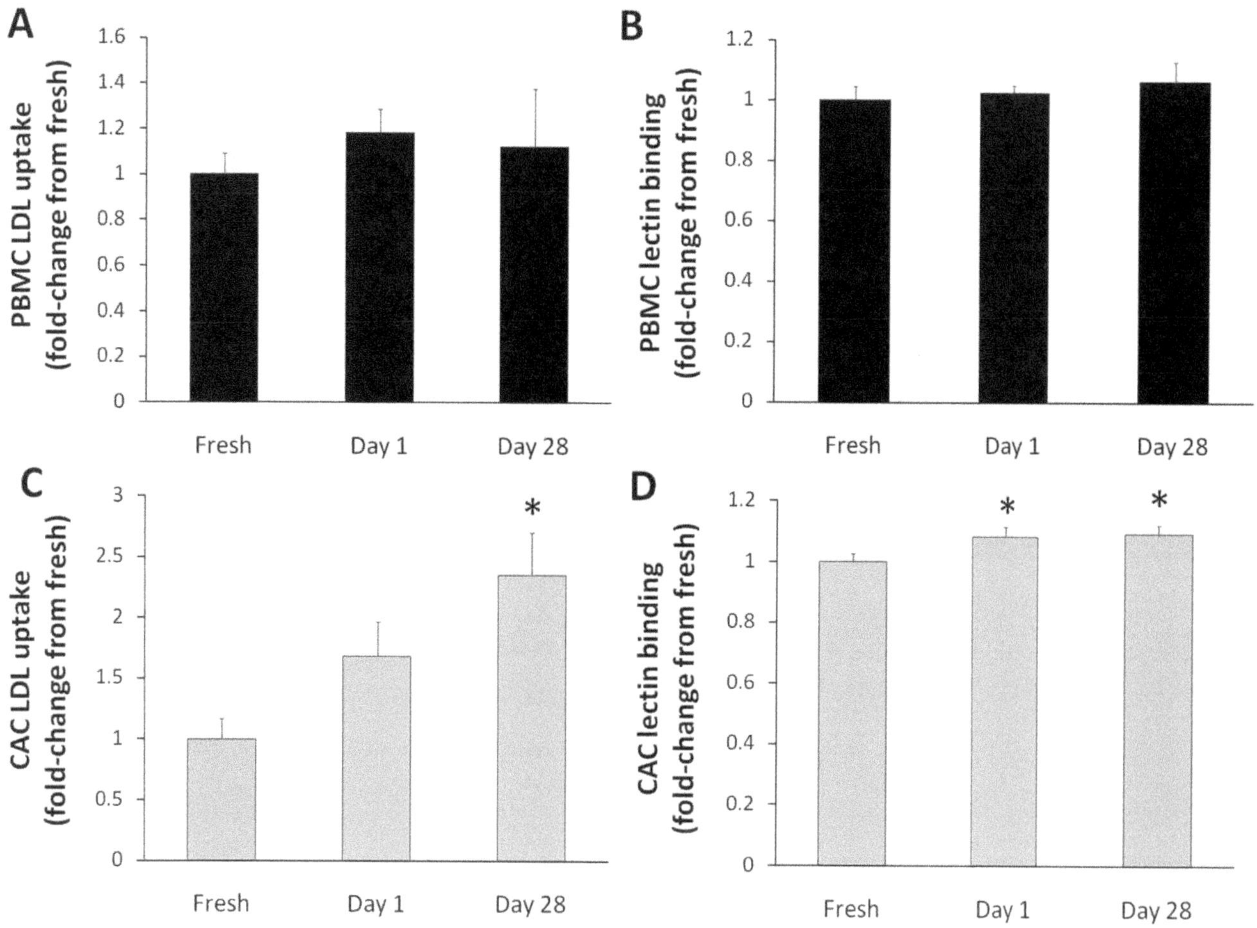

Figure 4. Cryopreservation affects LDL uptake and lectin binding of CACs. The numbers of positive PBMCs that took up fluorescent LDL (**A**) or that bound fluorescent lectin (**B**) were quantified using flow cytometry. The same procedure was conducted for CACs (**C, D**). Data is reported as fold-change from donor-matched fresh samples ± SE (n = 7). *$p<0.05$ compared to the fresh sample.

observed an increased expression following cryopreservation of PBMCs. Ketheesan *et al.* showed that the proportion of $CD34^+$ cells in frozen cord blood did not change after freezing [33], while Lanza *et al.* demonstrated that 90-day cryopreserved PBMCs derived from patients with non-Hodgkin's lymphoma had an increase in $CD34^+$ cells compared to fresh samples, attributed to a decrease in mature myeloid cells [34]. Many hypotheses exist to explain the changes in phenotypes after cryopreservation: 1) more cells could indeed shift towards expression of particular markers during the cryopreservation process; 2) due to cell death of other subpopulations, there is a relative increase of more cells expressing markers CD34, CD133 and VEGFR2 used to identify EPCs; or 3) an increase in false positives as perceived by the flow cytometry analyses. Given that the observed non-specific binding IgG controls did not differ among time points, it is plausible that the observed changes are due to one of the first two hypotheses; however, it is not yet clear whether one or both are predominant reasons for changes in stored cell phenotypes.

Although studies have shown that L-selectin decreases in immunomagnetically-purified $CD34^+$ CACs immediately after thawing, the cells do recover their expression of L-selectin after a short period of culture [35]. This may explain why, in our study, we observed a decrease in PBMC L-selectin expression, which were analyzed immediately, while the CACs were generated by a 4-day culture protocol before analysis, and demonstrated no change in L-selectin on day 1 cryopreserved cells and even an increased expression by day 28 cryopreserved cells. Surprisingly, the levels of CD31 were decreased in cryopreserved CACs in contrast to cryopreserved PBMCs where the expression of CD31 was unchanged or increased.

We also assessed the function of the populations' capacity for lectin-binding and LDL uptake, which are further characteristics used to assess CAC populations [1,2]. As expected, a higher percentage of cells in the CAC sample showed lectin binding and DiI-LDL uptake, compared to PBMCs. In PBMCs there was no significant difference between the number of fresh and cryopreserved cells staining positive for either one of these factors. However, for CACs a significant increase was seen for lectin binding and LDL uptake in 28 day cryopreserved cells. This is in contrast to an investigation by Mieno *et al.* that showed that LDL uptake and lectin-binding of EPCs is unaffected by cryopreservation [28].

The cells' function was further assessed by their adhesion and VEGF-mediated migration capabilities. The adhesion of PBMCs was unaffected by cryopreservation, which is somewhat contradictory to the observed decrease in L-selectin after cryopreservation. Binding of L-selectin inside the cell results in the activation of β1 and β2 integrins which promotes adhesion of leukocytes to fibronectin [36]. It is possible that while the loss of L-selectin in PBMCs is statistically significant, it is not, however, physiologically

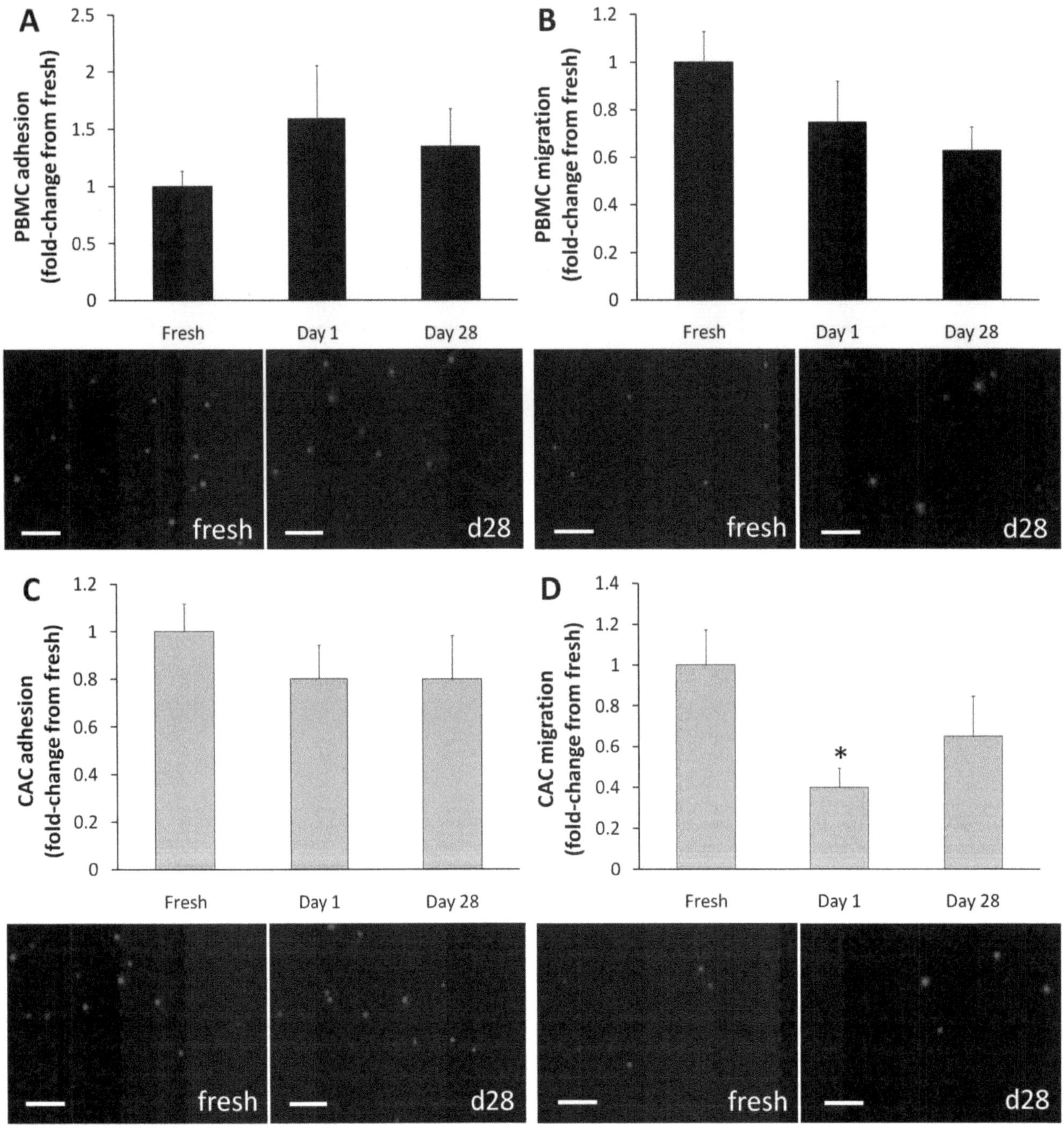

Figure 5. Cryopreservation does not affect adhesion and migration of PBMCs and CACs. PBMCs were pre-stained with DAPI for visualization by electron microscope for analysis of their adhesive (**A**) and migration (**B**) capabilities. The same procedure was performed for CACs (**C, D**). Five pictures were taken per well and the number of $DAPI^+$ cells (blue) were counted per FOV and representative pictures of each assay are shown. Data is reported as fold-change from donor-matched fresh samples ± SE (n = 7), scale bar = 80 µm. *$p<0.05$ compared to the fresh sample.

relevant, as compensatory mechanisms may be in action (e.g. other selectins may assume similar functions [37]). The migration potential of PBMCs using VEGF for chemotaxis was also unaffected by cryopreservation. The fact that there was no change in VEGF-mediated migration even though there was increased expression of one of its receptors (VEGFR2) was not further investigated. One theory could be attributed to the newly expressed receptor proteins being non-functional, and therefore being unable to contribute to the cells' migration.

CAC adhesion and migration capacities were mostly unaffected by cryopreservation. There is a significant decrease in CAC migration between fresh and 1-day frozen samples, but no significant difference between fresh and 28-day frozen samples. The loss of migratory function in the 1-day samples may be attributed to shock that the cells underwent during the short succession between freezing/thawing procedures; whereas the maintenance of migratory potential in the 28-day cells is in

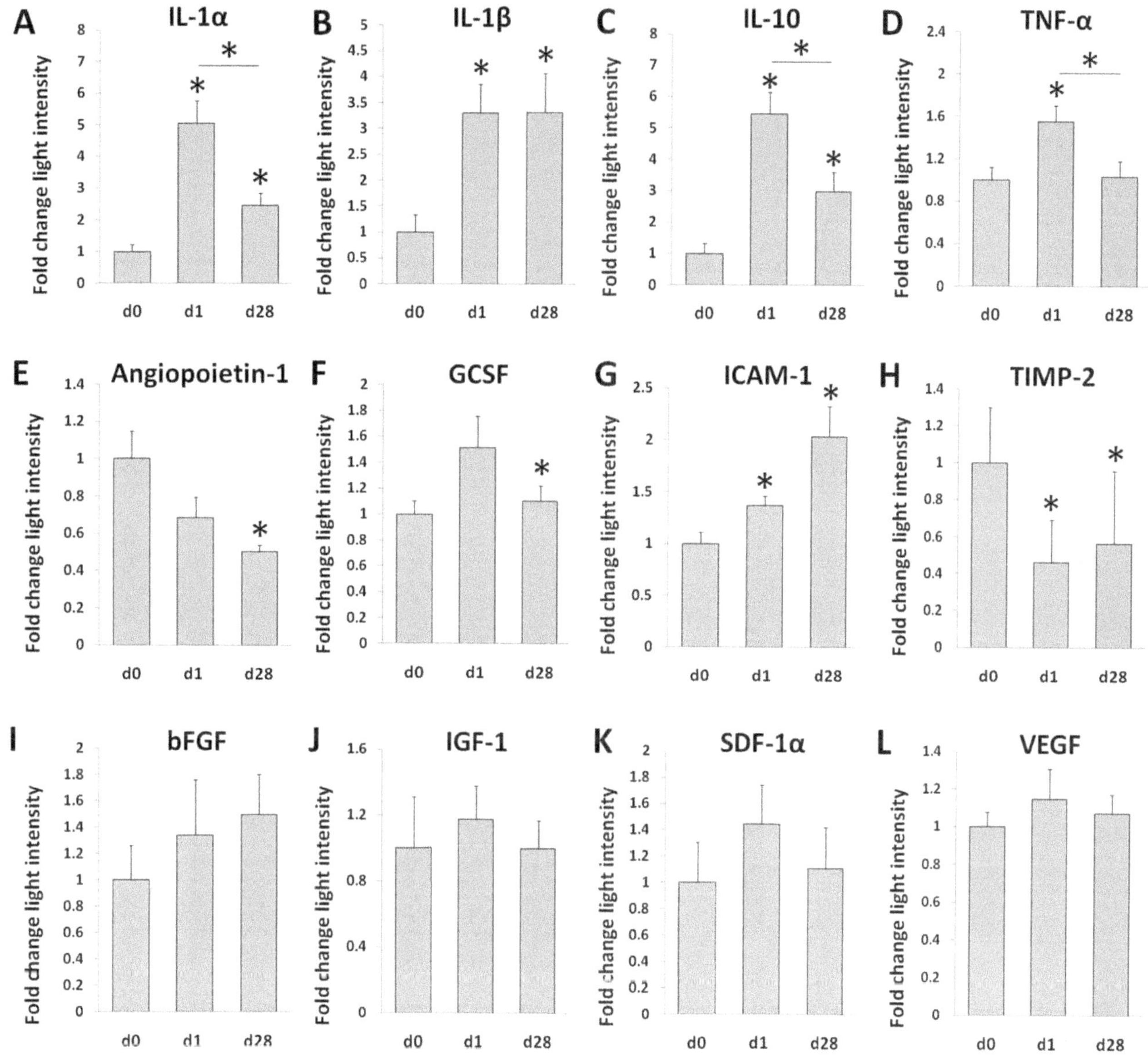

Figure 6. Cytokine secretion of CACs after cryopreservation. Fresh and cryopreserved PBMCs were cultured for 4 days to generate CACs. The media was collected and analyzed using a cytokine array (n = 5−6). Results of some of the cytokines are shown as a fold-change in light intensity ± SE; (**A–L**). *$p<0.05$ compared to the fresh sample, unless otherwise indicated.

accordance with other studies showing that migration is not affected by cryopreservation [28,38].

Secretion of cytokines is an important aspect of the CAC's ability to repair tissue after an ischemic injury [10]. Using an array we looked at a variety of cytokines released by cultured fresh and cryopreserved CACs. Inflammatory cytokines were increased a few fold in the cryopreserved cells; and this increase was greatest in day 1 cryopreserved cells, even compared to day 28 cells. Why the day 1 cells increase their secretion of inflammatory cytokines remains unknown. As the cryopreservation preparation procedures are the same, the difference between day 1 and day 28 might be explained by the short period of time between the freezing/thawing procedures. While changes in inflammatory cytokine production in cryopreserved CACs have not been reported, PBMCs have been previously shown in some studies to have a different cytokine secretion pattern post cryopreservation [39,40]. In one study, 12 months frozen PBMCs were found to spontaneously show an increase of inflammatory cytokines IL-6, IL-10, IL-13 and IFN-γ [39]. In our study, we found that the secretion of several other cytokines was also changed after cryopreservation. ICAM-1, an adhesion factor and, in its soluble form, an indicator of endothelial dysfunction [41], was found to have an increased secretion in cells after cryopreservation. However, this might be attributed to the increase in IL-1α, IL-β and TNF-α which have been found to up-regulate ICAM-1 in ECs [42]. Similarly, the observed angiopoietin-1 and TIMP-2 decrease in our cryopreserved cells may also be attributed to the effect of the increased IL-1β and TNF-α secretion by the cells [43,44]. While differences in the secretion of certain cytokines was observed in cryopreserved cells, some of the most important pro-regenerative and chemotactic cytokines such as IGF-1, bFGF, SDF-1α and VEGF remained unchanged between the fresh and

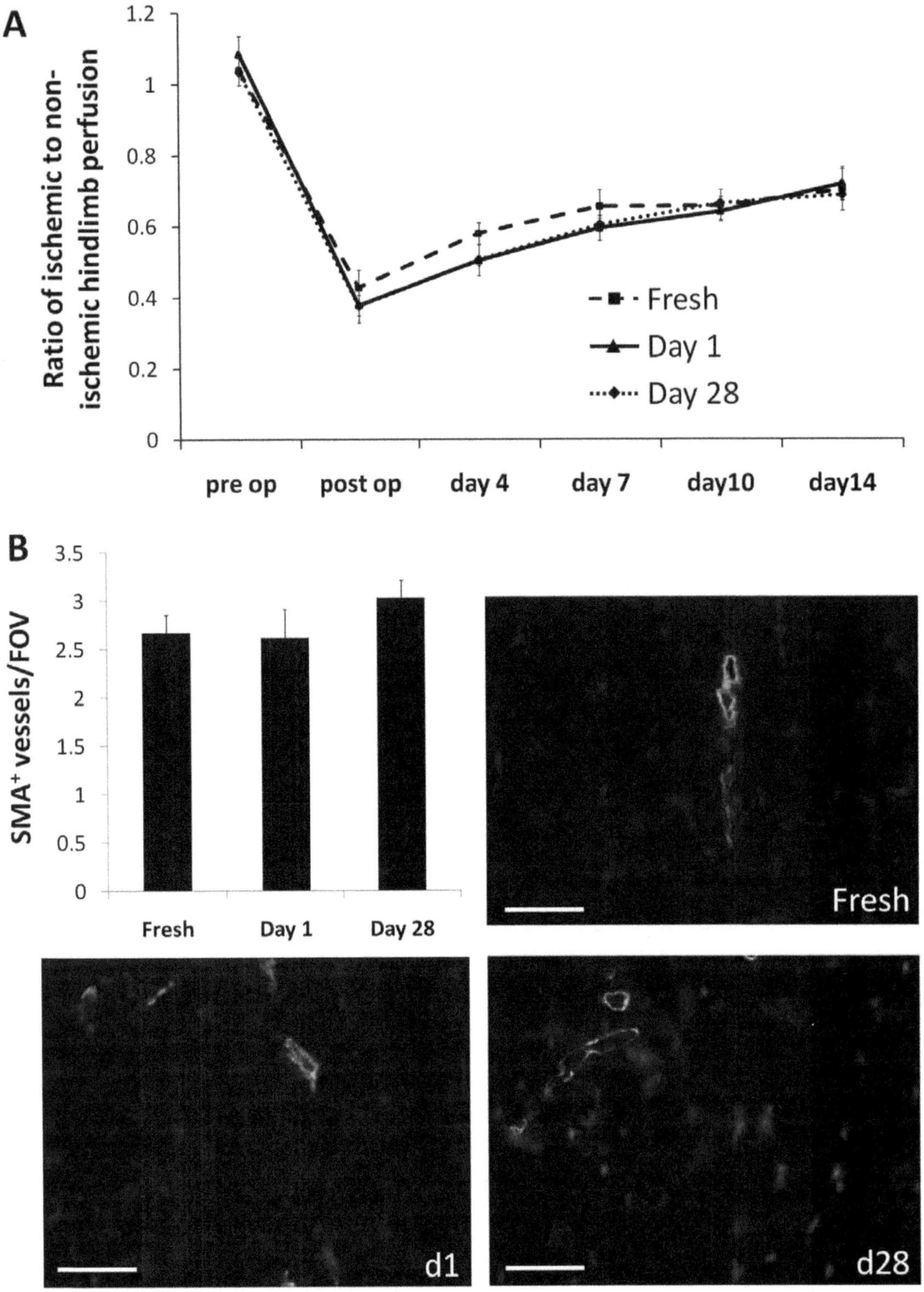

Figure 7. Cryopreservation does not affect the *in vivo* therapeutic function of CACs in a hindlimb ischemia model. Five million fresh, d1 or d28 cryopreserved-derived cells were injected into the skeletal muscle of mice that had undergone femoral artery ligation to induce hindlimb ischemia. Laser Doppler perfusion analysis was conducted over a period of 2 weeks (**A**). The tissue was harvested at 2 weeks and stained for SMA. The number of arterioles (per FOV ± SE) was determined (**B**), with representative images shown at 20× magnification (n = 5–6); scale bar = 50 μm.

cryopreserved CACs. Although some changes between fresh and cryopreserved cells were found, when their angiogenic potential was investigated *in vivo*, no differences in the recovery of perfusion or arteriole density were observed between the animals treated with cryopreserved cells vs. fresh cells, underscoring the conservation of therapeutic function after cryopreservation.

The extraction and cryopreservation of PBMCs is a well-established protocol, already in use for treatment of patients suffering with bone marrow cancers and other blood diseases. While the amount of serum used during our freezing step is 6% and other protocols have used storage solutions containing up to 90% serum [45,46], we did not observe a great loss of viable cells. This is an important finding, considering that a study investigating cryopreservation of MNCs reported that viability of <70% is associated with compromised proliferative responses to antigens and mitogens, and are not suitable for cytokine production studies, flow cytometric analyses, or immunomagnetic cell separation [47,48].

Overall, our data indicate that the function of therapeutic CAC populations is generally preserved after short- and longer-term cryopreservation, with the cells retaining their adhesion and migratory capabilities. Despite the more drastic phenotypic changes in PBMCs following cryopreservation, these cells can still generate CACs with no changes in the population of the potent vasculogenic $CD34^{+}VEGFR2^{+}CD133^{+}$ cells compared to their fresh counterparts. Furthermore, the secretion of important regenerative and chemotactic cytokines is unchanged in the cryopreserved CACs along with their angiogenic potential *in vivo*. However, the cryopreserved CACs do show a significant increase in secretion of inflammatory factors, which may affect the expression of additional cytokines such as angiopoietin-1, GCSF, TIMP-2 and ICAM-1. Our findings also have clinical implication, as more research is showing that PBMCs and CACs have a positive impact on neovascularization in ischemic and cardiovascular diseases. One important factor for successful cell therapy is generating enough therapeutic cells to achieve sufficient repair of the damaged tissue. Progenitor cells from the peripheral blood are much easier to obtain than the traditional bone marrow harvest, which is more painful and requires anesthesia. The PBMCs may be collected over a given period and stored each time until enough cells are collected and the patient is primed to receive the cell therapy, at which time the PBMCs may be cultured to generate the therapeutic CACs.

Conclusions

In conclusion, the CACs generated from the cryopreserved PBMCs show no significant difference in viability, in *in vitro* functions such as adhesion and migration, or in *in vivo* angiogenic potential; however, an increase in LDL uptake and lectin binding was observed. Also while there are some phenotypic changes, the therapeutically potent $CD34^{+}VEGFR2^{+}CD133^{+}$ population remains unchanged. Additionally, the secretion of important regenerative and chemotactic cytokines involved in tissue repair through paracrine mechanisms is unaffected in CACs after cryopreservation. The ability to store these therapeutic cells for a long period of time without large cell losses and disruption of function will offer more patients an opportunity to experience successful cell therapy.

Materials and Methods

A summary of the cell populations, methods and time points are presented in Figure 1.

Ethics Statement

The study was approved by the Human Research Ethics Board of the University of Ottawa Heart Institute, and informed consent was obtained from all donors.

Cell Isolation

Total PBMCs were isolated from 100 ml of fresh blood samples of seven healthy human donors both male and female (20–35 years of age) by Histopaque 1077 (Sigma-Aldrich, Oakville, Canada) density-gradient centrifugation as described previously [49]. The cells within the buffy coat were then either analyzed (baseline PBMCs), cultured to generate CACs, or cryopreserved.

Cryopreservation

PBMCs retrieved from the buffy coat were reconstituted in a 1 ml solution of Isocove's Modified Dulbecco's Medium (IMDM; Stem Cell Technologies, Vancouver, Canada) containing 5% dimethyl sulfoxide (DMSO; Sigma-Aldrich) and 6% donor cell-matched serum and placed into a cryovial (Corning Incorporated, Corning, USA). The DMSO was mixed with the IMDM prior to addition of blood serum and cells in order to prevent clump formation. The cells were cooled to −80°C over 3 hours before being transferred to liquid nitrogen for storage.

For rapid thawing and to prevent toxicity to cells, cryovial tubes were taken from the freezer and immediately placed in a 37°C water bath until ice crystals disappeared, after which the cells were diluted 5× with IMDM and centrifuged at 1400 rpm and immediately resuspended in fresh Endothelial Basal Media (EBM-2; Clonetics, Guelph, Canada).

CAC Culture

CACs were generated by seeding fresh PBMCs, or PBMCs thawed at days 1 or 28 post-freezing, on fibronectin (Sigma-Aldrich)-coated plates in EBM supplemented with EGM-2-MV-SingleQuots (Clonetics) containing 5% fetal bovine serum, human VEGF, human insulin-like growth factor 1, human epidermal growth factor and antibiotics for 4 days. After 4 days, the media was removed and the cells were lifted using gentle pipetting with PBS and prepared for subsequent assays.

Flow Cytometry Staining

Cells were counted using a Vi-Cell analyzer (Beckman Coulter, Mississauga, Canada). PBMCs or CACs (3×10^{5}) in 200 µl of EBM were stained for 30 minutes at 4°C with pre-conjugated antibodies against the following antigens: CD31-FITC (Beckman Coulter), CD34-PECy7 (BD Biosciences, Mississauga, Canada), KDR-PE (R&D Systems, Minneapolis, USA), CD133-APC (Miltenyl Biotec, Auburn, USA) and L-selectin-ECD (Beckman). Samples were also stained with appropriate IgG isotype-matched controls. Immediately prior to analysis, the viability stain 7-actinomycin D (7-AAD) (Invitrogen, Burlington, Canada) was added to samples to a final concentration of 8.3 µg/ml. Characterization also included incubating 3×10^{5} cells with 2 µg/ml of 1,1′-dioctadecyl-3,3,3′,3′-tetramethylindocarbocyanine–labeled acetylated low density lipoprotein (DiI-LDL) (Invitrogen) for one hour at 37°C after which the cells were pelleted, fixed for 10 minutes with fixation buffer (BD Biosciences) and further incubated for an hour at 37°C with 10 µg/ml of fluorescein isothiocyanate (FITC) labelled lectin from *Ulex europaeus* agglutinin-1 (Sigma-Aldrich). Cells were analyzed and quantified with a BD FACSaria cell sorter (BD Biosciences) until the number of events (cells) reached 200,000.

Functional Assays

Cells were suspended in 50 µg/ml of 4′,6-diamidino-2-phenylindole (DAPI; Sigma-Aldrich) in PBS for 30 minutes at 37°C, after which the cells were pelleted and counted. **Migration.** DAPI-stained PBMCs or CACs (2×10^{4}) were suspended in 100 µl of VEGF-free EBM and placed in the top compartment of a 24-well Boyden chamber (Corning Inc.) with the lower chamber coated with fibronectin and containing 350 µl of 0.05 µg/ml VEGF (Cedarlane, Burlington, Canada) in EBM. The cells were incubated for 24 hours in 5% CO_2 after which they were fixed with 4% paraformaldehyde and washed with PBS. **Adhesion.** DAPI-stained cells (2×10^4) were plated in 1 ml of EBM on fibronectin coated 24-well plates and incubated for 1 hour in 5% CO_2 after which they were fixed with 4% paraformaldehyde. The number of DAPI stained cells for both assays were counted in 5 random fields-of-view for each well at 20× magnification using Olympus B×60 fluorescent microscope (Olympus Canada Inc., Markham, Canada). Assays were performed in duplicate.

Cytokine Array

Three million PBMCs were plated on fibronectin-coated 6-well plates in 2 ml of EBM media. After culturing for 4 days to generate CACs, the media was collected. Cytokine arrays (RayBiotech, Norcross, USA) were incubated with 100 µl of media per sample according to manufacturer's protocol. Following the cytokine array procedures, the membranes were exposed and the light produced at each spot (proportional to the amount of cytokine bound) was quantified using AlphaEaseFC.

Hindlimb ischemia mouse model. Nude BALB/C mice (7–8 weeks old) underwent ligation of proximal femoral arteries under 2% isoflurane, as previously described [14]. Subsequently, the ischemic hindlimb was injected downstream of the ligation site with 2 equivolumetric injections (50 µl total, suspended in PBS) containing a total of 5×10^6 CACs that were derived from: i) fresh PBMCs; ii) day 1 cryopreserved PBMCs; or, iii) day 28 cryopreserved PBMCs, prepared as described above. Injections were performed with a 27 gauge needle. Prior to injection, cell viability was confirmed to be >85%.

Laser Doppler

Hindlimb perfusion was measured using laser Doppler analysis pre-operatively, post-operatively, and at days 4, 7, 10 and 14, as described previously [14]. Briefly, while mice were anaesthetized with 2% isoflurane, single point measurements were recorded (moorLD12; Moor Instruments, Axminster, UK) in both hindlimbs and used to evaluate perfusion. The data is expressed as a ratio of ischemic to non-ischemic hindlimb blood flow.

Immunohistochemistry

Two weeks post-CAC treatment, the mice were sacrificed and the ischemic hindlimb tissue was collected, frozen and sectioned. After fixing tissue sections with acetone for 10 minutes, the sections were washed for 3×5 minutes with PBS, blocked for an hour using 10% fetal bovine serum in PBS and stained with anti-SMA antibody (1:100; abcam, Cambridge, USA). Following 3 washes (5 minutes each) and secondary antibody application (1:600 anti-rabbit Alexa 488; Invitrogen,) the slides were mounted with DAPI and analyzed at 20× magnification with an Olympus BX50 microscope using Image Pro Plus. The number of arterioles was counted by 2 blind observers and expressed as the average number of SMA^+ vessels per field of view (FOV). Additionally, to visualize the injected CACs, hindlimb sections were stained with anti-human mitochondria antibody (1:100; Chemicon, Temecula, USA), following the manufacturers instruction.

Statistical Analysis

Results obtained for each phenotype and each cell type were analyzed for differences between days using a one-way Anova adjusted for donor. Sub-analyses were performed when a Bonferonni significant result was obtained using a paired t-test. Results were Bonferonni adjusted for the number of tests performed for each of viability, phenotype and function separately, to achieve an alpha of 0.05; *p*-values given for sub-analyses are not corrected. Statistical analyses were performed using R (R development core team, Vienna, Austria).

Acknowledgments

The authors would like to thank Branka Vulesevic for her assistance with animal procedures.

Author Contributions

Conceived and designed the experiments: TS ES DK. Performed the experiments: TS KM SC JM. Analyzed the data: TS KM RD DK. Contributed reagents/materials/analysis tools: TS KM SC JM RD. Wrote the paper: TS ES DK.

References

1. Yoder MC, Ingram DA (2009) Endothelial progenitor cell: ongoing controversy for defining these cells and their role in neoangiogenesis in the murine system. Curr Opin Hematol 16: 269–273.
2. Hirschi KK, Ingram DA, Yoder MC (2008) Assessing identity, phenotype, and fate of endothelial progenitor cells. Arterioscler Thromb Vasc Biol 28: 1584–1595.
3. Asahara T, Masuda H, Takahashi T, Kalka C, Pastore C, et al. (1999) Bone marrow origin of endothelial progenitor cells responsible for postnatal vasculogenesis in physiological and pathological neovascularization. Circ Res 85: 221–228.
4. Li B, Sharpe EE, Maupin AB, Teleron AA, Pyle AL, et al. (2006) VEGF and PlGF promote adult vasculogenesis by enhancing EPC recruitment and vessel formation at the site of tumor neovascularization. Faseb J 20: 1495–1497.
5. Kawamoto A, Asahara T, Losordo DW (2002) Transplantation of endothelial progenitor cells for therapeutic neovascularization. Cardiovasc Radiat Med 3: 221–225.
6. Fan CL, Gao PJ, Che ZQ, Liu JJ, Wei J, et al. (2005) Therapeutic neovascularization by autologous transplantation with expanded endothelial progenitor cells from peripheral blood into ischemic hind limbs. Acta Pharmacol Sin 26: 1069–1075.
7. Kalka C, Masuda H, Takahashi T, Kalka-Moll WM, Silver M, et al. (2000) Transplantation of ex vivo expanded endothelial progenitor cells for therapeutic neovascularization. Proc Natl Acad Sci U S A 97: 3422–3427.
8. Yamaguchi J, Kusano KF, Masuo O, Kawamoto A, Silver M, et al. (2003) Stromal cell-derived factor-1 effects on ex vivo expanded endothelial progenitor cell recruitment for ischemic neovascularization. Circulation 107: 1322–1328.
9. Scheubel RJ, Holtz J, Friedrich I, Borgermann J, Kahrstedt S, et al. (2008) Paracrine effects of CD34 progenitor cells on angiogenic endothelial sprouting. Int J Cardiol 139: 134–141.
10. Urbich C, Aicher A, Heeschen C, Dernbach E, Hofmann WK, et al. (2005) Soluble factors released by endothelial progenitor cells promote migration of endothelial cells and cardiac resident progenitor cells. J Mol Cell Cardiol 39: 733–742.
11. Britten MB, Abolmaali ND, Assmus B, Lehmann R, Honold J, et al. (2003) Infarct remodeling after intracoronary progenitor cell treatment in patients with acute myocardial infarction (TOPCARE-AMI): mechanistic insights from serial contrast-enhanced magnetic resonance imaging. Circulation 108: 2212–2218.
12. Schachinger V, Assmus B, Britten MB, Honold J, Lehmann R, et al. (2004) Transplantation of progenitor cells and regeneration enhancement in acute myocardial infarction: final one-year results of the TOPCARE-AMI Trial. J Am Coll Cardiol 44: 1690–1699.
13. Werner N, Kosiol S, Schiegl T, Ahlers P, Walenta K, et al. (2005) Circulating endothelial progenitor cells and cardiovascular outcomes. N Engl J Med 353: 999–1007.
14. Kuraitis D, Hou C, Zhang Y, Vulesevic B, Sofrenovic T, et al. (2011) Ex vivo generation of a highly potent population of circulating angiogenic cells using a collagen matrix. J Mol Cell Cardiol 51: 187–197.
15. Bonow RO, Smaha LA, Smith SC, Jr., Mensah GA, Lenfant C (2002) World Heart Day 2002: the international burden of cardiovascular disease: responding to the emerging global epidemic. Circulation 106: 1602–1605.
16. Khan SS, Solomon MA, McCoy JP Jr. (2005) Detection of circulating endothelial cells and endothelial progenitor cells by flow cytometry. Cytometry B Clin Cytom 64: 1–8.

17. Dimmeler S, Aicher A, Vasa M, Mildner-Rihm C, Adler K, et al. (2001) HMG-CoA reductase inhibitors (statins) increase endothelial progenitor cells via the PI 3-kinase/Akt pathway. J Clin Invest 108: 391–397.
18. Chen JZ, Zhang FR, Tao QM, Wang XX, Zhu JH, et al. (2004) Number and activity of endothelial progenitor cells from peripheral blood in patients with hypercholesterolaemia. Clin Sci (Lond) 107: 273–280.
19. Tepper OM, Galiano RD, Capla JM, Kalka C, Gagne PJ, et al. (2002) Human endothelial progenitor cells from type II diabetics exhibit impaired proliferation, adhesion, and incorporation into vascular structures. Circulation 106: 2781–2786.
20. Stockschlader M, Kruger W, Kroschke G, Zeller W, Hoffknecht M, et al. (1995) Use of cryopreserved bone marrow in allogeneic bone marrow transplantation. Bone Marrow Transplant 15: 569–572.
21. Haack-Sorensen M, Kastrup J (2011) Cryopreservation and Revival of Mesenchymal Stromal Cells. Methods Mol Biol 698: 161–174.
22. Motta JP, Gomes BE, Bouzas LF, Paraguassu-Braga FH, Porto LC (2010) Evaluations of bioantioxidants in cryopreservation of umbilical cord blood using natural cryoprotectants and low concentrations of dimethylsulfoxide. Cryobiology 60: 301–307.
23. Faint JM, Tuncer C, Garg A, Adams DH, Lalor PF (2011) Functional consequences of human lymphocyte cryopreservation: implications for subsequent interactions of cells with endothelium. J Immunother 34: 588–596.
24. Hulspas R, O'Gorman MR, Wood BL, Gratama JW, Sutherland DR (2009) Considerations for the control of background fluorescence in clinical flow cytometry. Cytometry B Clin Cytom 76: 355–364.
25. Seo DW, Li H, Guedez L, Wingfield PT, Diaz T, et al. (2003) TIMP-2 mediated inhibition of angiogenesis: an MMP-independent mechanism. Cell 114: 171–180.
26. Kleeberger CA, Lyles RH, Margolick JB, Rinaldo CR, Phair JP, et al. (1999) Viability and recovery of peripheral blood mononuclear cells cryopreserved for up to 12 years in a multicenter study. Clin Diagn Lab Immunol 6: 14–19.
27. Weinberg A, Zhang L, Brown D, Erice A, Polsky B, et al. (2000) Viability and functional activity of cryopreserved mononuclear cells. Clin Diagn Lab Immunol 7: 714–716.
28. Mieno S, Clements RT, Boodhwani M, Sodha NR, Ramlawi B, et al. (2008) Characteristics and function of cryopreserved bone marrow-derived endothelial progenitor cells. Ann Thorac Surg 85: 1361–1366.
29. Lu X, Proctor SJ, Dickinson AM (2008) The effect of cryopreservation on umbilical cord blood endothelial progenitor cell differentiation. Cell Transplant 17: 1423–1428.
30. Peichev M, Naiyer AJ, Pereira D, Zhu Z, Lane WJ, et al. (2000) Expression of VEGFR-2 and AC133 by circulating human CD34(+) cells identifies a population of functional endothelial precursors. Blood 95: 952–958.
31. Costantini A, Mancini S, Giuliodoro S, Butini L, Regnery CM, et al. (2003) Effects of cryopreservation on lymphocyte immunophenotype and function. J Immunol Methods 278: 145–155.
32. Griffin JD, Spertini O, Ernst TJ, Belvin MP, Levine HB, et al. (1990) Granulocyte-macrophage colony-stimulating factor and other cytokines regulate surface expression of the leukocyte adhesion molecule-1 on human neutrophils, monocytes, and their precursors. J Immunol 145: 576–584.
33. Ketheesan N, Whiteman C, Malczewski AB, Hirst RG, La Brooy JT (2004) Effect of cryopreservation on the immunogenicity of umbilical cord blood cells. Transfus Apher Sci 30: 47–54.
34. Lanza F, Moretti S, Castagnari B, Montanelli F, Latorraca A, et al. (1999) Assessment of distribution of CD34 epitope classes in fresh and cryopreserved peripheral blood progenitor cells and acute myeloid leukemic blasts. Haematologica 84: 969–977.
35. Hattori Y, Kato H, Nitta M, Takamoto S (2001) Decrease of L-selectin expression on human CD34+ cells on freeze-thawing and rapid recovery with short-term incubation. Exp Hematol 29: 114–122.
36. Hwang ST, Singer MS, Giblin PA, Yednock TA, Bacon KB, et al. (1996) GlyCAM-1, a physiologic ligand for L-selectin, activates beta 2 integrins on naive peripheral lymphocytes. J Exp Med 184: 1343–1348.
37. Sperandio M (2006) Selectins and glycosyltransferases in leukocyte rolling in vivo. Febs J 273: 4377–4389.
38. Vanneaux V, El-Ayoubi F, Delmau C, Driancourt C, Lecourt S, et al. (2010) In vitro and in vivo analysis of endothelial progenitor cells from cryopreserved umbilical cord blood: are we ready for clinical application? Cell Transplant 19: 1143–1155.
39. Axelsson S, Faresjo M, Hedman M, Ludvigsson J, Casas R (2008) Cryopreserved peripheral blood mononuclear cells are suitable for the assessment of immunological markers in type 1 diabetic children. Cryobiology 57: 201–208.
40. Shreffler WG, Visness CM, Burger M, Cruikshank WW, Lederman HM, et al. (2006) Standardization and performance evaluation of mononuclear cell cytokine secretion assays in a multicenter study. BMC Immunol 7: 29.
41. Chen Y, Osika W, Dangardt F, Gan LM, Strandvik B, et al. (2010) High levels of soluble intercellular adhesion molecule-1, insulin resistance and saturated fatty acids are associated with endothelial dysfunction in healthy adolescents. Atherosclerosis 211: 638–642.
42. McHale JF, Harari OA, Marshall D, Haskard DO (1999) TNF-alpha and IL-1 sequentially induce endothelial ICAM-1 and VCAM-1 expression in MRL/lpr lupus-prone mice. J Immunol 163: 3993–4000.
43. Fan F, Stoeltzing O, Liu W, McCarty MF, Jung YD, et al. (2004) Interleukin-1beta regulates angiopoietin-1 expression in human endothelial cells. Cancer Res 64: 3186–3190.
44. Alexander JP, Acott TS (2001) Involvement of protein kinase C in TNFalpha regulation of trabecular matrix metalloproteinases and TIMPs. Invest Ophthalmol Vis Sci 42: 2831–2838.
45. Makino M, Baba M (1997) A cryopreservation method of human peripheral blood mononuclear cells for efficient production of dendritic cells. Scand J Immunol 45: 618–622.
46. Zhou JY, Montefiori DC (1997) Antibody-mediated neutralization of primary isolates of human immunodeficiency virus type 1 in peripheral blood mononuclear cells is not affected by the initial activation state of the cells. J Virol 71: 2512–2517.
47. Weinberg A, Louzao R, Mussi-Pinhata MM, Cruz ML, Pinto JA, et al. (2007) Quality assurance program for peripheral blood mononuclear cell cryopreservation. Clin Vaccine Immunol 14: 1242–1244.
48. Reimann KA, Chernoff M, Wilkening CL, Nickerson CE, Landay AL (2000) Preservation of lymphocyte immunophenotype and proliferative responses in cryopreserved peripheral blood mononuclear cells from human immunodeficiency virus type 1-infected donors: implications for multicenter clinical trials. The ACTG Immunology Advanced Technology Laboratories. Clin Diagn Lab Immunol 7: 352–359.
49. Ruel M, Suuronen EJ, Song J, Kapila V, Gunning D, et al. (2005) Effects of off-pump versus on-pump coronary artery bypass grafting on function and viability of circulating endothelial progenitor cells. J Thorac Cardiovasc Surg 130: 633–639.

3

Cryopreserved Dental Pulp Tissues of Exfoliated Deciduous Teeth Is a Feasible Stem Cell Resource for Regenerative Medicine

Lan Ma[1,2], Yusuke Makino[1], Haruyoshi Yamaza[2], Kentaro Akiyama[3], Yoshihiro Hoshino[2], Guangtai Song[4], Toshio Kukita[1], Kazuaki Nonaka[2], Songtao Shi[3], Takayoshi Yamaza[1]*

1 Department of Molecular Cell Biology and Oral Anatomy, Graduate School of Dental Science, Kyushu University, Fukuoka, Japan, **2** Department of Pediatric Dentistry, Graduate School of Dental Science, Kyushu University, Fukuoka, Japan, **3** Center for Craniofacial Molecular Biology, Herman Ostrow School of Dentistry of USC, University of Southern California, Los Angeles, California, United States of America, **4** Department of Pedodontics, School of Stomatology, Wuhan University, Wuhan, China

Abstract

Human exfoliated deciduous teeth have been considered to be a promising source for regenerative therapy because they contain unique postnatal stem cells from human exfoliated deciduous teeth (SHED) with self-renewal capacity, multipotency and immunomodulatory function. However preservation technique of deciduous teeth has not been developed. This study aimed to evaluate that cryopreserved dental pulp tissues of human exfoliated deciduous teeth is a retrievable and practical SHED source for cell-based therapy. SHED isolated from the cryopreserved deciduous pulp tissues for over 2 years (25–30 months) (SHED-Cryo) owned similar stem cell properties including clonogenicity, self-renew, stem cell marker expression, multipotency, *in vivo* tissue regenerative capacity and *in vitro* immunomodulatory function to SHED isolated from the fresh tissues (SHED-Fresh). To examine the therapeutic efficacy of SHED-Cryo on immune diseases, SHED-Cryo were intravenously transplanted into systemic lupus erythematosus (SLE) model MRL/*lpr* mice. Systemic SHED-Cryo-transplantation improved SLE-like disorders including short lifespan, elevated autoantibody levels and nephritis-like renal dysfunction. SHED-Cryo amended increased interleukin 17-secreting helper T cells in MRL/*lpr* mice systemically and locally. SHED-Cryo-transplantation was also able to recover osteoporosis bone reduction in long bones of MRL/*lpr* mice. Furthermore, SHED-Cryo-mediated tissue engineering induced bone regeneration in critical calvarial bone-defect sites of immunocompromised mice. The therapeutic efficacy of SHED-Cryo transplantation on immune and skeletal disorders was similar to that of SHED-Fresh. These data suggest that cryopreservation of dental pulp tissues of deciduous teeth provide a suitable and desirable approach for stem cell-based immune therapy and tissue engineering in regenerative medicine.

Editor: Niels Olsen Saraiva Câmara, Universidade de Sao Paulo, Brazil

Funding: This work was supported by the grants from Ministry of Education, Culture, Sports, Science and Technology of Japan for Challenging Exploratory Research Project (no. 24659815 to TY and no. 23659967 to KN) and for Young Scientists (B) (no. 20790260 to HY) of Japan Society for Promotion of Science and from National Institute of Dental and Craniofacial Research, National Institutes of Health, Department of Health and Human Services, USA (R01DE17449 and R01DE019156 to SS). The funders had no role in study design, data collection and analysis, decision to publish, or preparation of the manuscript.

Competing Interests: The authors have declared that no competing interests exist.

* E-mail: yamazata@dent.kyushu-u.ac.jp

These authors contributed equally to this work.

Introduction

Mesenchymal stem cells (MSCs) have been isolated from a variety of fetal and adult tissues and considered as an ideal candidate source for cell-based therapy due to their unique properties such as multipotency and immunomodulatory functions [1]. Many researchers have investigated to apply MSCs as progenitors of osteoblasts for bone tissue engineering. Clinical evidences support the efficacy of MSC-based skeletal tissue regeneration [2,3]. On the other hand, MSCs exert striking regulatory effects on immune cells such as T- and B-lymphocytes, dendritic cells and natural killer cells [4,5]. This immunological traits of MSCs lead to take clinical advantages to immune diseases such as acute graft-versus-host-disease (GVHD) [4,6], hematopoietic stem cell (HSC) engraftment [7,8] and systemic lupus erythematosus (SLE) [9].

Recent discovery has evaluated that fresh dental pulp tissues of human exfoliated deciduous teeth preserve MSC population, termed SHED [10]. SHED display typical stem cell properties including clonogenicity, cell proliferation and multipotency to differentiate into odontoblast/osteoblast-, adipocyte-, and neural cell-like cells [10]. SHED also express a unique *in vivo* tissue regeneration capability of forming dentin/pulp and bone/bone marrow structures when subcutaneously transplanted into immunocompromised mice [10]. SHED implantation govern bone repair in critical-sized bone defects in mouse calvarias [11] and swine mandible [12]. Moreover, systemic SHED-transplantation exhibited effective improvement on SLE-like disorders including hyper-autoantibody levels, renal dysfunction and hyperactivity of interleukin 17 (IL-17)-producing helper T (Th17) cells, in MRL/*lpr* mice [13]. Therefore SHED are considered to be a feasible and promising cell source for cell-based tissue engineering and immune therapy in regenerative medicine.

Exfoliated deciduous teeth possess advantages of minimal invasiveness and easily accessible tissue source in comparison with other human tissues such as bone marrow and adipose tissue [10]. However, the effective preservation of deciduous teeth has remained a primary concern for clinical applications of SHED. In addition, SHED isolation is impractical immediately after the exfoliation of deciduous teeth because the opportunity of the exfoliation is unpredictable. Recently, cryopreservation of human cells and tissues is amenable to be a reliable and feasible approach for stem cell storage [14]. Herein, we cryopreserved dental pulp tissues of exfoliated deciduous teeth for over 2 years and investigated the effects of the long term cryopreservation on the recovering of SHED properties including *in vitro* and *in vivo* biological and immunological properties. Furthermore, we assessed the therapeutic efficacy of the recovered SHED from the cryopreserved deciduous dental pulp tissues on immune modulation and bone regeneration in SLE model-MRL/*lpr* and bone defect model-immunocompromised mice.

Materials and Methods

Ethics Statement

Procedures using human samples (exfoliated deciduous teeth and peripheral blood) were conducted in accordance with the Declaration of Helsinki and approved by the Kyushu University Institutional Review Board for Human Genome/Gene Research (Protocol Number: 393-01). We obtained written informed consent from all the children's parents on the behalf of the children participants involved in this study. All animal experiments were approved by the Institutional Animal Care and Use Committee of Kyushu University (Protocol Number: A21-044-1) and conformed to all the guidelines outlined in the Guide for the Care and Use of Laboratory Animals by the National Institutes of Health (NIH).

Human Subjects

Human exfoliated deciduous teeth were collected as discarded biological/clinical samples from children (5–7-year-old) at Department of Pediatrics of Kyushu University Hospital. Human peripheral blood mononuclear cells (PBMNCs) were collected from healthy volunteers (28–34 year-old).

Cryopreservation of Deciduous Dental Pulp Tissues of Exfoliated Deciduous Teeth and Isolation and Culture of SHED

A protocol for the cryopreservation of dental pulp tissues of exfoliated deciduous teeth was summarized in **Figure S1**. Dental pulp tissues were separated from exfoliated deciduous teeth. Half of the samples were mixed with a cryopreserved medium at 4°C and kept overnight at −80°C. The cryopreserved medium consisted of 10% dimethyl sulfoxide (DMSO) (Sigma, St Louis, MO) and 90% fetal bovine serum (FBS) (Equitech-Bio, Kerrville, TX). They were transferred into liquid nitrogen and stored for over 2 years (25–30 months). The other half of fresh deciduous dental pulp tissues were treated to isolate SHED (SHED-Fresh). SHED from the cryopreserved deciduous pulp tissues (SHED-Cryo), as well as SHED-Fresh, were isolated by an adherent colony-forming unit fibroblasts (CFU-F) method [10,13,15]. The cryopreserved tissues were quickly thawed at 37°C. Both cryopreserved and fresh dental pulp tissues were digested with 0.3% collagenase type I (Worthington Biochemicals, Lakewood, NJ) and 0.4% dispase II (Sanko Junyaku, Tokyo, Japan) in phosphate buffered saline (PBS, pH 7.4) for 60 min at 37°C. Single-cell suspensions were obtained through a 70-µm-cell strainer (BD Bioscience, San Jones, CA). The cells (1×10^6) were seeded on T-75 flasks, washed with PBS after 3 hours and cultured at 37°C in 5% CO_2 with a growth medium containing 15% FBS (Equitech-Bio), 100 µM L-ascorbic acid 2-phosphate (WAKO Pure Chemical, Osaka, Japan), 2 mM L-glutamine (Nacalai Tesque, Kyoto, Japan) and 100 U/ml penicillin and 100 µg/ml streptomycin (Nacalai Tesque) in alpha Modification of Eagle's Medium (alphaMEM) (Invitrogen, Carlsbad, CA) and subsequently cultured for 14–16 days until obtaining adherent colonies. The adherent colonies-forming cells were recognized as SHED as described before [10,13]. The colonies-forming cells were passed and sub-cultured in the growth medium. The medium was changed twice a week.

Antibodies

Antibodies used in this study are summarized in the **Table S1**.

Mice

C57BL/6J and Balb/cA-nu/nu mice (female, 6 week-old) were purchased from CLEA Japan. (Tokyo, Japan). C57BL/6J MRL/*lpr* mice (female, 6 week-old) were from Japan SLC. (Shizuoka, Japan). They were housed in temperature- and light-controlled environmental conditions with a 12-hour light and dark cycle, and permitted ad *libitum* consumption of water and standard pellet chow.

Histology of Cryopreserved Dental Pulp Tissues of Exfoliated Deciduous Teeth

Cryopreserved deciduous pulp tissues were fixed with 4% paraformaldehyde (PFA) in PBS and immersed in O.C.T. compound (Sakura Finetek Japan, Tokyo, Japan). The frozen specimens were cut into 6-µm thick sections. Some sections were stained with hematoxylin and eosin (H&E). The others were immunostained with anti-STRO-1 and anti-human CD146 antibodies by using SuperPicture kit (Invitrogen). Subclass-matched antibodies were used for negative controls for immunohistochemistry. The sections were observed under a microscope BIORVO BZ-9000 (Keyence Japan, Osaka, Japan).

CFU-F Assay

Cells (10×10^3) isolated from frozen/fresh deciduous dental pulp tissues were seeded on 100-mm culture dishes and cultured in the growth medium for 16 days. The flasks were treated with 4% PFA and 0.1% toluidine blue in PBS for 18 hours. Colonies containing >50 cells were recognized as single colony clusters under a microscope as previously [10,13]. The numbers of the colonies were counted.

Population Doubling Assay

Cells were seeded on T-75 culture flasks (BD Bioscience). When the cells reached at sub-confluent condition, they were passed. These steps were repeated until the cells lost dividing capability. The population doubling score was calculated at every passage according to the equation: $\log_2$ (number of final harvested cells/ number of initial seeded cells) and the total scores were determined by adding up each population doubling score in each sample as described in previous reports [10,13].

Bromodeoxyuridine (BrdU) Incorporation Assay

Passaged 3 (P3) SHED were seeded at 1×10^3 per well on 35-mm dishes and cultured in the growth medium for 2 days. BrdU reagent (Invitrogen) was added at 1:100 in the medium and subsequently cultured additionally for 24 hours. The cells were fixed in 70% ethanol for 15 min, treated with the BrdU staining kit (Invitrogen) and lightly stained with hematoxylin according to the company's instruction. The stained samples were observed under BIORVO microscope (BZ-9000, Keyence Japan, Osaka, Japan). Seven images were randomly selected to calculate BrdU-positive nuclei number in each sample. Cell proliferation capacity was shown as a percentage of BrdU-positive nuclei over total nucleated cells [13].

Telomerase Activity Assay

Telomerase activity was measured by a telomere repeat amplification protocol assay using the quantitative telomerase detection kit (Allied Biotech, Inc., Ijamsville, MD) applied with real-time PCR as referred to our previous reports [13]. As positive control, HEK293T cells were used. Some extracts from each cell were heated at 85°C for 10 min and used as negative control samples. The average starting quantity (SQ) of fluorescence units was used to compare the telomerase activity among the samples.

Flow Cytometric Analysis for SHED

Passaged 3 (P3) SHED were cultured at 50–60% confluent condition. Cells (100×10^3/100 μl) were stained with primary specific antibodies. Subclass-matched antibodies were used as negative controls. Flow cytometry was analyzed on FACSCalibur flow cytometer (BD Bioscience) [13]. The number (percentage) of positive cells was determined using CellQuest software (BD Bioscience) by comparison with the corresponding control cells stained with the subclass-matched antibody in which a false-positive rate of less than 1% was accepted.

Immunofluorescence for Cultured Cells

The cells were fixed with 4% PFA in PBS and blocked with PBS containing 10% normal serum matched to the secondary antibodies. The samples were incubated with the specific antibodies to cell surface markers or the subclass-matched antibodies overnight at 4°C and treated with CF 633-conjugated-secondary antibodies (Biotium, Hayward, CA). Finally, they were stained with 4', 6-diamidino-2-phenylindole (DAPI) (Dojindo, Kumamoto, Japan) and observed under the BIORVO microscope (Keyence Japan). Numbers of cells positive to the specific antibodies and nuclei stained with DAPI were counted and shown as a percentage of positive cells over total nucleated cells.

Semi-quantitative Reverse Transcription Polymerase Chain Reaction (RT-PCR)

Total RNA was isolated from the cultures using TRIzol (Invitrogen) and digested with DNase I. The cDNA was synthesized from 100 ng of total RNA using Revatra Ace (TOYOBO, Osaka, Japan). The specific primer pairs are listed in **Table S2**. PCR was performed using gene specific primers and RT-PCR Quick Taq HS DyeMix (TOYOBO) at 94°C for 2 min for one cycle and then react for 40 cycles with denature at 94°C for 45 sec, annealing at 56°C for 45 sec, extension at 72°C for 60 sec as one cycles, with a final 10-min extension at 72°C. Finally, 5 μl of each amplified PCR product was analyzed by 2% agarose gel electrophoresis and visualized by ethidium bromide staining. The intensity of bands was measured by using Image-J soft ware and normalized to an internal control gene, *glyceraldehyde 3-phosphate dehydrogenase* (*GAPDH*).

In vitro Multipotent Assay

***In vitro* dentinogenic/osteogenic induction assay.** SHED (P3, 5×10^3) were grown on 60-mm dishes in the growth medium until confluent condition and induced with an dentinogenic/osteogenic medium supplemented with 1.8 mM potassium dihydrogen phosphate (Sigma, St. Louis, MO) and 10 nM dexamethasone (Sigma) in the growth medium. The dentinogenic/osteogenic medium were changed twice a week. One week after the induction, dentinogenic/osteogenic markers were analyzed by semi-quantitative RT-PCR and alkaline phosphatase (ALP) activity test [13]. For mineralized nodule assay, the cultures were stained with 1% Alizarin Red-S (Sigma) at 4 weeks post the osteoinduction. The Alizarin red-positive area was analyzed using NIH image software Image-J and shown as a percentage of Alizarin red-positive area over total area [13].

***In vitro* chondrogenic induction assay.** SHED (P3, 2×10^6) were aggregated in a 15 mL tube and cultured with Dulbecco's modified Eagle's medium (Invitrogen) supplemented with 15% FBS (Equitech-Bio), 2 mM L-glutamine (Nacalai Tesque), 100 μM L-ascorbate-2-phosphate (Wako Pure Chemicals), 2 mM sodium pyruvate (Nacalai Tesque), 1% insulin-transferring-selene mixture (ITS) (BD Biosciences), 100 nM dexamethasone (Sigma), 10 ng/ml transforming growth factor $beta_1$ ($TGFbeta_1$) (PeproTech, Rocky Hill, NJ) and 100 U/ml penicillin and 100 μg/ml streptomycin (Nacalai Tesque). The chondrogenic medium were changed twice a week. After 3-week induction, chondrocyte-specific genes were analyzed by semi-quantitative RT-PCR.

***In vitro* adipogenic induction assay.** Cultured SHED (P3, 5×10^3/dish) were cultured until the confluent condition and induced in an adipogenic medium with the growth medium plus 500 μM isobutyl-methylxanthine (Sigma), 60 μM indomethacin (Sigma), 0.5 μM hydrocortisone (Sigma) and 10 μM insulin (Sigma). After 6-week induction, the cultures were stained with 0.3% Oil red O (Sigma) to detect lipid droplets. The samples stained with Oil red O were treated with isopropanol and the absorbance of the extracts were measured at 520 nm. Adipocyte-specific genes were also analyzed by semi-quantitative RT-PCR.

***In vitro* endothelial cell induction assay.** P3 SHED (5×10^3 cells) were seeded on fibronectin-coated 35-mm dishes (BD Biosciences) and cultured by using a commercial available endothelial cell growth media kit, EGM-2, (Lonza, Basel, Switzerland) with 100 U/ml penicillin and 100 μg/ml streptomycin (Nacalai Tesque) according to the manufacture's instruction. The endothelial growth medium was changed every 2 days. The cultures were fixed 7 days after the induction. Endothelial cell differentiation was determined by immunofluorescence with anti-CD31 and CD34 antibodies.

***In vitro* neuronal cell induction assay.** P3 SHED (5×10^3 cells/well) were plated in laminin-coated 35-mm dishes (BD Biosciences) and cultured in a neurogenic medium containing supplemented with 1xN2 supplement (Invitrogen), 10 ng/ml basic fibroblast growth factor (bFGF) (PeproTech), 10 ng/ml epidermal growth factor (EGF) (PeproTech), 100 U/ml penicillin and 100 μg/ml streptomycin (Nacalai Tesque) in Neurobasal A (Invitrogen) for 21 days according to the recent study [10]. The medium was changed with 50% of fresh medium twice a week. Neural cell differentiation was determined by immunofluorescence with anti-*glial fibrillary acidic protein*, anti-neural filament M and anti-betaIII tubulin antibodies.

***In vitro* hepatic induction assay.** P3 SHED (P3, 5×10^3) were seeded on fibronectin-coated 35-mm dishes (BD Biosciences) and cultured with Iscove's modified Dulbecco's medium (Invitrogen) supplemented with 10 nM dexamethasone (Sigma), 1% ITS

(BD Biosciences), 20 ng/ml EGF (PeproTech), 10 ng/ml bFGF (PeproTech), 20 ng/ml hepatocyte growth factor (PeproTech), 20 ng/ml oncostain M (PeproTech), 100 U/ml penicillin and 100 μg/ml streptomycin (Nacalai Tesque) for 4 weeks. Hepatocyte-specific gene albumin was assayed by semi-quantitative RT-PCR.

In vivo Tissue Regeneration Assay

Xenogenic transplantation was performed as previously [13,16]. Cultured SHED (P3) in the growth medium (4×10^6) were mixed with a carrier, hydroxyapatite tricalcium phosphate (HA/TCP) (40 mg, Zimmer Inc., Warsaw, IN). The mixtures were subcutaneously transplanted into the dorsal surface of 8–10-week-old Balb/c *nude/nude*. The implants were harvested 8 weeks after the surgery, fixed with 4% PFA and decalcified with 10% EDTA. Frozen sections were cut and used for H&E staining or immunofluorescence observed under a BIORVO microscope (BZ-9000, Keyence Japan).

In vivo Self-renewal Assay

In vivo self-renewal assay with a sequential transplantation was referred to a recent study [17]. SHED (P3) (4×10^6) were implanted with HA/TCP carrier (40 mg) (Zimmer) into primary Balb/c *nu/nu* mice. Eight weeks after the transplantation, the primary implants were harvested and treated with 0.4% dispase II (Sanko Junyaku) for 60 min at 37°C to obtain single cells from the implants. The cells were stained with R-PE-conjugated anti-human CD146 and magnetic-beads-conjugated anti-R-PE antibodies (Miltenyi Biotec, Bergisch Gladbach, Germany) and sorted magnetically to obtain human CD146-positive cells. The purity was confirmed by flow cytometry with anti-human CD146 or anti-mouse CD146 antibody. The sorted cells were seeded at low density and cultured to obtain CFU-F. Expanded CFU-F-forming cells (P1) (4×10^6) were subcutaneously transplanted with HA/TCP carriers (40 mg) (Zimmer) into secondary immunocompromised mice for 8 weeks and the secondary implants were assayed by H&E staining and immunofluorescence.

Histological Analysis of Implant Tissues

Implant tissues were fixed with 4% PFA in PBS overnight and decalcified with 5% EDTA solution (pH 7.4). Frozen sections were cut into 8-μm thickness and treated with H&E staining and immunofluorescence with anti-human mitochondria, anti-STRO-1 or anti-human CD146 antibody. The sections were observed under BIORVO microscope (BZ-9000, Keyence Japan). To measure newly-formed area of mineralized tissue, seven fields were randomly selected and the newly-formed area was calculated by NIH image-J software, and shown as a percentage over total tissue area as described before [13,16].

Single Colonies-derived Cell Assay

As referred to a previous study [18], cells isolated from cryopreserved deciduous dental pulp tissues were seeded at 1, 2 or 4 cells per well on 24-well multiplates with the growth medium. The wells containing more than two attached cells were excluded from further analysis. The single cell-attached wells were cultured for 14–16 days and obtained CFU-F-forming cells. The single colony-forming cells were used for population doubling, BrdU incorporation and dentinogenic/osteogenic analyses.

Induction Assay of Th17 cells

Induction of Th17 cells was performed as previously [13]. Human $CD4^+CD25^-$ T cells were magnetically sorted from human PBMNCs using $CD4^+CD25^+$ regulatory T cell isolation kit (Miltenyi Biotec). They (1×10^6/well) were activated by plate-bounded anti-CD3 (5 μg/ml) and soluble anti-CD28 (1 μg/ml) antibodies for 3 days and loaded on SHED (20×10^3/well) with recombinant human $TGFbeta_1$ (2 μg/ml, PeproTech) and IL-6 (50 μg/ml, PeproTech) for 3.5 days. T cells were stained with anti-PerCP-conjugated CD4 and FITC-conjugated anti-CD8a antibodies, treated with R-PE-conjugated anti-IL-17 and APC-conjugated anti-interferon gamma (IFNgamma) antibodies using Foxp3 staining buffer kit (eBioscience, San Diego, CA) and analyzed on FACSCalibur (BD Biosciences). IL-17 levels in the culture supernatants were analyzed by enzyme linked immunosorbent assay (ELISA) with a commercial available kit (R&D Systems) according to the manufacture's instruction.

Systemic Transplantation of SHED into MRL/lpr Mice

The protocol for systemic transplantation of SHED into MRL/*lpr* mice was referred to our previous reports [9,13]. Briefly, SHED (0.1×10^6/10 g body weight) or PBS were systemically infused into MRL/*lpr* mice via the tail vein at the age of 16 weeks old. Their survival was inspected daily until died. Peripheral blood, urine, kidney, axial lymph nodes and long bones were collected from MRL/*lpr* mice at the age of 20 weeks old according to our previous studies [9,13].

Assay for Autoantibodies, Immunoglobulins, Biomarkers and Cytokines in Peripheral Blood Serum, Urine and Tissue Samples

Biological factors in mouse biological samples were measured by ELISA with commercial available kits (anti-double strand DNA [dsDNA] IgG and IgM, anti-nuclear antibody [ANA], albumin and complement C3: alpha diagnostic [San Antonio, TX]; IL-17 and nuclear factor kappaB ligand [sRANKL] [R&D Systems]; C-terminal telopeptides of type I collagen [CTX]: Nordic Bioscience [Herlev, Denmark]) according to the manufactures' instructions. Creatinine (R&D Systems) and urine protein (Bio Rad, Hercules, CA) were also assayed by colorimetry according to the manufactures' instructions.

Flow Cytometric Analysis for Mouse Peripheral Blood Th17

To measure Th17 cells in mouse peripheral blood, mouse PBMNCs were incubated with PerCP-conjugated anti-CD4, FITC-conjugated anti-CD8a, followed by the treatment with R-PE-conjugated anti-IL-17 and APC-conjugated anti-IFNgamma antibodies using a Foxp3 staining buffer kit (eBioscience). The stained cells were analyzed on FACSCalibur (BD Bioscience).

Histopathological Analysis of Mouse Kidney

Kidney samples were fixed with 4% PFA in PBS overnight. After washed with PBS, some samples were processed for paraffin embedding and cut into 6-μm thick paraffin sections. The others were cut into 8-μm thick cryosections. The paraffin sections were treated with H&E, Gomori trichrome or Periodic Acid Schiff (PAS) staining. The cryosections were stained with anti-mouse complement C3 antibody and treated with CF 633-conjugated-secondary antibodies (Biotium). Finally, they were stained with DAPI (Dojindo) and observed under a BIORVO microscope (BZ-9000, Keyence Japan).

In vivo Tracing Assay of SHED

To trace the trafficking of SHED transfused in MRL/*lpr* mice, SHED were labeled with carboxyfluorescein diacetate succinimidyl ester (CFSE) labeling. Single suspended SHED population (10×10^6/ml) was incubated with CFSE solution (Invitrogen) for 10 min at 37°C. The labeled cells (1×10^6) were intravenously injected into the tail vein of MRL/*lpr* mice. Lymph nodes, kidneys and femurs of the mice were harvested either 24 h or 1 week after the infusion. The tissue samples were fixed with 4% PFA in PBS overnight. Only bone samples were treated with 10% EDTA. All of the samples were cut into 6-µm frozen sections, stained with DAPI (Dojindo) and observed under a microscope.

Bone Phenotype Analysis

The femoral bone samples were fixed with 4% PFA in PBS overnight and analyzed by micro-computed tomography (microCT) with Skyscan 1076 scanner (Skyscan, Kontich, Belgium). Bone mineral density (BMD) and bone structural parameters including bone volume/trabecular volume (BV/TV), trabecular thickness (Tb.Th), trabecular number (Tb.N), trabecular separation (Tb.Sp) and trabecular space (Tb.Spac) were calculated using CT Analyzer software (Skyscan) according to the manufacture's instruction. After microCT analysis, the bone samples were decalcified with 10% EDTA and cut into 6-µm paraffin sections. The paraffin sections were treated with H&E staining. Some sections were stained with a tartrate-resistant acid phosphate (TRAP) staining solution containing 0.01% naphthol AS-MX phosphate (Sigma), 50 mM tartrate (Sigma) and 0.06% fast red violet LB salt (Sigma) in 0.1 M acetate buffer (pH 5.0) for 10 min. TRAP-positive cells were counted and quantified as previously [19].

In vitro Osteoclastic and Osteogenic Assays

Bone marrow were flashed out from the bone cavity of femurs and tibias with heat-inactivated 3% FBS (Equitech-Bio) in PBS. For osteoclastic assay, the bone marrow nucleated cells (BMCs) were seeded at 1×10^6 per well on the 24-well culture plates with 10 ng/ml macrophage colony stimulating factor (M-CSF) (R&D Systems) and 25 ng/ml sRANKL (PeproTech) in alphaMEM for 5–6 days. The medium was changed on Day 3. The cultures were treated with TRAP staining [19] and the TRAP-positive cells with muitinuclei (n>3) were counted [19]. The BMCs (1×10^6) were also seeded on 35-mm culture dishes and cultured in an osteogenic medium containing 10% FBS (Equitech-Bio), 2 mM L-glutamine (Nacalai Tesque), 100 µM L-ascorbic acid 2-phosphate (WAKO pure chemicals), 2 mM beta-glycerophosphate (Sigma), 10 nM dexamethasone (Sigma), 100 U/ml penicillin and 100 µg/ml streptomycin (Nacalai Tesque) in alphaMEM [20]. Four weeks after the induction, the osteogenic cultures were stained with 1% Alizarin Red (Sigma). The mineralized area was quantified by using NIH Image-J [13].

Bone Regeneration in Calvarial Bone Defects

The bone defect models in calvarial bone were generated on immunocompromised mice as described before [11]. Briefly, P3 SHED (4×10^6) were mixed with HA/TCP carriers (40 mg) (Zimmer). Calvarial bones, especially parietal bones, were removed to make a critical defect area. The mixtures of SHED and HA/TCP were placed on the defect area and the calvarial skins were sutured. Twelve weeks after the implantation, the calvariae were fixed with 4% PFA in PBS overnight and imaged by Skyscan 1076 scanner (Skyscan). Then, the samples were decalcified with 10% EDTA and cut into 6-µm thick paraffin sections. The sections were treated with H&E staining and TRAP staining. Some of sections were used for immunofluorescence with anti-human CD146 antibody.

Statistical Analysis

Student's *t*-test was used to analyze significance between 2 groups. A *P* value of less than 0.05 was considered as a significant difference.

Results

SHED-Cryo Possess MSC Properties

SHED have been elucidated to possess MSC characteristics including clonogenicity, self renew, multipotency, *in vivo* regeneration and *in vitro* immunomodulatory functions [10,13]. Here, we demonstrated the effects of cryopreservation of deciduous pulp tissues on these SHED properties. Dental pulp tissues were removed from exfoliated deciduous teeth and frozen in the freezing medium (10% DMSO and 90% FBS) at −80°C overnight followed by the preservation in a liquid nitrogen-filled tank for over 2 years (**Figure S1**). The remaining tooth bodies were returned to the donor children. The long-term cryopreserved tissues were quickly thawed at 37°C before used (**Figure S1**). Histological analysis confirmed that the cryopreserved tissues showed dense connective tissues containing blood vessels and nerve fibers (**Figure 1A**), but not odontoblastic cells. The odontoblast layer may be lost because of mechanical damage or freezing fracture of the pulp samples. Early MSC markers STRO-1 and CD146-positive cells were detected at the perivascular area (**Figure 1B**) as dental pulp stem cells (DPSCs) were localized around the blood vessels in adult dental pulp tissues [21]. These data suggested a possibility of the remaining of MSCs in the cryopreserved deciduous pulp tissues.

Next, we isolated MSCs from the cryopreserved dental pulp tissues of exfoliated deciduous teeth, SHED-Cryo, by CFU-F approach (**Figure S1**), which is a classical and standard method [15]. When seeded at low density, single cells were adhered to the plastic culture dishes and then divided to generate cell clusters (**Figure 1C**). The clonogenic cell clusters exhibited different size and varied density (**Figure 1D**). The colony forming efficiency of SHED-Cryo showed similar to that of SHED-Fresh (**Figure 1E**). BrdU incorporation assay demonstrated that the proliferation capacity of SHED-Cryo maintained at a high level (**Figure 1F**) similar to that of SHED-Fresh (**Figure 1G**). Flow cytometric analysis verified that SHED-Cryo were positive to STRO-1, CD146, CD73 and CD105 but negative to hematopoietic cell markers CD34, CD45 and CD14 (**Figure 1H**). SHED-Cryo were also positive to CD90 (over 95%) and the positive level in SHED-Cryo was similar to that in SHED-Fresh (data not shown). SHED-Cryo shared the immunophenotype with SHED-Fresh (**Figure 1I**). RT-PCR demonstrated that SHED-Cryo expressed genes for embryonic stem cells, *NANOG* and *octamer 4*, and for neural crest cells, *NOTCH1*, *NESTIN* and *low-affinity neural growth factor* (**Figure 1J**), as seen in SHED-Fresh (**Figure 1K**). Expression of a neural crest cell marker, Nestin, was also detected in SHED-Cryo by flow cytometry (**Figure 1L**) and the expression in SHED-Cryo showed a similar level to that in SHED-Fresh (data not shown). Recent studies [22,23] demonstrated the effect of cryopreservation on the expression of NANOG, octamer 4 and Nestin in deciduous teeth stem cells and may support, at least in partially, our cryopreserved effect of deciduous dental pulp tissues on SHED properties. Taken together, these data suggested that SHED-Cryo retained primitive MSC properties likely to SHED-Fresh.

SHED-Cryo Own Multipotency

Four weeks after dentinogenic/osteogenic induction, SHED-Cryo were capable of forming Alizarin Red-positive nodules

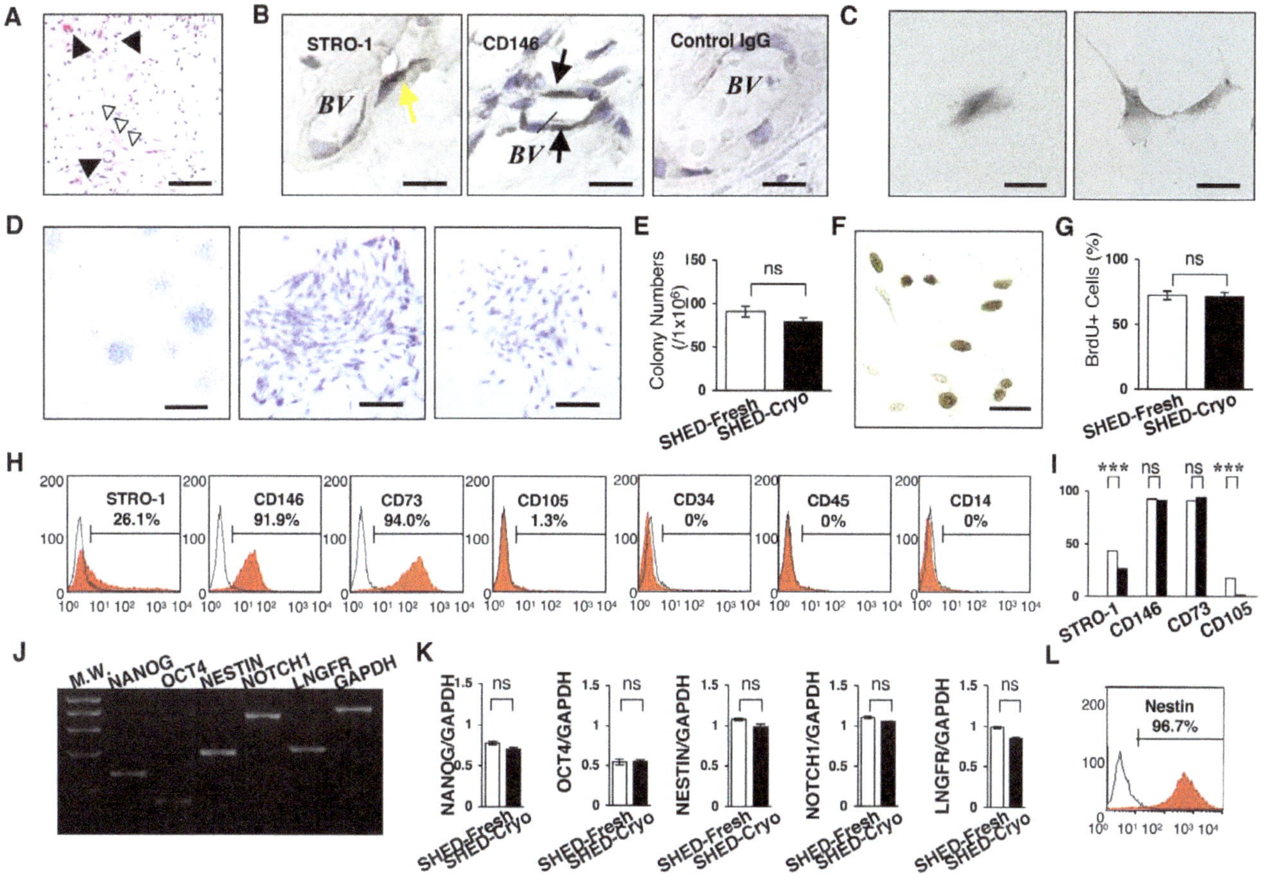

Figure 1. Clonogenicity, cell proliferation capacity and stem cell marker expression of SHED-Cryo. (**A**) Histology of cryopreserved dental pulp tissue of exfoliated deciduous teeth. Black arrowheads: blood vessel, white arrowheads: nerve fibers. H&E staining. (**B**) Localization of MSC markers in the cryopreserved deciduous pulp tissues. Yellow arrows: STRO-1-positive cells, black arrows: CD146-positive cells. *BV*: blood vessel, Control: subclass-matched antibody staining. (**C–E**) CFU-F assay. Formation of a clonogenic cell cluster from a single attached cell (**C**). Images of attached colonies of SHED-Cryo. Toluidine blue staining (**D**). Comparison of CFU-F number (**E**). (**F, G**) Cell proliferation assay. Immunostaining of BrdU-positive nuclei. (**F**). Comparison of cell proliferation (**G**). (**H, I**) Flow cytometry of MSC markers in SHED-Cryo. Representative histograms (**H**). Comparison of STRO-1, CD146, CD73 and CD105. Black columns: SHED-Cryo, white columns: SHED-Fresh (**I**). (**J, K**) Gene expression of embryonic stem and neural crest cell markers. MW: molecular weight markers (**J**). Comparative analysis of *NANOG, octamer 4* (*OCT4*), *NESTIN, NOTCH1* and *low affinity nerve growth factor receptor* (*LNGFR*) (**K**). (**L**) Flow cytometry of Nestin in SHED-Cryo. **A, B:** n = 3. **C–L:** n = 5 for all group. **A–E:** Bar = 30 μm (**A**), 5 μm (**B, C, F**), 1 mm (**D**, left) 25 μm (**D**, middle and right). **G, I, K:** ****P*<0.005. ns: no significance. The graph bars represent mean±SD.

(**Figure 2A**). SHED-Cryo showed a high ALP activity (**Figure 2B**) and expressed odontoblast/osteoblast-specific genes *runt-related gene 2*, *ALP*, *osteocalcin*, and *dentin sialophosphoprotein* (**Figure 2C**) after the 1-week induction. SHED-Cryo expressed chondrocyte-specific genes for *SOX9*, *aggrecan* and *type X collagen* 3 weeks after chondrogenic culture (**Figure 2D**). SHED-Cryo expressed adipocyte-like phenotypes including accumulation of Oil red-O-positive droplets (**Figure 2E**) and expression of adipocyte-specific genes *lipoprotein lipase* and *peroxisome proliferator activated receptor-gamma2* (**Figure 2F**) 6 weeks after adipogenic induction. SHED-Cryo expressed albumin gene, one of hepatocyte-specific genes, 4 weeks after hepatogenic induction (**Figure 2G**). Immunofluorescence revealed that SHED-Cryo expressed endothelial cell markers CD31 and CD34 1 week after endothelial cell-induction (**Figure 2H**) and neural cell markers neurofilament M and tubulin betaIII 3 weeks after neural cell-induction (**Figure 2I**). SHED-Cryo also exhibited similar capabilities of differentiating into odontoblasts/osteoblasts, adipocytes, chondrocytes, hepatocytes, endothelial cells and neural cells to SHED-Fresh (**Figure 2**). These data indicated that SHED-Cryo maintained multipotency as MSCs.

SHED-Cryo Showed *in vivo* Mineralized Tissue Regeneration

Eight weeks after subcutaneous transplantation of SHED-Cryo with HA/TCP carriers into immunocompromised mice (**Figure S2**), dentin/pulp-like complex and bone/bone marrow-like structures were formed around the surface of HA/TCP by histological analysis (**Figure 3A**). Immunofluorescence showed that human mitochondria-positive cells were arranged on mineralized matrix (**Figure 3C**). Human specific STRO-1 and CD146 antibodies-positive cells were also detectable on the regenerated mineralized matrix (**Figure 3D**). On the other hand, control transplant-tissues that implanted only HA/TCP carriers without SHED-Cryo did not express any mineralized tissue and human specific antibody-positive cells (**Figure S3**). Therefore these results suggested that SHED-Cryo were responsible cells for mineralized tissue formation in the implant tissues. SHED-Cryo formed similar amount of the regenerated mineralized tissues when compared to

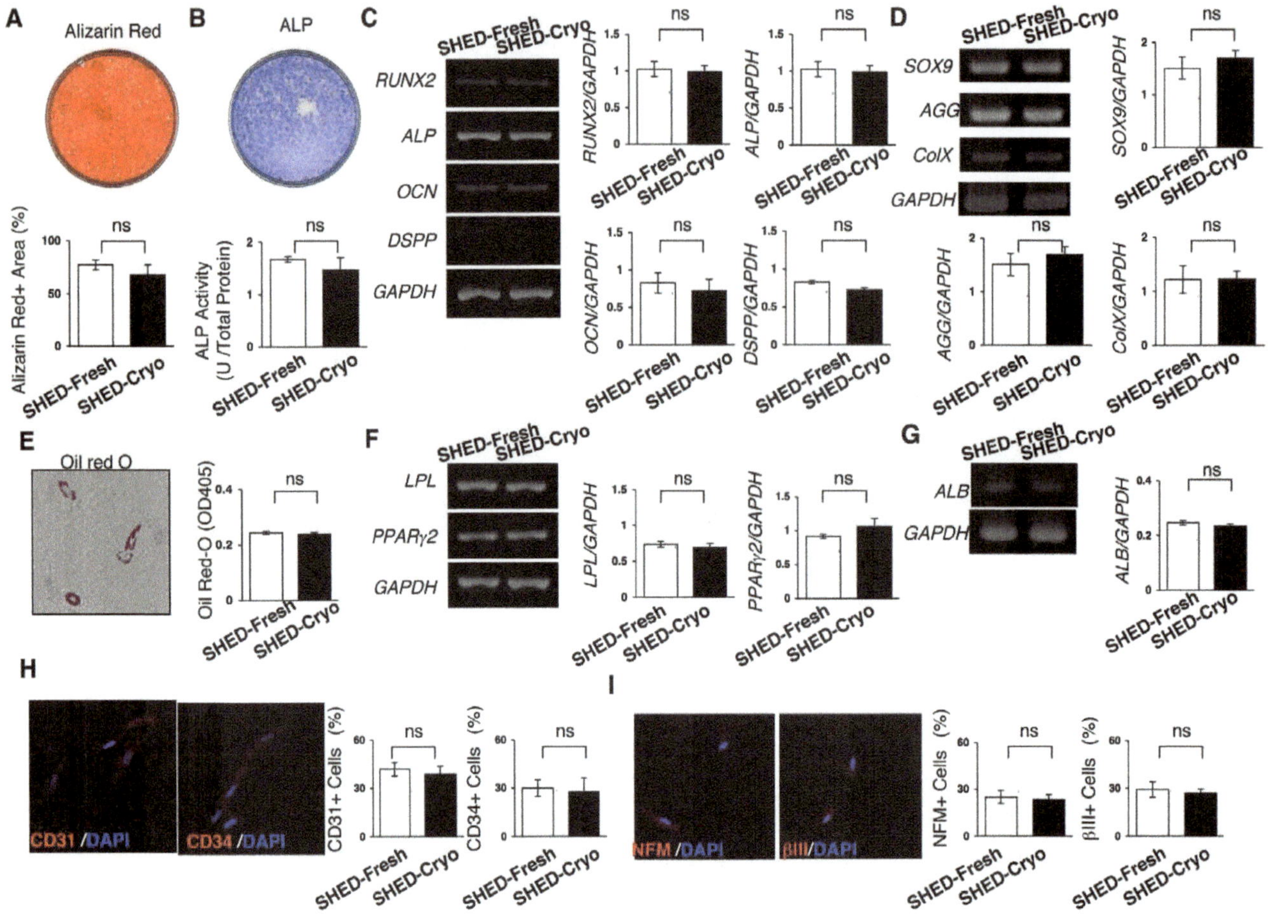

Figure 2. Multipotency of SHED-Cryo. (**A–C**) Dentinogenic/osteogenic differentiation capacity. Images of Alizarin Red staining (**A**) and alkaline phosphatase (ALP) activity (**B**) of SHED-Cryo. Comparison of Alizarin Red-positive (Alizarin Red+) area (**A**), ALP activity (**B**) and odontoblast/osteoblast-specific genes, *runt-related gene 2 (RUNX2), ALP, osteocalcin (OCN), and dentin sialophosphoprotein (DSPP)* (**C**). (**D**) Chondrogenic differentiation capacity. Comparison of chondrocyte-specific genes, *SOX9, aggrecan (AGG)* and *type X collagen (ColX)*. (**E, F**) Adipogenic differentiation assay. A representative image of Oil Red-O staining and comparison of Oil Red-O accumulation (**E**). Comparison of adipocyte-specific genes *lipoprotein lipase (LPL)* and *peroxisome proliferator activated receptor-gamma2 (PPARgamma2)* (**F**). (**G**) Hepatogenic differentiation capacity. Comparison of hepatocyte-specific gene *albumin (ALB)*. (**H**) Endothelial cell differentiation assay. Comparison of endothelial cell markers CD31 and CD34. (**H**) Neural cell differentiation assay. Comparison of neural cell markers neurofilament M (NFM) and tubulin betaIII (betaIII). **A–I:** n = 5 for all group. ns: no significance. The graph bars represent mean±SD.

SHED-Fresh (**Figure 3B**). These data indicated that SHED-Cryo retained a unique *in vivo* regenerative capacity likely to SHED-Fresh. Although the origin of bone marrow cells are host cells in bone marrow MSC-implants [24,25], the origin of cells in bone marrow and dental pulp in SHED-implants have not been elucidated. Furthermore study will be needed to clarify the origin of the dental pulp cells and bone marrow cells in the SHED-implants in future.

SHED-Cryo Retain Self-renewal Capability

To evaluate self-renewal capability of SHED-Cryo, sequential transplantation, which is one of traditional and gold standard methods to identify the self-renewal capability of stem cells [17], was performed (**Figure S2**). CD146 is considered to be a critical cell surface marker for human MSCs [18]. Cell population was isolated from the primary implants and stained with human CD146 antibody. Then human CD146 antibody-positive cells were purified by a magnet sorting system. Flow cytometric analysis confirmed that the sorted cells were positive to human CD146 (>95%) but negative to mouse CD146 (0%) (**Figure 3E**), evaluating the high purity of the sorted cells as human stem cells. When the sorted cells were seeded at low density, they were capable of forming CFU-Fs that exhibited positive to human CD146 by immunostaining (**Figure 3F**), indicating that the CD146-positive human cells maintained as intact MSCs in the transplant tissues after the long-term implantation. After the colony-forming cells were secondarily transplanted into immunocompromised mice for 8 weeks, the secondary implants contained similar dentin/pulp complex-like structures (**Figure 3G**) to the mineralized structural complexes in the primary transplants. Cells positive to anti-human CD146 or anti-human specific mitochondria antibodies were localized on the mineralized matrix (**Figure 3H**). Population doubling and telomerase activity are associated with self-renewal potential of stem cells [26]. Population doubling assay indicated a prolonged and time-limited cell proliferation in SHED-Cryo (**Figure 3I**). Telomerase activity test revealed the lower activity in SHED-Cryo (**Figure 3J**). Collectively, these results verified that SHED-Cryo represented a self-renewal capacity.

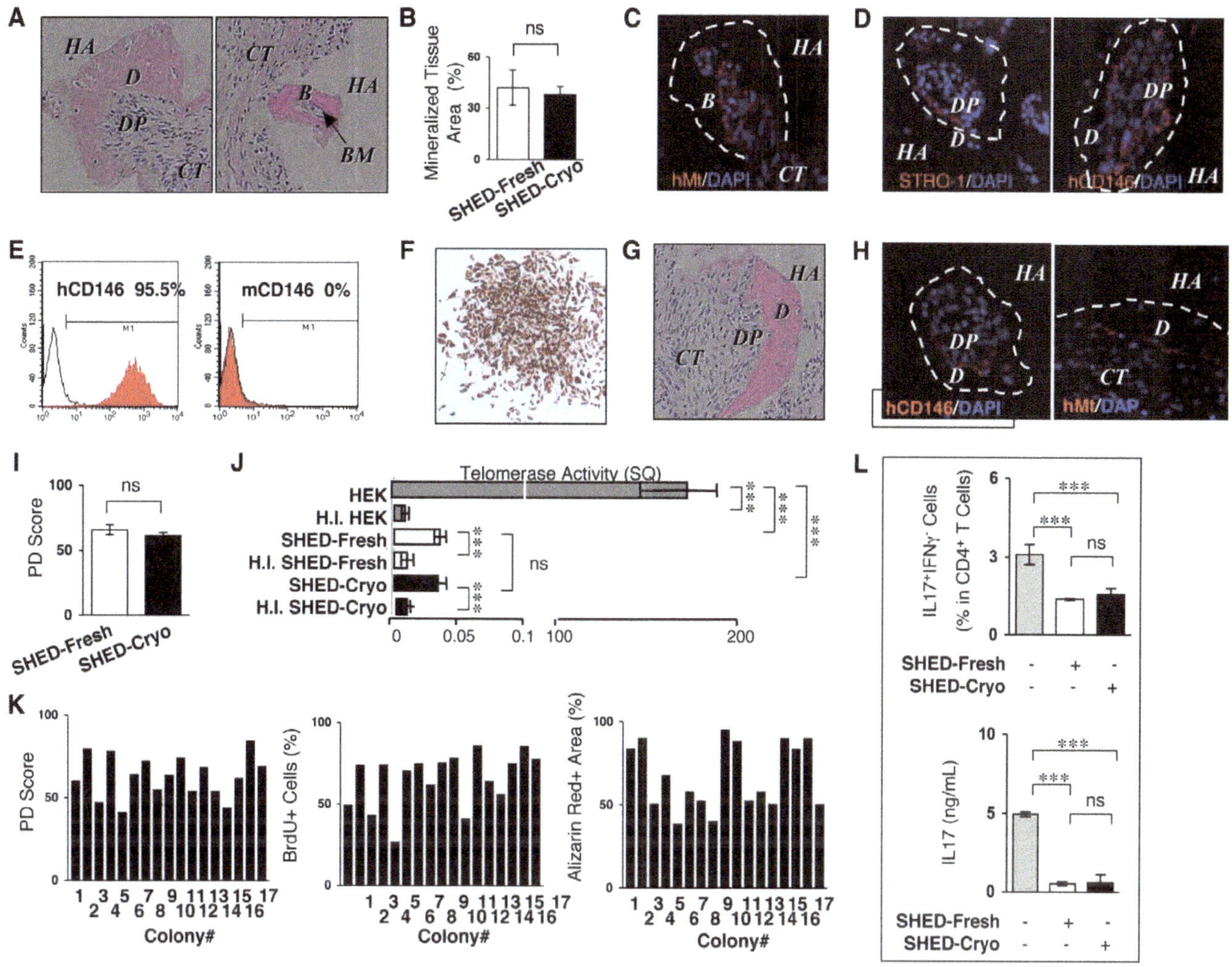

Figure 3. Tissue regeneration capability, self-renewal potency, heterogeneity and *in vitro* immunomodulatory functions of SHED-Cryo. (**A–D**) Images of primary transplant tissues of SHED-Cryo. H&E staining (**A**). Comparison of newly formed-mineralized tissue (B). Immunofluorescence with anti-human specific mitochondria (hMt) (**C**) and anti-STRO-1/human CD146 (hCD146) (**D**) antibodies. (**E, F**) Purity of hCD146 antibody-sorted cells from primary transplants. Flow cytometry with hCD146 and mouse CD146 (mCD146). (**E**). Immunocytochemistry with hCD146 antibody of sorted cell-derived CFU-F (F). (**G, H**) *in vivo* self-renewal assay. Images of secondary transplant tissues. H&E staining (**G**). Immunofluorescence with anti-hCD146/anti-Mt antibodies. (**H**). (**I**) Comparison of population doubling (PD) scores. (**J**) Comparison of telomerase activity. (**K**) Single-colony-derived cell assay with 17 single cell colonies from a cryopreserved deciduous pulp tissues. (**L**) *In vitro* direct immunosuppressive effects of SHED-Cryo on human Th17 cells. **A–J, L**: $n = 5$ for all group. **A, C, D, F, H**: *B*: bone, *BM*: bone marrow, *CT*: connective tissue, *D*: dentin, *DP*: dental pulp, *HA*: HA/TCP, **I**: HEK: HEK293 cells, H.I. HEK: heat inactivated HEK, H.I. SHED-Cryo: heat inactivated SHED-Cryo, H.I. SHED-Fresh: heat inactivated SHED-Fresh. **C, D, H**: Dot lined areas: mineralized tissue. Nuclei are counterstained with DAPI. **B, H, I, L**: $^{***}P<0.005$, ns: no significance. The graph bars represent mean±SD.

SHED-Cryo are Heterogeneous Population

To identify the heterogeneity of SHED-Cryo, total 17 clonogenic single-colonies were acquired from a cryopreserved dental pulp tissue of exfoliated deciduous tooth. Each single-colony-derived SHED-Cryo showed various population-doubling score, diverse cell proliferation capacity and varied Alizarin red-positive area (**Figure 3K**). These findings indicated that SHED-Cryo displayed heterogeneous as seen in SHED-Fresh [10].

Immunomodulatory Properties of SHED-Cryo *in vitro*

To explore whether SHED-Cryo display immunomodulatory capacity to Th17 cells as seen in SHED-Fresh (**Figure 3L**) [13], SHED-Cryo were co-cultured with anti-CD3 and anti-CD28 antibodies-activated human $CD4^{+}CD25^{-}$ T cells under the stimulation with IL-6 and $TGFbeta_1$. SHED-Cryo were able to inhibit both the differentiation of $CD4^{+}IL17^{+}IFNgamma^{-}$ Th17 cells and the secretion of IL-17 (**Figure 3L**), suggesting that SHED-Cryo maintained *in vitro* inhibitory effect on Th17 cells.

SHED-Cryo Transplantation Prolongs the Life Span and Improves Autoimmune Disorder and Renal Dysfunction in MRL/*lpr* Mice

SLE-like autoimmune disorders appear around age 7–8 weeks in MRL/*lpr* mice. Peripheral levels of autoimmune antibodies increase extremely from about 12 weeks of age in MRL/*lpr* mice. To evaluate the therapeutic potency of SHED-Cryo in severed SLE condition as seen in previous studies [13], SHED-Cryo were intravenously transfused to human SLE model MRL/*lpr* mice at the age of 16 weeks old (**Figure S4**). Kaplan-Meyer curve demonstrated that systemic SHED-Cryo-transplantation significantly prolonged the lifespan of MRL/*lpr* mice (**Figure 4A**). Mantel-Haenszel test evaluated that the median survival time was 155.0, 219.0 and 212.5 days in control, SHED-Fresh-transplanted and SHED-Cryo-transplanted MRL/*lpr* mice, respectively. Elevated serum levels of autoantibodies to ANA and anti-dsDNA IgG and IgM antibodies, which are critical clinical markers in human SLE therapy, were markedly decreased in SHED-Cryo transplanted MRL/*lpr* mice (**Figure 4B**). Increased peripheral immunoglobulins including IgG_1, IgG_{2a}, IgG_{2b} and IgM were also significantly reduced in SHED-Cryo-transplanted mice (**Figure S5**). Histopathological analysis demonstrated that SHED-Cryo-transplantation prevented renal nephritis associated with hypercellularity, mesangial matrix hyperplasia and basal membrane disorder in MRL/*lpr* mice (**Figure 4C**). Immunofluorescence showed that complement C3-deposition in the glomeruli of the kidney in MRL/*lpr* mice was disappeared after SHED-Cryo-transplantation (**Figure 4C**). SHED-Cryo-transplantation elevated serum albumin level and reduced urine protein (**Figure 4D**), meanwhile, decreased serum creatinine in MRL/*lpr* mice (**Figure 4D**). SHED-Cryo-transplantation displayed similar therapeutic effects on the lifespan, autoantibody levels and renal function in MRL/*lpr* mice with SHED-Fresh-transplantation (**Figures 4 and S5**). These findings provided that SHED-Cryo retained therapeutic efficacy on MRL/*lpr* mice.

Transplantation of SHED-Cryo Suppresses Peripheral Th17 Cells in MRL/*lpr* Mice

Th17-cell regulation is a critical therapeutic strategy for SLE treatment [27]. The present study showed that the levels of peripheral Th17 cells and IL-17 were remarkably reduced in SHED-Cryo-received MRL/*lpr* mice at the age of 20 weeks old (**Figures 5A–5C**). This Th17-cell suppressive effect was similar to the effect of SHED-Fresh-transplantation on MRL/*lpr* mice (**Figures 5B and 5C**) [13], suggesting that SHED-Cryo maintained *in vivo* immunomodulatory functions.

Systemically Infused SHED-Cryo Home to Lymph Node and Kidney in MRL/*lpr* Mice and Regulate the Local Immune Microenvironment

To assess the homing ability of SHED-Cryo into sites of injured tissues, CFSE-labeled SHED-Cryo were intravenously injected into MRL/*lpr* mice at the age of 16 weeks old. The high frequency of CFSE-positive SHED-Cryo was observed in the lymph nodes and kidneys, particularly in the glomeruli, on one day after the

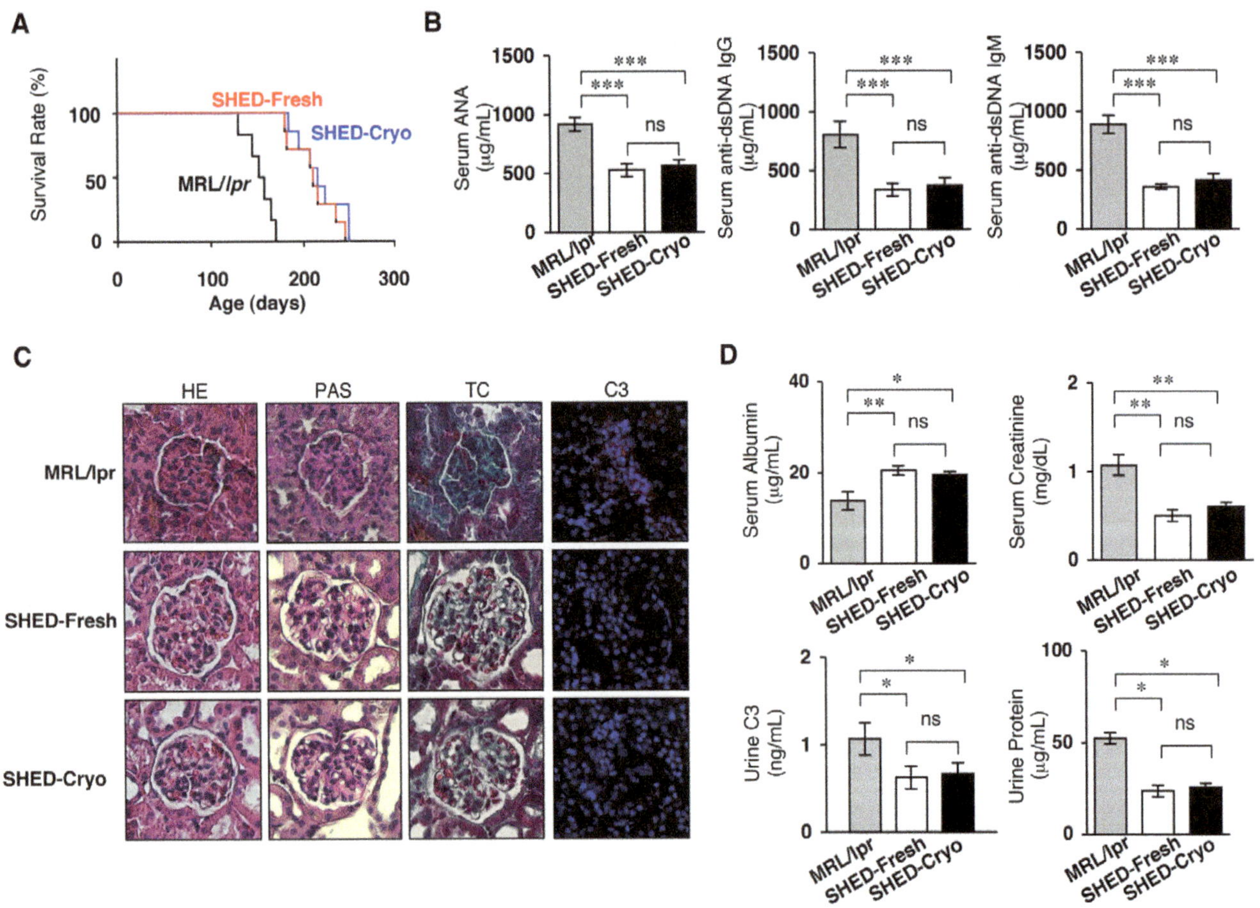

Figure 4. Systemic SHED-Cryo-transplantation improves lifespan and SLE-like disorders in MRL/*lpr* mice. (**A**) Kaplan-Meier survival curve of MRL/*lpr* mice. (**B**) ELISA of serum levels of autoantibodies ANA and anti-dsDNA IgG and IgM antibodies. (**C**) Histopathology of kidneys. *G* and dot-circled area: glomerular. HE: H&E staining, TC: Gomori trichrome staining, PAS: Periodic acid-Schiff staining, C3: Immunofluorescence of Complement C3. DAPI staining. (**D**) Levels of serum albumin and creatinine and urine C3 and protein. **A–D**: MRL/*lpr*: control group, SHED-Cryo: SHED-Cryo-transplant group, SHED-Fresh: SHED-Fresh-transplant group. **A**: n = 7, **B–D**: n = 5 for all group. **B, D**: *$P<0.05$, **$P<0.01$, ***$P<0.005$, ns: no significance. The graph bars represent mean±SD.

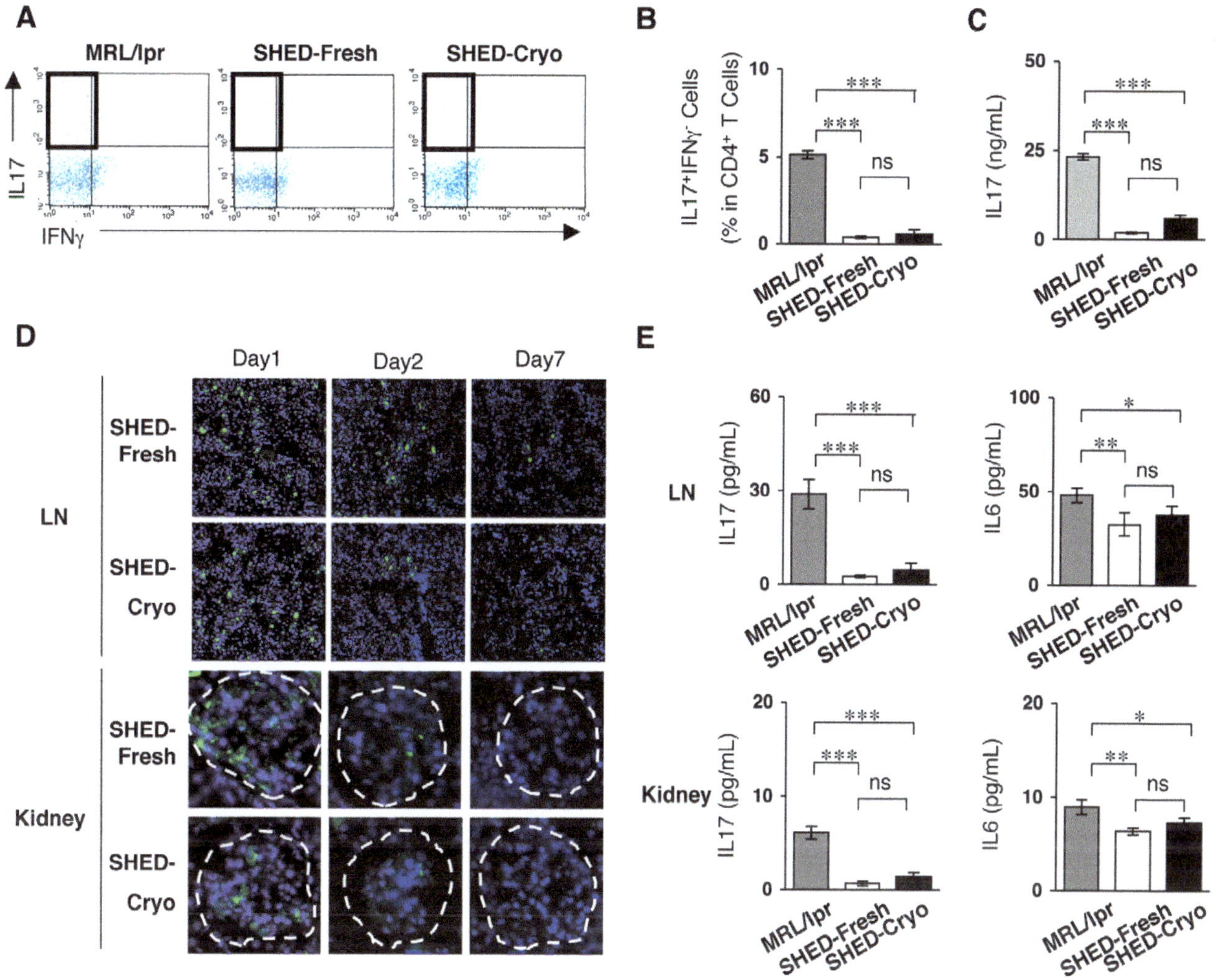

Figure 5. SHED transplantation suppresses circulating and local levels of Th17 cells in MRL/*lpr* mice. (**A, B**) Flow cytometry of peripheral $CD4^{+}IL17^{+}IFNgamma^{-}$ Th17 cells. (**C**) Serum levels of IL-17. (**D**) Homing of systemically infused CFSE-labeled SHED-Cryo and SHED-Fresh to lymph node (LN) and kidney of MRL/*lpr* Mice after 1- (Day 1), 2- (Day 2) or 7- (Day 7) day transplantation. Dot-circled area: glomerular. (**E**) ELISA of IL-17 an IL-6 levels in lymph node and kidney. **A–E**: n = 5 for all group. MRL/*lpr*: control group, SHED-Cryo: SHED-Cryo-transplant group, SHED-Fresh: SHED-Fresh-transplant group. **B, C, E**: **P*<0.05, ***P*<0.01, ****P*<0.005, ns: no significance. The graph bars represent mean±SD.

transplantation (**Figures 5D and S6**). The frequency of CFSE-positive SHED-Cryo decreased gradually in both tissues from the 2nd day to 7th day after the transplantation (**Figures 5D and S6**). The localization of SHED-Cryo were similar to that of SHED-Fresh (**Figures 5D and S6**). The levels of inflammatory cytokines IL-17 and IL-6 in the lymph nodes and kidneys were significantly reduced in SHED-Cryo-transfused MRL/*lpr* mice, as well as SHED-Fresh-transfused mice, when compared to control MRL/*lpr* mice (**Figure 5E**). Taken together, these *in vivo* studies suggested that SHED-Cryo were capable of homing to lesional sites and might improve the pathological environments of damaged tissues.

Systemic SHED-Cryo-transplantation Improves Osteoporotic Skeletal Disorder in MRL/*lpr* Mice

The homing ability of SHED-Cryo into bones was analyzed 1 day after the infusion into MRL/*lpr* mice at the age of 16 weeks old. The CFSE-positive SHED-Cryo was sparsely observed in the bone marrow space of MRL/*lpr* mice 7 days after the infusion (**Figure S7**). The frequency of CFSE-positive SHED-Cryo was similar to that of SHED-Fresh (**Figure S7**). MRL/*lpr* mice expressed a remarkable osteoporotic bone-loss in their long bones (**Figures 6A–6D**). Systemic SHED-Cryo-transplantation was capable of increasing BMD and recovering trabecular bone structures in the long bones of MRL/*lpr* mice (**Figures 6A and 6B**). TRAP-positive cells were significantly reduced in the long bones of SHED-Cryo-transplanted group compared to the non-transplanted control group (**Figure 6E**). SHED-Cryo-transplantation markedly reduced serum levels of sRANKL and CTX in MRL/*lpr* mice (**Figure 6F**). Flow cytometry revealed that $CD4^{+}IL\text{-}17^{+}IFNgamma^{-}$ Th17 cells were significantly reduced in bone marrow cells of SHED-Cryo- and SHED-Fresh-transplanted MRL/*lpr* mice compared to the control MRL/*lpr* mice (**data not shown**), suggesting that immunomodulatory functions of SHED-Cryo and SHED-Fresh may contribute to reduce the bone reduction in MRL/*lpr* mice. To examine *ex vivo* osteoclastogenesis and osteogenesis, BMCs were isolated from control, SHED-Fresh-transplanted and SHED-Cryo-transplanted MRL/*lpr* mice, termed control-, SHED-Fresh- and SHED-Cryo-BMCs, respectively. When BMCs were stimulated with M-CSF and sRANKL, the number of TRAP-positive multinucleated cells was significantly reduced in SHED-Cryo-BMCs than in control-BMCs (**Figure 6G**). Osteogenic analysis showed that Alizarin Red-positive area was shared larger in SHED-Cryo-BMCs than control-BMCs (**Figure 6H**). SHED-Cryo-transplanted MRL/*lpr* mice expressed similar bone regenerative effects to SHED-Fresh-transplanted MRL/*lpr* mice (**Figure 6**). These data indicated that SHED-Cryo-transplantation improved osteoporotic disorder in MRL/*lpr* mice. Furthermore studies will be needed to evaluate the therapeutic mechanism of SHED to osteoporotic disorder in immune diseases likely SLE.

SHED-Cryo-implantation Repairs the Calvarial Bone Defects in Immunocompromised Mice

From the present *in vivo* tissue regeneration capability of SHED-Cryo, we hypothesized that SHED-Cryo could regenerate bone tissues in bone defects, as seen in SHED-Fresh (**Figure 7A–C**) [11]. We generated critical calvarial bone defects on immunocompromised mice and implanted SHED-Cryo with HA/TCP carrier onto the defect area (**Figure S8**). SHED-Cryo-implantation was able to regenerate the calvarial defects with a large amount of bone-like structures and bone-marrow-like components compared to implantation with only HA-TCP (**Figure 7A–C**). The amount of regenerated bone and bone marrow at SHED-Cryo-implanted sites was similar to that at SHED-Fresh implanted sites (**Figure 7A–C**). Immunofluorescence with anti-human CD146 antibody revealed that SHED-Cryo were responsible cells for bone regeneration in SHED-Cryo- and SHED-Fresh-implanted group, but not in HA/TCP-implanted group (**Figure 7B**). These findings suggested that SHED-Cryo and SHED-Fresh could be differentiated into bone-forming cells to contribute to repair the bone defect. A large number of TRAP-positive osteoclast-like cells were found in the regenerated bone tissues in SHED-Cryo-implanted sites, as well as in that of SHED-Fresh-implanted sites, but less in HA/TCP-implanted sites (**Figure 7D**), suggesting that the regenerated bone tissues might indicate

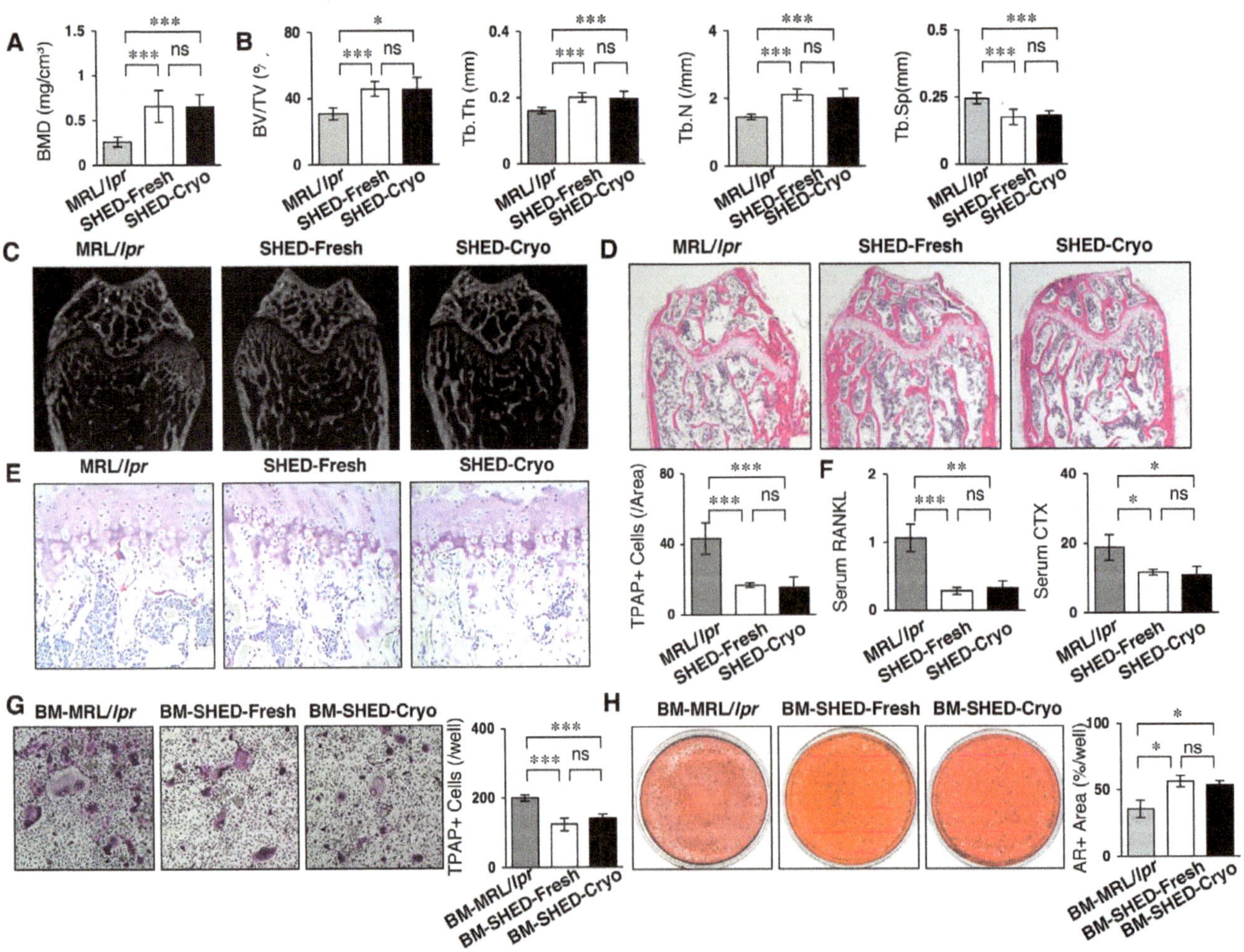

Figure 6. SHED-Cryo transplantation ameliorates osteoporotic bone disorder in MRL/*lpr* mice. (A, B) MicroCT analysis of tibiae. BMD (**A**). Trabecular parameters, bone volume ratio to tissue volume (BV/TV), trabecular thickness (Tb.Th), and trabecular number (Tb.N) along with increased trabecular separation (Tb.Sp) (**B**). (**C, D**) MicroCT (**C**) and histological (**D**) images of trabecular bone structures of tibiae. H&E staining (**D**). (**E, F**) *In vivo* osteoclast activity. TRAP staining (**E**). ELISA of serum sRANKL and C-terminal telopeptides of type I collagen (CTX) (**F**). (**G**) *Ex vivo* sRANKL-induced osteoclastogenesis. TRAP+ cells: TRAP-positive osteoclast-like cells. (**H**) *Ex vivo* osteogenic capacity. Alizarin red-positive (AR+) area after four-week induction. **A–H:** n = 5 for all groups. **A–F:** MRL/*lpr*: control group, SHED-Cryo: SHED-Cryo-transplant group, SHED-Fresh: SHED-Fresh-transplant group, **G, H:** BM-MRL/*lpr*: control MRL/*lpr* mice-derived bone marrow cells, BM-SHED-Cryo: SHED-Cryo-transplanted mice-derived bone marrow cells, BM-SHED-Fresh: SHED-Fresh-transplanted mice-derived bone marrow cells. **A, B, E, F:** *$P<0.05$, ***$P<0.005$ (vs. MRL/*lpr*). **G, H:** *$P<0.05$, **$P<0.01$, ***$P<0.005$ (vs. BM-MRL/*lpr*), ns: no significance. **A, B, E–H:** The graph bars represent mean±SD.

a physiological bone-remodeling ability by osteoclasts and osteoblasts. These data implied that SHED-Cryo were a useful cell source for tissue engineered bone regeneration. Implanted MSCs are impaired by host lymphocytes through secreting the pro-inflammatory cytokines IFNgamma and TNFalpha to suppress MSC-mediated bone regeneration [28]. On the other hand, deciduous teeth stem cells are capable of regenerating bone tissues in the calvarial defects of non-immunosuppressed rats [29]. Furthermore study will be necessary to clarify the kinetics of SHED under immunocompetent conditions for future clinical tissue engineering.

Discussion

Since SHED has been identified in dental pulp of deciduous tooth and characterized as MSCs [10], deciduous dental pulp tissues have been considered a promising stem cell source. SHED or deciduous teeth stem cells express multipotency into several lineage cells including of dentin/bone-forming cells [10,30], endothelial cells [30], neural cells [10,31] and myocytes [22] *in vitro* and *in vivo*. SHED or deciduous teeth stem cells were also applied for tissue-engineering in large animal models including bone defects, muscular dystrophy and dentin defects [32–34], as well as small animal models including bone defect and spinal cord injury [11,29,35]. Recent study demonstrates *in vitro* immunomodulatory functions of SHED and evaluates the immune therapeutic efficacy on SLE-like model mice [13]. Herein, we demonstrated the feasibility of cryopreserved dental pulp tissue of human deciduous teeth in SHED-based bone tissue engineering and immune therapy. Our results indicate that long-term cryopreservation of human dental pulp tissues of deciduous teeth provides a great potential in future translational researches and clinical applications.

Postnatal stem cells have offered great promise to care diverse diseases. Besides HSCs have acquired outstanding and extensive success in a variety of human diseases such as leukemia, aplastic anemia and autoimmune diseases over half century [36,37]. To date, MSCs also admit to regenerative medicine for GVHD [6], skeletal reconstruction [2,3] and SLE [9]. Several challenges have remained to concern quality and safety in respective processes of MSC transplantation such as the cell processing and preserving *ex vivo*. Traditional bone marrow MSCs significantly reduced their frequency and multipotency donor-age-dependently [38–41]. Aspiration of bone marrow might accompany sever invasion to the donors [42]. As these significant disadvantages promote to seek alternative resources with accessible and least invasive approaches, novel MSC populations have been identified from various sources

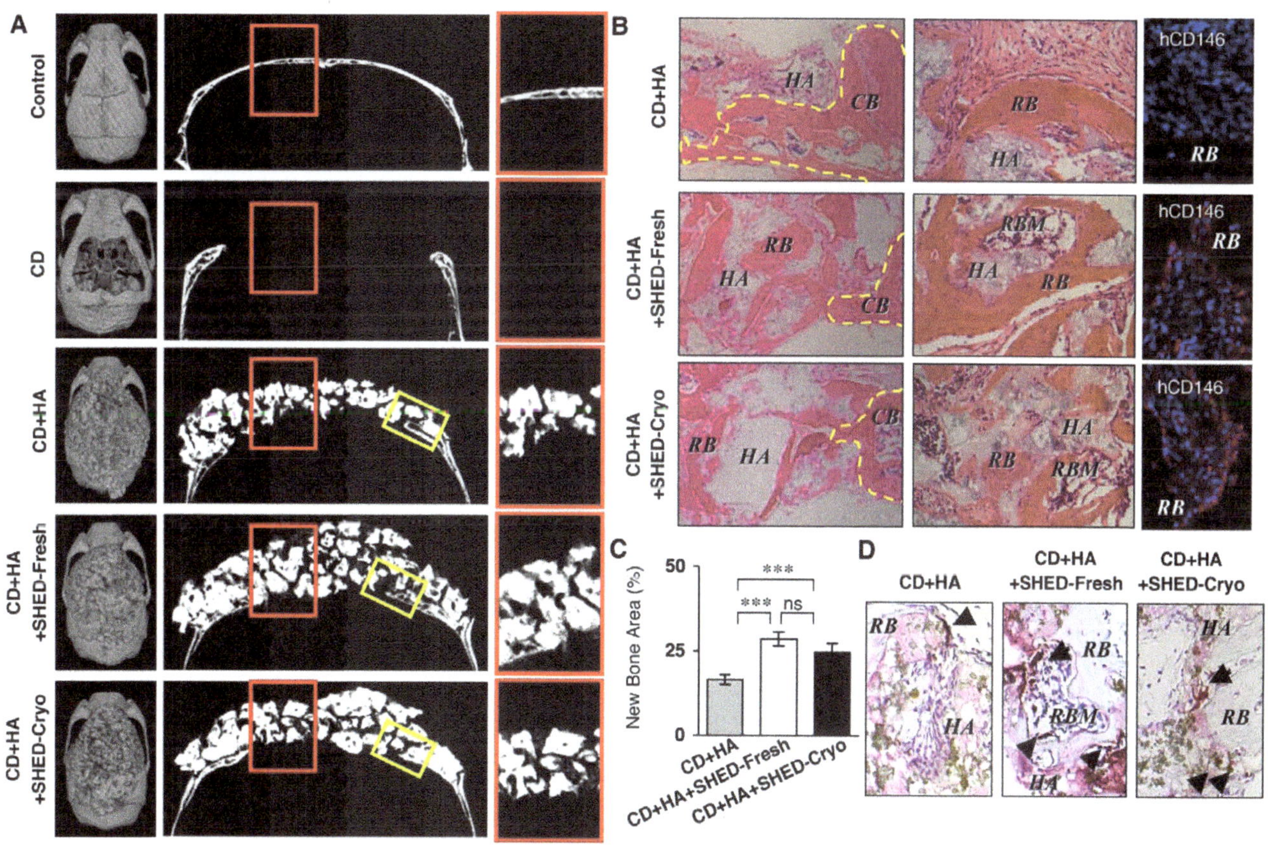

Figure 7. SHED-Cryo are capable of repairing critical calvarial bone defects in immunocompromised mice. (**A**) MicroCT images of mouse calvariae. Left panels: cranial images, middle panels: saggital images, right panels: images of red-bowed area in riddle panels. (**B**) Histology of bone regeneration in mouse calvariae. Left panels: edge parts of the defect area (yellow-boxed area in Figure 7A). H&E staining; middle panels: middle parts of the defect area (red red-bowed area in Figure 7A). H&E staining; right panels: immunofluorescence with anti-human CD146 antibody (hCD146). DAPI staining. CB, yellow dot-circled area: calvarial bone, HA: HA/TCP, RB: regenerated bone, RBM: regenerated bone marrow. (**C**) Regenerated bone area in the defect area. (**D**) Distribution of osteoclasts. Arrowheads: TRAP-positive cells. TRAP staining. **A–D:** n = 5 for all groups. Control: control (non-defect) group, CD: calvarial defect group, CD+HA: HA/TCP-implanted group, CD+HA+SHED-Fresh: SHED-Fresh-implanted group, CD+HA+SHED-Cryo: SHED-Cryo-implanted group. **C:** ***$P<0.005$, ns: no significance. The graph bars represent mean±SD.

such as adipose tissue, cord blood and dental pulp [43–45]. While the isolation process of MSCs is intricate, clinically applicable processing for storing and banking of the resources can offer great advantages for MSC-based therapy as well as the reduction of the number of staffing required for cell processing. Cryopreservation of stem cells has provided several utilities such as long-term storage, adjusting a therapeutic cell dose, reducing contamination for safety and quality in the clinical applications [34,46]. This approach has been used widely and successfully in bone marrow transplantation [47] and HSC transplantation [48]. Therefore, cryopreserved store and banking of MSC resources would be a variable, indispensable and practical approach for stem cell-based therapy.

SHED have been considered to be a primary promising source for regenerative medicine [10]. Exfoliated deciduous teeth represent the most easily, least invasively accessible and feasible resource because of their natural fate (exfoliation) and clinical abolition. SHED offered profound therapeutic efficacy to skeletal defects [11,12] and autoimmune disease [13]. On the other hand, several tasks have remained to be solved in SHED processing and banking. It is generally hard to expect a chance of the exfoliating of deciduous teeth and to maintain the tissue activity of the interests for a while after the harvesting. In addition, isolation process of SHED accompanies with several complicated, arduous and time-consuming steps and attentive operations. Recent discoveries about functional MSCs recovered from cryopreserved intact dental pulp and periodontal ligament (PDL) tissues of adult human third molars suggest that the least minimal processing may be adequate for the banking of samples [49,50]. Recovered MSCs after the cryopreservation can still maintain the immunomodulatory capacity *in vitro* [51,52]. Deciduous teeth stem cells is known to maintain the stem cell property and multipotency after the long-term cryopreservation [22,23]. In the present study, we firstly demonstrate that stem cells are capable of recovering from human dental pulp tissues of exfoliated deciduous teeth after long-term (over 2 years) cryopreservation. The recovered SHED retained superior MSC properties such as self-renew, multipotency, *in vivo* dentin/bone-regeneration and *in vitro* immunomodulatory function. In addition, transplantation of SHED-Cryo showed critical therapeutic efficiency on both immune and skeletal disorders in MRL/*lpr* mice and bone defects in the calvariae of immunocompromised mice. Taken together, these data suggest that long-term cryopreservation of dental pulp tissues of deciduous teeth is an innocuous and designable approach for clinical banking of stem cells and gives great advantages in immune therapy and bone tissue engineering of regenerative medicine. Furthermore, the present pulp tissue banking system could allow children and their parents to return the "baby teeth" bodies as the most precious souvenirs to the children's growing.

Current studies indicate that the biological recovery of the PDL stem cells and adult DPSCs are inferior to the freshly isolated stem cells after the cryopreservation of PDL tissues [50] and adult teeth [49]. Whereas, cryopreserved stem cells from apical papilla (SCAP) show a similar biological and immunomodulatory functions to freshly isolated SCAP [51]. Apical papillae have a responsibility to form and extend tooth roots at the developing stage [53,54], indicating that apical papillae are an active tissue biologically. Cryopreserved deciduous teeth stem cells can also maintain the stem cell property and multipotency [22,23]. The present study demonstrated that cryopreservation of deciduous dental pulp did not affect the biological and immunological properties of SHED. Moreover, SHED show higher cell proliferation capability than adult DPSCs [10], suggesting that deciduous pulp tissues maintain greater biological activity than adult dental pulp tissues. The discrepancy of the functionally recovery efficiency among deciduous dental pulps, adult PDLs and adult dental pulps after the cryopreservation might depend on the age and/or potential activity of donor samples, supporting the feasibility of the cryopreservation of deciduous dental pulp tissues.

In conclusion, the present cryopreservation of dental pulp tissues of human exfoliated deciduous teeth does not affect on the biological, immunological and therapeutic functions of SHED. Therefore, cryopreserved approach of deciduous dental pulp tissues not only serve as a most clinically desirable banking approach, but also provide sufficient number of SHED for critical therapeutic benefits to stem cell-based immune therapy and tissue engineering in regenerative medicine.

Supporting Information

Figure S1 A scheme of the cryopreservation and isolation of mesenchymal stem cells (MSCs) from dental pulp tissues of exfoliated deciduous teeth. Deciduous dental pulp tissues in the remnant crown (yellow-dot circled region) were removed *en bloc* mechanically, stored in a freezing medium and preserved in a liquid nitrogen tank over 2 years. The frozen tissues were thawed at 37°C and treated with an enzyme solution. SHED from the cryopreserved deciduous dental pulp tissues (SHED-Cryo), as well as SHED from fresh deciduous dental pulp tissues (SHED-Fresh), were obtained by colony forming units fibroblasts (CFU-F) method.

Figure S2 A scheme of *in vivo* tissue regeneration and self-renewal assays of SHED-Cryo. SHED-Cryo were subcutaneously transplanted with HA/TCP carrier into immunocompromised mice. Eight weeks after the implantation, the primary transplants were harvested. Some transplants were used for histological and immunofluorescent analyses. Cells were isolated from the other primary transplants and stained with human-specific CD146 antibody. Human CD146-positive cells were magnetically sorted. The purity of the cells was confirmed by flow cytometry as described in Materials and Methods. The CD146-positive cells were seeded at low density to obtain CFU-F-forming cells. The colony-forming cells were expanded and transplanted secondarily into immunocompromised mice. The secondary transplants were harvested 8 weeks after the implantation and analyzed morphologically.

Figure S3 Images of primary transplant tissues with HA/TCP alone. (**A**) H&E staining (HE). (**B-D**) Immunofluorescence with anti-human specific mitochondria (hMt) (**B**), anti-STRO-1 (**C**) and human CD146 (hCD146) (**D**) antibodies.

Figure S4 A scheme of the transplantation of SHED-Cryo into MRL/*lpr* mice (*lpr*). SHED-Cryo were infused into MRL/*lpr* mice via the tail vein at the age of 16 weeks. The mice were maintained until died for the survival assay. At 20-week-old, some mice were harvested and biological samples were collected to assess the therapeutic efficacy.

Figure S5 Systemic SHED-Cryo-transplantation improves levels of serum immunoglobulins in MRL/*lpr* mice. n = 5 for all group. $^{*}P<0.05$, $^{**}P<0.01$, $^{***}P<0.005$, ns: no significance. The graph bars represent mean±SD. MRL/lpr: non-transplanted group, SHED-Fresh: SHED-Fresh-transplanted group, SHED-Cryo: SHED-Cryo-transplanted group.

Figure S6 Homing of systemically infused SHED-Cryo to lymph node and kidney of MRL/*lpr* Mice. Images of CFSE-labeled SHED-Cryo and SHED-Fresh in lymph nodes (LN) and kidneys of MRL/*lpr* mice 1 (Day 1) or 7 (Day 7) days after the transplantation. CFSE: CSFE image, DAPI: DAPI image, SHED-Fresh: SHED-Fresh-infused group, SHED-Cryo: SHED-Cryo-infused group.

Figure S7 Homing of systemically infused SHED-Cryo to bone of MRL/*lpr* Mice. Images of CFSE-labeled cells in bone of MRL/*lpr* mice 7 days after the transplantation. CFSE: CFSE image, DAPI: DAPI image, CFSE/DAPI: Merged image of CFSE and DAPI images, SHED-Fresh: SHED-Fresh-infused group, SHED-Cryo: SHED-Cryo-infused group.

Figure S8 A scheme of the transplantation of SHED-Cryo into calvarial bone defect of immunocompromised mice. SHED-Cryo were expanded and mixed with HA/TCP carriers. Calvarial bones, especially parietal bone area (P), were removed to generate a bone defect on immunocompromised mice. SHED & HA/TCP mixture were implanted to cover over the defect area. Twelve weeks after the implantation, the samples were harvested and analyzed by microCT and histology.

Table S1 The list of antibodies.

Table S2 The list of primer pairs for RT-PCR.

Author Contributions

Conceived and designed the experiments: TY SS. Performed the experiments: TY LM YM HY KA YH. Analyzed the data: TY LM YM HY KA YH GS KN TK SS. Contributed reagents/materials/analysis tools: TY HY. Wrote the paper: TY.

References

1. Porada CD, Almeida-Porada G (2010) Mesenchymal stem cells as therapeutics and vehicles for gene and drug delivery. Adv Drug Deliv Rev 62: 1156–1166.
2. Kwan MD, Slater BJ, Wan DC, Longaker MT (2008) Cell-based therapies for skeletal regenerative medicine. Hum Mol Genet 17: R93–98.
3. Panetta NJ, Gupta DM, Quarto N, Longaker MT (2009) Mesenchymal cells for skeletal tissue engineering. Panminerva Med 51: 25–41.
4. Aggarwal S, Pittenger MF (2005) Human mesenchymal stem cells modulate allogeneic immune cell responses. Blood 105: 1815–1822.
5. Nauta AJ, Fibbe WE (2007) Immunomodulatory properties of mesenchymal stromal cells. Blood 110: 3499–3506.
6. Le Blanc K, Rasmusson I, Sundberg B, Götherström C, Hassan M, et al. (2004) Treatment of severe acute graft-versus-host disease with third party haploidentical mesenchymal stem cells. Lancet 363: 1439–1441.
7. Koç ON, Gerson SL, Cooper BW, Dyhouse SM, Hayneswothr SE, et al. (2000) Rapid hematopoietic recovery after coinfusion of autologous-blood stem cells and culture-expanded marrow mesenchymal stem cells in advanced breast cancer patients receiving high-dose chemotherapy. J Clin Oncol 18: 307–316.
8. Noort WA, Kruisselbrink AB, in't Anker PS, Kruger M, van Bezooijen RL, et al. (2002) Mesenchymal stem cells promote engraftment of human umbilical cord blood-derived CD34 cells in NOD/SCID mice. Exp Hematol 30: 870–878.
9. Sun L, Akiyama K, Zhang H, Yamaza T, Hou Y, et al. (2009) Mesenchymal Stem Cell Transplantation Reverses Multi-Organ Dysfunction in Systemic Lupus Erythematosus Mice and Humans. Stem Cells 27: 1421–1432.
10. Miura M, Gronthos S, Zhao M, Lu B, Fisher LW, et al. (2003) SHED: Stem cells from human exfoliated deciduous teeth. Proc Natl Acad Sci USA 100: 5807–5812.
11. Seo BM, Sonoyama W, Yamaza T, Coppe C, Kikuiri T, et al. (2008) SHED repair critical-size calvarial defects in immunocompromised mice. Oral Dis 14: 428–434.
12. Zheng Y, Liu Y, Zhang CM, Zhang HY, Li WH, et al. (2009) Stem cells from deciduous tooth repair mandibular defect in swine. J Dent Res 88: 249–254.
13. Yamaza T, Akiyama K, Chen C, Liu Y, Shi Y, et al. (2010) Immunomodulatory properties of stem cells from human exfoliated deciduous teeth. Stem Cell Res Ther 1:5.
14. Wood EJ, Benson JD, Agca Y, Crister JK (2004) Fundamental cryobiology of reproductive cells and tissues. Cryobiology 48: 146–156.
15. Friedenstein AJ (1980) Stromal mechanisms of bone marrow: cloning in vitro and retransplantation in vivo. Haematol Blood Transfus 25: 19–29.
16. Shi S, Gronthos S, Chen S, Reddi A, Counter CM, et al. (2002) Bone formation by human postnatal bone marrow stromal stem cells is enhanced by telomerase expression. Nat Biotechnol 20: 587–591.
17. Bi Y, Ehirchiou D, Kilts TM, Inkson CA, Embree MC, et al. (2007) Identification of tendon stem/progenitor cells and the role of the extracellular matrix in their niche. Nat Med. 13: 1219–1227.
18. Gronthos S, Zannettino AC, Hay SJ, Shi S, Graves SE, et al. (2003) Molecular and cellular characterisation of highly purified stromal stem cells derived from human bone marrow. J Cell Sci 116: 1827–1835.
19. Yamaza T, Miura Y, Bi Y, Liu Y, Akiyama K, et al. (2008) Pharmacologic stem cell based intervention as a new approach to osteoporosis treatment in rodents. PLoS ONE 3: e2615.
20. Danjo A, Yamaza T, Kido MA, Shimohira D, Tsukuba T, et al. (2007) Cystatin C stimulated the differentiation of mouse osteoblastic cells and bone formation. Biochem Biophys Res Commun 360: 199–204.
21. Shi S, Gronthos S (2003) Perivascular Niche of Postnatal Mesenchymal Stem Cells Identified in Human Bone Marrow and Dental Pulp. J Bone Miner Res 18: 696–704.
22. Kerkis I, Kerkis A, Dozortsev D, Stukart-Parsons GC, Gomes Massironi SM, et al. (2006) Isolation and characterization of a population of immature dental pulp stem cells expressing OCT-4 and other embryonic stem cell markers. Cells Tissues Organs. 184: 105–116.
23. Lizier NF, Kerkis A, Gomes CM, Hebling J, Oliveira CF, et al. (2012) Scaling-up of dental pulp stem cells isolated from multiple niches. PLoS One 7: e39885.
24. Miura Y, Gao Z, Miura M, Seo BM, Sonoyama W, et al. (2006) Mesenchymal stem cell-organized bone marrow elements: an alternative hematopoietic progenitor resource. Stem Cells 24: 2428–2436.
25. Yamaza T, Miura Y, Akiyama K, Bi Y, Sonoyama W, et al. (2009) Mesenchymal stem cell-mediated ectopic hematopoiesis alleviates aging-related phenotype in immunocompromised mice. Blood 113: 2595–2604.
26. Morrison SJ, Prowse KR, Ho P, Weissman IL (1996) Telomerase activity in hematopoietic cells is associated with self-renewal potential. Immunity 5: 207–216.
27. Yang J, Yang X, Zou H, Chu Y, Li M (2011) Recovery of the immune balance between Th17 and regulatory T cells as a treatment for systemic lupus erythematosus. Rheumatology 50: 1366–1372.
28. Liu Y, Wang L, Kikuiri T, Akiyama K, Chen C, et al. (2011) Mesenchymal stem cell-based tissue regeneration is governed by recipient T lymphocytes via IFN-gamma and TNF-α. Nat Med 17: 1594–1601.
29. de Mendonça Costa A, Bueno DF, Martins MT, Kerkis I, Kerkis A, et al. (2008) Reconstruction of large cranial defects in nonimmunosuppressed experimental design with human dental pulp stem cells. J Craniofac Surg 19. 204–210.
30. Sakai VT, Zhang Z, Dong Z, Neiva KG, Machado MA, et al. (2010) SHED differentiate into functional odontoblasts and endothelium. J Dent Res 89: 791–796.
31. Nourbakhsh N, Soleimani M, Taghipour Z, Karbalaie K, Mousavi SB, et al. (2011) Induced in vitro differentiation of neural-like cells from human exfoliated deciduous teeth-derived stem cells. Int J Dev Biol 55: 189–195.
32. Zheng Y, Liu Y, Zhang CM, Zhang HY, Li WH, et al. (2009) Stem cells from deciduous tooth repair mandibular defect in swine. J Dent Res 88: 249–254.
33. Kerkis I, Ambrosio CE, Kerkis A, Martins DS, Zucconi E, et al. (2008) Early transplantation of human immature dental pulp stem cells from baby teeth to golden retriever muscular dystrophy (GRMD) dogs: Local or systemic? J Transl Med 6: 35.
34. Zheng Y, Wang XY, Wang YM, Liu XY, Zhang CM, et al. (2012) Dentin regeneration using deciduous pulp stem/progenitor cells. J Dent Res 91: 676–682.
35. Sakai K, Yamamoto A, Matsubara K, Nakamura S, Naruse M, et al. (2012) Human dental pulp-derived stem cells promote locomotor recovery after complete transection of the rat spinal cord by multiple neuro-regenerative mechanisms. J Clin Invest 122: 80–90.
36. Korbling M, Estrov Z (2003) Adult stem cells for tissue repair–a new therapeutic concept? N Engl J Med 349: 570–582.
37. Woods EJ, Pollok KE, Byers MA, Perry BC, Purtteman S, et al. (2007) Cord blood stem cell cryopreservation. Transfus Med Hemother 34: 276–285.
38. D'Ippolito G, Schiller PC, Ricordi C, Roos BA, Howard GA (1999) Age-related osteogenic potential of mesenchymal stromal stem cells from human vertebral bone marrow. J Bone Miner Res 14: 1115–1122.
39. Stenderup K, Justesen J, Clausen C, Kassem M (2003) Aging is associated with decreased maximal life span and accelerated senescence of bone marrow stromal cells. Bone 33: 919–926.
40. Stolzing A, Jones E, McGonagle D, Scutt A (2008) Age-related changes in human bone marrow-derived mesenchymal stem cells: consequences for cell therapies. Mech Age Dev 129: 163–173

41. Garvin G, Connie F, Sharp JG, Berger A (2007) Does the number or quality of pluripotent bone marrow stem cells decrease with age? Clin Orthod Relat Res 465: 202–207.
42. Muschler GF, Boehm C, Easley K (1997) Aspiration to obtain osteoblast progenitor cells from human bone marrow: The influence of aspiration volume. J Bone Joint Surg Am 79: 1699–1709.
43. Zuk PA, Zhu M, Mizuno H, Huang J, Futrell JW, et al. (2001) Multilineage cells from human adipose tissue: implications for cell-based therapies. Tissue Eng 7: 211–228.
44. Erices A, Conget P, Minguell JJ (2000) Mesenchymal progenitor cells in human umbilical cord blood. Br J Haematol 109: 235–242.
45. Gronthos S, Mankani M, Brahim J, Robey PG, Shi S (2000) Postnatal human dental pulp stem cells (DPSCs) in vitro and in vivo. Proc Natl Acad Sci USA 97: 13625–13630.
46. Hubel A (1997) Parameters of cell freezing: implications for the cryopreservation of stem cells. Transfus Med Rev 11: 224–233.
47. Areman E, Sacher R, Deeg H (1990) Processing and storage of human bone marrow: A survey of current practices in North America. Bone Marrow Transplant 6: 203–209.
48. Watt SM, Austin E, Armitage S (2007) Cryopreservation of hematopoietic stem/ progenitor cells for therapeutic use. Methods Mol Biol 368: 237–259.
49. Perry BC, Zhou D, Wu X, Yang FC, Byers MA, et al. (2008) Collection, cryopreservation, and characterization of human dental pulp-derived mesenchymal stem cells for banking and clinical use. Tissue Eng Part C. 14: 149–156.
50. Seo BM, Miura M, Sonoyama W, Coppe C, Stanyon R, et al. (2005) Recovery of stem cells from cryopreserved periodontal ligament. J Dent Res 84: 907–912.
51. Ding G, Wang W, Liu Y, An Y, Zhang C, et al. (2010) Effect of cryopreservation on biological and immunological properties of stem cells from apical papilla. J Cell Physiol 223: 415–422.
52. Zhao ZG, Li WM, Chen ZC, You Y, Zou P (2008) Hematopoiesis capacity, immunomodulatory effect and ex vivo expansion potential of mesenchymal stem cells are not impaired by cryopreservation. Cancer Invest 26: 391–400.
53. Sonoyama W, Liu Y, Fang D, Yamaza T, Seo BM, et al. (2006) Mesenchymal stem cell-mediated functional tooth regeneration in swine. PLoS One 1: e79.
54. Sonoyama W, Liu Y, Yamaza T, Tuan RS, Wang S, et al. (2008) Characterization of the Apical Papilla and Its Residing Stem Cells from Human Immature Permanent Teeth: A Pilot Study. J Endod 34: 166–171.

Cryopreserved Reticulocytes Derived from Hematopoietic Stem Cells Can Be Invaded by Cryopreserved *Plasmodium vivax* Isolates

Florian Noulin[1,3*], Céline Borlon[1], Peter van den Eede[1], Luc Boel[2], Catherine M. Verfaillie[3], Umberto D'Alessandro[1,4], Annette Erhart[1]

1 Unit of Malariology, Institute of Tropical Medicine Antwerp, Belgium, **2** Unit of Immunology, Institute of Tropical Medicine Antwerp, Belgium, **3** Stem cells institute, Catholic University Leuven, Belgium, **4** Medical Research Council Unit, Fajara, The Gambia

Abstract

The development of a system for the continuous culture of *Plasmodium vivax in vitro* would benefit from the use of reticulocytes derived from differentiated hematopoietic stem cells (HCS). At present, the need to use both fresh reticulocytes and fresh *P. vivax* isolates represents a major obstacle towards this goal, particularly for laboratories located in non-endemic countries. Here, we describe a new method for the cryopreservation of HSC-derived reticulocytes to be used for both *P. falciparum* and *P. vivax* invasion tests. Cryopreserved *P. falciparum* and *P. vivax* isolates could invade both fresh and cryopreserved HSC-derived reticulocytes with similar efficiency. This new technique allows the storage of HSC-derived reticulocytes which can be used for later invasion tests and represents an important step towards the establishment of a continuous *P. vivax* culture.

Editor: Ivo Mueller, Walter & Eliza Hall Institute, Australia

Funding: Funding was provided by ITM Secondary Research Funding (SOFI-B), http://www.itg.be/itgtool_v2/Projecten/Project.asp?PNr=755023. The funders had no role in study design, data collection and analysis, decision to publish, or preparation of the manuscript.

Competing Interests: The authors have declared that no competing interests exist.

* E-mail: fnoulin@itg.be

Introduction

Outside sub-Saharan Africa, *Plasmodium vivax* is the most common human plasmodium species [1]. Though it has long been considered as less dangerous than *P. falciparum*, severe and fatal *P. vivax* cases as well as drug resistant parasites are now increasingly reported [2,3,4,5]. Understanding of the biology and the transmission dynamics of *P. vivax* lags behind that of *P. falciparum*, partly because the maintenance of a continuous line of this parasite *in vitro* has not been established. The availability of an *in vitro* culture system would improved our understanding of the biology of *P. vivax*, e.g. invasion and relapse mechanisms, resistance markers, and would contribute to the development of new treatments and vaccines.

P. vivax preferentially invades reticulocytes [6] and thus in order to achieve a continuous *in vitro* culture system the availability of large amounts of these young red blood cells, which circulate in the peripheral blood at low concentration (1% of total red blood cells) and for a very short time (24 h), is required. Russell *et al.* [7] recently published a reliable invasion assay protocol using fresh *P. vivax* schizonts from fresh clinical isolates and enriched cord blood reticulocytes. It has been previously shown that reticulocytes can also be successfully produced through the differentiation of hematopoietic stem cells (HSC) [8] and that such HSC-derived reticulocytes may be used for *P. vivax* culture, though both the reticulocyte production and the parasite densities obtained were extremely low [9]. The contribution of this paper is to report an improved method to produce and cryopreserve HSC-derived reticulocytes to be later invaded by *P. vivax*, allowing research on *P. vivax* culture to be carried out outside endemic areas, increasing the number of teams potentially working on this subject and hence the chances for major discoveries.

Materials and Methods

Ethics Statement

***P. vivax* sample collection:** MUTM 2008-15 from the ethics Committee of the faculty of Tropical Medicine, Mahidol University, Bangkok, Thailand.

Cord blood sample collection: blood was collected anonymously and patients were informed orally with a possibility of opting-out. Each Patient was notified on the hospital admission form of this opting-out possibility.

Procedure was accepted by ethic committee of UZA and ITM.

Study was approved by the ITM review board, number: SBB.219.2007/1410.

Hematopoietic Stem Cell (HSC) Isolation

Umbilical cord blood samples (40 ml each) were collected from pregnant women delivering at the University hospital, Antwerp (UZA), after obtaining an individual informed consent. Mononuclear cells (MNC) were separated by Ficoll-Isopaque (GE Healthcare) centrifugation (250 g, 10 min) and enriched for $CD34^+$ cells by supermagnetic microbead selection using Mini-MACS columns (Miltenyi Biotech) according to a previously published procedure [10].

HSC Culture

The amplification procedure was adapted from a three-step expansion of $CD34^+$ cells by sequential supply of the culture with specific combination of cytokines and growth factors [11]. HSCs isolated from cord blood were cultured at 37°C, 5% CO_2 in a modified serum-free media (IMDM, Biochrom) supplemented with L- Glutamine (4 mM, Sigma), Penicilline/Streptomycine (1%, Invitrogen), Inositol (40 µg/ml, Sigma), Folic acid (10 µg/ml, Sigma), Monothioglycerol (1.6 10^{-4} M, Sigma), Transferrin (120 µg/ml, Sigma), insulin (10 µg/ml, Sigma), Bovine Serum Albumin detoxified by beads resin AG501-X8 (Biorad) (BSA, 100 mg/ml, PAA) [12].

Step 1 (day 0 to 8): $CD34^+$ cells were cultured with Stem Cell Factor (SCF, 100 ng/ml, Bioke), IL-3 (5 ng/ml, R&D System), Hydrocortisone (HDS, 10^{-6} M, Sigma), and Erythropoietin (EPO, 3 IU/ml, R&D System). At day 4, cells were diluted 1:2 in IMDM medium completed with all the four above mentioned growth factors and incubated at 37°C for 4 additional days. At day 7, cells were re-suspended (10^6 cells per vial) in culture medium (IMDM) and an equal volume of 80% Foetal Calf Serum (FCS)/20% DMSO solution was added drop by drop to obtain an IMDM/40% FCS/10% DMSO solution before progressively freezing them at −80°C using a Mr Frosty [13].

Step 2 (day 8–11): at day 8, 250 000 cells were added to each 25 cm^2 flask and incubated in 5 ml of IMDM medium supplemented with only EPO (3 IU/mL).

Step 3 (day 11–20): the culture was maintained in IMDM without growth factors or cytokines and the medium was changed every 3 days. The culture was stopped at day 14 corresponding to the peak of reticulocytes counts determined by microscopic examination of thin films done by cytospin (Thermo scientific): 200 000 cells were washed with PBS once and re-suspended in 50 µL of PBS. 50 µL of Cresyl Blue(Merck)(previously diluted 1/100) were added to the tube and incubated 30 minutes. FCS (30 µL) was added to protect cells during cytospin centrifugation and the cells were placed in a cytospin funnel. After a centrifugation (700 rpm, 3 minutes), slides were removed from the funnel and stained with Giemsa. A reticulocyte scoring positive would contain two or more blue-stained RNA granules.

HSC Cryopreservation

HSC derived reticulocytes at day 14 in the culture were subjected to three different cryopreservation protocols:

Glycerolyte solution [14]: 100 µL of Glycerolyte (Baxter) were drawn up to the cell pellet using an insulin syringe. Firstly, 20% of the volume of Glycerolyte was added to the blood cell suspension drop by drop while continuously agitating the tube to mix the content. This suspension was incubated 5 minutes at room temperature before adding the rest of Glycerolyte. RBC-Glycerolyte mixture was then aliquoted into cryovials and frozen at −80°C overnight before being stored in liquid nitrogen.

Glycerol & Sorbitol solution [15]: Cells were frozen in a medium composed of 28% glycerol, 3% sorbitol, and 0,9% NaCl diluted in distillate water. This solution was added drop by drop to packed cells, at room temperature. Cells were immediately transferred at −80°C.

IMDM/10% DMSO/40% FCS solution [16]: Cells were frozen according to a standard method with medium containing 10% DMSO/40% FCS as described above for the HSC culture at day 7.

Reticulocytes were kept in liquid nitrogen up to 1 year. Cells were thawed following the same protocol as for *Plasmodium falciparum*: the volume (V) of the thawed content was measured and transferred to a 50 ml centrifuge tube. Then 0.1 V of pre-warmed (37°C) 12% NaCl solution was added drop by drop to the parasitized red blood cells (pRBC), which were allowed to stay for 5 min at room temperature. Subsequently 10 V of 1.6% NaCl solution (pre-warmed at 37°C) were carefully added to the cells, which were then centrifuged for 10 min at 650 g without applying a break. The supernatant was discarded and 10 ml of 0.9% NaCl solution pre-warmed at 37°C was carefully added to the pellet, followed by a 5 min centrifugation at 650 g [17]. The cryopreservation protocols were evaluated according to the following criteria: viability of the cells after thawing, preservation of the reticulocyte population, presence or absence of clots, and presence of invaded reticulocytes by *Plasmodium*.

FACS Analysis

HSCs cultures were analysed by flow cytometer at day 8, 11, 14 and 17 of their in-vitro maturation. After thawing the cryopreserved reticulocytes they were analysed by FACS, to assess the expression of stage specific markers: CD36, CD45, CD71 and CD235a. Briefly, 200 000 cells were centrifuged for 5 min at 650 g. The pellet was re-suspended in 100 µL of PBS buffer. 5 µl of CD45-PE (BD bioscience), 10 µL of CD71-APC (BD bioscience), 5 µL of CD36-FITC (BD bioscience), and 5 µL of CD235-PerCP (BD bioscience), were added to the cells and incubated for 15 min at room temperature in the dark. One tube containing only antibody isotypes coupled to fluorochromes was used as a negative control: 10 µL of IgG1γ-PE (BD bioscience), 5 µL of IgM-FITC (BD bioscience), 5 µL of IgG1 K isotype-PerCP (BD bioscience), and 5 µL of IgG1 K isotype-APC (BD bioscience). After incubation cells were washed with FACS buffer (Miltenyi Biotech) and re-suspended in 500 µL of buffer before analysis. FACS analysis were carried out on a FACScalibur 4 color cytometer (BD bioscience).

Hemoglobin Content

The haemoglobin (Hb) content (foetal or adult haemoglobin) of the HSC derived reticulocytes was analyzed using the Kleihauer method [18]. Umbilical cord blood (≥95% of foetal Hb) and adult peripheral blood (≥95% of adult Hb) were used as controls. Cells containing foetal haemoglobin will be stained in red whereas cells containing adult haemoglobin will look empty and only their membranes will be observed.

Plasmodium Falciparum

P. falciparum D10 strains, cultured and cryopreserved in the Institute of Tropical Medicine, Antwerp, Belgium, (ITMA) cryobank were used for this study. Parasite isolates were thawed in a 37°C incubator until no ice was left in the tube. The thawing protocol was the same as the one used for reticulocytes. To check for the invasion of new RBCs (RBC derived from HSC maturation) by P. falciparum, pRBC were concentrated on a Percoll gradient (90%, 70%, and 40%; Sigma) [19] and centrifuged for 30 min at 2600 g without applying a break. After centrifugation, the layer between the 70% and 40% Percoll was kept, transferred into a new tube and washed with RPMI (Lonza). 1 µl of purified pRBC (≥90% mature forms) was mixed with 4 µL of the HSCs derived reticulocytes and 94 µL of RPMI medium supplemented with HEPES (25 mM, Lonza), L-glutamine (1 mM), gentamycin (40 µg/ml, Sigma), D-glucose (2%, Sigma), hypoxanthine (50 mg/L, Sigma) diluted in NaOH 1 M and 10% heat inactivated Human Serum (final volume: 100 µL, haematocrit 5%). The culture was incubated in a hermetic chamber flushed gas (90% N2, 5% CO2, 5% O_2) at 37°C for 24 h. Parasite invasion was determined by microscopic examination of Giemsa stained thin films. The number of newly invaded cells by

P. falciparum ring stages were counted and divided by the total number of cells to obtain the parasite density (% of invaded RBC). For each experiment, two technicians independently examined the slides and the mean parasite density was taken as the final value. If reading results were discordant (parasite counting difference ≥30%), both technicians re-read the slide until an agreement was found.

Plasmodium Vivax

P. vivax isolates were collected from patients with acute *P. vivax* malaria (mono-infection with a density of >1 parasite/1000 red blood cells) attending the clinics of the Shoklo Malaria Research Unit (SMRU, Mae Sot, North-West Thailand), after obtaining their individual written informed consent. *P. vivax* isolates contained more than 80% early ring forms. White blood cells were removed after filtration on CF11 column (Whatmann) [20]; Parasites were frozen in Glycerolyte solution [21], as described above, and then transferred and kept in liquid nitrogen at the ITMA. Subsequently, *P. vivax* isolates were thawed following the same protocol as described above for *P. falciparum*, and were re-suspended in pre-warmed 37°C McCoy's 5A (Gibco) medium supplemented with 20% heat inactivated human serum from normal AB group donors and 0.5% D-glucose. The medium containing *P. vivax* was then transferred to a 25 cm^2 culture flask, flushed with gas (90% N2, 5% CO2, 5% O_2), and placed in a 37°C incubator for 36 to 44 hours to allow the maturation into schizonts. Parasite stage was determined by microscopic examination of Giemsa stained thick and thin films. Mature schizonts were concentrated in 45% Percoll by centrifugation for 15 minutes at 1580 g without applying a break (Sigma) [22]. Purified schizonts (≥90% mature forms) were mixed with HSCs derived reticulocytes resulting in a starting parasite density between 0.5 and 1%. Complete Mc Coy's medium was added in order to obtain a 5% haematocrit level and the culture was incubated at 37°C in a hermetic chamber flushed for 24 h with gas (90% N_2, 5% CO_2, 5% O_2) [7]. The efficiency of invasion was checked as described above for *P. falciparum*.

Data Analysis

During HSC maturation, reticulocytes and RBC were counted for each selected day during the maturation process. Median value and interquartile range (IQR) were calculated for the reticulocytes counts (6 independent experiments) as well as for the expression of the surface markers measured by FACS (3 independent experiments). Parasite densities for *P. vivax* invasion assays were calculated per 500 counted cells. The number of infected cells was divided by the total number of counted cells and multiplied by 100 to obtain a percentage.

Results

Amplification of Erythroid Cells

Microscopic examination of thin films of the erythroid cells (Figure 1) confirmed a normal differentiation pattern and the commitment of the HSC culture to the erythroid lineage as defined elsewhere [23]. A total of 30 independent experiments were successfully carried out, and median reticulocyte and RBC population counts were computed at regular intervals (every 3 days between day 8 and day 20) for 6 experiments (Figure 2). The first reticulocytes were observed at day 12 of differentiation, and their number dramatically increased between days 13 and 14 (from 5% to 18%), corresponding to a peak of enucleation. Subsequently, the percentage of reticulocytes decreased steadily (10% at day 16) and almost disappeared at day 19 of differentiation. The haemoglobin content measured at day 14 of maturation was mainly foetal (about 90%, data not shown). Mature red blood cells (RBC) showed an opposite trend as compared to reticulocytes as they started to increase following the reticulocyte peak (0.4% at day 15), outnumbered the reticulocytes at day 17 and reached a plateau of above 20% at day 20. After cryopreservation, the cells viability was up to 70% and the proportion of reticulocytes remained stable compared to the one before freezing (except for glycerolyte which was not tested due to the presence of clots during the thawing process).

FACS Analysis of HSC Maturation

The proportion of positive cells expressing different surface markers was calculated at different time points (Figure 3a). The CD45, a specific marker for young cells (pre-proerythroblast stage), decreased dramatically from 90% at day 8 to 3% at day 11, while CD235a (Glycophorin A marker) followed the opposite trend. Both CD71 and CD36 (both markers of stages between the

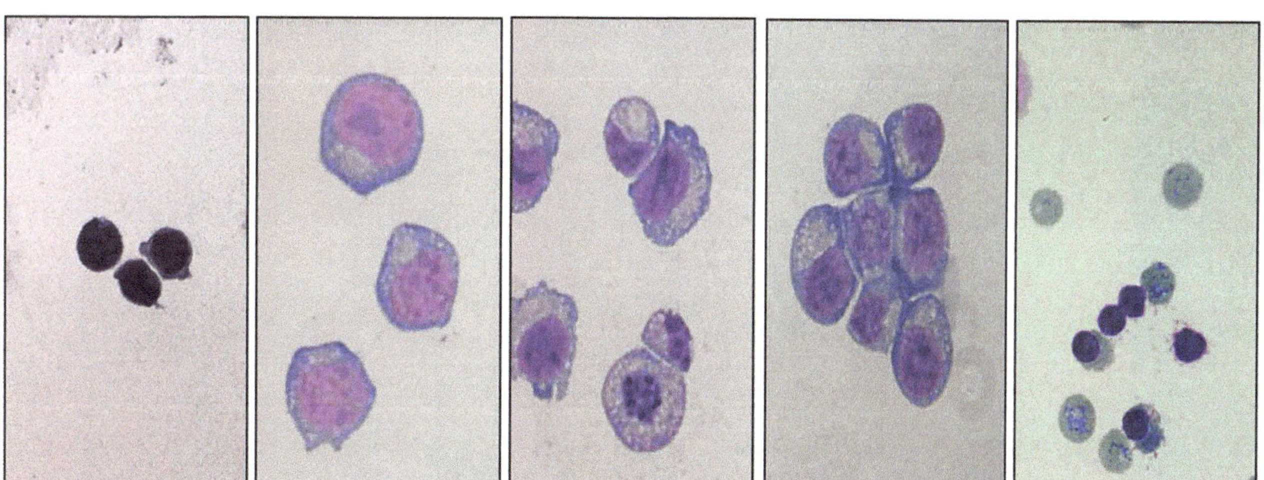

Figure 1. Production of reticulocytes from HSCs, morphologic analysis after Giemsa staining. (A) Human CD34+ cells isolated from cord blood (B) Day 4 of culture, (C) Day 8- (D) Day 11- (E) Day 14 of culture, reticulocytes are highlighted by cresyl blue staining (arrows).

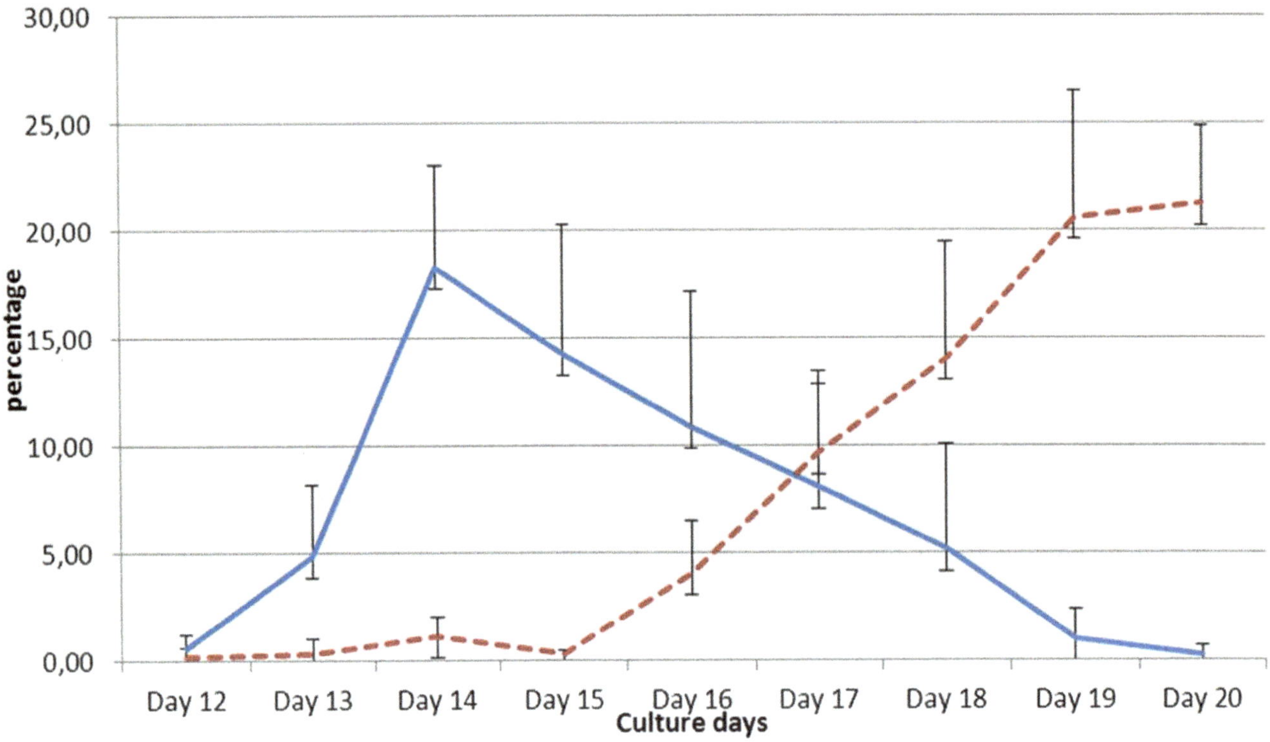

Figure 2. Evolution of reticulocytes (continuous line) and red blood cells (dotted line) counts from day 12 to day 20 of HSCs maturation. Each value represents the median a total of 6 independent experiments, and vertical bars represent the interquartile range (IQR). Reticulocytes were identified by cresyl blue staining.

proerytroblast and mature RBC) were highly expressed at day 8, 100% and 80% respectively, and decreased following a different pattern until day 17: the CD71 decreased slightly to 65% at day 17, while the CD36 dramatically decreased to less than 10% at day 17. A freezing cycle (up to 1 year in the frame of this work) of the HSC derived reticulocytes which were matured till day 14 did not affect the expression of specific reticulocytes markers for any of the three protocols tested. Indeed the comparison with the freshly produced HSC derived reticulocytes gave similar results: (Figure 3b) a high expression of CD71 (between 95% and 98% according to the freezing protocol *versus* 90% for fresh reticulocytes at day 14), and CD235a (between 90 and 98% *versus* 90% in fresh reticulocytes), a quasi-absence of CD45 (around 0.5% *versus* 1%), and the expression of CD36 reaching 30% for the 3 protocols tested (*versus* 40%).

Invasion Assays

Both *P. falciparum* and *P. vivax* could successfully invade freshly HSC derived reticulocytes (Figure 4a and 4b). Parasite density after invasion of fresh reticulocytes was around 0.5% for *P. vivax* and 4% for *P. falciparum*. Parasite invasion was then tested on the cryopreserved reticulocytes derived from HSC maturation. *P. falciparum* could invade cryopreserved reticulocytes subjected to any of the 3 cryopreservation methods (parasite density: 4.2% for glycerol solution, 4.1% for Glycerolyte and 8% for the IMDM/10% DMSO/40% FCS; 3 independent assays). However, the method using the Glycerolyte solution was discarded as clots were observed during the thawing process. Therefore, for *P. vivax* invasion, only reticulocytes cryopreserved with the Glycerol solution and the Medium/10% DMSO/40% FCS solutions were used. These were successfully invaded with the identification of ring stages 24 h post invasion and the parasite density was counted for all 8 independent experiments (Figure 5). *P. falciparum* and *P. vivax* could invade the erythroblasts stages, with both fresh and cryopreserved HSC derived cells (Figure 4c).

Parasite density obtained after invasion of cryopreserved HSC derived reticulocytes varied substantially (from 0.19% to 2.43%) according to the *P. vivax* isolate used. However, parasite density for each single parasite isolate did not vary when using different batches of cryopreserved reticulocytes. The range of parasite densities after invasion was between 0.19% and 2.43% for reticulocyte cryopreserved in the IMDM/10% DMSO/40% FCS solution. In comparison, invasion rate of reticulocytes cryopreserved in glycerol solution was 30 to 50% less efficient in 4 replicates) (Figure 5).

Discussion

This is the first report showing a reliable cryopreservation protocol for HSC-derived reticulocytes derived that can be successfully invaded by both cryopreserved *P. falciparum* and *P. vivax* isolates. This will contribute to the establishment of a continuous *P. vivax* culture, since the latter requires a continuous supply of reticulocytes in a large amount. Until now, availability of adequate quantities of reticulocytes was a bottleneck, as they represent only 1% of all circulating cells in the blood stream [24]. Several approaches have been tested to tackle this problem. One used blood from hematochromatosis patients [25] as they need to be regularly bled and have a higher percentage of reticulocytes. This method allowed for short-term *P. vivax in vitro* cultures for up to 15 days. However, the limited access to hemochromatosis patients is a drawback. A second source of reticulocytes is cord blood, in which the reticulocyte concentration is slightly higher (3–

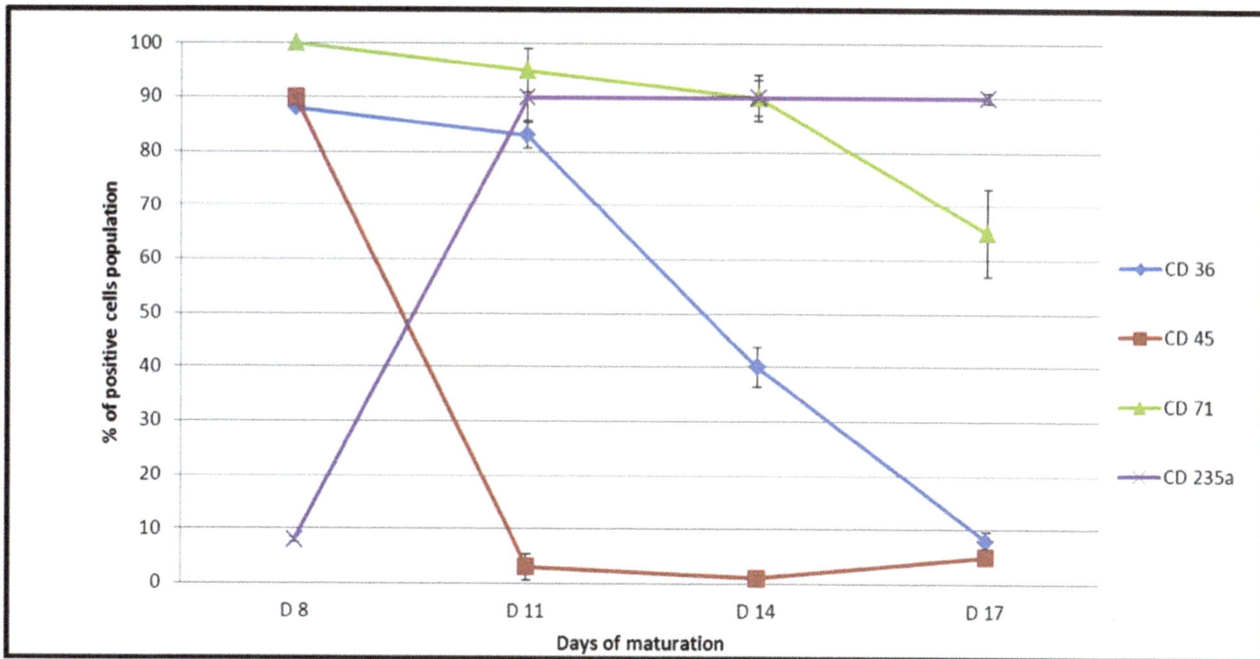

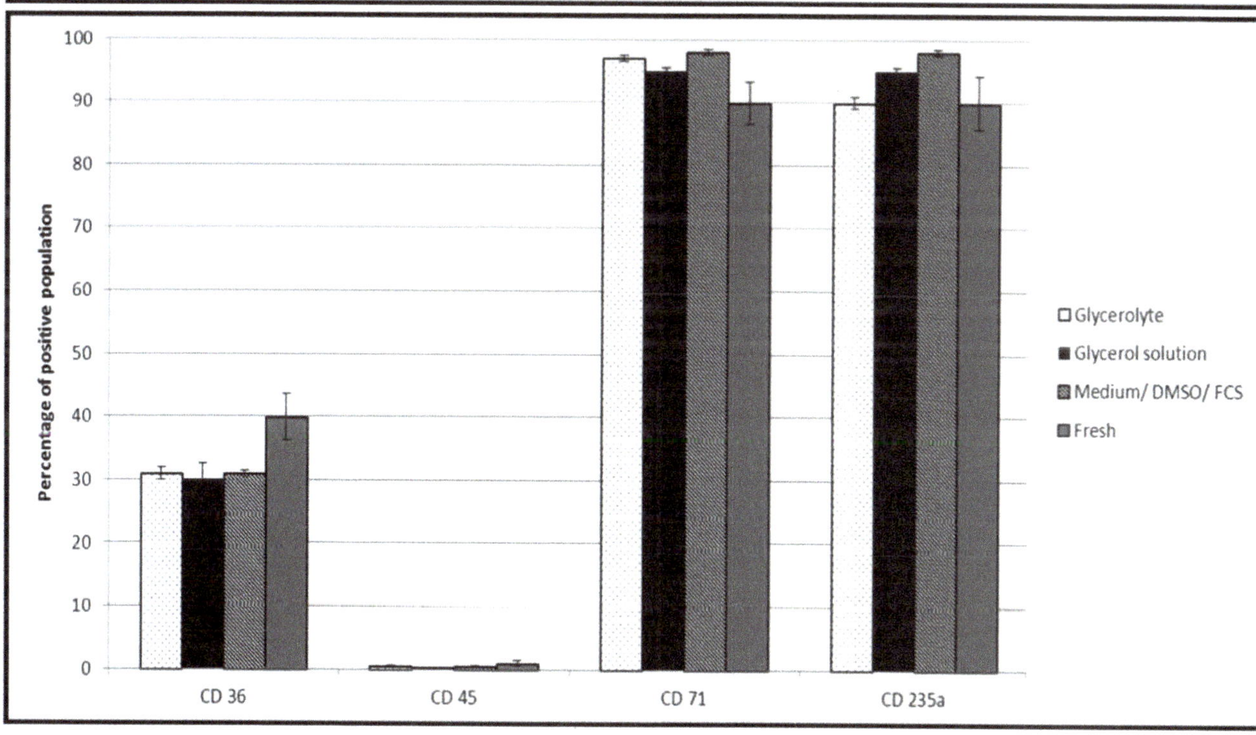

Figure 3. Expression of reticulocyte surface markers by flow cytometry. 3a: Expression of HSCs surface markers at different maturation stages characterized by flow cytometry. Cells were collected, respectively at 8 days, 11 days, 14 days and 17 days of maturation and stained with specific antibodies against CD36, CD45, CD71 and CD325a receptors, respectively. Each value represents the median percentage for 3 independent experiments (vertical bars represent the extreme values in this case). 3b: Comparison of the surface markers expression (%) at day 14 after thawing for 3 different cryopreservation protocols compared to fresh HSC derived reticulocytes (median % values of 3 independent experiments; vertical bars represent the extreme values in this case).

8%) than in adult blood; this source has been previously used permitted maintaining *P. vivax* fresh isolates in culture for about one month [26]. However, the low amount of reticulocytes obtained did not allow for an optimal parasite multiplication (parasite density <0.7%). Moreover, substantial variation of parasite densities obtained makes this protocol unsuitable for studying *P. vivax* biology.

Cord blood HSCs represent a valuable source, as reticulocytes can be produced in substantial quantities [9,27]. The differentiation protocol presented here was adapted from the one published by Giarratana *et al.* [8,28], which originally matured HSCs into

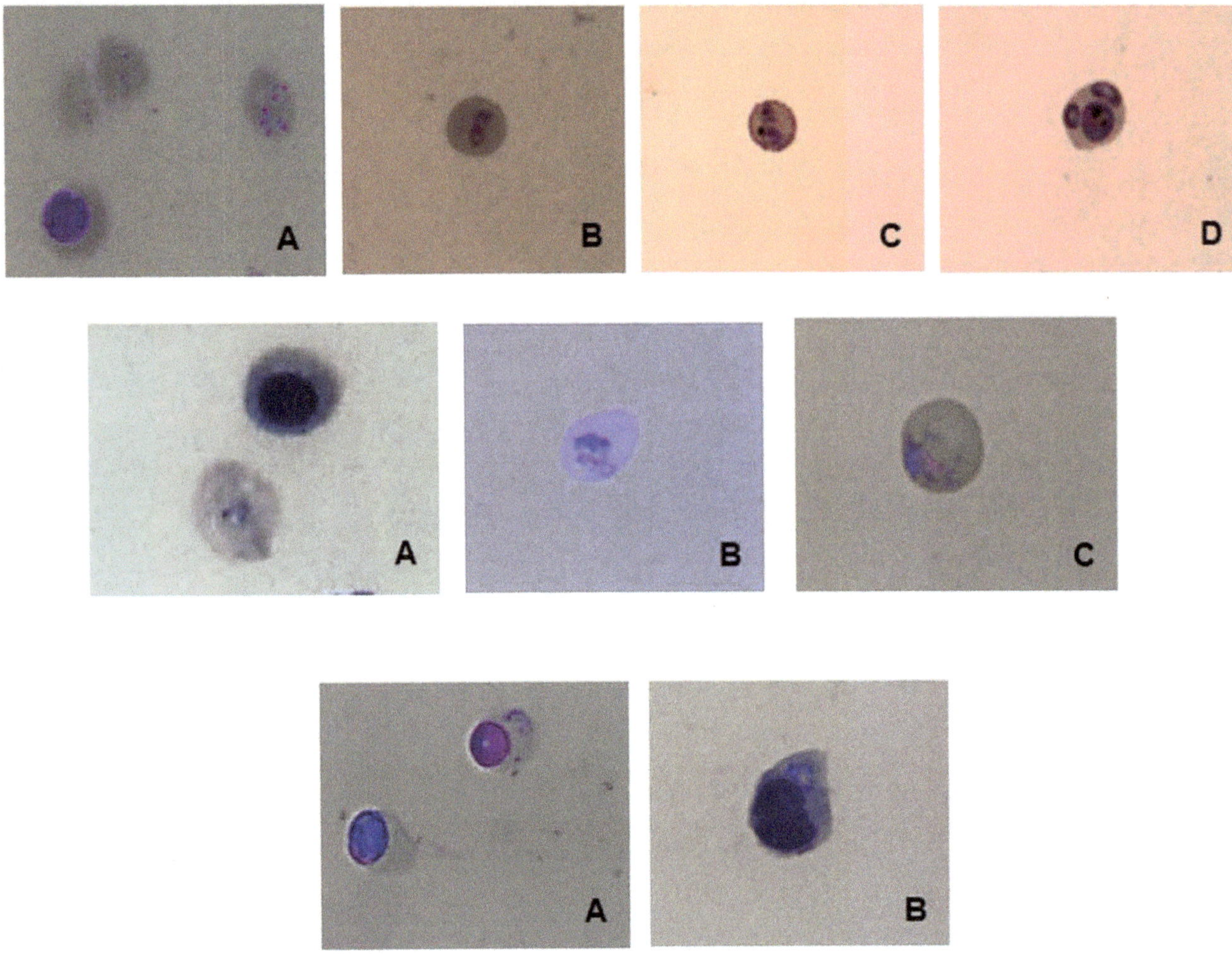

Figure 4. Pictures of HSC-derived reticulocyte invaded by *Plasmodium*. 4a: Infected enucleated cells by *P. falciparum* 24 h after invasion. (A) Fresh reticulocytes derived from HSCs maturation, (B) Cryopreserved reticulocytes with glycerolyte, (C) Cryopreserved reticulocytes with glycerol solution, (D) Cryopreserved reticulocytes with Medium/40% FCS/10% DMSO (3 replicates). 4b: Infected reticulocytes by *P. vivax* 24 h after invasion. (A) Fresh reticulocytes derived from HSCs maturation, (B) Cryopreserved reticulocytes with glycerol solution, (C) Cryopreserved reticulocytes with Medium/40% FCS/10% DMSO (8 replicates). 4c: Infected erythroblast (A) By *P. falciparum*. (B) By *P. vivax* (8 replicates).

erythroid lineage by using a co-culture with stromal cells. The latter are mainly used to retain nucleated cells, free nuclei and debris so that the culture appears cleaner. We have modified this technique by omitting the co-culture with stromal cells, which resulted in a less pure population of enucleated cells in our model (20% enucleated/80% nucleated cells *versus* 100% enucleated cells when using stromal cells) since the nucleated cells were not retained by the MS5 layer. Nevertheless, while Giarratana aimed at obtaining RBCs with a lifespan of about 120 days, achieved during the later stages of RBC maturation, our purpose was to obtain reticulocytes whose lifespan is 2 days. This explains why maturation was stopped at the peak of the reticulocyte concentration (D14). Moreover, reticulocytes obtained after maturation on a MS5 layer gave similar result in term of invasion than those produced without stromal cells (data not shown). A previous attempt of producing HSC-derived reticulocytes without using stromal cells in co-culture attained a maximum reticulocyte concentration of 0.5% [9]. This is in contrast with the results of our methods which produced a 40-fold (20%) higher concentration of reticulocytes. The low efficiency of the previous method could be explained by several factors such as the use of non-detoxified BSA (a key factor for the last steps of maturation and enucleation) [27], a very high (lethal?) concentration of monothioglycerol, or the absence of inositol or folic acid. It is also important to mention that the parasite density in culture supported by HSC-derived reticulocytes obtained with our method was substantially higher (2.5% vs 0.0015%) than that reported for the previous one [9], providing sufficient biological material to possibly keep the culture for a longer period.

During the invasion assays we performed, the cell population was not purely composed by reticulocytes but contained also erythroblasts. Both *P. falciparum* and *P. vivax* were able to invade these cells, as previously reported [29]. No information on the quality of the parasite development in this type of cells is currently available.

FACS analyses of both fresh and cryopreserved reticulocytes showed the expression of CD235a marker (Glycophorin A), an important receptor for *P. falciparum* invasion [30]. Its role in *P. vivax* invasion has not been described yet but its homology with the Duffy Binding Protein (DBP) [31,32] suggests a potential role in the *P. vivax* invasion process of the reticulocytes. Thus the

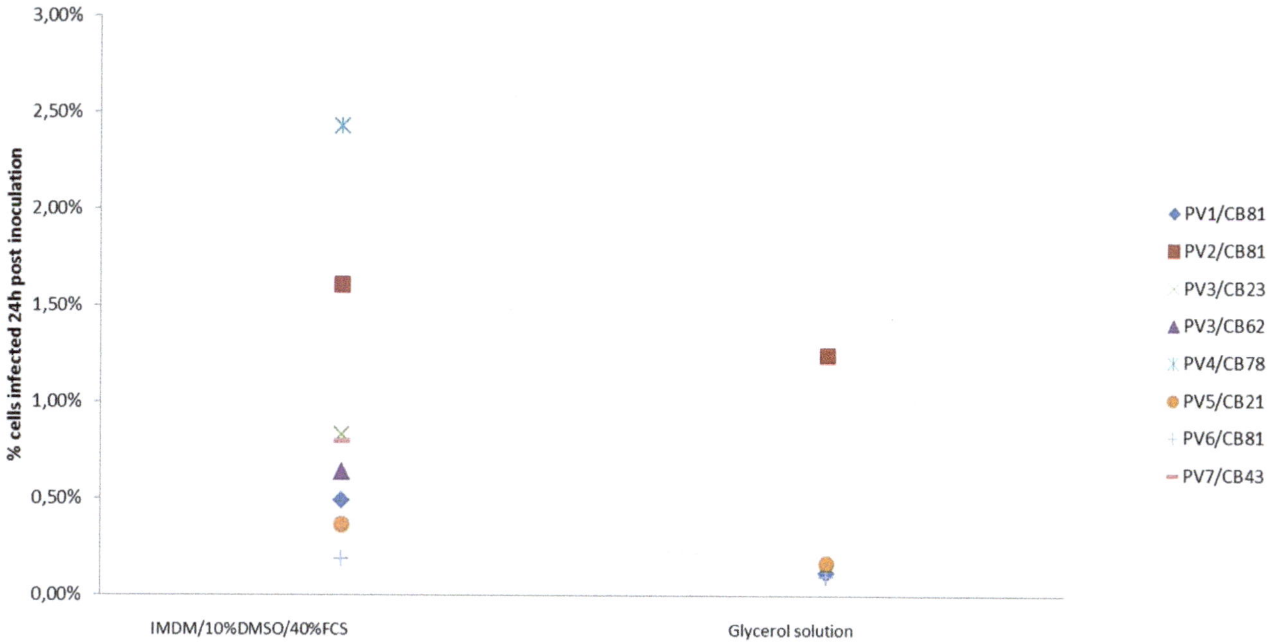

Figure 5. Parasitemia of infected HSC derived cells for the 2 protocols of cryopreservation (Medium/10%DMSO/40%FCS and Glycerol solution) 24 h after invasion assay with *P. vivax* cryopreserved isolates.

cryopreservation protocol kept the important receptors on the reticulocyte membrane.

The added value of the work presented is the set up of a cryopreservation technique for the long-term storage of cord blood HSC-derived reticulocytes, providing a continuous source for the performance of invasion assays and possibly the maintenance of a *P. vivax in vitro* culture. The technique is robust as it has been tested on several HSCs batches of that gave comparable results with a single *P. vivax* isolates (Fig.5, PV3/CB 81 and 23). The reticulocyte cryopreservation using Glycerolyte was abandoned due to the presence of clots during thawing. It is unclear why this occurred but a possible hypothesis would be the harmful effect of glycerolyte on nucleated cells.

The source of reticulocytes for a *P. vivax* invasion assay is not limited to HSC-derived cells. Borlon *et al* [32] were able to maintain a short-term (up to 10 days) *P. vivax* culture using reticulocytes directly concentrated from cord blood. This was done with similar efficacy with both fresh and frozen parasites and reticulocytes (using Glycerolyte 57 as cryopreservant). The advantage of concentrating reticulocytes from cord blood is that these are almost immediately available (pending availability of cord blood) while HSC-derived ones need 14 days to mature. Nevertheless, with the former technique the concentration process (percoll centrifugation, application of several washing/concentration steps) could damage the cells and affect the invasion efficiency while using HSC-derived reticulocytes may guarantee a more homogeneous and standardised cell population. Eventually, using both methods: concentrated from cord blood and HSC-derived reticulocytes could substantially increase the amount of available reticulocytes from a single cord blood sample.

In conclusion, the cryopreservation method described in this paper provides a continuous and substantial source of reticulocytes, possibly a first step to achieve the continuous *in vitro* culture of *P. vivax*. In addition, the possibility of cryopreserving both reticulocytes and parasites offers the opportunity of investigating *P. vivax* invasion assays/culture in non-endemic countries, opening this field of research to more research groups thus increasing the probability of fast advances in knowledge.

Acknowledgments

We would like to thank Prof. François Nosten from the Shoklo Malaria Research Unit in Mae Sot (Thailand) for providing the *P. vivax* samples. We also would like to thank Dr. R. Udomsangpetch from Mahidol University, Bangkok (Thailand) for providing the initial training on the HSCs maturation, as well as Dr. M-C. Giarratana from INSERM U417, Paris (France) for her technical advises concerning HSCs maturation. We thanks Dr Bruce Russell (A-STAR Singapore) for training on the *P. vivax* invasion method. Cord blood samples collection was possible thanks to the support of Prof. Dr. Y. Jacquemin and Dr. I. Beckstedde from UZ Antwerp (Belgium). We are very grateful to P[r] J. Van Den Abbeele and Dr R. Fitzhenry for their comments on this manuscript.

Author Contributions

Conceived and designed the experiments: FN CB UDA AE. Performed the experiments: FN. Analyzed the data: FN. Contributed reagents/materials/analysis tools: CB LB PVDE. Wrote the paper: FN. Reviewed the manuscript: CB CMV UDA AE.

References

1. Guerra CA, Howes RE, Patil AP, Gething PW, Van Boeckel TP, et al. (2010) The international limits and population at risk of Plasmodium vivax transmission in 2009. PLoS neglected tropical diseases 4: e774.
2. Tjitra E, Anstey NM, Sugiarto P, Warikar N, Kenangalem E, et al. (2008) Multidrug-resistant Plasmodium vivax associated with severe and fatal malaria: a prospective study in Papua, Indonesia. PLoS medicine 5: e128.
3. Price RN, Douglas NM, Anstey NM (2009) New developments in Plasmodium vivax malaria: severe disease and the rise of chloroquine resistance. Current opinion in infectious diseases 22: 430–435.
4. Prajapati SK, Joshi H, Dev V, Dua VK (2011) Molecular epidemiology of Plasmodium vivax anti-folate resistance in India. Malaria journal 10: 102.

5. Yohannes AM, Teklehaimanot A, Bergqvist Y, Ringwald P (2011) Confirmed vivax resistance to chloroquine and effectiveness of artemether-lumefantrine for the treatment of vivax malaria in Ethiopia. The American journal of tropical medicine and hygiene 84: 137–140.
6. Mons B (1990) Preferential invasion of malarial merozoites into young red blood cells. Blood cells 16: 299–312.
7. Russell B, Suwanarusk R, Borlon C, Costa FT, Chu CS, et al. (2011) A reliable ex vivo invasion assay of human reticulocytes by Plasmodium vivax. Blood.
8. Giarratana MC, Kobari L, Lapillonne H, Chalmers D, Kiger L, et al. (2005) Ex vivo generation of fully mature human red blood cells from hematopoietic stem cells. Nature biotechnology 23: 69–74.
9. Panichakul T, Sattabongkot J, Chotivanich K, Sirichaisinthop J, Cui L, et al. (2007) Production of erythropoietic cells in vitro for continuous culture of Plasmodium vivax. International journal for parasitology 37: 1551–1557.
10. Jaatinen T, Laine J (2007) Isolation of hematopoietic stem cells from human cord blood. Current protocols in stem cell biology Chapter 2: Unit 2A 2.
11. Giarratana MC KL, Lapillonne H, Chalmers D, Kiger L, Cynober T, Marden MC, Wajcman H, Douay L. (2005) Ex vivo generation of fully mature human red blood cells from hematopoietic stem cells. Nat Biotechnol 1: 69–74.
12. J B, Nienhuis AW (1979) A method for the deionization of bovine serum albumin. Methods in Cell Science Volume 1: 111–112.
13. Goldman JM, Th'ng KH, Park DS, Spiers AS, Lowenthal RM, et al. (1978) Collection, cryopreservation and subsequent viability of haemopoietic stem cells intended for treatment of chronic granulocytic leukaemia in blast-cell transformation. British journal of haematology 40: 185–195.
14. Kosaisavee V, Suwanarusk R, Nosten F, Kyle DE, Barrends M, et al. (2006) Plasmodium vivax: isotopic, PicoGreen, and microscopic assays for measuring chloroquine sensitivity in fresh and cryopreserved isolates. Experimental parasitology 114: 34–39.
15. Laforest J, Combrisson A (1968) [Preservation of erythrocytes by freezing in the presence of glycerol and sorbitol. II. Dosage of glycerol and sorbitol in the blood. Behavior of these polyalcohols at contact with the erythrocytes]. Revue francaise de transfusion 11: 23–32.
16. Loretz LJ, Li AP, Flye MW, Wilson AG (1989) Optimization of cryopreservation procedures for rat and human hepatocytes. Xenobiotica; the fate of foreign compounds in biological systems 19: 489–498.
17. Diggs C, Joseph K, Flemmings B, Snodgrass R, Hines F (1975) Protein synthesis in vitro by cryopreserved Plasmodium falciparum. The American journal of tropical medicine and hygiene 24: 760–763.
18. Kleihauer E, Braun H, Betke K (1957) [Demonstration of fetal hemoglobin in erythrocytes of a blood smear]. Klinische Wochenschrift 35: 637–638.
19. Rivadeneira EM, Wasserman M, Espinal CT (1983) Separation and concentration of schizonts of Plasmodium falciparum by Percoll gradients. The Journal of protozoology 30: 367–370.
20. Sriprawat K, Kaewpongsri S, Suwanarusk R, Leimanis ML, Lek-Uthai U, et al. (2009) Effective and cheap removal of leukocytes and platelets from Plasmodium vivax infected blood. Malaria journal 8: 115.
21. Rossan RN (1985) Cryopreservation of the blood stages of Plasmodium falciparum and Plasmodium vivax for in vivo studies. The American journal of tropical medicine and hygiene 34: 207–208.
22. Carvalho BO, Lopes SC, Nogueira PA, Orlandi PP, Bargieri DY, et al. (2010) On the cytoadhesion of Plasmodium vivax-infected erythrocytes. The Journal of infectious diseases 202: 638–647.
23. Tsiftsoglou AS, Vizirianakis IS, Strouboulis J (2009) Erythropoiesis: model systems, molecular regulators, and developmental programs. IUBMB life 61: 800–830.
24. Mechetner EB, Sedmak DD, Barth RF (1991) Heterogeneity of peripheral blood reticulocytes: a flow cytometric analysis with monoclonal antibody HAE9 and thiazole orange. American journal of hematology 38: 61–63.
25. Golenda CF, Li J, Rosenberg R (1997) Continuous in vitro propagation of the malaria parasite Plasmodium vivax. Proceedings of the National Academy of Sciences of the United States of America 94: 6786–6791.
26. Udomsangpetch R, Somsri S, Panichakul T, Chotivanich K, Sirichaisinthop J, et al. (2007) Short-term in vitro culture of field isolates of Plasmodium vivax using umbilical cord blood. Parasitology international 56: 65–69.
27. Douay L, Giarratana MC (2009) Ex vivo generation of human red blood cells: a new advance in stem cell engineering. Methods in molecular biology 482: 127–140.
28. Giarratana MC, Rouard H, Dumont A, Kiger L, Safeukui I, et al. (2011) Proof of principle for transfusion of in vitro generated red blood cells. Blood.
29. Tamez PA, Liu H, Fernandez-Pol S, Haldar K, Wickrema A (2009) Stage-specific susceptibility of human erythroblasts to Plasmodium falciparum malaria infection. Blood 114: 3652–3655.
30. Pasvol G, Jungery M, Weatherall DJ, Parsons SF, Anstee DJ, et al. (1982) Glycophorin as a possible receptor for Plasmodium falciparum. Lancet 2: 947–950.
31. Rayner JC, Vargas-Serrato E, Huber CS, Galinski MR, Barnwell JW (2001) A Plasmodium falciparum homologue of Plasmodium vivax reticulocyte binding protein (PvRBP1) defines a trypsin-resistant erythrocyte invasion pathway. The Journal of experimental medicine 194: 1571–1581.
32. Adams JH, Sim BK, Dolan SA, Fang X, Kaslow DC, et al. (1992) A family of erythrocyte binding proteins of malaria parasites. Proceedings of the National Academy of Sciences of the United States of America 89: 7085–7089.

5

On-Chip Cryopreservation: A Novel Method for Ultra-Rapid Cryoprotectant-Free Cryopreservation of Small Amounts of Human Spermatozoa

Yujie Zou[1], Tailang Yin[1,2]**, Shijing Chen[3]**, Jing Yang[1]*, Weihua Huang[3]***

1 Reproductive Medicine Center, Renmin Hospital of Wuhan University, Wuhan, China, 2 Center of Reproductive Medicine, Department of Obstetrics and Gynecology, Peking University Third Hospital, HaiDian District, Beijing, China, 3 Key Laboratory of Analytical Chemistry for Biology and Medicine (Ministry of Education), College of Chemistry and Molecular Sciences, Wuhan University, Wuhan, China

Abstract

Cryopreservation of human spermatozoa free from cryoprotectant can avoid toxicity caused by highly concentrated cryoprotectant and a series of specific carriers have been previously explored, except for PDMS chip. Our study is aimed at exploring a novel device for ultra-rapid cryopreservation of small numbers of spermatozoa without cryoprotectant based on polydimethylsiloxane (PDMS) chips. Spermatozoa from 25 healthy men were involved in this study, comparing on-chip cryopreservation with different micro-channel height (group A: 10 μm height, group B: 50 μm height, group C: 100 μm height) and conventional freezing (group D) in liquid nitrogen for 72 h. The viability, motility, DNA integrity by comet assay and acrosome integrity by fluorescein isothiocyanate-conjugated peanut agglutinin (FITC-PNA) staining of frozen-thawed spermatozoa of each group were compared. The motility and viability of post-thawed spermatozoa was significantly decreased than that of pre-freezing spermatozoa. There was no difference of viability and motility of frozen-thawed spermatozoa between group A and D, while viability and motility of group B and C decreased compared to group A. Comet assay showed that no matter for group A or D, there was no difference of CR, TL, TD and OTM between pre-frozen and post-thawed spermatozoa. There was no difference of CR, TL, TD and OTM of post-thawed spermatozoa between group A and group D neither, while spermatozoa DNA damage was more serious in group B and group C with increasing height of micro-channel compared with group A. The proportion of intact acrosome of post-thawed spermatozoa in group A was the highest when compared with group B and group C, though similar to that of group D. In conclusion, PDMS chip with 10 μm height micro-channel is ideal for ultra-rapid cryopreservation of small quantity of spermatozoa without cryoprotectant.

Editor: Wei Yan, University of Nevada School of Medicine, United States of America

Funding: This work was supported by the National Natural Science Foundation of China (No. 30973196, 20975077, 31070995), the Science Fund for Creative Research Groups (No. 20921062), and the Program for New Century Excellent Talents in University (NCET-10-0611). The funders had no role in study design, data collection and analysis, decision to publish, or preparation of the manuscript.

Competing Interests: The authors have declared that no competing interests exist.

* E-mail: DrYangQing@hotmail.com (JY); whhuang@whu.edu.cn (WH)

These authors contributed equally to this work.

Introduction

Male fertility has declined during the past several decades and the number of patients with azoospermia and oligospermia is increasing [1]. Testicular spermatozoa extraction (TESE) is the common way to obtain spermatozoa from patients with azoospermia, which is expensive, time-consuming and destructive to blood-testicular barrier. In particular, the patients have to bear the repeated surgery procedure after each ICSI failure. Thus, it's necessary to find a method for effective cryopreservation of small amounts of human spermatozoa.

Microfluidic technology is an emerging subject manipulating miniaturized fluids based on chips with micro-fabricated channels and chambers. Microfluidic technology has been utilized in numerous biological applications specifically for miniaturization and simplification of laboratory techniques [2]. Polydimethylsiloxane (PDMS) is the commonest material for making a chip, and it has been proved to be safe and friendly to reproductive cells [3]. Here we just develop a novel method for ultra-rapid cryopreservation of small amounts of human spermatozoa without cryoprotectant using a PDMS chip for the first time.

Materials and Methods

Device design and fabrication

We strictly obeyed the Declaration of Helsinki for Medical Research involving Human Subjects during the project and written consent was obtained from all subjects. This study was approved by the Ethics Committee of Renmin Hospital, Wuhan University. As shown in Figure 1a, 1b and 1c, the chip was comprised of three regions: spermatozoa inlet, spermatozoa store channel and spermatozoa outlet. To find the best parameter for channel design, we designed three channels of different heights: 10 μm, 50 μm and 100 μm. The width and length of the three channels were separately 10 μm and 5 cm. The volume of 10 μm, 50 μm and 100 μm height channel was separately 5×10^{-3} μL, 25×10^{-3} μL and 50×10^{-3} μL. The chip was fabricated with two layers of PDMS. Figure 2a showed the chip compared with a coin.

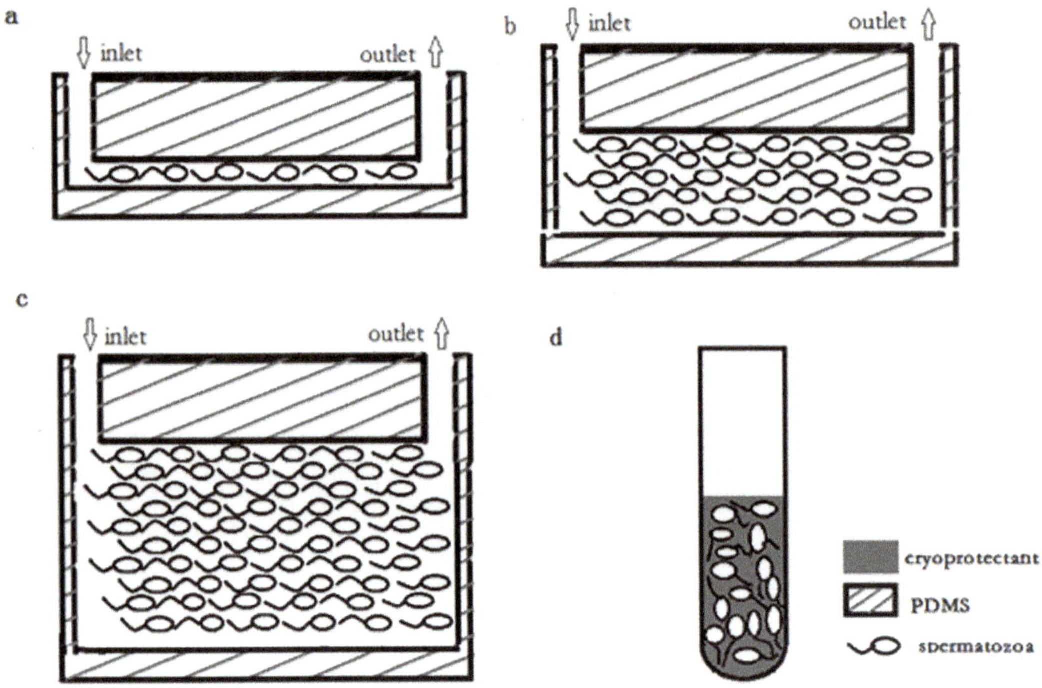

Figure 1. Schema of spermatozoa store in PDMS chip and freezing tube. a. spermatozoa cryopreserved in micro-channel of which height is 10 μm (group A) one by one; **b.** spermatozoa cryopreserved in micro-channel of which height is 50 μm (group B); **c.** spermatozoa cryopreserved in micro-channel of which height is 100 μm (group C); **d.** spermatozoa cryopreserved disorderly in a 1.8 ml freezing tube (group D).

The size of the chip was about 4 cm (length)×3 cm (width)×0.5 cm (height). The upper layer had a micro-channel with an inlet and an outlet punched separately at each ends. The lower layer was smooth without any structures at the surface and plasma bonded together with the upper layer to hold the whole system.

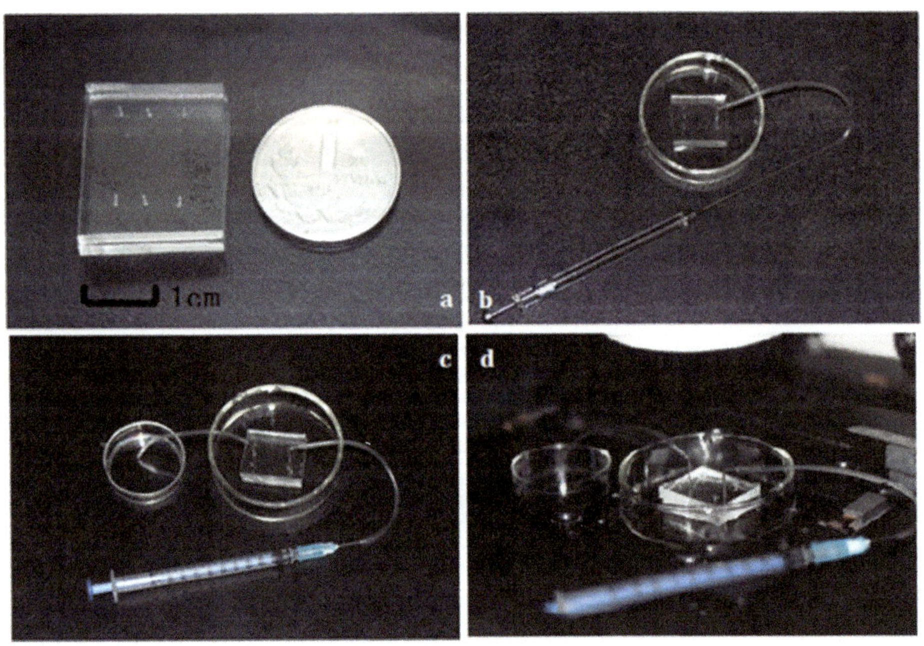

Figure 2. Workflow of PDMS chip. a. a PDMS chip compared with a coin; **b.** sample loading with a micro-injector; **c.** thawing of frozen spermatozoa; **d.** push the syringe and thawed spermatozoa in micro-channel is observed to be transferred into the tissue culture dish under the microscope.

Validation of the method

A chip was stored in the liquid nitrogen for 72 h, and then taken out and recovered in 37°C water bath for 10 min. To detect the tolerance of PDMS to very low temperature in liquid nitrogen, we examined the planeness of surface and elasticity of whole device. Then we loaded some deionized water with a syringe into the channel of chip, and observed whether the channel became deformed or the fluid went forward without any difficulty in the channel.

Collection of spermatozoa sample

Ejaculates were obtained from 25 healthy men by masturbation after at least 72 h of sexual abstinence. 12 samples were used for detection of spermatozoa motility, viability and DNA integrity, and 13 samples were used for acrosome integrity evaluation. The ejaculates were required to contain 20 million or more spermatozoa/mL and to show at least 50% progressive spermatozoa motility and 14% morphologically normal spermatozoa. Informed consent was obtained from each donor. Semen analysis was performed according to published guidelines of the World Health Organization [4].

PureSperm gradient centrifugation

The PureSperm gradient centrifugation was performed according to the PureCeption™ Sperm Separation Media protocol. To prepare the single bilayered gradient for spermatozoa purification, 2 mL of Lower Phase (PureCeption™ 80%) was pipetted into the bottom of the centrifuge tube and 2 mL of Upper Phase (PureCeption™ 40%) was carefully layered over the Lower Phase. 2.5 mL of liquefied semen was gently placed onto the Upper Phase of the single bi-layered gradient and centrifuged at 350 g for 20 min. Remove all layers except the lowest portion (approximately 0.3 mL), add 2 mL of Spermatozoa Washing Medium(PureCeption™, SAGE,USA, 5 mg/ml HEPES-buffered Human Tubal fluid with Human Serum Albumin) and resuspend the pellet. The resulting pellet, placed in the bottom of the tube, was centrifuged at 250 g for 6 min, aspirated and diluted in 0.5 mL Spermatozoa Washing Medium. After estimate of motility, viability, DNA integrity and acrosome integrity, the sample was divided into four equal parts for on-chip cryopreservation (group A: 10 μm; group B: 50 μm; group C: 100 μm) and conventional slow freezing (group D).

On-chip cryopreservation and conventional freezing of human spermatozoa

Before use, the chip was sterilized in 75% ethanol for 30 min, dried, and illuminated under UV for 30 min. Then the sample contacting surfaces of the chip were hydrophilized by plasma treatment. Before on-chip cryopreservation of group A, B and C, the density of spermatozoa sample was adjusted to 10^5 cells/μL so that we can guarantee a high number of spermatozoa stored in the micro-channel. Then the spermatozoa was transferred into a soft tube connected with a needle by pulling a 5 μL micro-injector (Gaoge Industry Trade Co., Ltd., Shanghai, China) (Figure 2b). Next we inserted the needle into the inlet of chip and pushed the micro-injector slowly so that the spermatozoa were administrated into the channel (Figure 2c). Take the 10 μm height channel for example, 5×10^{-3} μL washing medium containing 1000 spermatozoa was thus stored in the micro-channel (Figure 3). For 50 μm and 100 μm height channel, the volume distribution in micro-channel just increased by 5 folds and 10 folds. All the performance was observed under a microscope at the same time, so that we could stop pushing the syringe when the whole channel was filled with spermatozoa (Figure 2d). Wrap the whole chip with silver paper as a protective layer from direct contact with liquid nitrogen. Keep it in the liquid nitrogen for 72 h. Conventional freezing of group D: Spermatozoa sample was mixed with cryoprotectant (Quinn's Advantage@ spermatozoa Freezing Medium, SAGE,USA) of equal volume size and the mixture was allowed to equilibrate for 3 minutes at room temperature. Quinn's Advantage@ spermatozoa Freezing Medium was a HEPES-buffered salt solution containing 10 mg/mL human serum albumin, glycerol and sucrose as the cryoprotective agent, and gentamicin as an antibiotic. The 1.8 mL freezing tube (Corning Costar, USA) was first placed at 4°C for 60 min, and then transferred quickly to liquid nitrogen vapor (−80°C) which was about 10 cm above the surface of the liquid nitrogen for 30 min. Finally we quickly plunged the freezing tube into liquid nitrogen (−196°C) and stored for 72 h.

Thawing procedure of frozen spermatozoa

Group A, B, C: Take the frozen chip out of liquid nitrogen, remove silver paper and recover in 37°C incubator for 10 min. After the spermatozoa were completely thawed, we slowly pushed the syringe connected to the inlet and observed until the spermatozoa were completely transferred into the tissue culture dish (BD FALCON, USA) under the soft tube connected to outlet. Take 10 μm micro-channel for example, about 800–900 out of 1000 spermatozoa could be finally transferred out in 1 mL fertilization medium(Quinn's Advantage@ Fertilization HTF Universal Medium,SAGE,USA) during this process. Group D: Take the freezing tube out of liquid nitrogen and warm in 37°C water bath for 10 min. After centrifugating at 400 g for 5 min, the pellet was collected and resuspended in 100 μL fertilization medium.

Viability of frozen-thawed spermatozoa by eosin Y staining

Spermatozoa viability was assessed using eosin Y staining. The viability of cryopreserved spermatozoa in each group was determined before freezing and after 10 min thawing by 37°C incubation in group A, B and C or melting in 37°C water bath in group D. Considering a low density of post-thawed spermatozoa in group A, B and C, we directly used the 0.5 mL centrifugate tube instead of tissue culture dish to hold the thawed spermatozoa. After centrifugating at 400 g for 5 min, we collected and resuspended the spermatozoa in 5 μL fertilization medium. Then 2 μL drop of stain and 2 μL spermatozoa was placed on microscope slide. A coverslip was applied and the slide was examined with a 400× lens. The slides were scored for living and dead spermatozoa under light microscope (Olympus). Intact spermatozoa excluded the eosin and appeared white, while dead spermatozoa with damaged membrane appeared red. A minimum of 200 spermatozoa under a magnification of 400× was analyzed from each specimen.

Motility of frozen-thawed spermatozoa

The motility of pre-frozen spermatozoa was assessed immediately after PureSperm gradient centrifugation. The motility of cryopreserved spermatozoa in each group was determined immediately after instant thawing by melting in 37°C water bath in group D or 37°C incubation in group A, B and C as described above. A Makler Counting Chamber was used for motility scoring. This index was estimated under the light microscope at a magnification of 400×. Only spermatozoa showing progressive motility were assessed.

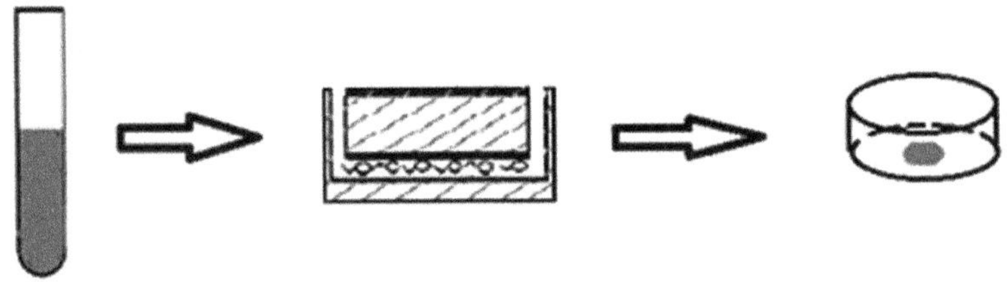

Figure 3. Schematic illustration of the volume distribution of sperm suspension in each step when cryopreserved in 10 μm height channel. The density of spermatozoa sample was adjusted to 10^5/μL before loading. 5×10^{-3} μL medium containing 1000 spermatozoa can be stored in 10 μm height channel and 800–900 spermatozoa can be finally transferred out in 1 mL fertilization medium.

Assessment of spermatozoa DNA integrity by comet assay

We performed the comet assay using OxiSelect™ Comet Assay Kit. 1× Lysis Buffer(containing 3.65 mol/L NaCl, 50 mM/LEDTA), Alkaline Solution(300 mM/L NaOH, 1 mM/L EDTA), Electrophoresis Running Solution (300 mM NaOH, 1 mM EDTA, pH>13) were prepared following the protocol. Post-thawed spermatozoa was washed by centrifugation at 700 g for 2 min, and the pellet was washed with ice-cold PBS(without Mg^{2+} and Ca^{2+}) giving a final concentration at 1×10^5 cells/mL. 10 μL cell sample was mixed with 100 μL OxiSelect™ Comet Agarose and 75 μL/well was immediately pipetted onto the OxiSelect™ Comet Slide. The slide was maintained horizontally at 4°C for 15 min, then immersed in the pre-chilled Lysis Buffer (25 mL/slide) for 30 minutes at 4°C, and finally in the pre-chilled Alkaline Solution (25 mL/slide) for 30 min at 4°C. The above procedures were all performed in the dark. After draining off the fluid, slides were placed in a horizontal electrophoresis unit and were equilibrated in cold Alkaline Electrophoresis Solution for 20 min before being electrophoresed at 15 V (1 volt/cm) and 300 mA for 30 min at room temperature. After electrophoresis, the slides were placed in 70% ethanol for 5 min, air-dried and subsequently stained with 100 μL/well of 1× Vista Green DNA Dye for 15 min at room temperature. View slides by epifluorescence microscopy (Olympus) using a FITC filter at 400× magnification. Images were analyzed by Comet Assay Software (Comet Assay Internet Group, http://www.casp.of.pl). For each sample, a minimum of 100 cells were simultaneously scored for comet rate (CR, %), tail length (TL, length of the comet tail measured from head area to end of tail, μm), tail DNA percentage (TD, percent of DNA in the comet tail) and Oliver tail moment (OTM, tail DNA% × tail length).

Assessment of acrosome integrity by FITC-PNA staining method

Acrosome integrity was evaluated using FITC-PNA (fluorescein isothiocyanate-conjugated peanut agglutinin) staining method, following the protocol of Sperm Morphology FITC-PNA staining kit (GENMED SCIENTIFICS. Shanghai, China) specific to stain the acrosome. The spermatozoa sample was centrifugated at 600 g for 5 min, collected and resuspended in BWW medium adjusting to a final concentration of 1×10^5 cells/mL. 10 μL spermatozoa sample was smeared on a microscope slide, air dried, and then fixed in ice–cold methanol for 2 min. After drying of the fixed smears, 200 μL FITC-PNA stain (25 μg/mL) was gently spread over the smear and incubate in the dark for 30 min at room temperature. The slides were then rinsed with 200 μL of phosphate–buffered saline (PBS) solution to remove excess stain. The acrosome status of spermatozoa was monitored and photographed with an epifluorescence microscope (Olympus) using an excitation wavelength of 480 nm and emissions of 530 nm. According to Reid [5], acrosomes were classified as I = intact, II and III are intermediate forms, IV = reacted. The percentage of fluorescent acrosome-intact spermatozoa (I) was counted in at least 200 sperm cells under the fluorescence microscope at 1000× magnification.

Statistical analysis

Results are presented as mean values ± SD. To identify significant differences, two groups were compared by the independent t-test using SPSS 11.0. Differences were considered statistically significant at $P<0.05$.

Results

Validation of the method

After recovery after cryopreservation in the liquid nitrogen for 72 h, we found that the surface of PDMS was still smooth and whole chip was elastic. After we loaded deionized water with a syringe into the channel of the chip, no deformation of the channel was detected and the fluid went forward without any difficulty in the channel. Thus, PDMS was believed to tolerate very low temperature in liquid nitrogen and the whole device can work well after thawing.

Motility and viability of frozen-thawed spermatozoa

As shown in Table 1, the motility and viability of thawed spermatozoa was significantly decreased than that of pre-freezing spermatozoa. There is no significant difference of motility and viability between group A and group D, while the motility and viability of group B and group C is significantly lower than that of group A. Namely, the wider the channel is, the lower viability and motility could be.

Assessment of spermatozoa DNA integrity by comet assay

Spermatozoa with undamaged DNA did not form a 'comet' and was considered as a viable sperm (Figure 4a). Sperms with low (Figure 4c) and high (Figure 4d) degree of DNA damage due to frozen–thawed process displayed increasing migration of the DNA from the nucleus towards the anode. No matter for group A or D, there was no difference of CR, TL, TD and OTM between pre-frozen and post-thawed spermatozoa. There was no difference of CR, TL, TD and OTM of post-thawed spermatozoa between group A and D neither. Compared with group A, spermatozoa

Table 1. Comparison of viability and motility of frozen-thawed spermatozoa among the groups.

Group	Number	Pre-freezing		Post-thawing	
		viability(%)	motility(%)	viability(%)	motility(%)
A(10 μm height)	12	88.6±3.0[a]	87.1±3.1[a]	51.3±2.8	49.5±2.8
B (50 μm height)	12	88.6±3.0[a]	87.1±3.1[a]	34.2±2.5[b]	33.5±2.6[b]
C (100 μm height)	12	88.6±3.0[a]	87.1±3.1[a]	13.2±2.0[b]	12.1±2.1[b]
D(conventional freezing)	12	88.6±3.0[a]	87.1±3.1[a]	50.9±2.2	49.1±2.5

[a]indicates a statistical difference when compared with post-thawing spermatozoa from Group A to Group D ($P<0.05$);
[b]indicates a statistical difference when compared with Group A ($P<0.05$).

DNA damage was more serious in group B and group C with increasing height of micro-channel (Table 2).

Assessment of acrosome integrity by FITC-PNA staining

The acrosome status of frozen-thawed spermatozoa among the groups was evaluated by FITC-PNA staining. Human spermatozoa stained with FITC-PNA (Figure 5) were classified into four catergories: I (Figure 5a), spermatozoa displaying intensively bright fluorescence of the acrosomal cap, indicating an intact acrosome; II (Figure 5b) and III (Figure 5c) are intermediate forms, displaying disrupted fluorescence of the acrosomal cap or fluorescence only shown in equatorial plane of spermatozoa respectively; IV(Figure 5d), spermatozoa displaying little or no fluorescence, indicating reacted acrosome. As shown in Table 3, the proportion of intact acrosome of post-thawed spermatozoa in group A was 72.8±3.6, while in group B and C, the proportion of intact acrosome was markedly reduced to 61.1±4.0 and 45.3±3.1 respectively. However, proportion of intact acrosome of post-thawed spermatozoa in group A showed no difference when compared with that of spermatozoa in group D (72.8±3.6 vs. 72.6±3.2), indicating a similar acrosome integrity between on-chip cryopreservation in 10 μm height channel and conventional freezing method.

Discussion

There are two potentially fatal mechanisms during the cryopreservation process, excessive osmosis and intracellular ice damages. Ice formation occur firstly in the extracellular aqueous solution during the freezing process and results in high solute concentration surrounding the living cells, which causes the osmosis effect to drive water out of the living cells. If the cooling rate is too slow, osmosis can remove water from cells excessively, leading to a harmful high electrolyte concentration inside cells. If the cooling rate is too fast, intracellular ice could form and become the recrystallization sites of large ice crystals later during the thawing process and could damage the cell membrane [6]. The commonest way for spermatozoa cryopreservation is conventional slow freezing and the ideal cooling rate is expected neither to be too fast nor too slow, which is actually a fragile balance to keep.

In theory, the rate of cooling/warming and the concentration of cryoprotectants are inversely related. That is to say, the faster cooling/warming rate is undertaken, the lower concentration of cryoprotectant is needed [7]. One way to increase the rate of cooling is to minimize both the sample volume and amount of cryoprotectant solution, which means a very large surface area-to-volume ratio. Moreover, the use of advanced and specific devices and materials as supports can markedly increase the rate of cooling.

The present study proposes a new methodology for ultra-rapid cryoprotectant-free cryopreservation of small amounts of human spermatozoa, using a PDMS chip with 10 μm height channel, and leads to a satisfactory cryopreservation effect. To find the best parameter of channel for cryopreservation of spermatozoa, we design channels with a height of 10 μm, 50 μm, 100 μm separately

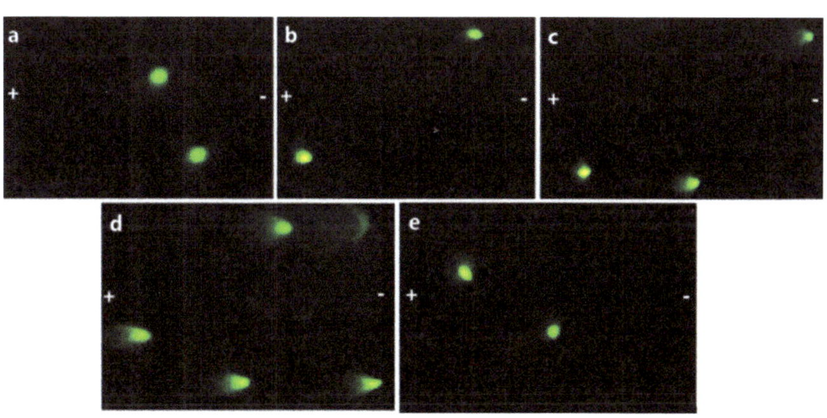

Figure 4. Photographs showing DNA integrity of frozen-thawed spermatozoa in each group at a magnification of 400×. The symbol "−" and "+" represent cathode and anode respectively during electrophoresis of negatively charged DNA. **a.** pre-frozen spermatozoa; **b.** post-thawed spermatozoa of Group A; **c.** comet tail of post-thawed spermatozoa in group B; **d.** obvious long comet tail in post-thawed spermatozoa of Group C; **e.** post-thawed spermatozoa of Group D.

Table 2. Comparison of DNA integrity of frozen-thawed spermatozoa among groups.

Group		Number	CR(%)	TL(μm)	TD(%)	OTM
Pre-freezing		12	11.4±2.9	15.2±2.6	25.9±8.0	3.91±1.0
Post-thawing	A(10 μm height)	12	12.3±2.6	16.0±2.2	26.6±7.4	4.19±0.7
	B(50 μm height)	12	22.8±3.0[a]	19.5±3.2[a]	27.4±4.9	5.35±0.9[a]
	C(100 μm height)	12	45.8±3.1[a]	23.6±5.3[a]	33.0±5.2[a]	7.39±0.7[a]
	D(conventional freezing)	12	12.8±2.2	16.9±3.4	27.8±7.6	4.46±1.0

[a]indicates a statistical difference when compared with Group A ($P<0.05$).

(Figure 1a, 1b, 1c). We found that spermatozoa cryopreserved in channel with a height of 10 μm showed a higher viability and motility compared with the other two groups. If the fluid space around spermatozoa in channel is overmuch, it would be hard to achieve a ideal cooling rate with the increasing sample volume; if the channel space is too narrow, it would be very easy to be obstructed when spermatozoa passes by. The size of spermatozoa head is estimated to be 6 μm, so the spermatozoa just advances in the 10 μm height channel one by one (Figure 1a) and the channel won't be too narrow to be obstructed at the same time.

When compared with conventional freezing, on-chip cryopreservation in 10 μm height channel just displayed a similar viability and motility of frozen-thawed spermatozoa. However, our design has some advantages over conventional freezing in application. First, the fluid volume is small and precise. The volume of the 10 μm height channel is minimized about 5×10^{-3} μL, which is the smallest and most precise volume calculated ever. This means a very large surface area-to-volume ratio, which ensures a very fast and mean cooling/warm rate. Second, no cryoprotectant is needed for our method. Human spermatozoa are too sensitive to tolerate the high concentrations of cryoprotectant conventionally used. Thus high concentrations of permeable cryoprotectants (30%–50%) with the consequent cytotoxic effects due to osmotic stress cannot be applied to spermatozoa cryopreservation. On one hand, our method can avoid the toxicity cryoprotectant caused to cells and decrease the following risk such as abortion and pregnancy loss. On the other hand, our method can simplify the thawing procedure and avoid removing cryoprotectant from spermatozoa. It's reported that removal of cryoprotectant in multiple steps could lead to a loss of spermatozoa motility and membrane disruption to different extents [8]. Third, PDMS is well biocompatible and safe to germ cells. Fourth, PDSM is transparent so that we can get whole control of our operation on spermatozoa with observation under microscope.

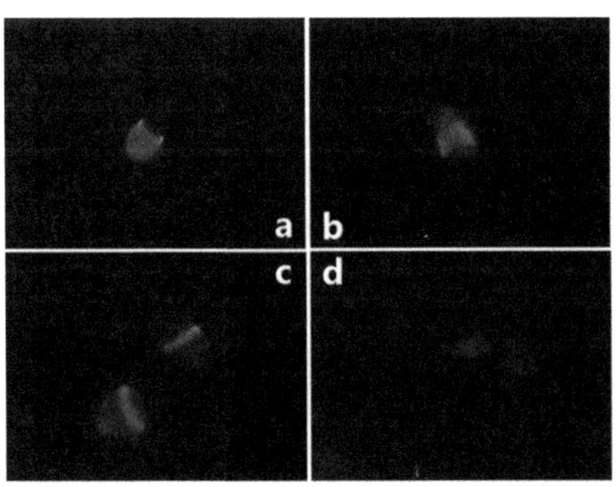

Figure 5. Photographs of four categories of acrosome status evaluated by FITC-PNA staining at a magnification of 1000×. a. I-intact acrosome; **b.** II-intermediate form of minimal acrosome reactivity; **c.** III-intermediate form of severe acrosome reactivity; **d.** IV-reacted acrosome.

Since Nawroth [9] succeeded in cryopreservation of human spermatozoa by direct plunging into liquid nitrogen without cryoprotectants for the first time in 2002, many researchers [10,11] have made a lot of efforts in cryopreservation for spermatozoa in the absence of cryoprotectants. They commonly believe that as long as we can find a carrier with least volumes of the samples and largest surface area-to-volume ratios, we can realize the cryoprotectant-free cryopreservation of spermatozoa by an ultra-rapid cooling rate. A series of specially designed carriers [11–13] have been developed till now, such as human or mouse empty zona pellucidae, cryoloop and the cryotop. But these carriers have their own limitations. Micromanipulation is required for zona pellucidae evacuation and spermatozoa insertion into ZP using human or mouse empty zona pellucidae for cryopreservation [12,14]. Moreover, as the exogenous protein, mouse zona pellucidae may cause pollution to frozen human spermatozoa. For cryoloop cryopreservation [15], a big minus is that we need to be very careful with the thin and fragile liquid film during whole operation [10]. However, on-chip cryopreservation can overcome these disadvantages above. First, on-chip cryopreservation just loads the fluid without any delicate micromanipulation. Second, there is no need to worry about the exogenous protein pollution in our design. Third, the hand-made PDMS chip is strong enough to store the sample and keep stable during freezing and thawing procedure.

Good-quality spermatozoa DNA is of paramount importance for the correct conveyance of genetic material from one generation to the next, the additional chromosomal damage or strand breaks

Table 3. Comparison of acrosome integrity of frozen-thawed spermatozoa among groups.

Group		Number	Intact acrosome(%)
Pre-freezing		13	75.4±3.9
Post-thawing	A(10 μm height)	13	72.8±3.6
	B(50 μm height)	13	61.1±4.0[a]
	C(100 μm height)	13	45.3±3.1[a]
	D(conventional freezing)	13	72.6±3.2

[a]indicates a statistical difference when compared with Group A ($P<0.05$).

in DNA caused by freezing and thawing may have detrimental consequences [16]. It is, therefore, crucial to ensure that spermatozoa are cryopreserved in a way as to minimize DNA injury. In the current investigation, our results suggest that the DNA integrity detected by comet assay was similar between pre-frozen and post-thawed spermatozoa no matter by cryopreservation in 10 μm height channel or by the conventional freezing, which means for semen samples of high quality, on-chip cryopreservation and slow freezing were equally effective in terms of recovery of spermatozoa from healthy and fertile population. These results are in accordance with previous studies recruiting such healthy volunteers [17,18]. However, in contrast to Thomson's study using slow, controlled-rate freezing [19], the amount of DNA damage was increased in the post-thaw spermatozoa samples, suggesting the susceptibility of spermatozoa to damage in the freeze-thaw process, which may due to a specified infertility population. Studies have shown that DNA fragmentation and the percentage of apoptotic spermatozoa significantly increase after cryopreservation [20], possibly owing to formation of reactive oxygen species (ROS), which eventually causes chromatin damage [21].

Another important aspect in the evaluation of semen quality is acrosome integrity. The acrosome is a structure attached to the nucleus of a matured spermatozoa, which contains proteolytic enzymes released to dissolve zona pellucida proteins during fertilization [22]. Acrosome reaction is associated to the activation and exocytotic release of proteolytic enzymes from the acrosome and is thought to play a role in the penetration of the egg investments [23]. FITC-PNA can be used as a reliable probe for detecting acrosome reactions in spermatozoa and the signal representing PNA binding is mainly limited to the acrosomal cap of spermatozoa at the level of fluorescence microscopy. In our study, evaluated by FITC-PNA staining method, frozen-thawed spermatozoa cryopreserved in the 10 μm height channel displayed a higher proportion of intact acrosome compared with that of spermatozoa cryopreserved in the 50 μm or 100 μm height channel, though similar to that of post-thawed spermatozoa by conventional freezing.

Microbial or viral cross-contamination in the liquid nitrogen is a critical issue we are concerned about. However, it is arguable whether the spermatozoa would be infected with bacteria or pathogens mediated by liquid nitrogen, which have leaked into the PDMS chip during storage in a cryotank. According to Endo [24] and Vajta [25], the risk of contamination of germ plasma during cryopreservation and cryobanking in IVF units is small and no report mentions LN2 as a probable vehicle for disease transmission in assisted reproduction treatment. Our study failed to detect the contaminants in the cryotanks, which have been managed in a clean room at 23–28°C with air conditioning for over 5 years of continuous use (data not shown). Further, using an open device, Gimenez et al [26] and Cobo et al [27] have provided the negative evidence of cross-contamination mediated by liquid nitrogen between human pathogens and oocytes/embryos, even from HIV-, HBV-, and HCV-seropositive patients. All in all, information above may suggest that the risk of the cross-contamination between gametes and LN2 in our study is vanishingly small.

In conclusion, this is the first report about successful ultra-rapid on-chip cryopreservation of human spermatozoa without cryoprotectant by direct plunging into liquid nitrogen. Small number of human spermatozoa can be thus successfully cryopreserved without significant loss of important physiological parameters. Ultra-rapid cryopreservation of human spermatozoa without permeable cryoprotectants is a prospective direction for investigations. Anyhow, the final goal of this PDMS chip is to be applied in the cryopreservation of individual spermatozoa. Therefore, in later stage of this project, it is planned to use spermatozoa by TESE as experimental subjects. Another limitation of the micro-device is that we have to spend some time making the chip. All in all, our report of successful recovery of spermatozoa following on-chip cryopreservation is encouraging for further research and clinical implementation.

Author Contributions

Conceived and designed the experiments: YZ. Performed the experiments: YZ TY SC. Analyzed the data: YZ WH. Contributed reagents/materials/analysis tools: TY JY. Wrote the paper: YZ WH.

References

1. ESHRE Capri Workshop Group (2010) Europe the continent with the lowest fertility. Hum Reprod Update 16(6): 590–602.
2. Yi CQ, Li CW, Ji SL, Yang MS (2006) Microfluidics technology for manipulation and analysis of biological cells. Analytica Chimica Acta 560(1): 1–23.
3. Wheeler MB, Walters EM, Beebe DJ (2007) Toward culture of single gametes: The development of microfluidic platforms for assisted reproduction. Theriogenology (Suppl 1): 178–189.
4. World Health Organization (1999) Laboratory manual for the examination of human semen and semen-cervical mucus interaction. 4th ed. Cambridge, United Kingdom: The Press Syndicate of the University of Cambridge.
5. Reid CE, Hermes R, Blottner S, Goeritz F, Wibbelt G, et al. (2009) Split-sample comparison of directional and liquid nitrogen vapour freezing method on post-thaw semen quality in white rhinoceroses (Ceratotherium simum simum and Ceratotherium simum cottoni) Theriogenology 71(2):275–291
6. Li S, Liu W, Lin LW (2010) On-Chip Cryopreservation of Living Cells. JALA 15(2):99–106.
7. Isachenko E (2003) Vitrification of mammalian spermatozoa in the absence of cryoprotectant: from past practical difficulties to present success. Reproductive Biomedicine online 6(2):191–200.
8. Pukazhenthi B, Spindler R, Wildt D, Bush LM, Howard JG (2002) Osmotic properties of spermatozoa from felids producing different proportions of pleiomorphisms: influence of adding and removing cryoprotectant. Cryobiology 44(3): 288–300.
9. Nawroth F, Isachenko V,Dessole S (2002) Successful cryopreservation of human spermatozoa by direct plunging into liquid nitrogen (vitrification) without cryoprotectants (abstract). Fertil Steril, Suppl 78:S129.
10. Isachenko V, Isachenko E, Montag M, Zaeva V, Krivokharchenko I, et al. (2005) Clean technique for cryoprotectant-free vitrification of human spermatozoa. Reproductive BioMedicine online 10(3): 350–354.
11. Desai NN, Blackmon H, Goldfarb J (2004) Sperm cryopreservation on cryoloops: an alternative to hamster zona for freezing individual spermatozoa. Reproductive Biomedicine Online 9(1):47–53
12. Hsieh YY, Tsai HD, Chang CC, Lo HY (2000) Cryopreservation of human spermatozoa within human or mouse empty zona pellucidae. Fertil Steril 73(4): 694–698.
13. Endo Y, Fujii Y, Shintani K, Seo M, Motoyama H, et al. (2011) Single spermatozoon freezing using cryotop. J Mamm Ova Res 28(1): 47–52
14. Ye YH, Xu CM, Qian YL, Jin F, Huang HF (2009) Evaluation of human spermatozoa function after being cryopreserved within the zona pellucidae. Fertil Steril 92(3): 1002–1008.
15. Desai NN, Blackmon H, Goldfarb J (2004) Single spermatozoa cryopreservation on cryoloops: an alternative to hamster zona for freezing individual spermatozoa. Reproductive Biomedicine Online 9(1): 47–53.
16. Donnelly ET, Steele EK, McClure N, Lewis SE (2001) Assessment of DNA integrity and morphology of ejaculated spermatozoa from fertile and infertile men before and after cryopreservation. Hum Reprod 16(6):1191–1199.
17. Gandini L, Lombardo F, Lenzi A, Spano M, Dondero F (2006) Cryopreservation and sperm DNA integrity. Cell Tissue Bank 7(2): 91–98.
18. Kalthur G, Adiga SK, Upadhya D, Rao S, Kumar P (2008) Effect of cryopreservation on sperm DNA integrity in patients with teratospermia. Fertil Steril 89(6): 1723–1727.
19. Thomson LK, Fleming SD, Schulke L, Barone K, Zieschang JA, et al. (2009) The DNA integrity of cryopreserved spermatozoa separated for use in assisted reproductive technology is unaffected by the type of cryoprotectant used but is related to the DNA integrity of the fresh separated preparation. Fertil Steril 92(3): 991–1001.
20. Paasch U, Sharma RK, Gupta AK, Grunewald S, Mascha EJ, et al. (2004) Cryopreservation and thawing is associated with varying extent of activation of

apoptotic machinery in subsets of ejaculated human spermatozoa. Biol Reprod 71(6):1828–1837.

21. Baumber J, Ball BA, Linfor JJ, Meyers SA (2003) Reactive oxygen species and cryopreservation promote DNA fragmentation in equine spermatozoa. J Androl 24(4): 621–628.
22. Yan W (2009) Male infertility caused by spermiogenic defects: Lessons from gene knockouts. Molecular and Cellular. Endocrinology 306(1–2):24–32.
23. Valcircel A, Heras de las MA, Pkrez L, Moses DF, Baldassarre H (1997) Assessment of the acrosomal status of membrane-intact ram spermatozoa after freezing and thawing, by simultaneous lectin/Hoechst 33258 staining. Animal Reproduction, Science 45(4):299–309
24. Endo Y, Fujii Y, Shintani K, Seo M, Motoyama H, et al. (2012) Simple vitrification for small numbers of human spermatozoa. Reproductive BioMedicine Online 24(3): 301–307.
25. Vajta G, Reichart A (2011) Risk of contamination of germ plasma during cryopreservation and cryobanking in IVF units. Hum Reprod (Suppl 1): i42–43.
26. Gimenez JR, Cabal AC, Santos MJ, Bellver J, Mifsu A, et al. (2011) Viral screening of spent culture media and liquid nitrogen samples of oocytes or embryos coming from HIV, HCV, and HBV chronically infected women undergoing IVF cycles. Hum Reprod (Suppl 1): i45.
27. Cobo A, Bellver J, Santos MJ, Remohí J (2012) Viral screening of spent culture media and liquid nitrogen samples of oocytes and embryos from hepatitis B, hepatitis C, and human immunodeficiency virus chronically infected women undergoing in vitro fertilization cycles. Fertil Steril 97(1): 74–78

6

Towards Gene Banking Amphibian Maternal Germ Lines: Short-Term Incubation, Cryoprotectant Tolerance and Cryopreservation of Embryonic Cells of the Frog, *Limnodynastes peronii*

Bianca Lawson, Simon Clulow, Michael J. Mahony, John Clulow*

School of Environmental and Life Sciences, University of Newcastle, Callaghan, New South Wales, Australia

Abstract

Gene banking is arguably the best method available to prevent the loss of genetic diversity caused by declines in wild populations, when the causes of decline cannot be halted or reversed. For one of the most impacted vertebrate groups, the amphibians, gene banking technologies have advanced considerably, and gametes from the male line can be banked successfully for many species. However, cryopreserving the female germ line remains challenging, with attempts at cryopreserving oocytes unsuccessful due to their large size and yolk content. One possible solution is to target cryopreservation of early embryos that contain the maternal germ line, but consist of smaller cells. Here, we investigate the short term incubation, cryoprotectant tolerance, and cryopreservation of dissociated early embryonic cells from gastrulae and neurulae of the Striped Marsh Frog, *Limnodynastes peronii*. Embryos were dissociated and cells were incubated for up to 24 hours in various media. Viability of both gastrula and neurula cells remained high (means up to 40–60%) over 24 hours of incubation in all media, although viability was maintained at a higher level in Ca^{2+}-free Simplified Amphibian Ringer; low speed centrifugation did not reduce cell viability. Tolerance of dissociated embryonic cells was tested for two cryoprotectants, glycerol and dimethyl sulphoxide; dissociated cells of both gastrulae and neurulae were highly tolerant to both—indeed, cell viability over 24 hours was higher in media containing low-to-medium concentrations than in equivalent cryoprotectant-free media. Viability over 24 hours was lower in concentrations of cryoprotectant higher than 10%. Live cells were recovered following cryopreservation of both gastrula and neurula cells, but only at low rates. Optimal cryodiluents were identified for gastrula and neurula cells. This is the first report of a slow cooling protocol for cryopreservation of amphibian embryonic cells, and sets future research directions for cryopreserving amphibian maternal germ lines.

Editor: Nei Moreira, Federal University of Parana (UFPR) – Campus Palotina, Brazil

Funding: This work was supported by the University of Newcastle. The funders had no role in study design, data collection and analysis, decision to publish, or preparation of the manuscript.

Competing Interests: The authors have declared that no competing interests exist.

* E-mail: john.clulow@newcastle.edu.au

Introduction

There is a growing interest in cryobanking amphibian cells and tissues as a conservation tool [1–6] to address the catastrophic collapse in amphibian biodiversity that has been well documented since 1990 [7–13]. The most obvious target for cryopreservation is the embryo; however, this is not likely to be achieved in the near term, given the failure of progress in cryopreserving fish oocytes and embryos, which have similar structural and biochemical properties such as large cell size and high yolk composition [14–18]. This is unfortunate, as external fertilisation and development is the norm in anurans, and the resumption of development of cryopreserved embryos would complete a simple procedure to store, retrieve and restore the diploid genome (mitochondrial and nuclear) without the need for additional complicated post thaw procedures such as implantation. The cryopreservation of embryonic cells is an alternative, but more circuitous path, to storing and retrieving the diploid genome, and will require the reconstitution of embryos through either nuclear transfer or the generation of chimeras. These procedures have been documented for various fish species with reports of the cryopreservation of blastomeres from dissociated embryos [19–27] and the generation of chimeras from unfrozen [28–31] and cryopreserved blastomeres [32–35]. There is also one report of the generation of nuclear transfer embryos in an anuran species from cryopreserved blastomeres [36].

Despite the approach being investigated for fish, other than the single report above, there are no reports of the cryopreservation of early embryonic cells from anuran amphibians.

This is despite the existence of a large literature and long history of nuclear transfer in amphibians [37]. The development of procedures for the isolation and cryopreservation of embryonic cells is a necessary step towards the routine generation of nuclear transfer or chimeric embryos from cryopreserved anuran embryonic cells. This paper reports a study of the isolation, incubation, tolerance to cryoprotectants and cryopreservation of cells from early embryos of the frog, *Limnodynastes peronii* (the Striped Marsh Frog, a common temperate species of south east Australia). The recovery of live cells from gastrulae and neurulae, following cryopreservation with the penetrating cryoprotectants dimethyl sulphoxide (DMSO) and glycerol,was investigated. The study also provided the opportunity to investigate the effect of developmental stage, with the associated changes in yolk content and cell size, on recovery following cryopreservation.

Materials and Methods

Ethics Statement

Collection of spawn and larvae of *Limnodynastes peronii* were approved under NSW NPWS scientific licence S10382 and the research protocol 706 0607 was approved by the University of Newcastle Animal Care and Ethics Committee.

Collection of embryos

Spawn of Striped Marsh Frogs (*Limnodynastes peronii*) were collected from the wild or from animals spawning in captivity. *L. peronii* is a foam-nesting species that deposits its fertilised eggs into a floating foam nest produced from oviducal secretions that the female beats into a froth as eggs are deposited in pond water during amplexus.

L. peronii spawn were held in disposable plastic containers with a small volume of pond water in an incubator at 8°C until required, but for no longer than 5 days (temperature/interval for which the embryos could be held without adverse effects, whilst holding development at desired stages). Embryos were used in experiments at two developmental stages [38]: mid to late gastrula (Gosner stages 10–12) or neurula (Gosner stages 14–16); Fig. 1a, 1b.

Preparation of embryos for experimental protocols

Embryos were removed from refrigeration and allowed to equilibrate at room temperature 30 minutes prior to dissociation. During each experiment, some embryos were withheld and maintained at room temperature in petri dishes overnight to verify viability of embryos after the holding period (Fig. 1c).

A stereoscopic microscope (Model SZH-ILLB, Olympus) was used on low magnification with an external fibre optic light source to view embryos for determining stage of development and removing egg jelly. The outer, sticky jelly layer was removed for all experimental procedures. This was performed under the stereoscopic microscope whilst resting the embryo on a sheet of plastic coated absorbent paper (Benchkote, Whatman). The embryo was gently rolled on the absorbent paper to remove residual nest mucous "foam", before the egg jelly layers were removed with fine forceps, leaving the embryo encapsulated within the vitelline membrane.

Dissociation of Embryos

Individual de-jellied embryos were transferred into droplets on the surface of plastic petri dishes containing 25 μL Ca^{2+}-free Simplified Amphibian Ringer (SAR) with 4 mM ethylenediaminetetraacetic acid (EDTA, Sigma) (referred to herein as Ca^{2+}-free SAR/EDTA; see below). The droplets containing the embryos were left for 15 mins after which a further 10 μL of the same solution was added, and a 0.5–250 μL yellow pipette tip (Finntip 250 μL universal, Thermo Fisher Scientific) with ~1.5 mm cut off the tip was used to transfer the embryos (within the ~35 μL droplets) into 1500 μL Eppendorf tubes. A gentle up and down pipetting action was used to dissociate the embryos within the

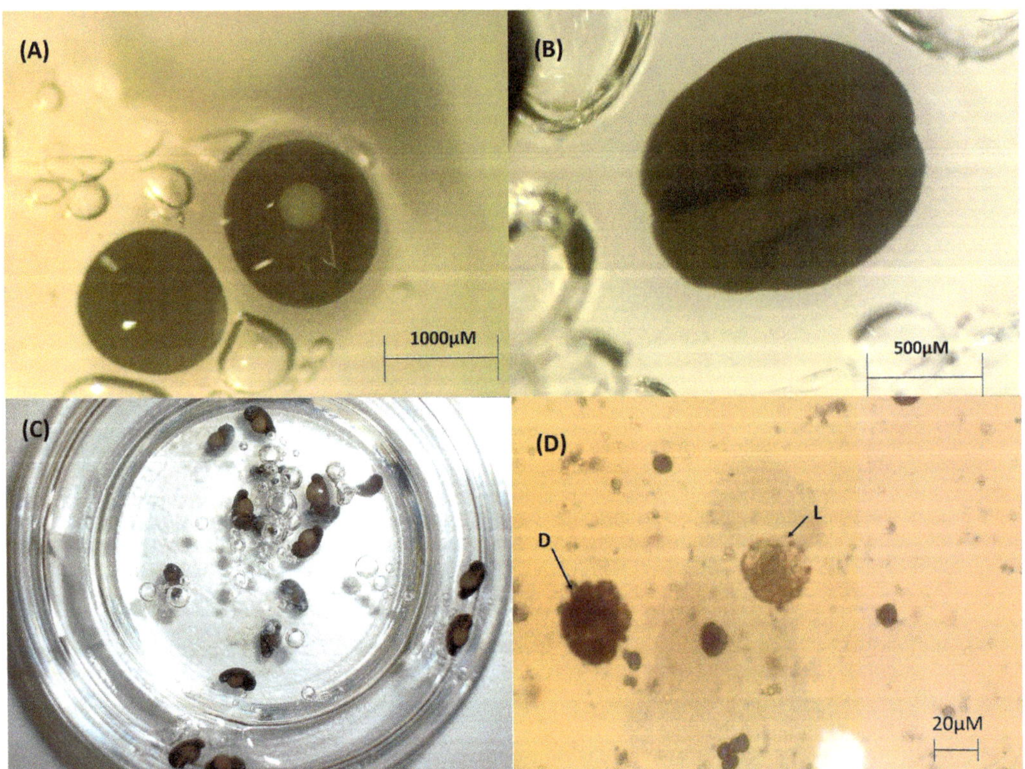

Figure 1. *Limnodynastes peronii* embryos and embryonic cells. (**A**) Late gastrula (Gosner Stage 12). (**B**) Neurula (Gosner Stage 15). (**C**) *L. peronii* embryos continuing normal development 24 hours after removal from holding at 8°C. (**D**) Post thaw viable and non-viable neurula cells determined with trypan blue live (L) - dead (D) staining; 400× magnification.

Eppendorf tube. The resulting cell suspensions were left for a further 10 mins prior to being used in experimental protocols.

Solutions for dissociating and culturing embryos

The following solutions were used to dissociate and culture embryos in experiments for up to 24 hours: Simplified Amphibian Ringer (containing calcium) consisting of 113 mM NaCl, 2 mM KCl, 1.35 mM $CaCl_2$,1.2 mM $NaHCO_3$ [39,40]; calcium-free SAR with 4 mM EDTA (Ca^{2+}-free SAR/EDTA); culture medium (CM) consisting of 0.67× Dulbecco's Modified Eagle's Medium (Gibco) v/v, 10 mM HEPES, 3 mM $NaHCO_3$, 0.1% w/v bovine serum albumin (BSA, CSL), 2% v/v Penicillin-Streptomycin (Sigma), 0.8% v/v Fungizone (Thermo Scientific), and 5 µL/100 ml Tween 80 (Sigma).

Culturing and assessing dissociated embryos

The effect of culture in various solutions on viability of dissociated embryos over 24 hours was investigated by diluting the dissociated embryos (described above) in 3 types of solution: Ca^{2+}-free SAR/EDTA, SAR or CM. As an additional treatment to test the effect of centrifugation on dissociated cell viability, some dissociated embryos were centrifuged at 400×g for 3 minutes, and resuspended in CM. Each treatment was replicated in N = 6 embryos (2 each from 3 different spawns).

All 24 hour incubation solutions were added stepwise (total of 250 µl) to the ~35 µL starting volumes to minimise osmotic disruption as follows: 0 min, 5 µl; 5 min, 15 µl, 10 min, 30 µl, 15 min, 200 µl. Tubes were incubated at room temperature. 20 µl sub-samples were removed at 0.5, 1.5, 2.5 and 24 hours to assess the proportion of viable cells by the addition of 10 µl of 0.4% w/v trypan blue (Sigma). Cells were examined in the chamber of an improved Neubauer haemocytometer under an Olympus BH-2 compound microscope at 400× magnification and at least 100 cells were counted in each replicate; cells excluding the trypan blue dye (Fig. 1d) were scored as viable; cells taking the dye were scored as non-viable. The use of trypan blue to assess cell viability was tested against an alternative fluorescent staining technique (propidium iodide, SYBR® 14, Invitrogen, Oregon, USA; [41]) with no significant difference in results (data not shown).

Cryoprotectant Toxicity

The cytotoxicity of the cryoprotectants dimethylsulphoxide (DMSO, Ajax Chemicals) and glycerol (BDH Chemicals) to cells at room temperature were tested by the stepwise addition of 250 µl of SAR or CM containing 5, 10, 15 or 20% v/v of cryoprotectant to individual embryos dissociated (as described previously) in 35 µl of Ca^{2+} free SAR. The 250 µl cryoprotectant solutions (or SAR or CM controls with no cryoprotectant) were added stepwise to the 35 µl of dissociated embryo in Eppendorf tubes as 5 µl at t = 0 mins, 15 µl at 10 mins, 30 µl at 15 mins, 200 µl at 20 mins and these were inverted at the time of addition to mix solutions; aliquots were removed to assess the proportion of viable cells using trypan blue (as described above) at 30 mins (after the first of the stepwise additions of cryoprotectant), and at 1.5, 2.5 and 24 hours after the first addition of cryoprotectant to each dissociated embryo. Each treatment (and control) was replicated in 6 embryos (2 embryos from each of 3 spawn).

Cryopreservation of cells from dissociated embryos

The same cryoprotectant/media configurations described above for cryoprotectant toxicity tests were used in a cryopreservation experiment to determine the rate of post-thaw recovery of viable cells following cryopreservation. 150 µL of dissociated cell preparations (prepared as for the cryoprotectant tolerance experiment) were loaded into 250 µL semen straws (IMV technologies, France) and sealed with poly vinyl alcohol powder. Samples were cooled and frozen in a controlled cryochamber (Cryologic, Melbourne, Australia). The following cooling protocol was applied: straws prepared from samples held at room temperature were equilibrated at 10°C for 10 minutes, cooled to −7°C at a rate of −1°C/min, held at −7°C for 10 mins, cooled at −1°C/min to −30°C, held at −30°C for 10 mins before being allowed to go into free fall. When the temperature of the chamber was <−100°C, the straws were removed and plunged into liquid nitrogen where they were subsequently stored. The effect of seeding of straws at −7°C (using forceps cooled in liquid nitrogen) showed no effect on recovery rates (data not shown), and was not used in the cooling protocol.

Straws were thawed in air at room temperature and equilibrated for 30 minutes. The contents were emptied into 1.5 mL Eppendorf tubes, and 10 µL aliquots were stained with trypan blue to determine viability as previously described. Counts were performed at 0.5, 1.5, 2.5 and 24 hours post thaw; each treatment was replicated in 6–10 embryos. Due to low recovery of viable cells, the number of cells counted in replicates was increased wherever possible, with more than 500 cells counted in 56% of replicates (up to 1020 in gastrulae, 2400 in neurulae).

Analyses of Data

Data for embryonic cell culture and cryoprotectant cytotoxicity experiments were converted to percentages, arcsine transformed, and subjected to repeated measures analysis of variance using Statistica 5.5 with media, cryoprotectant type, cryoprotectant concentration and embryonic stage as the other main effects. *Post hoc* tests (Duncan's multiple range and Planned Contrasts within the ANOVA routine in Statistica 5.5) were used to test hypotheses of specific differences between groups of means, and for quadratic effects within the cryoprotectant toxicity data. Due to the high number of zero values and the data following a poisson distribution, the cryopreservation data were analysed using the log likelihood function within the Statistica 5.5 Visual GLZ procedure.

Results

Dissociation of embryos and culture of embryonic cells

The viability of cells recovered from dissociated embryos was initially high for both gastrulae and neurulae, but declined in all media over 24 hours. The mean viability of dissociated cells taken from gastrulae (Fig. 2a) were high and within a narrow range (range of means 83.4 to 86.7% at 0.5 hr, the highest being in Ca^{2+}free SAR/EDTA); mean viability for neurula cells at 0.5 hr was also highest in Ca^{2+}free SAR/EDTA (83.0%, Fig. 2b), but ranged lower in other media (58.5 to 74.8%).

There were significant effects of embryonic stage (gastrula versus neurula; $p<0.01$), media ($p<0.05$) and time ($p<0.001$) on the proportion of viable cells following dissociation, as well as significant interactions between embryonic stage and time ($p<0.001$). The data indicated that the recovery of viable cells after dissociation was generally higher in gastrulae than neurulae, but not in all equivalent treatments at equivalent times; in cells of both embryonic stages there was a significant decline over 24 hour of culture (Fig. 2).

There was a benefit to cells dissociated and maintained in Ca^{2+}free SAR with EDTA compared to SAR containing calcium ($p<0.02$), or to culture media ($p<0.05$) which also contained calcium. The significant differences between media were largely

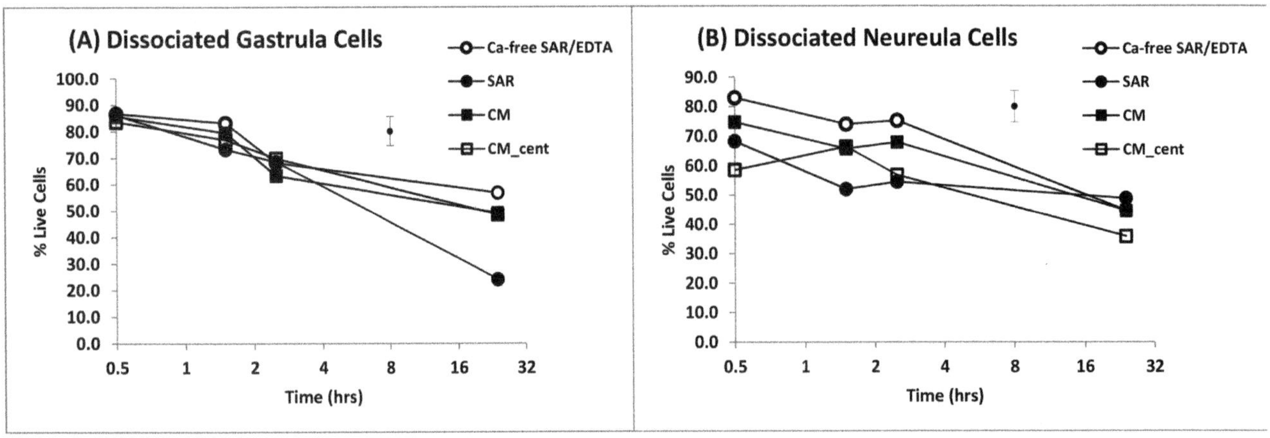

Figure 2. Mean viability of cells from gastrulae and neurulae dissociated in Ca^{2+}-free SAR/EDTA and incubated in various media for 24 hours. (**A**) gastrulae; (**B**) neurulae. Ca^{2+}-free-SAR/EDTA = Ca^{2+}-free-SAR containing 4 mM EDTA; SAR = Simplified Amphibian Ringer; CM = culture medium; CM_cent = centrifuged and resuspended in CM. Each mean, n = 6 (2 embryos from each of 3 spawns); error bar = pooled standard error.

due to differences between means from neurulae (SAR/EDTA 83.0±2.1%, SAR 68.2±3.3%, CM 74.8±2.1% at t = 0.5 hr; $p<0.05$; Fig. 2b); within the gastrulae (Fig. 2a), the media were not significantly different (e.g. SAR/EDTA 86.7±4.4%, SAR 85.7±3.7%, CM 83.6±3.9% at t = 0.5 hr) except for SAR at t = 24 hrs, which was significantly lower ($p<0.01$) than the other media.

Viability of dissociated cells was not significantly reduced as a result of concentrating cells by centrifugation and resuspending in CM for either gastrulae ($p = 0.967$) or neurulae ($p = 0.135$).

Cryoprotectant Toxicity (unfrozen cells)

The viability of gastrula and neurula cells from dissociated embryos cultured for 24 hours in CM or SAR in the presence of DMSO or glycerol (Figs. 3 and 4) showed significant main effects for time, media, cryoprotectant type and cryoprotectant concentration (all $p<0.001$) but no significant main effect of embryonic stage (gastrula versus neurula, $p = 0.416$). There were also significant interaction effects including cryoprotectant type×cryoprotectant concentration ($p<0.01$), cryoprotectant type×time ($p<0.01$) cryoprotectant concentration×time ($p<0.001$), and some higher order effects. The lack of a significant main effect for embryonic stage was indicative of a similar overall response of gastrula and neurula cells to cryoprotectant and other treatments (visible by inspection of Fig. 3), although there were some significant higher order interaction effects involving embryonic stage e.g. stage×media×concentration×time ($p<0.01$).

The presence of cryoprotectants in the media over 24 hours of incubation had complex effects on the viability of cells related both to cryoprotectant type and concentration (Figs. 3 and 4), as indicated by the significant interaction effects described above. These included higher survival of cells in media containing cryoprotectants at low concentrations than in cryoprotectant free controls ($p<0.05$), but lower survival at higher cryoprotectant concentrations ($p<0.001$). DMSO and glycerol both significantly ($p<0.05$) improved viability over controls at low concentrations. The significantly higher overall viability (main effect; $p<0.001$) of cells in DMSO than glycerol was due to the toxicity of glycerol at high concentrations, rather than at low concentrations where it was generally no more toxic, or in some cases, less toxic than DMSO.

Overall, trends were similar in SAR and CM media. However, beneficial effects of cryoprotectant on viability were maintained to higher concentrations in CM than SAR ($p<0.001$), an effect that was most obvious after 24 hours of incubation (Fig. 4). SAR/glycerol were the most toxic media/cryoprotectant combinations at higher concentrations after 24 hours, indicating that SAR was less effective than CM at reducing toxic effects of higher concentrations of glycerol (Fig. 4).

Beneficial effects of cryoprotectants on survival after 24 hours resulted in high mean 24 hr values for many of the 5% and 10% v/v treatments compared to the cryoprotectant free controls e.g. in gastrula cells: 68.5±2.9% in CM/5% glycerol and 56.7±8.2% in CM/5% DMSO versus 36.2±3.4% in CM control; in neurula cells: 62.7±4.2%, CM/5% glycerol and 63.2±4.3% in CM/5% DMSO versus 41.4±4.3% in CM controls. The beneficial effect of 5–10% v/v cryoprotectant concentrations on viability in comparison to media with no cryoprotectant is evident in Figure 4, where there was a statistically significant quadratic effect of concentration on viability at 24 hours of incubation ($p<0.001$) due to viability rising significantly with cryoprotectant concentration initially before declining at higher concentrations. For gastrulae (Fig. 4a), 5 and 10% w/v glycerol and DMSO means were higher than cryoprotectant free controls with the exception of 10% w/v glycerol in SAR. For neurula cells (Fig. 4b), all 5 and 10% w/v cryoprotectant concentrations were equal to or greater than controls; and 15% w/v DMSO means were higher than controls as well.

Recovery of gastrula and neurula cells after cryopreservation

Recovery of viable cells following cryopreservation was low in all treatments for both gastrula (Table 1) and neurula cells (Table 2). As an indicator of the low rates of recovery post-thaw, across all treatments and replicates, only 22% of gastrula replicates and 32% of neurula replicates recovered more than 1% viable cells, and only 7.5% of gastrula and neurula replicates had greater than 2% recovery. The highest cell recovery in any gastrula replicate was 4.4% (CM/10% DMSO) and in any neurula replicate was 5.1% (CM/15% glycerol) assessed at 0.5 hours post thaw. The highest mean recovery for gastrula cells was 1.1±0.8 (n = 6) in CM/10% DMSO and for neurula cells was 1.9%±0.7% (n = 8) in CM/20% glycerol.

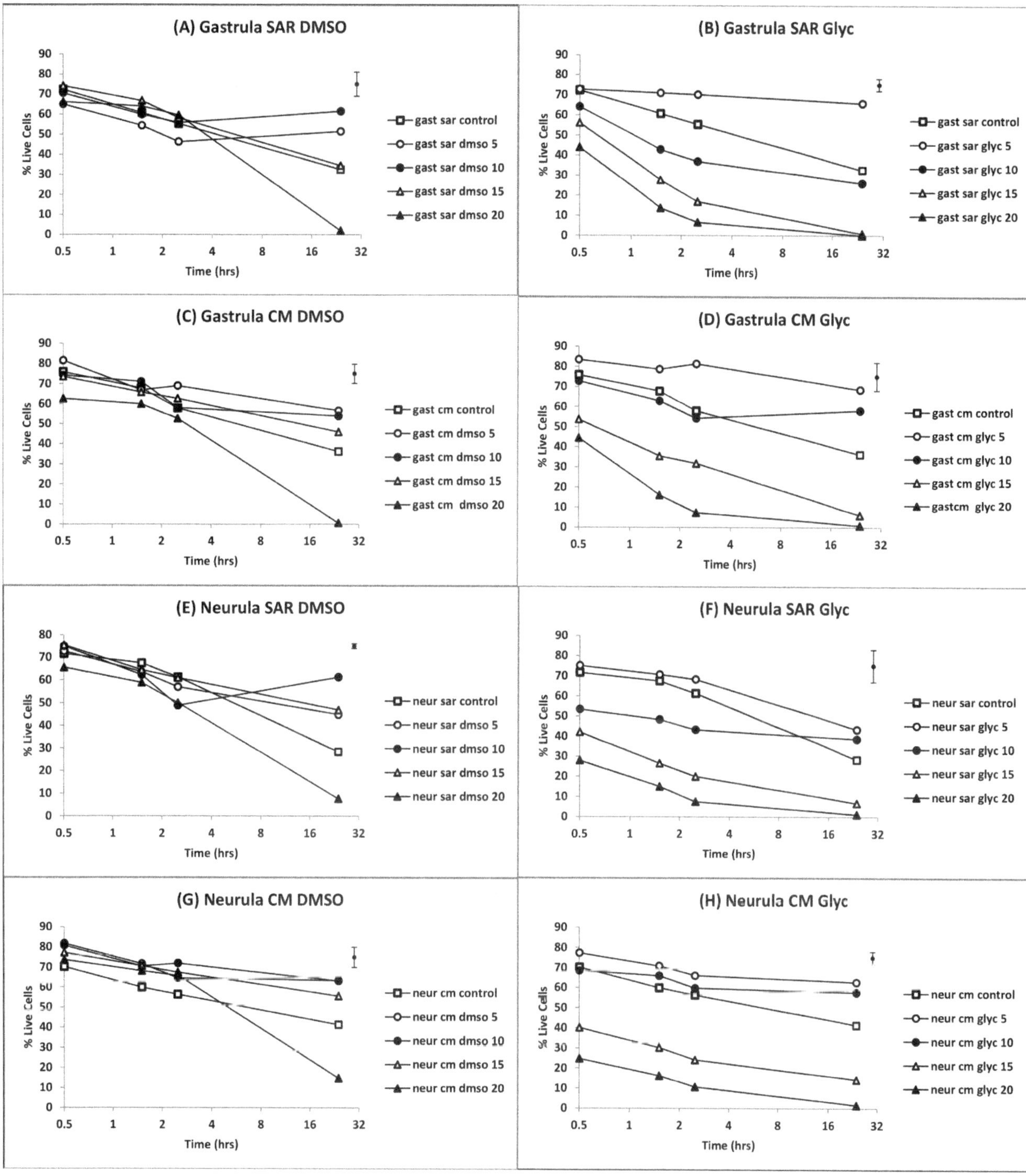

Figure 3. Mean viability of cells from gastrulae and neurulae incubated in various media/cryoprotectant combinations for 24 hours. (**A–D**) gastrulae; (**E–H**) neurulae. CM = culture medium; SAR = Simplified Amphibian Ringer; dmso = DMSO; glyc = glycerol; numbers 5–20 indicate % v/v concentration of cryoprotectant in media; control = 0% v/v cryoprotectant. Each mean, n = 6 (2 embryos from each of 3 spawns); error bar = pooled standard error.

When all data at 0.5 hours post thaw were analysed together, the mean recovery of viable cells was higher in neurula (6.5 cells per thousand; cpt) than gastrula cells (4.6 cpt), $p<0.001$; in CM (6.2 cpt) than SAR (5.3 cpt), $p<0.01$; in glycerol (8.2 cpt) than DMSO (4.5 cpt, $p<0.001$) and there was a significant mean increase in recovery with increasing concentration of cryoprotectant (2.6, 5% v/v; 7.7 cpt, 10% v/v, 7.3 cpt, 15% v/v, 8.1, 20% v/v; $p<0.001$). The higher recovery of cryopreserved cells from neurulae at 0.5 hours persisted with time, and only thawed neurula cells were still viable after 24 hours of incubation (Table 2).

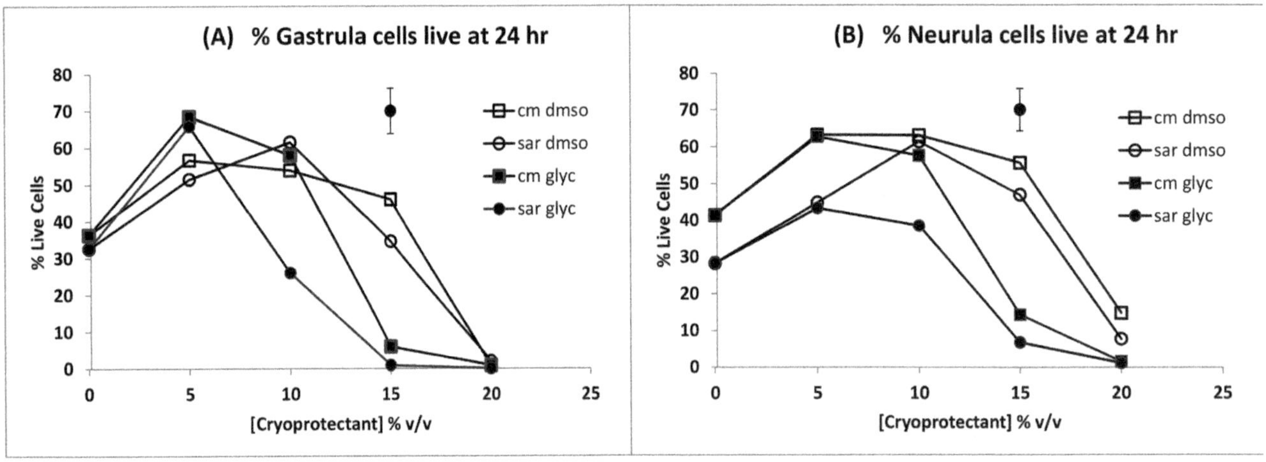

Figure 4. Mean viability of cells from gastrulae and neurulae after 24 hours of incubation in various media/cryoprotectant combinations. (**A**) gastrulae; (**B**) neurulae. cm dmso = CM containing DMSO, sar dmso = SAR containing DMSO, cm glyc = CM containing glycerol, sar glyc = SAR containing glycerol. Each mean, n = 6 (2 embryos from each of 3 spawns); error bar = pooled standard error.

Nevertheless, there were major, significant interaction effects in mainly the gastrula, but also the neurula data that complicate simple identification of the most effective cryopreservation protocols. These included significant media×cryoprotectant type ($p<0.01$), media×cryoprotectant concentration ($p<0.001$), cryoprotectant type×cryoprotectant concentration ($p<0.001$) and even media×cryoprotectant type×cryoprotectant concentration ($p<0.001$) effects for gastrula cells. Cryoprotectant (glycerol 6.0 cpt vs DMSO 4.4 cpt) and cryoprotectant concentration, but not media, main effects were significant ($p<0.01$) in gastrula cells. For neurula cells, interaction effects weren't as strong with only cryoprotectant type×cryoprotectant concentration interactions significant ($p<0.001$), while cryoprotectant (glycerol 10.4 cpt vs DMSO 4.6 cpt) and cryoprotectant concentration ($p<0.001$), but not media, main effects were significant.

Notwithstanding the complexities of the interactions, reasonable interpretations of the data in Tables 1 and 2 lead to the conclusions that: (1) recovery of cells from cryopreservation is higher and persists longer in neurula cells; (2) media effects on recovery are weak or only apparent as complex interaction effects; (3) recovery is higher overall in glycerol than in DMSO, but DMSO recovery can be comparable in favourable media/cryoprotectant concentration combinations; (4) recovery in both glycerol and DMSO is generally higher in concentrations above 5% v/v. These data allow the identification of 10–15% DMSO/CM and 10–15% glycerol/SAR as favourable cryodiluents for gastrula cells and 15–20% glycerol in CM or SAR as favourable cryodiluents for neurula cells (taking into account the persistence of viability at 2.5 hours post thaw).

Discussion

This study found that cells from dissociated embryos of *L. peronii* can be cryopreserved by slow cooling, although the rate of recovery of viable cells, as assessed by live/dead staining, is very low. The optimal cryodiluents identified in this study were 10–15% DMSO/CM and 10–15% glycerol/SAR for gastrula cells and 15–20% glycerol in CM or SAR as favourable cryodiluents for neurula cells. There are no other reports of amphibian embryonic cell cryopreservation against which to benchmark this study with the exception of the report of Uteshev and colleagues [36,42], in which cells from blastulas of *Rana temporaria* were vitrified in a cryoprotectant medium containing 10% sucrose and 10% DMSO. In that study, a high recovery rate of 87% of cryopreserved blastomeres was reported. The reasons for such a large difference in recovery rates between the two studies will require further investigation but may be due to many factors including species differences, embryonic stage (blastula versus gastrula/neurula), handling procedures, cryodiluent composition and cooling rate (slow cooling versus vitrification). The availability of only two reports on amphibian embryonic cell cryopreservation leaves many questions unanswered. In this study, the viability of unfrozen cells, even in the presence of high levels of cryoprotectants, for up to 24 hours indicates that the cryopreservation process, not the handling procedures were the cause of the loss of viability.

Unfrozen cells of *L. peronii* embryos were shown to have a high tolerance to cryoprotectants at room temperature, and even to benefit from their presence in the media over extended periods of incubation (with many combinations of cryoprotectant and medium maintaining above 60% viability after 24 hours). Interestingly, cell viability was maintained at a higher level in the presence of low to medium levels of penetrating cryoprotectant (5–10% v/v DMSO and glycerol) than in cryoprotectant free controls, with increases of more than 20–25% in survival at 24 hours of incubation. Determining the reasons for the beneficial effect of cryoprotectant in the media were beyond the scope of this study, but may derive from a wide range of mechanisms such as anti-oxidant or membrane stabilising effects. Comparable studies on isolated fish blastomeres have also reported high levels of tolerance to DMSO including rainbow trout (85% intact in 8.7% v/v DMSO; [26]), chum salmon (90% survival after 30 mins in 5 or 10% v/v DMSO, and 65% survival in 20% v/v DMSO; [24]), zebrafish (no reduction in hatching of embryos after 2 hours in 7% v/v DMSO; [23]) and medaka, whiting and pejerrey (greater than 80–90% survival after more than 5 hours in up to 9% v/v DMSO [43]). Higher concentrations above those described were associated with significant toxic effects in a number of these species. In contrast, much lower levels of glycerol have been reported as toxic to fish blastomeres: [24] reported only 65% survival in 5% v/v glycerol, and less than 20% in 10% v/v, and less than 10% in 20% v/v glycerol in chum salmon blastomeres following 30 minutes of glycerol exposure; [44] abandoned attempts to investigate permeability of zebrafish embryos to glycerol because of its toxicity to early embryos. None of those fish studies reported the

Table 1. Recovery of gastrula cells after cryopreservation.

		Time (hours)							
		0.5		1.5		2.5		24	
Treatment	[CP] %v/v	$\bar{x}$	# Viable	$\bar{x}$	# Viable	$\bar{x}$	# Viable	$\bar{x}$	# Viable
CM control	0	0.0	0/5	0.0	0/5	0.0	0/5	0.0	0/5
CM/DMSO	5	0.0	0/6	0.6	1/6	1.4	1/6	0.0	0/6
	10	11.1	2/6	0.0	0/6	3.7	1/6	0.0	0/6
	15	8.9	4/6	0.5	1/6	3.1	2/6	0.0	0/6
	20	1.6	1/6	0.7	1/6	0.0	0/6	0.0	0/7
CM/Glycerol	5	1.7	1/6	0.0	0/6	1.2	2/6	0.0	0/6
	10	5.2	2/5	2.6	2/6	0.0	0/6	0.0	0/6
	15	6.4	3/5	0.0	0/6	1.5	1/6	0.0	0/6
	20	9.8	4/6	1.9	2/6	1.5	2/6	0.0	0/6
SAR control	0	0.0	0/6	0.0	0/6	0.0	0/6	0.0	0/6
SAR/DMSO	5	3.5	2/6	0.0	0/6	0.0	0/6	0.0	0/5
	10	8.3	3/6	1.7	1/6	0.0	0/6	0.0	0/6
	15	1.4	1/6	1.8	1/6	0.0	0/6	0.0	0/6
	20	0.0	0/6	0.0	0/6	0.0	0/6	0.0	0/6
SAR/Glycerol	5	5.9	2/6	2.1	2/6	3.5	2/6	0.0	0/6
	10	9.9	5/6	3.9	4/6	4.3	3/6	0.0	0/6
	15	2.7	1/6	5.0	1/6	4.4	3/6	0.0	0/6
	20	6.5	2/6	8.0	2/6	7.4	2/6	0.0	0/6

Post-thaw recovery of cell viability (live/dead staining) expressed as mean number of viable cells per thousand counted (left hand column at each time) and the number of replicates with viable cells (right hand column at each time) for each treatment group. [CP] = cryoprotectant concentration.

increase in survival found with low concentrations of cryoprotectant reported here for *L. peronii* embryonic cells. The data indicate that there is no need to remove cryoprotectants from media containing *L. peronii* embryonic cells, and that these cryoprotectants may even be retained in media for purposes of cell stabilisation. It also worth noting that combinations of cryoprotectants and culture medium generally supported cell viability at a higher level than cryoprotectants with Ringer.

The data from this study suggest that dissociated cells benefited from a low calcium environment (at least, in the absence of cryoprotectants in the media) since survival over 24 hours was higher in Ca^{2+} free SAR than SAR with Ca^{2+}. This effect was only significant in neurula cells. Ca^{2+}-free media was used to dissociate embryos in this study, as in other studies involving fish and frog embryos [19,30,45] as this reduces damage to cells from mechanical disruption of cell to cell junctions (Ca^{2+} is required to maintain intercellular junctions and cell adhesion, [46–48]). However, since all replicates were first dissociated in Ca^{2+} free media, the lower viability of neurula cells in Ca^{2+} -SAR cannot be attributed to the presence of Ca^{2+} during dissociation, but must be due to other effects post-dissociation when Ca^{2+} was added back. Interpretation of these data may be complicated by the observation of re-aggregation of neurula cells in some replicates in the presence of Ca^{2+} (data not shown), which may have increased the relative proportion of non-viable cells in suspension by selective re-aggregation of viable cells.

The recovery of post thaw viability of cryopreserved cells did not follow the same pattern of response to cryoprotectant concentration as the unfrozen cells. In unfrozen cells, viability was significantly decreased by higher concentrations, but in the cryopreservation treatments, there was a trend for higher recovery rates in higher concentrations of glycerol and DMSO, although the effects were complicated by complex interactions. It would appear that the protective effects of higher cryoprotectant concentrations during cryopreservation outweigh the toxic effects during incubation at room temperature. The recovery of cryopreserved *L. peronii* cells following cryopreservation was low compared to reports of the recovery of fish blastomeres after cryopreservation; in this study mean recovery rates were less than 2% with the highest recovery in a gastrula replicate 4.4%, and neurula replicate 5.1%. Most reports of fish blastomere cryopreservation are for cryodiluents employing DMSO. Depending on the species, and the method of freezing, the results are variable, but in all cases higher than values for *L. peronii*: zebrafish 70–85% recovery by slow cooling in 7–10% v/v DMSO [23,25] and 93% with vitrification in 35% v/v DMSO [21]; rainbow trout 36% recovery in 8.7% DMSO [26], up to 95% recovery in 1.4 M propanediol [19]; chum salmon 59% recovery in 10% v/v DMSO/10% [24]; pejerrey 67% recovery in 18% DMSO, but lower recovery rates for whiting (20%) and medaka (34%) in 9% v/v DMSO [43]. Nevertheless, even with fish blastomeres, large variations in recovery rates occur with variations in cryoprotectant type and concentration [24,43], and some treatments yield recovery rates as low as values found in this study. It is difficult to make meaningful comment on differences in recovery rates between fish and amphibian embryonic cells, given this is the only study of cryopreservation by slow cooling with no comparable data from other studies of amphibians, and the only other report by [36] involving vitrification. Factors that may contribute to the differences include the developmental stage (most fish studies involve blastomeres from 50% epiboly blastula versus gastrula and neurula in this study), methodological differences, and potentially

Table 2. Recovery of neurula cells after cryopreservation.

		Time							
		0.5		1.5		2.5		24	
Treatment	[CP] %v/v	$\bar{x}$	# Viable	$\bar{x}$	# Viable	$\bar{x}$	# Viable	$\bar{x}$	# Viable
CM control	0	0.0	0/8	0.0	0/8	0.0	0/8	0.0	0/8
CM/DMSO	5	2.8	5/13	2.3	5/13	0.0	0/13	0.0	0/13
	10	4.8	6/10	2.3	4/10	0.3	1/10	0.2	1/10
	15	3.6	8/10	1.4	5/10	0.0	0/10	0.2	1/10
	20	6.8	11/14	1.4	8/14	0.2	1/14	0.0	0/14
CM/Glycerol	5	2.2	3/8	1.0	3/8	1.0	2/8	0.3	1/8
	10	9.3	6/8	5.7	6/8	4.0	4/8	0.0	0/8
	15	17.2	7/8	8.9	7/8	6.1	7/8	0.6	2/8
	20	19.1	7/8	6.5	7/8	4.9	8/8	0.5	1/8
SAR control	0	0.0	0/8	0.0	0/8	0.0	0/8	0.0	0/8
SAR/DMSO	5	2.0	4/8	0.3	1/8	0.0	0/8	0.0	0/8
	10	5.8	8/8	1.2	4/8	1.2	4/8	0.0	0/8
	15	4.9	6/8	1.8	3/8	0.6	3/8	0.0	0/8
	20	5.8	6/8	2.7	4/8	1.2	3/8	0.0	0/8
SAR/Glycerol	5	2.7	5/8	2.0	4/8	1.5	1/8	0.5	1/8
	10	8.7	8/8	3.9	6/8	2.4	5/8	0.0	0/8
	15	11.4	8/8	6.0	6/8	3.2	5/8	0.4	1/8
	20	12.7	7/8	4.8	5/8	3.2	4/8	0.3	2/8

Post-thaw recovery of cell viability (live/dead staining) expressed as mean number of viable cells per thousand counted (left hand column at each time) and the number of replicates with viable cells (right hand column at each time) for each treatment group. [CP] = cryoprotectant concentration.

fundamental differences between fish and amphibians in the response to cryopreservation. The yolk content of amphibian oocytes and embryos may be even higher than that of fish in some cases (compare [18,49] to values from [44]).

One response that was similar between this study and studies reporting the cryopreservation of fish blastomeres was the trend for recovery rates to be higher in later stages of embryo development in fish (across varying stages of blastula [19,43], and in this study higher recovery of neurula than gastrula cells). This may be due to many factors, but could include more favourable biophysical properties of the smaller cells generated as cleavage proceeds in pre-hatching embryos [45,50]. In the current study, gastrula cells of *L. peronii* were mostly in the order of 30–70 μm and neurula cells in the range of 20–30 μm. [14] also reports a number of fish species in which chill sensitivity in whole embryos declines as embryonic development proceeds, paralleling the results with blastomere cryopreservation.

It should be recognised that even though the rate of recovery of intact cryopreserved *L. peronii* embryonic cells was low, this may not indicate that use of cryopreserved cells will not be successful in the generation of nuclear transfer embryos. Although cell viability assessed by the recovery of intact cells with sufficient membrane integrity to exclude vital stains was the measure of response in this study, the integrity of the cell nuclei may still be high, and they may be capable of supporting nuclear transfer. In amphibian nuclear transfer (although not the case with chimeras), cell membranes are deliberately disrupted when cell nuclei are transferred [37].

This study is a step forward in investigating the cryopreservation of amphibian embryonic cells, but poses many questions that require further investigation if cryopreserved embryonic cells are to generate viable embryos for the use in amphibian genome banking and species management for conservation. Further studies need to investigate whether recovery can be boosted by changing factors such as cooling and thawing rates, and cryodiluent composition e.g. improvements in fish blastomere cryopreservation by including foetal calf serum as reported by [24].

Author Contributions

Conceived and designed the experiments: BL SC MM JC. Performed the experiments: BL SC JC. Analyzed the data: BL JC. Wrote the paper: JC SC.

References

1. Clulow J, Mahony MJ, Browne RK, Pomering M, Clark AK (1999) Applications of assisted reproductive technologies (ART) to endangered anuran amphibians. In: Campbell A, editor. Declines and Disappearances of Australian Frogs. Canberra: Environment Australia. pp. 219–225.
2. Lermen D, Blömeke B, Browne R, Clarke ANN, Dyce PW, et al. (2009) Cryobanking of viable biomaterials: implementation of new strategies for conservation purposes. Molecular Ecology 18: 1030–1033.
3. Lloyd R, McGregor Reid G, Gibson R, Holt W, Sainsbury A, et al. (2012) Abstracts from the Amphibian Ark Biobanking Advisory Committee Workshop Towards a Biobanking Strategy for Amphibian Conservation September 6–8, 2010 London and Portsmouth, UK. Biopreservation and Biobanking 10: 62–69.
4. Mahony MJ, Clulow J (2011) Appendix 2. Cryopreservation and Reconstitution Technologies: A Proposal to Establish A Genome Resource Bank For Threatened Australian Amphibians. In: Murray K, Skerratt LF, Marantelli G, Berger L, Hunter D, et al., editors. Guidelines for minimising disease risks

associated with captive breeding, raising and restocking programs for Australian frogs. Canberra: Australian Government, Department of Sustainability, Environment, Water, Population and Communities.

5. Maya-Soriano JM, Holt WV, Lloyd RE (2012) Biobanked amphibian samples confirmed to species level using 16S rRNA DNA barcodes. Biopreservation and Biobanking 10: 22–28.
6. Clulow J, Clulow S, Guo J, French AJ, Mahony MJ, et al. (2012) Optimisation of an oviposition protocol employing human chorionic and pregnant mare serum gonadotropins in the Barred Frog *Mixophyes fasciolatus* (Myobatrachidae). Reproductive Biology and Endocrinology 10: 60.
7. Bishop PJ, Angulo A, Lewis JP, Moore RD, Rabb GB, et al. (2012) The Amphibian Extinction Crisis - what will it take to put the action into the Amphibian Conservation Action Plan? SAPIENS 5.2.
8. Murray K, Retallick R, McDonald KR, Mendez D, Aplin K, et al. (2010) The distribution and host range of the pandemic disease chytridiomycosis in Australia, spanning surveys from 1956–2007. Ecology 91: 1557–1558.
9. Murray KA, Retallick RWR, Puschendorf R, Skerratt LF, Rosauer D, et al. (2011) Assessing spatial patterns of disease risk to biodiversity: implications for the management of the amphibian pathogen, *Batrachochytrium dendrobatidis*. Journal of Applied Ecology 48: 163–173.
10. Skerratt L, Berger L, Speare R, Cashins S, McDonald K, et al. (2007) Spread of chytridiomycosis has caused the rapid global decline and extinction of frogs. EcoHealth 4: 125–134.
11. Stuart SN, Chanson JS, Cox NA, Young BE, Rodrigues ASL, et al. (2004) Status and trends of amphibian declines and extinctions worldwide. Science 306: 1783–1786.
12. Wake DB (1991) Declining amphibian populations. Science 253: 860.
13. Shoo LP, Olson DH, McMenamin SK, Murray KA, Van Sluys M, et al. (2011) Engineering a future for amphibians under climate change. Journal of Applied Ecology 48: 487–492.
14. Hagedorn M, Hsu E, Kleinhans FW, Wildt DE (1997) New Approaches for Studying the Permeability of Fish Embryos: Toward Successful Cryopreservation. Cryobiology 34: 335–347.
15. Hagedorn M, Kleinhans FW, Artemov D, Pilatus U (1998) Characterization of a major permeability barrier in the Zebrafish embryo. Biology of Reproduction 59: 1240–1250.
16. Hagedorn M, Peterson A, Mazur P, Kleinhans FW (2004) High ice nucleation temperature of zebrafish embryos: slow-freezing is not an option. Cryobiology 49: 181–189.
17. Kleinhans FW, Guenther JF, Roberts DM, Mazur P (2006) Analysis of intracellular ice nucleation in *Xenopus oocytes* by differential scanning calorimetry. Cryobiology 52: 128–138.
18. Wallace RA (1963) Studies on amphibian yolk III. A resolution of yolk platelet components. Biochimica Biophysica Acta 74: 494–504.
19. Calvi SL, Maisse G (1998) Cryopreservation of Rainbow Trout (*Oncorhynchus mykiss*) blastomeres: influence of embryo stage on postthaw survival rate. Cryobiology 36: 255–262.
20. Calvi SL, Maisse G (1999) Cryopreservation of carp (*Cyprinus carpio*) blastomeres. Aquatic Living Resources 12: 71–74.
21. Cardona-Costa J, García-Ximénez F (2007) Vitrification of zebrafish embryo blastomeres in microvolumes. Cryoletters 28: 303–309.
22. Dash SN, Routray P, Dash C, Guru BC, Swain P, et al. (2008) Use of the non-toxic cryoprotectant trehalose enhances recovery and function of fish embryonic stem cells following cryogenic storage. Current Stem Cell Research & Therapy 3: 277–287.
23. Harvey B (1983) Cooling of embryonic cells, isolated blastoderms, and intact embryos of the zebra fish *Brachydanio rerio* to −196°c. Cryobiology 20: 440–447.
24. Kusuda S, Teranishi T, Koide N (2002) Cryopreservation of chum salmon blastomeres by the straw method. Cryobiology 45: 60–67.
25. Lin C, Zhang T, Rawson DM (2009) Cryopreservation of Zebrafish (Danio rerio) Blastomeres by Controlled Slow Cooling. Cryoletters 30: 132–141.
26. Nilsson EE, Cloud JG (1993) Cryopreservation of rainbow trout (*Oncorhynchus mykiss*) blastomeres. Aquatic Living Resources 6: 77–80.
27. Routray P, Dash C, Dash SN, Tripathy S, Verma DK, et al. (2010) Cryopreservation of isolated blastomeres and embryonic stem-like cells of Leopard danio, Brachydanio frankei. Aquaculture Research 41: 579–589.
28. Hong N, Chen S, Ge R, Song J, Yi M, et al. (2012) Interordinal Chimera Formation Between Medaka and Zebrafish for Analyzing Stem Cell Differentiation. Stem Cells and Development 21: 2333–2341.
29. Lin S, Long W, Chen J, Hopkins N (1992) Production of germ-line chimeras in zebrafish by cell transplants from genetically pigmented to albino embryos. Proceedings of the National Academy of Sciences 89: 4519–4523.
30. Nilsson EE, Cloud JG (1992) Rainbow trout chimeras produced by injection of blastomeres into recipient blastulae. Proceedings of the National Academy of Sciences 89: 9425–9428.
31. Yamaha E, Mizuno T, Hasebe Y, Yamazaki F (1997) Chimeric fish produced by exchanging upper parts of blastoderms in goldfish blastulae. Fisheries Science 63: 514–519.
32. Al-Anzi B, Chandler DE (1998) A sperm chemoattractant is released from *Xenopus* egg jelly during spawning. Developmental Biology 198: 366–375.
33. Higaki S, Eto Y, Kawakami Y, Yamaha E, Kagawa N, et al. (2010) Production of fertile zebrafish (Danio rerio) possessing germ cells (gametes) originated from primordial germ cells recovered from vitrified embryos. Reproduction 139: 733–740.
34. Kusuda S, Teranishi T, Koide N, Nagai T, Arai K, et al. (2004) Pluripotency of cryopreserved blastomeres of the goldfish. Journal of Experimental Zoology Part A: Comparative Experimental Biology 301A: 131–138.
35. Yasui GS, Fujimoto T, Sakao S, Yamaha E, Arai K (2011) Production of loach (Misgurnus anguillicaudatus) germ-line chimera using transplantation of primordial germ cells isolated from cryopreserved blastomeres. Journal of Animal Science 89: 2380–2388.
36. Uteshev VK, Melnikova EV, Kaurova SA, Nikitin VA, Gakhova EN, et al. (2002) Fluorescent analysis of cryopreserved totipotent cells of amphibian embryos Biofizika 47: 539–545.
37. McKinnell RG (1978) Cloning. Nuclear Transplantation in Amphibia. Minneapolis, MN: University of Minnesota Press.
38. Gosner KL (1960) A simplified table for staging anuran embryos and larvae with notes on identification. Herpetologica 16: 183–190.
39. Edwards DL, Mahony MJ, Clulow J (2004) Effect of sperm concentration, medium osmolality, and oocyte storage on artificial fertilisation success in a myobatrachid frog (*Limnodynastes tasmaniensis*) Reproduction, Fertility and Development 16: 347–354.
40. Rugh R (1962) Induced Breeding. Experimental Embryology Techniques and Procedures. Minneapolis, MN: Burgess Publishing Company. pp. 91–103.
41. Fitzsimmons C, McLaughlin EA, Mahony MJ, Clulow J (2007) Optimisation of handling, activation and assessment procedures for Bufo marinus spermatozoa. Reproduction, Fertility and Development 19: 594–601.
42. Uteshev VK, Gakhova EN (2005) Gene cryobanks for conservation of endangered amphibian species. In: Ananjeva N, Tsinenko O, editors. Russian J of Herpetology, vol 12 (Suppl), Proceedings of the 12th Ordinary General Meeting of the Societas Europaea Herpetologica, Saint-Petersburg, Aug 12–16, 2003 pp. 233–234.
43. Strussmann CA, Nakatsugawa H, Takashima F, Hasobe M, Suzuki T, et al. (1999) Cryopreservation of Isolated Fish Blastomeres: Effects of Cell Stage, Cryoprotectant Concentration, and Cooling Rate on Postthawing Survival. Cryobiology 39: 252–261.
44. Hagedorn M, Kleinhans FW, Wildt DE, Rall WF (1997) Chill Sensitivity and Cryoprotectant Permeability of Dechorionated Zebrafish Embryos, *Brachydanio rerio*. Cryobiology 34: 251–263.
45. Wang P, Hayden S, Masui Y (2000) Transition of the blastomere cell cycle from cell size-independent to size-dependent control at the midblastula stage in *Xenopus laevis*. Journal of Experimental Zoology 287: 128–144.
46. Armitage WJ, Juss BK (2000) Assembly of intercellular junctions in epithelial cell monolayers following exposure to cryoprotectants. Cryobiology 41: 58–65.
47. Faure S, Cau J, de Santa Barbara P, Bigou S, Ge Q, et al. (2005) *Xenopus* p21-activated kinase 5 regulates blastomeres' adhesive properties during convergent extension movements. Developmental Biology 277: 472–492.
48. Nomura K, Uchida M, Kageura H, Shiokawa K, Yamana K (1986) Cell to Cell Adhesion Systems in *Xenopus laevis*, the South African Clawed Frog I. Detection of Ca2+ Dependent and Independent Adhesion Systems in Adult and Embryonic Cells. Development, Growth & Differentiation 28: 311–319.
49. Guenther JF, Seki S, Kleinhans FW, Edashige K, Roberts DM, et al. (2006) Extra- and intra-cellular ice formation in Stage I and II *Xenopus laevis* oocytes. Cryobiology 52: 401–416.
50. Sze LC (1953) Changes in the amount of desoxyribonucleic acid in the development of *Rana pipiens*. Journal of Experimental Zoology 122: 577–601.

7

Inhibition of Ice Growth and Recrystallization by Zirconium Acetate and Zirconium Acetate Hydroxide

Ortal Mizrahy, Maya Bar-Dolev*, Shlomit Guy, Ido Braslavsky

The Institute of Biochemistry, Food Science, and Nutrition, The Robert H. Smith Faculty of Agriculture, Food and Environment, The Hebrew University of Jerusalem, Rehovot, Israel

Abstract

The control over ice crystal growth, melting, and shaping is important in a variety of fields, including cell and food preservation and ice templating for the production of composite materials. Control over ice growth remains a challenge in industry, and the demand for new cryoprotectants is high. Naturally occurring cryoprotectants, such as antifreeze proteins (AFPs), present one solution for modulating ice crystal growth; however, the production of AFPs is expensive and inefficient. These obstacles can be overcome by identifying synthetic substitutes with similar AFP properties. Zirconium acetate (ZRA) was recently found to induce the formation of hexagonal cavities in materials prepared by ice templating. Here, we continue this line of study and examine the effects of ZRA and a related compound, zirconium acetate hydroxide (ZRAH), on ice growth, shaping, and recrystallization. We found that the growth rate of ice crystals was significantly reduced in the presence of ZRA and ZRAH, and that solutions containing these compounds display a small degree of thermal hysteresis, depending on the solution pH. The compounds were found to inhibit recrystallization in a manner similar to that observed in the presence of AFPs. The favorable properties of ZRA and ZRAH suggest tremendous potential utility in industrial applications.

Editor: Boris Rubinsky, University of California at Berkeley, United States of America

Funding: This work was supported in part by grants from The Lady Davis Fellowship Trust and The Valazzi-Pikovsky Fellowship Fund (to MBD), the Israel Science Foundation (ISF) grants #1279/10 and #1281/10, the European Research Council (ERC) grant #281595, and the Marie Curie International Reintegration grant #256364. The funders had no role in study design, data collection and analysis, decision to publish, or preparation of the manuscript.

Competing Interests: The authors have declared that no competing interests exist.

* E-mail: mayabar00@gmail.com

Introduction

Ice growth and nucleation play a significant role in life on earth, for example, in the freezing of environmental water resources, as well as the freezing of water in plant and animal tissue, both of which can result in fatal damage. Organisms living in cold environments must respond appropriately to survive when temperatures drop below the freezing point. Such organisms have evolved a series of mechanisms for controlling ice growth and protecting themselves from freezing damage, either by avoiding freezing or by tolerating the freezing process. Control over ice growth is crucial in industrial settings as well. Processes such as ice templating are used in the engineering of porous materials [1,2]. In cryopreservation, control over ice growth is crucial to maintaining the viability of cells, tissues, and organs [3,4].

Ice growth may be controlled by inhibiting ice recrystallization, in which the growth of large ice crystals is favored at the expense of small ones (Otswald ripening [5]). Ice recrystallization is lethal in tissues, for example in crops at times of a frost. This process produces mechanical damage to cell membranes during the freeze–thaw cycle of cryopreservation [3,4,6]. In the frozen food industry, ice recrystallization can avert food texture [7]. Substances that can inhibit ice recrystallization and control ice growth are, therefore, attractive for industrial and medical applications.

Ice binding proteins (IBPs) are a set of natural proteins that may be used to control ice growth and recrystallization. This expanding group of proteins interacts with ice for various biological purposes such as freeze avoidance, ice recrystallization inhibition (IRI), and ice nucleation promotion, thereby protecting organisms from ice damage [8,9]. A subset of this group is the antifreeze proteins (AFPs), which were found to control ice shape, growth and melting in a non-colligative manner. AFPs can depress the freezing temperature of seed ice crystals below the melting point, resulting in a thermal hysteresis gap. AFPs were also shown to have IRI activities. AFPs are expressed in a variety of organisms that require survival in cold environments, including insects, fish, bacteria, plants, fungi, and plankton (reviewed in [8]). These proteins are excellent candidates for use in cryopreservation, the frozen food industry, and other fields in which control over ice growth and recrystallization are important. Although AFPs are currently used in industry, for example, in ice cream [10], the difficulties associated with producing AFPs in large quantities preclude them from use in other industrial applications. Furthermore, the lifetimes of AFPs are limited compared to other cryoagents. An ideal synthetic alternative cryoprotectant would be low in cost, readily available, and simple to use.

Non-protein cryoprotective agents (CPAs) that act in a colligative manner such as glycerol, dimethyl sulfoxide (DMSO), and certain sugars are widely used in cryopreservation. These materials can protect cells from damage caused by the formation of intracellular ice during freezing and thawing. CPAs require high concentrations to be effective and some display high cytotoxic effects [11]. Few additives have been identified as exhibiting non-colligative effects similar to those of AFPs upon crystal growth and recrystallization. These include block co-polymers based on poly(ethylene oxide) and polyethyleneimine modified with hydroxyl groups [12], polyvinyl alcohol (PVA) [13], synthetic peptides [14] and peptoids [15], and xylomannan, a substance containing mostly polysaccharides and fatty acids which was isolated from a freeze-tolerant beetle [16,17]. Recently, the inorganic material zirconium(IV) acetate (ZRA) was shown to induce six-fold symmetry in cavities of materials prepared by ice templating over the pH range 3.5–4.5. These shapes are similar to ice shapes observed in the presence of a variety of AFPs [18]. Such shaping was not apparent in similar materials, such as yttrium acetate, barium acetate, or zirconium(IV) acetate hydroxide (ZRAH)] [19].

Here, we have expanded the study of ZRA and ZRAH by measuring their potential to act as cryoprotectants through direct ice shaping, modulating the growth rate of ice, and inhibiting ice recrystallization. ZRA and ZRAH induced hexagonal ice shaping, and both materials significantly reduced the rate of ice crystal growth compared to buffer, depending on the solution pH. ZRA and ZRAH were also shown to inhibit ice recrystallization in a manner similar to AFPs.

Materials and Methods

Solution Preparation

ZRAH powder and a ZRA solution were purchased from Sigma-Aldrich Ltd. ZRAH was dissolved in 100 mM acetate buffer to a final concentration of 30 g/L. This concentration provides 100–150 mM of Zr in the solution, as calculated using the empiric formula $(CH_3CO_2)_xZr(OH)_y$, x+y~4. The pH of the ZRAH solution was adjusted using 100 mM acetic acid for pH values of 3.3 and 4.2, and 100 mM sodium acetate for pH 4.7. The ZRA is provided by the manufacturer as 16% solution in dilute acetic acid with the formula $Zr^{x+}(CH_3COOH)_x$. Using the given density of 1.279 g/ml, this solution contains 200 g/L or 2.2 M Zr. The solution was diluted either with pure water (for pH of 3.3) or with 100 mM sodium acetate (for pH values of 4.2 or 4.7) to obtain a final Zr concentration of 13.3 g/L (150 mM). The pH was adjusted using HCl. A sample of a recombinant type III AFP from Ocean pout was kindly provided by Peter L. Davies (Queen's University) [20]. The AFP solution contained 40 μM protein in 100 mM ammonium bicarbonate at pH 8.

Growth Rate Measurements

The ice growth rate was measured using a custom-designed nanoliter osmometer controlled by a LabView interface developed in our lab. The system and procedure are described in detail elsewhere [21,22]. The temperature in this apparatus can be controlled with a precision of 2–3 millidegrees [22,23]. Briefly, sub-microliter volumes of the solutions were injected into a temperature-controlled sample holder using a micropipette (prepared with a pipette puller device). The sample solution was flash frozen and the temperature was slowly increased toward the melting point until a small (10–20 μm diameter) single ice crystal remained. Once the single crystal was formed, its melting temperature (T_m) was determined as the lowest temperature at which melting was observed. The temperature was then reduced to the desired supercooling temperature and held at this temperature while growth was recorded. The experiment was repeated at least three times on three different samples for each temperature.

Growth rates were determined from recorded images using an ImageJ analysis routine [24]. To determine the growth rate of each point, at least five snapshots were taken, one at the beginning of the cooling period and the others during the growth course. For each pair of snapshots, the distance between the crystal edges across the *c* and/or the *a* axis was measured and the difference between the measured distances divided by a factor of 2 was taken as the elongation of the crystal along the measured axis. The growth rate was determined as the division of the elongation by the time that passed during the growth course, and a mean value was calculated from the different snapshots for each experiment. The directionality of the ice crystals was determined according to the crystal symmetry (twofold symmetry in the *c* direction and six-fold symmetry in the *a*-direction). ZRA and ZRAH were tested for their effects on ice crystal growth over the pH range 3.3–4.7. pH 4.7 was the highest pH at which measurements could be collected. At higher pH values, the solutions polymerized and turned into gels. Even at pH 4.7, the solution became viscous and gelated after 4–6 days.

Ice Recrystallization Inhibition (IRI) Activity

IRI experiments were performed using a Nikon optical microscope equipped with a LC-PolScope (Abrio IM™ Imaging System) and a Linkam MDBCS 196 cold stage (Linkam Scientific Instruments Ltd, UK). A sample of 5 μL was inserted into a 15 mm inner diameter quartz sample holder and covered with a circular glass coverslip. The quartz dish was then placed on a temperature-controlled silver block inside the cold stage. A flow of dry nitrogen was subjected on top to prevent humidity in the ambient air from condensing on the upper window. The experimental procedure was as follows: Samples were cooled from room temperature to −20°C at a cooling rate of 50°C/min. During this cooling phase, ice crystals spontaneously nucleated from the supercooled solution. Samples were kept at −20°C for 2 min, and during this time, images of the frozen samples were collected. The temperature was then warmed to −3°C at a constant rate of 5°C/min, and thereafter the temperature was maintained at −3°C for 10–60 min to permit recrystallization. During this time period the development of the ice crystals was recorded. With the LC-PolScope image system, multiple images with different polarization settings are recorded by a digital camera. These images were used to determine the optical axis in the sample fractions. A color gradation is used to indicate the orientations of the optical axis in each ice crystal. Three replicate experiments were performed, at a minimum, for each determination.

Results

Influence of ZRA and ZRAH on Ice Shaping

Ice in ZRAH solutions at pH 4.2 and 4.7 and ice in ZRA solutions at all pH values developed shapes that appeared to be faceted with hexagonal symmetry in the *a*-direction and twofold symmetry in the *c* direction (Fig. 1). These shapes were similar to those observed in the presence of several mutated AFPs [25,26], AFP analogues [27], or AFPs at low concentrations [26], in accordance with the results of [18]. Crystals assumed the shapes of elongated hexagonal tubes and retained these shapes throughout the growth process. For ZRAH, some shrinking in the basal planes was occasionally detected during growth, leading to the development of truncated pyramidal shapes (Fig. S1). We did not detect the formation of full bipyramidal shapes in which basal planes are not exposed. Ice crystals grown in the ZRAH solution at pH 3.3 assumed flat disc shapes (Fig. 1b), similar to the shape observed in a buffer solution (Fig. 1a) or in pure water [28]. This circular shape resulted in the more rapid growth of the crystal in the directions perpendicular to the *c*-axis. Interestingly, at high supercooling levels, the crystals developed columnar hopper

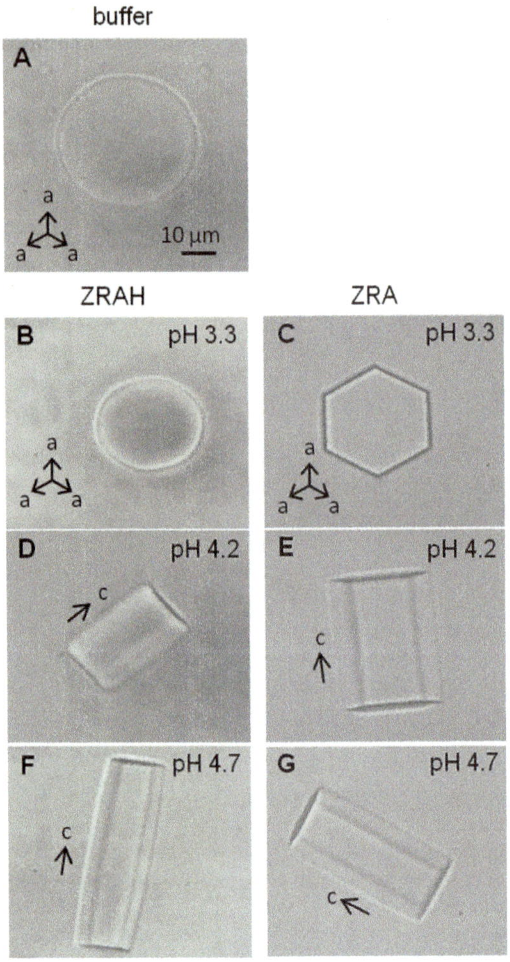

Figure 1. Ice crystal shaping during growth in the presence of (A) 100 mM sodium acetate (pH 4.2), (B, D, F) ZRAH (30 g/L) and (C, E, G) ZRA (13.3 g/L) solutions. The arrows indicate the crystal orientation in the sample.

shapes with hollow ends (Fig. S2). Hopper crystal shapes have been described by Deville et al, albeit under different growth conditions [19]. In all cases, the faceting structures that were apparent during growth were lost upon melting, as has been observed in the presence of moderately active AFPs [29].

Growth Rate Reduction and TH Activity

Crystals in ZRAH and ZRA solutions at pH 3.3 did not significantly affect the ice growth rate along the *a*-axis compared to buffer. However, at pH values of 4.2 and 4.7 the growth rate of the ice crystals was dramatically decreased (Fig. 2–3). Growth inhibition was significantly affected by the solution pH, with a higher activity observed in solutions with a higher pH. ZRA was significantly more active than ZRAH. Over a certain range of temperatures, the growth rate along the *a*-axis was less than 0.02 μm/s (Fig. 2b). This slow growth rate is 10 times slower than the rate previously used as the threshold for the freezing temperature in thermal hysteresis (TH) measurements, marking the boundary between no growth and growth conditions [26]. In such thermal hysteresis measurements, the melting temperature of a single ice crystal is determined, and then the crystal is cooled down in a slow rate until growth is detected, and the difference between the melting and freezing temperatures is determined as the TH activity of the sample. Supercooling temperatures at which the growth rates were below 0.2 μm/s were considered within the TH gap [26]. The accuracy of our equipment allows detection of slower growth rates and we defined the threshold for TH activity to be 0.02 μm/s. According to this definition, ZRA and ZRAH exhibited TH activity along the *a*-axis, as indicated by the arrow in Fig. 2b. Nevertheless, observation of an ice crystal in ZRAH at pH 4.2 over long periods of time (up to two hours) at temperatures within the semi-TH gap (0.060°C below T_{m}) revealed slow but steady growth at a rate of 0.01 μm/s along the *a*-axis. Notably, this growth rate was about 600 times slower than the growth rate observed in a buffer solution for the same supercooling level.

The growth of ice crystals along the *c*-axis in the presence of ZRA and ZRAH was measured. The relatively high growth velocities along the *a*-axis in the solutions at pH 3.3 resulted in flat shapes, which forced the crystals to float such that the *c*-axis was parallel to the view direction. This orientation limited our ability to measure the growth rate along the *c*-axis (Fig. 3). Note that in

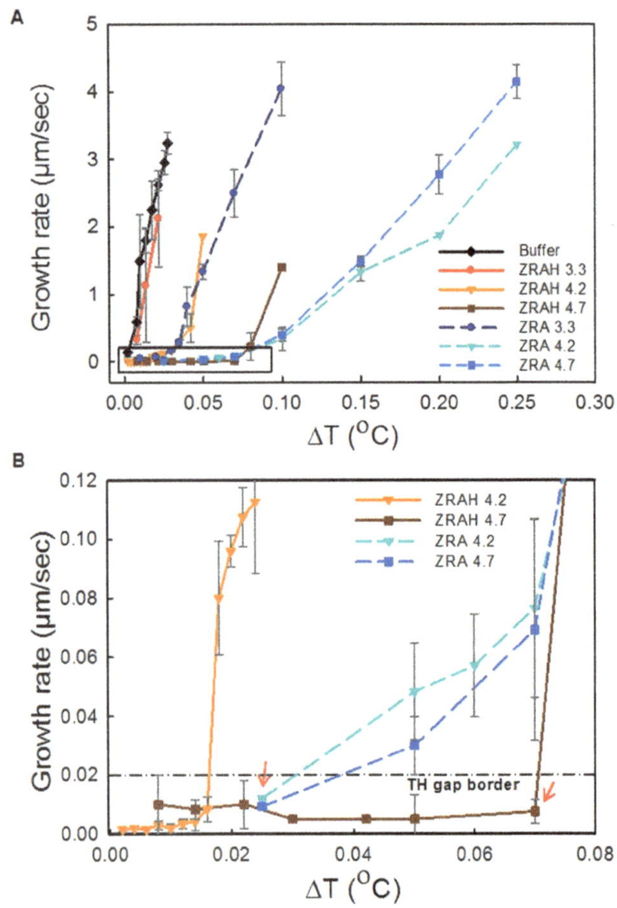

Figure 2. The effect of ZRA and ZRAH on the growth rate of ice crystals along the *a*-axis. The X-axis represents the temperature below T_{m}. The Y-axis represents the rate of growth of the radius of the ice crystal. Each point represents the average of at least 3 measurements. (B) A magnified view of the area enclosed by the black rectangle in A. The horizontal dash-dot line indicates a growth rate of 0.02 μm/sec, which is the threshold value used to define the TH gap. The arrows indicate the semi-TH activity according to this modified definition. Note that according to the old threshold definition (0.2 μm/sec) the TH activity in the *a*-axis direction would be >0. 1°C.

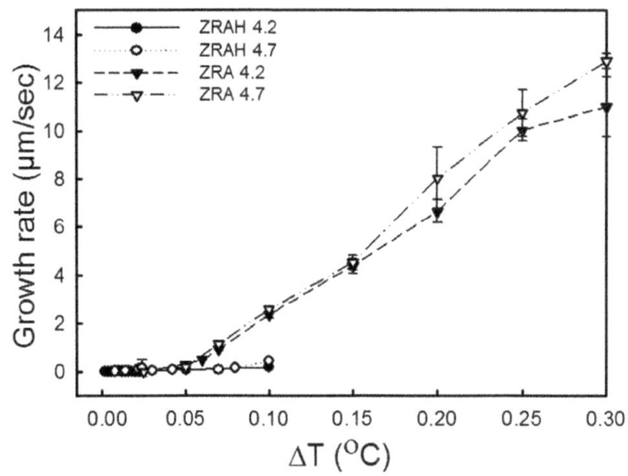

Figure 3. Comparison of the effects of ZRAH and ZRA on the ice growth rates along the *c*-axis. The X-axis represents the temperature below T_m. The Y-axis represents the rate of growth of the radius of a seed ice crystal 20 μm in length. For ZRAH solutions, growth at ΔT >0.1°C could not be measured due to crystal orientation.

ZRAH, the growth velocity along the *c*-axis was lower than in ZRA under a supercooling of 0.05–0.1°C. This behavior may have resulted from latent heat release during the fast growth in the *a*-axis direction, which increased the temperature at the basal plane.

IRI Activity

ZRA and ZRAH were examined qualitatively for their IRI activities relative to the acetate buffer negative control or fish AFP type III positive control. The frozen samples were warmed to the annealing temperature (–3°C), and recrystallization appeared immediately in the acetate buffer sample (Fig. 4), but not in the ZRAH or ZRA samples. Delicate signs of change were apparent in the case of ZRA at pH 4.7, as observed in Fig. 4a. ZRAH efficiently prevented recrystallization of ice at all pH values studied. The ice observed during annealing did not change over a time course of 60 minutes, similar to the behavior observed in solutions containing the AFP type III. Ice recrystallization in the acetate buffer was clearly observed after this time period, as indicated by the pattern change in the images. Notably, the acetate buffer appeared to contain more ice crystals in the initial freezing state than in the ZRAH or AFP type III solutions (Fig. 4c, first column).

Discussion

TH Activity

The findings of this study showed that the inorganic synthetic materials ZRA and ZRAH exhibited certain traits that are characteristic of AFPs. A comparison of the growth of ice in a buffer solution or in the presence of ZRA or ZRAH indicated a hundred-fold reduction in the growth rates along the *a*-axis compared to the growth in buffer over the given temperature range. Nevertheless, whereas AFPs can completely halt ice growth over a range of temperatures, ZRA and ZRAH could not. The observation of continuous growth along the direction of the *c*-axis indicated that ZRA and ZRAH did not produce true TH activity, as is observed in the presence of antifreeze proteins. We note that although the hyperactive AFPs inhibit ice growth along the basal plane [29,30,31,32,33], moderate AFPs do not block basal plane growth; however, moderate AFPs clearly induce ice crystals to grow such that the prism plane is tilted, resulting in a bi-pyramidal shape that eliminates the presence of a basal plane altogether. Both Moderate and hyperactive AFPs completely halt crystal growth within the TH gap. These properties were not observed in the ZRA or ZRAH solutions in which ice crystals assumed columnar shapes with continuous basal growth. Still, over a defined temperature range, ice growth along the direction of the *a*-axis was reduced to less than 0.02 μm/s, which is 10 times below the rate that was previously considered to be "no growth" and, therefore, considered to be TH activity [26]. One can claim that ZRA and ZRAH have directionally-limited TH activity. Nevertheless, in our experimental resolution we were able to note that after a few hours, slow but continuous growth of the crystal to the *a* direction was observed also within this temperature range, indicating that the interactions between ZRA or ZRAH and ice are not as strong as the interactions of AFPs. The mechanism by which ZRA inhibits growth must, therefore, differ from the mechanism by which AFPs interact with ice, as noted by Deville et al. [19].

In addition to growth rate reduction, melting inhibition was observed. ZRAH at pH 4.7 produced a melting rate of 0.2 μm/s for ice at a temperature of 0.01°C above the T_m, which is ten times slower than the melting rate of ice in an acetate buffer at the same temperature (2.1 μm/s). Such melting effects are characteristic of AFPs, as have been demonstrated by Celik et al. [34] and Bar-Dolev et al. [29]. Stabilization of the ice front against melting indicates a direct interaction between the molecules of the ZRA (and ZRAH) complexes and the receding ice front.

IRI Activity

The presence of ZRAH or ZRA dramatically slowed or completely stopped ice recrystallization compared to the buffer. Similar IRI results were obtained for the AFP type III positive control. Although recrystallization was observed for only a relatively short period of time, the effects of the compounds on IRI were clear.

Ice Shaping Activity

The ice shaping properties of ZRA and ZRAH, which led to hexagonal columnar growth patterns, along with the growth inhibition activities to the *a*-axis direction, suggested that these compounds were active on surfaces parallel to the ice prism planes or some other planes that were slightly tilted with respect to the prism planes. Moderate AFPs have been shown to interact with the prism planes of ice and surfaces slightly tilted with respect to these planes [35,36,37,38]. In some cases (AFP type I), interactions with the pyramidal planes have been proposed [37,38], and in others, multiple binding sites appear plausible (AFP III) [36,39,40]. In either case, no direct interactions between the moderate AFPs and the basal planes of ice are observed; however, the AFPs induced bipyramidal ice shapes with minimal or no exposed basal planes. Seed ice crystals grown in a solution containing moderate AFPs display initial growth along the *c*-axis. As growth along the *c*-axis progresses, the basal planes shrink until they are reduced and the crystal eventually acquires a bipyramidal shape and is protected from growth in any direction. Despite the fact that these AFPs do not directly interact with the basal planes [29,41], the crystals are protected from all directions, and growth is arrested throughout the TH gap. Ice assumes hexagonal columnar shapes in the presence of ZRA or ZRAH solutions. The basal planes are, therefore, exposed and not well protected, leading to continuous growth along the *c*-axis. The differences between the growth patterns of ice in the presence of AFPs, ZRA, or ZRAH can be explained according to differences in the

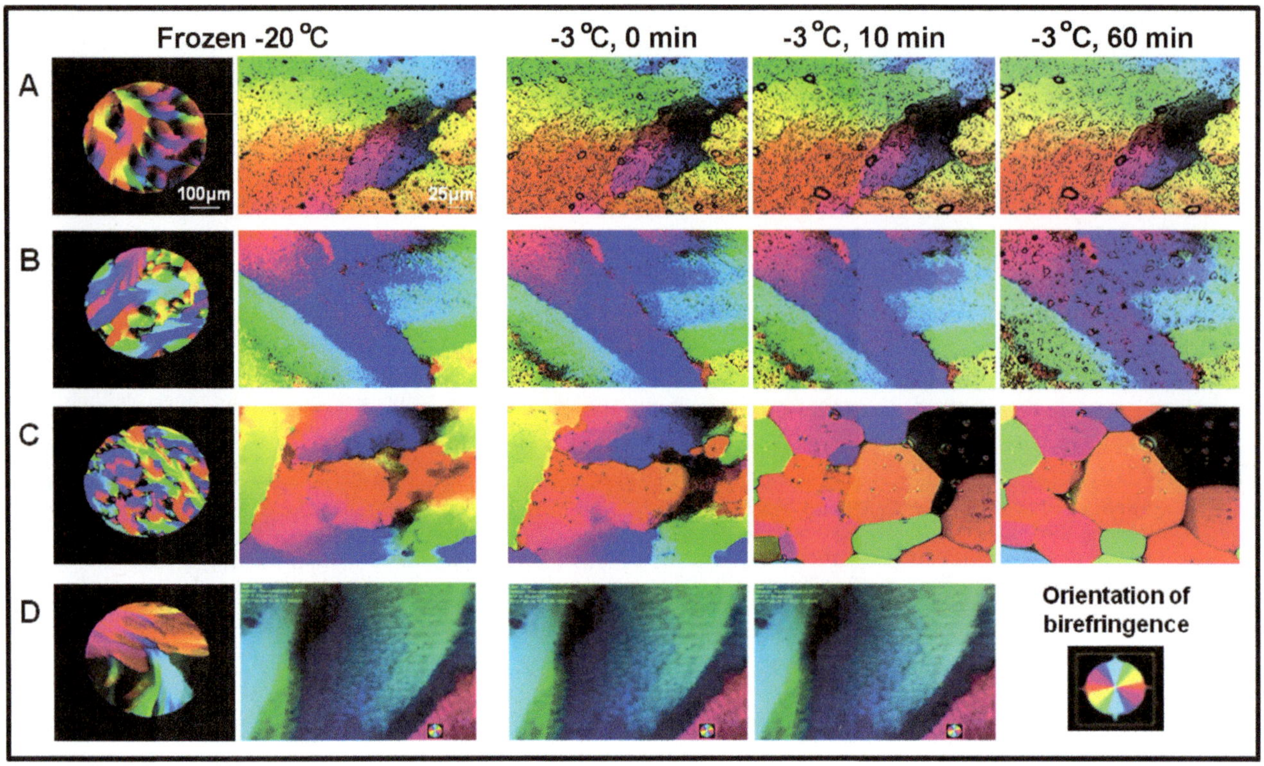

Figure 4. Ice recrystallization inhibition by ZRA and ZRAH. LC-PolScope images for (A) ZRA at pH 4.2, (B) ZRAH at pH 4.2, (C) 100 mM sodium acetate pH 4.2 (negative control), and (D) AFP III (40 μM solution). The colors represent the optical axis orientations of the ice which was determined by the Abrio birefringence detection system. The color map in the bottom right corner corresponds to the respective birefringence orientation. The images in the first and second columns show the initial frozen state at 4× and at 50× magnifications, respectively. Images in the last three columns were collected at an annealing temperature of −3°C, at times 0, 10, and 60 minutes. Note the grain boundaries that appeared during annealing only in the buffer sample. These boundaries shifts indicate recrystallization process.

adsorption planes or sites on the ice surfaces. Whereas AFPs may bind to more than one plane [40], ZRA and ZRAH may not. One possible mechanism for AFP binding relies on irreversible binding [23,34,42]. The binding of ZRA and ZRAH may well be reversible, which can explain the slow growth along the *a*-direction observed after hours of incubation.

Hexagonal column-shaped crystals and continuous slow growth were previously observed in the presence of mutated AFPs [25,26] and antifreeze glycoprotein (AFGP) analogs [27]. In the latter, the slow growth was explained by a reduction in the adsorption constants of the protein analogues to ice, an explanation that may be valid for ZRA and ZRAH. Our data, nevertheless, does not rule out the possibility that the observed activity stemmed from the effects of the compounds on the water structure. ZRA and ZRAH may act at the quasi-liquid layer and interact with the water molecules there. This mechanism was proposed previously as underlying the interactions in carbohydrate or glycopeptide solutions, which displayed IRI without influencing the dynamics of ice shaping [43]. Recent studies have showed that AFGPs affect water molecules in the far hydration shell and, therefore, support this possible mechanism [44].

Differences between the ZRA and ZRAH Activities

While little is known about the actual structure of ZRA in an aqueous solution [45], it is known that ZRA tends to polymerize upon aging, heating, or a pH increase [45,46]. The monomeric units of ZRA prior to polymerization may consist of four zirconium atoms arranged in a tetrameric form, and these tetramers can arrange themselves in organized rod-like structures linked by double hydroxyl bridges [46]. Indeed, we noted that some of our solutions, mainly those at high pH, gelatinized after a few hours or days at room temperature. Studies of zircolyl chloride solutions have shown that oligomerization of a tetramer and an octamer zirconyl species occurs by heating, aging, or reduction of solution acidity [47,48]. Similar polymerization processes may occur in solutions obtained either by dissolving ZRAH in an acetate buffer or by dissolving ZRA in an acetic acid and water solution. The active species in both ZRA and ZRAH may, therefore, be the same or very similar oligomeric forms. The lower activity of ZRAH relative to ZRA stems from the lower concentration of the active species in solution. Deville et al. claimed that ZRAH solutions displayed no ice shaping activities [19]. We showed that ZRAH produced meaningful ice recognition, as indicated by its IRI and ice growth inhibition activity.

pH Dependence of Ice Growth Retardation and IRI

The pH dependence of ice growth inhibition by ZRA and ZRAH solutions may be due to differences in the complex structures, stoichiometry, or oligomerization in solution, as indicated above. Solutions containing a Zr:acetate ratio of 1:2 (at concentrations of 0.01 or 0.02 M) were found to exhibit a dramatic change in the main complex species between pH 3 and 4 [49]. The authors attributed this change to a complex transition from one acetate group per Zr atom to two acetate groups per Zr. Studies of Zr in aqueous solutions revealed that the dominant species at low pH is the tetrameric form $[Zr_4(OH)_8(H_2O)_{16}]^{8+}$

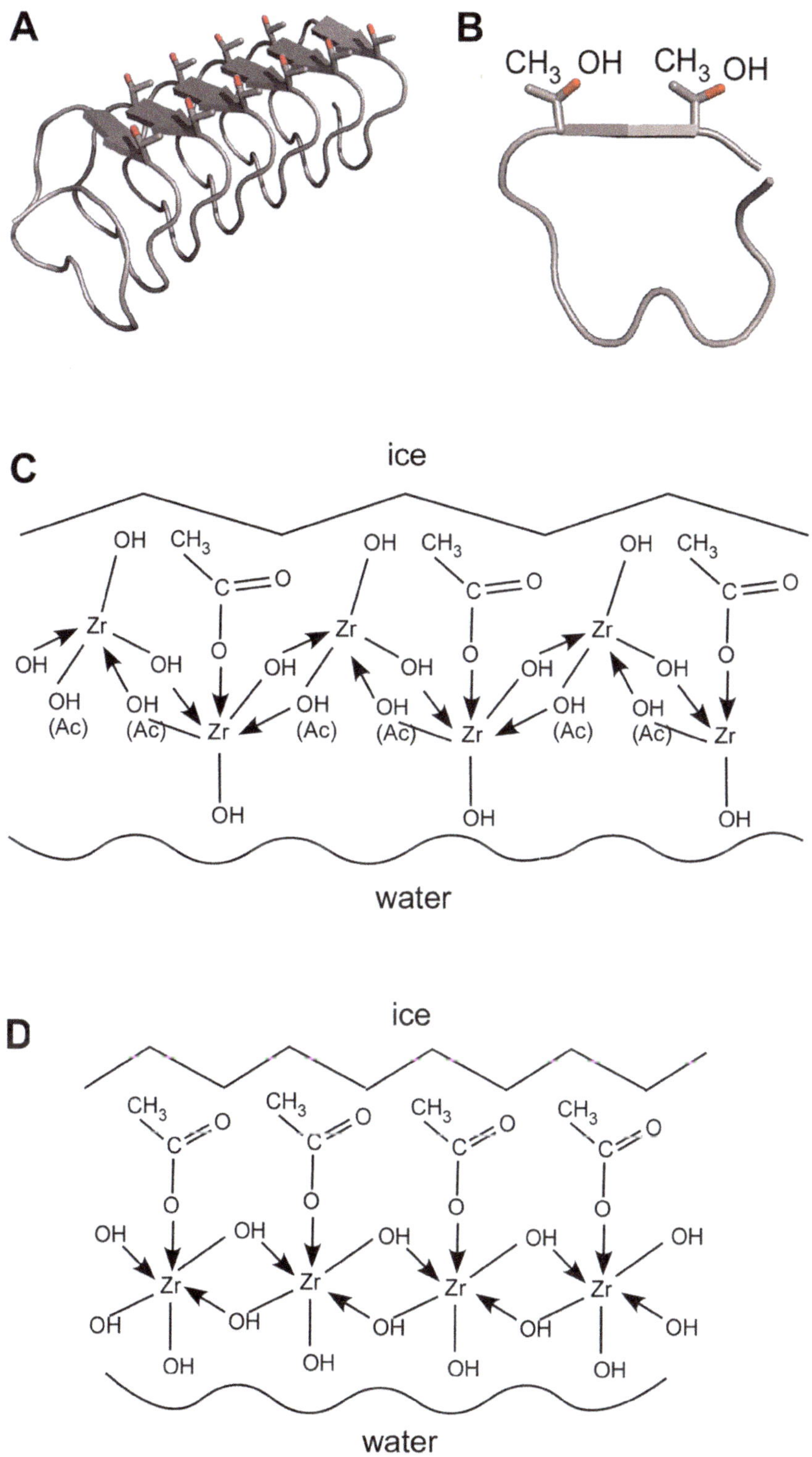

Figure 5. Proposed mechanism for ice recognition by ZRA and ZRAH. A–B. Representation of the Structure of AFP from mealworm beetle (*Tenebrio molitor*) (*Tm*AFP), showing the backbone as a ribbon and the threonine residues forming the ice-binding face in a ball-and-stick model. Oxygen atoms are presented in red and carbon atoms in gray. (A) Side view of the protein. (B) Cross-section of the structure. (C–D) Models suggested for the interactions of ZRA and ZRAH with ice. See text for details.

[50]. Later on, it has been demonstrated that an octameric species of Zr is in equilibrium with the tetramer, and that the octamer is predominant in solutions of lower acidity [47]. Other studies revealed that pH increase as well as aging of Zr(IV) aqueous solutions result in larger oligomeric species [51]. A different study described that pH increase or reflux treatment of acidic zirconyl

solutions resulted in oligomerization of the tetrameric units into two-dimensional sheets that later on formed gels and percipitate [48]. We associate the correlation between the increase in ice growth inhibition activity and the decrease in acidity of the solutions to an enhancement of the oligomerization at higher pH levels, thereby increasing the amount of the active species in solution. Active oligomeric zirconium acetate may either be in the form of chains or in the form of sheets. The slight reduction in the growth inhibition activity of ZRA at pH 4.7 relative to 4.2 may have been due to over-polymerization. Partial oligomerization may be important for ice recognition because more ZRA units per molecule increase the number of interactions with ice; however, over-polymerization can also disturb the ice recognition activity due to slow diffusion of the heavy molecules or to an increase in the presence of non-active species in solution. An extensive spectroscopic study of Zr(IV) complex formation in acetic acid or aqueous solutions showed that the stoichiometry of an acetate complex of Zr(IV) is independent of the Zr/acetic acid ratio, the Zr concentration, or the solution pH [52]. This study supports the notion that the main difference between the samples prepared at different pH values was the oligomerization state.

Deville et al. observed hexagonal shapes in the ice crystals at pH 4 but not at pH 4.5 or 3.3. The author suggested that the decrease in ice shaping activity at these pH values arose from surface charge repulsion between the ice and ZRA [18]. Contrasting with these findings, we observed ice recognition activity by ZRA at both pH 3.3 and 4.7. The activity of ZRAH was higher at pH 4.7 than at pH 4.2, precluding the possibility of charge repulsion at this pH range. The differences between our findings and those of Deville et al. may stem from the different procedures of sample preparation as well as differences in the assays. Notably, we found that ZRAH at a pH as low as 3.3 was sufficient for IRI but not for growth inhibition or ice shaping. Excellent IRI, along with a low TH activity, have been observed in solutions of the AFP derived from ryegrass [53].

Ice Recognition by ZRA and ZRAH

The functional groups that may participate in ice recognition by the hydroxyl-bridged ZRA oligomer are the acetate and hydroxyl moieties. The acetate group is linked to the Zr(IV) ion through an oxygen atom, leaving a methyl group and another oxygen or hydroxyl group exposed to the solvent. This arrangement is similar to the side chain of the threonine amino acid, which plays a significant role in the ice binding faces of many AFPs [30,31,54]. The acetate groups in a ZRA polymer are thought to be exposed on one side of the chain, along with hydroxyl groups, whereas the other side of the chain exposes only hydroxyl groups. The ZRA polymer is thought to be neutral in charge [55]. The part of the polymer that presents the acetate groups may, to a certain extent, mimic the ice binding sites of several ice-binding proteins. The crystal structures [30,31,54] as well as mutagenesis studies [56,57] of several AFPs have revealed that the ice binding faces of these proteins consist of an array of threonine residues with side chains exposed on the outer surface of the protein and arranged in such a way as to match certain planes of the ice crystals nearly perfectly. Furthermore, it has been shown that the hydrophobic part of the ice-binding faces is important for ice recognition while the more hydrophilic portions of the proteins are solvent-exposed [56,58,59]. A similar ice-binding mechanism can be envisioned for ZRA, as illustrated in Fig. 5. Here, a much less perfect match is achieved, offering a possible explanation for why ZRA only incompletely arrests ice growth.

On this basis, we propose two structural arrangements by which ZRA can bind to an ice surface (Fig. 5). In the first model, a methyl group from the acetate and an adjacent hydroxyl group act as the ice recognition unit. In this model, the acetate groups of ZRA are oriented in multiple directions, which is possible for short-chain oligomers or small two dimensional clusters. In the second model, adjacent acetate groups act as the ice recognition site. The interactions between ZRA and ice are then mainly governed by van der Waals forces involving the methyl group, similar to the interactions observed in a heavily mutated AFP in which most of the threonine residues on the ice binding face were replaced with valine, thereby exposing only methyl groups to the solvent. This mutant showed a low TH activity but was very efficient in ice shaping and ice growth inhibition [25]. Although it is unlikely that ZRA interacts with ice via an anchored clathrate mechanism [54], ZRA may interact with ice through a mechanism similar to that employed by AFP mutants. Deville et al. recently stated that it is more likely that ZRA interacts with ice through its hydrophilic side through a mechanism that differs from that observed among AFPs; however, the authors did not rule out the possibility of interactions with the hydrophobic side [19].

Note that although the models presented here consist of chains and assume that preliminary polymerization steps have occurred, the interactions are valid for other ZRA structures that may be present in solution, such as a tetramer in which four zirconium atoms are bound by hydroxo [46] or oxo [52] bridges with a ratio of one [52] or two [49] acetate groups per zirconium atom.

The effects of the ZRA compound on the ice crystal growth or melting may be advantageous for cryopreservation. The efficient inhibition of ice recrystallization and the induction of smooth growth patterns without spontaneous "burst" growth, as are observed in the presence of AFPs, could prevent fatal damage to cells, tissues, and organs. For example, the addition of AFP type I [60,61] or type III [61] was accompanied by extensive growth of ice around the cell during cooling, which damaged the cells, presumably due to the spicular growth pattern.

Conclusions

This work examined the extent to which ice growth could be controlled in the presence of ZRA and ZRAH. These compounds were shown to share many of the properties of natural AFPs, including IRI activities, ice-shaping activities, and a reduction in the ice growth and melting rates. The structures of ZRA complexes and polymers suggested that ZRA interacts directly at the ice:water interface, although further studies are needed to elucidate the mechanism of action underlying the ZRA effects. While ZRA do not have high TH activity, it presents several practical advantages over AFPs, including a low production cost, good stability, and abundance in nature. ZRA is an excellent candidate agent that warrants further study for possible use in applications in which ice growth control remains a challenge.

Supporting Information

Figure S1 Ice crystal growth as truncated pyramidal shapes in a solution containing ZRAH at pH 4.7. The images were taken at a temperature 0.040°C below the melting point.

Figure S2 Hopper shapes induced by ZRAH. Ice crystal grown at a high supercooling rate in the presence of ZRAH at pH 4.7 developed hopper shape with increased basal planes that appeared to be hollow. The sequence of images was taken during the growth course. Frame A was collected before the temperature had been reduced, at the melting point. The other three images were taken during temperature decline. The actual temperature below the melting point and the time lapse is presented. The arrows in A denote the crystal orientation in the sample.

Acknowledgments

We thank Rivka Elbaum (The Hebrew University) for the LC-PolScope (Abrio IM™ Imaging System) and for assistance in operating the device. We acknowledge Peter Davies (Queen's University, Kingston, Ontario, Canada) for providing the AFP type III, and Liat Bahari for help with the IRI experiments.

Author Contributions

Conceived and designed the experiments: MBD OM IB. Performed the experiments: OM SG. Analyzed the data: MBD OM SG. Wrote the paper: MBD OM IB.

References

1. Deville S, Saiz E, Nalla RK, Tomsia AP (2006) Freezing as a path to build complex composites. Science 311: 515–518.
2. Munch E, Launey ME, Alsem DH, Saiz E, Tomsia AP, et al. (2008) Tough, Bio-Inspired Hybrid Materials. Science 322: 1516–1520.
3. Mazur P (1984) Freezing of living cells: mechanisms and implications. Am J Physiol 247: C125–142.
4. Rubinsky B, Arav A, Devries AL (1992) The cryoprotective effect of antifreeze glycopeptides from antarctic fishes. Cryobiology 29: 69–79.
5. Boistelle R, Astier JP (1988) Crystallization mechanisms in solution. Journal of Crystal Growth 90: 14–30.
6. Knight CA, Duman JG (1986) Inhibition of recrystallization of ice by insect thermal hysteresis proteins - A possible cryoprotective role. Cryobiology 23: 256–262.
7. Hui YH, Lim M-h, Nip W-k, Smith JS, Yu PHF (2008) Principles of Food Processing. Food Processing: Blackwell Publishing. 1–29.
8. Venketesh S, Dayananda C (2008) Properties, Potentials, and Prospects of Antifreeze Proteins. Critical Reviews in Biotechnology 28: 57–82.
9. Duman JG (2001) Antifreeze and ice nucleator proteins in terrestrial arthropods. Annual Review of Physiology 63: 327–357.
10. Lindner MN, Oldroyd RJ, Sztehlo A, Towell JD (2006) Frozen confectionery product comprising ice structuring proteins. In: N.V U, Plc U, editors. USA, Europe.
11. Fahy GM (1986) The relevance of cryoprotectant "toxicity" to cryobiology. Cryobiology 23: 1–13.
12. Baruch E, Mastai Y (2007) Antifreeze Properties of Polyglycidol Block Copolymers. Macromolecular Rapid Communications 28: 2256–2261.
13. Budke C, Koop T (2006) Ice Recrystallization Inhibition and Molecular Recognition of Ice Faces by Poly(vinyl alcohol). ChemPhysChem 7: 2601–2606.
14. Wierzbicki A, Knight CA, Rutland TJ, Muccio DD, Pybus BS, et al. (2000) Structure–Function Relationship in the Antifreeze Activity of Synthetic Alanine–Lysine Antifreeze Polypeptides. Biomacromolecules 1: 268–274.
15. Huang ML, Ehre D, Jiang Q, Hu C, Kirshenbaum K, et al. (2012) Biomimetic peptoid oligomers as dual-action antifreeze agents. Proceedings of the National Academy of Sciences.
16. Walters KR, Serianni AS, Sformo T, Barnes BM, Duman JG (2009) A nonprotein thermal hysteresis-producing xylomannan antifreeze in the freeze-tolerant Alaskan beetle Upis ceramboides. Proceedings of the National Academy of Sciences 106: 20210–20215.
17. Ishiwata A, Sakurai A, Nishimiya Y, Tsuda S, Ito Y (2011) Synthetic Study and Structural Analysis of the Antifreeze Agent Xylomannan from Upis ceramboides. Journal of the American Chemical Society 133: 19524–19535.
18. Deville S, Viazzi C, Leloup J, Lasalle A, Guizard C, et al. (2011) Ice Shaping Properties, Similar to That of Antifreeze Proteins, of a Zirconium Acetate Complex. PLoS ONE 6: e26474.
19. Deville S, Viazzi C, Guizard C (2012) Ice-Structuring Mechanism for Zirconium Acetate. Langmuir.
20. Chao H, Davies PL, Sykes BD, Sönnichsen FD (1993) Use of proline mutants to help solve the NMR solution structure of type III antifreeze protein. Protein Science 2: 1411–1428.
21. Pertaya N, Celik Y, DiPrinzio CL, Wettlaufer JS, Davies PL, et al. (2007) Growth-melt asymmetry in ice crystals under the influence of spruce budworm antifreeze protein. Journal of Physics: Condens Matter 19: 412101 (412112pp)
22. Braslavsky I, Drori R (2012) LabVIEW-operated novel nanoliter osmometer for ice binding protein investigations. The Journal of Visualized Experiments 72: e4189.
23. Celik Y, Drori R, Pertaya-Braun N, Altan A, Barton T, et al. (2013) Microfluidic experiments reveal that antifreeze proteins bound to ice crystals suffice to prevent their growth. Proc Natl Acad Sci U S A 110: 1309–1314.
24. Schneider CA, Rasband WS, Eliceiri KW (2012) NIH Image to ImageJ: 25 years of image analysis. Nature Methods 9: 671–675.
25. Bar M, Celik Y, Fass D, Braslavsky I (2008) Interactions of beta-helical antifreeze protein mutants with ice. Crystal Growth & Design 8: 2954–2963.
26. DeLuca CI, Chao H, Sonnichsen FD, Sykes BD, Davies PL (1996) Effect of type III antifreeze protein dilution and mutation on the growth inhibition of ice. Biophys J 71: 2346–2355.
27. Peltier R, Evans CW, DeVries AL, Brimble MA, Dingley AJ, et al. (2010) Growth Habit Modification of Ice Crystals Using Antifreeze Glycoprotein (AFGP) Analogues. Crystal Growth & Design 10: 5066–5077.
28. Hobbs PV (1974) *Ice Physics*: Clarendon Oxford.
29. Bar-Dolev M, Celik Y, Wettlaufer JS, Davies PL, Braslavsky I (2012) New insights into ice growth and melting modifications by antifreeze proteins. J R Soc Interface 9: 3249–3259.
30. Graether SP, Kuiper MJ, Gagne SM, Walker VK, Jia Z, et al. (2000) Beta-helix structure and ice-binding properties of a hyperactive antifreeze protein from an insect. Nature 406: 325–328.
31. Liou YC, Tocilj A, Davies PL, Jia Z (2000) Mimicry of ice structure by surface hydroxyls and water of a beta-helix antifreeze protein. Nature 406: 322–324.
32. Pertaya N, Marshall CB, Celik Y, Davies PL, Braslavsky I (2008) Direct Visualization of Spruce Budworm Antifreeze Protein Interacting with Ice Crystals: Basal Plane Affinity Confers Hyperactivity. Biophys J 95: 333–341.
33. Scotter AJ, Marshall CB, Graham LA, Gilbert JA, Garnham CP, et al. (2006) The basis for hyperactivity of antifreeze proteins. Cryobiology 53: 229–239.
34. Celik Y, Graham LA, Mok YF, Bar M, Davies PL, et al. (2010) Superheating of ice crystals in antifreeze protein solutions. Proceedings of the National Academy of Sciences of the United States of America 107: 5423–5428.
35. Antson AA, Smith DJ, Roper DI, Lewis S, Caves LSD, et al. (2001) Understanding the mechanism of ice binding by type III antifreeze proteins. Journal of Molecular Biology 305: 875–889.
36. Jia Z, DeLuca CI, Chao H, Davies PL (1996) Structural basis for the binding of a globular antifreeze protein to ice. Nature 384: 285–288.
37. Knight CA, Cheng CC, DeVries AL (1991) Adsorption of alpha-helical antifreeze peptides on specific ice crystal surface planes. Biophys J 59: 409–418.
38. Wierzbicki A, Taylor MS, Knight CA, Madura JD, Harrington JP, et al. (1996) Analysis of shorthorn sculpin antifreeze protein stereospecific binding to (2–1 0) faces of ice. Biophys J 71: 8–18.
39. Sönnichsen FD, Sykes BD, Chao H, Davies PL (1993) The nonhelical structure of antifreeze protein type III. Science 259: 1154–1157.
40. Jia Z, Davies PL (2002) Antifreeze proteins: an unusual receptor-ligand interaction. Trends in Biochemical Sciences 27: 101–106.
41. Pertaya N, Marshall CB, DiPrinzio CL, Wilen L, Thomson ES, et al. (2007) Fluorescence Microscopy Evidence for Quasi-Permanent Attachment of Antifreeze Proteins to Ice Surfaces. Biophys J 92: 3663–3673.
42. Knight CA, Wierzbicki A (2001) Adsorption of Biomolecules to Ice and Their Effects upon Ice Growth. 2. A Discussion of the Basic Mechanism of "Antifreeze" Phenomena. Crystal Growth & Design 1: 439–446.
43. Tam RY, Rowley CN, Petrov I, Zhang T, Afagh NA, et al. (2009) Solution Conformation of C-Linked Antifreeze Glycoprotein Analogues and Modulation of Ice Recrystallization. Journal of the American Chemical Society 131: 15745–15753.
44. Ebbinghaus S, Meister K, Born B, DeVries AL, Gruebele M, et al. (2010) Antifreeze Glycoprotein Activity Correlates with Long-Range Protein–Water Dynamics. Journal of the American Chemical Society 132: 12210–12211.
45. Geiculescu AC, Spencer HG (1999) Thermal decomposition and crystallization of aqueous sol-gel derived zirconium acetate gels: Effects of the precursor solution pH. Journal of Sol-Gel Science and Technology 16: 243–256.
46. Geiculescu AC, Rack HJ (2002) X-ray scattering studies of polymeric zirconium species in aqueous xerogels. Journal of Non-Crystalline Solids 306: 30–41.
47. Singhal A, Toth LM, Lin JS, Affholter K (1996) Zirconium(IV) Tetramer/Octamer Hydrolysis Equilibrium in Aqueous Hydrochloric Acid Solution. Journal of the American Chemical Society 118: 11529–11534.
48. Clearfield A (1990) The Mechanism of Hydrolytic Polymerization of Zirconyl Solutions. Journal of Materials Research 5: 161–162.
49. Ts.B.Konunova M.S.Popov, Kuang C (1969) The formation of Zirconium complexes of monocarboxylic acids in aqueous solutions. Russian Journal of Inorganic Chemistry 14(8): 1084–1085.
50. Muha GM, Vaughan PA (1960) Structure of the Complex Ion in Aqueous Solutions of Zirconyl and Hafnyl Oxyhalides. J Chem Phys 33: 194–199.
51. Jutson JA, Richardson RM, Jones SL, Norman C (1990) Small Angle X-ray Scattering Studies of Polymeric Zirconium Species in Aqueous Solution. MRS Online Proceedings Library 180: 123.
52. Hannane S, Bertin F, Bouix J (1994) NMR, Infrared and raman-spectroscopy study of Zr(IV) complexation by acetic-acid in aqueous-solution. Bulletin de la Societe chimique de France 131(3): 262–270.
53. Yu SO, Brown A, Middleton AJ, Tomczak MM, Walker VK, et al. (2010) Ice restructuring inhibition activities in antifreeze proteins with distinct differences in thermal hysteresis. Cryobiology 61: 327–334.
54. Garnham CP, Campbell RL, Davies PL (2011) Anchored clathrate waters bind antifreeze proteins to ice. Proc Natl Acad Sci U S A 108: 7363–7367.

55. Moles PJ (2002) The use of zirconium in Surface Coatings. Data sheet 117776–1780: MEL Chemicals.
56. Baardsnes J, Kondejewski LH, Hodges RS, Chao H, Kay C, et al. (1999) New ice-binding face for type I antifreeze protein. FEBS Letters 463: 87–91.
57. Marshall CB, Daley ME, Graham LA, Sykes BD, Davies PL (2002) Identification of the ice-binding face of antifreeze protein from Tenebrio molitor. FEBS Letters 529: 261–267.
58. Sönnichsen FD, DeLuca CI, Davies PL, Sykes BD (1996) Refined solution structure of type III antifreeze protein: hydrophobic groups may be involved in the energetics of the protein–ice interaction. Structure 4: 1325–1337.
59. Chao H, Houston ME, Hodges RS, Kay CM, Sykes BD, et al. (1997) A Diminished Role for Hydrogen Bonds in Antifreeze Protein Binding to Ice†. Biochemistry 36: 14652–14660.
60. Carpenter JF, Hansen TN (1992) Antifreeze protein modulates cell-survival during cryopreservation - Mediation through influence on ice crystal-growth Proceedings of the National Academy of Sciences of the United States of America 89: 8953–8957.
61. Ishiguro H, Rubinsky B (1994) Mechanical Interactions between Ice Crystals and Red Blood Cells during Directional Solidification. Cryobiology 31: 483–500.

8

Effect of a Magnetic Field on Drosophila under Supercooled Conditions

Munekazu Naito[1], Shuichi Hirai[1*], Makoto Mihara[2], Hayato Terayama[1], Naoyuki Hatayama[1], Shogo Hayashi[3], Masayuki Matsushita[4], Masahiro Itoh[1]

1 Department of Anatomy, Tokyo Medical University, Tokyo, Japan, **2** Department of Plastic and Reconstructive Surgery, Graduate School of Medicine, University of Tokyo, Tokyo, Japan, **3** Medicinal Education Center, School of Medicine, Aichi Medical University, Aichi, Japan, **4** Molecular and Cellular Physiology, University of the Ryukyus, Okinawa, Japan

Abstract

Under subzero degree conditions, free water contained in biological cells tends to freeze and then most living things die due to low temperatures. We examined the effect of a variable magnetic field on Drosophila under supercooled conditions (a state in which freezing is not caused even below the freezing point). Under such supercooled conditions with the magnetic field at 0°C for 72 hours, −4°C for 24 hours and −8°C for 1 hour, the Drosophila all survived, while all conversely died under the supercooled conditions without the magnetic field. This result indicates a possibility that the magnetic field can reduce cell damage caused due to low temperatures in living things.

Editor: Efthimios M. C. Skoulakis, Alexander Flemming Biomedical Sciences Research Center, Greece

Funding: This research was supported by a Grant-in-Aid for General Science Research from the Ministry of Education, Science, Sports, and Culture, Japan. The funders had no role in study design, data collection and analysis, decision to publish, or preparation of the manuscript.

Competing Interests: The authors have declared that no competing interests exist.

* E-mail: shinamon611@gmail.com

These authors contributed equally to this work.

Introduction

Generally, water changes from liquid to solid at 0°C. However, when the temperature of water is gradually lowered using a program freezer, etc., a "supercooling phenomenon" occurs in which the water does not change phase from liquid to solid even when the temperature falls below 0°C. Furthermore, it has been clarified that this supercooling phenomenon is stably caused by artificially applying pressure and voltage to water [1–3]. We recently invented a program freezer that causes supercooling conditions using a magnetic field. It is known that when the temperature is further reduced from the supercooled condition using this device in order to freeze the internal organs, the shape of the tissues maintains a better form compared to the group that used liquid nitrogen [4]. In physical experiments, it has been clarified that uniform ice crystals with small cracks are formed when a variable magnetic field is added to the coagulation process of water, thus suggesting that this caused a reduction in damage to tissues [5]. However, the detailed mechanism of this action has not yet been clarified.

On the other hand, it is known that the metabolism of organisms declines when low temperatures are reached. In theory, when a cell is frozen at very low temperatures, the metabolic activities of the cells stop, thus allowing for the semipermanent maintenance of the state before freezing. To date, cryopreservation of cells such as sperm [6], red blood cells [7], etc. has become possible by adding glycerol, etc. Moreover, regarding organs, cryopreservation has been successfully achieved in mouse and pig livers [8] as well as in human ovaries [9]. However, the cryopreservation of an organism is difficult at this point due to severe tissue damage following thawing.

It is becoming clear that in humans, therapeutic hypothermia increases the resuscitation rate following cardiopulmonary arrest [10]. In addition, in the natural world, several organisms live or hibernate by controlling their metabolism to below 0°C. Frogs can live at −6°C with glucose as antifreeze [11], fish inhabiting Antarctica with antifreezing protein can live at −2°C [12], and arctic squirrels can live at −2.9°C by bringing their internal bodies to a supercooled condition [13]. Therefore, in this study, we prepared a supercooled condition (a state in which freezing is not caused even below the freezing point) using a program freezer and examined how a magnetic field affects organisms using Drosophila.

Materials and Methods

Drosophila

Drosophila adult flies (Oregon-R: wild type) were purchased from the Kagaku club (Shiga, Japan) and they were raised on a standard cornmeal medium at 25°C.

Program freezer with a variable magnetic field

An aluminum cooling chamber (inner diameter: 125.5×93 mm; depth: 40 mm) was installed in the cooling unit of a stirling engine, which is a cooling device, and a hole with an inner diameter of 12.8 mm and depth of 32 mm was made inside to insert the specimens. The chamber was covered with a heat insulating material, with a teflon-coated coil of wire coiled around the outside thereof. Alternating current was applied using a specialized function generator, the coil was electrified with 30 Hz, 2 V,

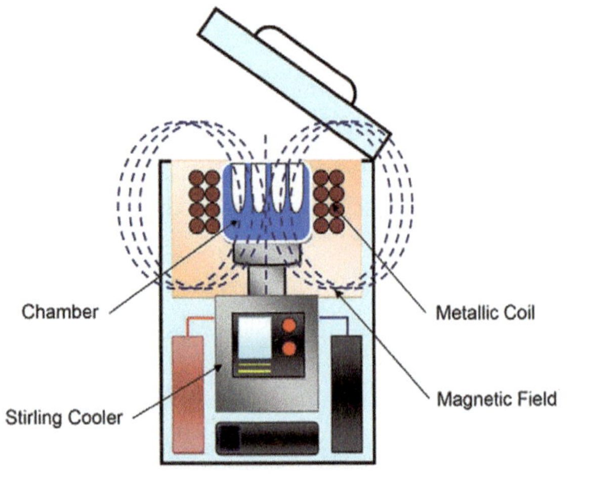

Figure 1. System of program freezer using magnetic field. Function: Cooling while generating a 0.5 mT magnetic field inside the chamber is possible by electrifying the coil from the generator.

0.8 A, and a variable magnetic field of 0.5 mT (tesla) was caused inside the chamber. The entire device is shown in Fig. 1.

Evaluation of the cooling process on distillated water and saline with or without a magnetic field

1 ml distilled water (freezing point: 0°C) and 1 ml saline (freezing point: −0.65°C) were put in a plastic tube (Nunc cryotube, Thermo Fisher Scientific Inc, Waltham, USA), a magnetic field with an intensity of 0.5 mT was added, and it was cooled at a regular speed of −3°C/min. Those without a magnetic field added were classified as the control group. A type K thermocouple sensor was installed inside the chamber, and the temperature and the time were measured using a thermometer (PC card-type data acquisition system, type: NR-250, KEYENCE CORPORATION "Keyence", Japan, Osaka). To clarify whether a variable magnetic filed affects the temperature, the experiments were repeated fifteen times, respectively, and the data were analyzed by the non-paired *t*-test and F-test for the time required to reach 0, −2, −4, −6, −8 and −10°C.

Supercooling preservation experiment on Drosophila with or without a magnetic field

Five to ten Drosophila were put into 2.5 ml plastic tubes respectively. This tube was set in a chamber and it was cooled from room temperature at −3°C/min. The point at which 5 minutes elapsed after the target temperature had been reached was determined to be 0 hours. The Drosophila were preserved at 0°C for 1 hour, 6 hours, 12 hours, 24 hours, 48 hours, 72 hours and 96 hours; at −2 and −4°C for 1 hour, 6 hours, 12 hours, 24 hours and 48 hours; at −6 and −8°C for 1 hour and 6 hours; and at −10°C for 1 hour. The experimental group was exposed to a variable magnetic field with an intensity of 0.5 mT from the beginning of the cooling. Subsequently, the tube was taken from the freezer and re-warmed to room temperature. The Drosophila were observed for 24 hours to check for survival. Those that were not exposed to a variable magnetic field added were classified as the control group. To clarify whether the survival rate of Drosophila is related to a magnetic filed, the data were analyzed using a three-parameter logistic regression model of the survival rate by time, temperature and with/without a variable magnetic field.

Results

The temperature change of distilled water and saline is shown in Fig. 2. When distilled water was put in a tube, the temperature gradually dropped by 3°C per minute from a room temperature of 20°C, and the supercooled condition continued until −10°C was reached (Fig. 2a). When saline was put in a tube, the temperature gradually dropped by 3°C per minute in the same manner as when distilled water was put in; however, the supercooled condition continued until −12°C was reached. In both distilled water and saline, the time to reach 0, −2, −4, −6, −8 or −10°C were equivalent in averages and variables weather a magnetic field was added or not.

Next, the outcome of preserving Drosophila is shown in Table 1. When Drosophila were preserved for 72 hours at 0°C, 13 of 60 survived in the experimental group while all 47 Drosophila died in the control group. Under conditions at −2 and −4°C for 24 hours, 25 of 54 and 15 of 45 survived in the experimental group, respectively, while all 106 Drosophila died in the control group. Under the condition at −8°C for 1 hour, 57 of 67 survived in experimental group, while in the control group, all 66 Drosophila died. Under a condition of −10°C, all 48 Drosophila died following 1 hour of preservation for both groups. When the Drosophila were observed under a microscope immediately following preservation under a supercooled condition of −2°C or less, there was no body movement at all. However, the surviving Drosophila gradually commenced body movement with re-warming and then widely fluttered about several hours later.

In a logistic regression model, the survival rate was significantly affected by time (chi square = 904.4, $p<0.001$), temperature (chi square = 645.2, $p<0.001$) and weather magnetic field are added or not (chi square = 224.5, $p<0.001$) respectively. Especially, the odds ratio associated with the magnetic field was 12.6, indicating that Drosophila in a magnetic field had 12.6 times higher survival than without a magnetic field. The interactions between them were also significant ($p<0.001$) but much smaller (chi-square = 27.3–67.2).

Discussion

In this experiment, it was shown that Drosophila survive when they are preserved at 0°C for 72 hours, at −4°C for 24 hours, and at −8°C for 1 hour under a supercooled condition with a 0.5 mT variable magnetic field added. As far as we know, this is the first report investigating how magnetic fields affect organisms under supercooled conditions.

In previous reports, several experiments examined whether or not a supercooled condition is useful for preserving rat organs [1–3]. Takahashi et al. [1] reported that when livers were preserved under a supercooled condition of −2°C with 5MPa pressure added, the shape of the tissue was maintained and the success rate of liver transplantation improved. Monzen et al. [2] clarified that when the heart, liver and kidneys are preserved under a supercooled condition of −4°C with 100 Volt and 500 Volt added, the biomarker of organ preservation declines compared to preservation at 4°C. Moreover, Okamoto et al. [3] reported that when the lungs are preserved under a supercooled condition of −2°C with 3000 volt added, reperfusion injury clearly declines compared to when preserving at 4°C. In these articles, the authors conclude that the supercooled condition lowered the metabolism of cells and decreased cell damage. In our experiment, when Drosophila were preserved under supercooled conditions at 0°C for 72 hours, −4°C for 24 hours and −8°C for 1 hour without a magnetic filed, the Drosophila all died. Perhaps while the metabolism of the cells declined when low temperatures were

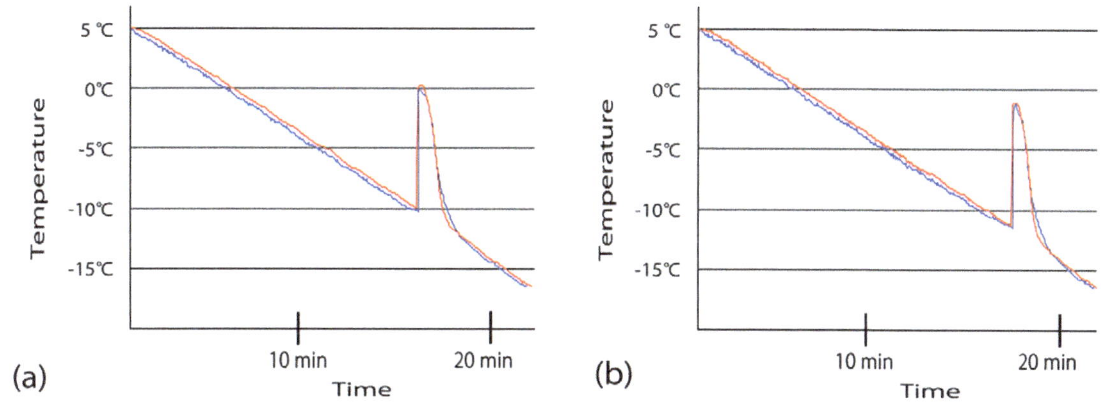

Figure 2. Supercooling phenomenon in distilled water (a) and normal saline (b) under magnetic field (Red line) or not (Blue line). Each line shows the average temperature of fifteen experiments every 10 second. Supercooling is an unstable phenomenon in which the temperature is brought below the freezing point but a phase change from liquid to solid does not occur.

reached, cell damage was caused due to the low temperatures. However, when Drosophila were preserved under supercooled conditions with a 0.5 mT magnetic field added, the some Drosophila survived. Especially, fifty-seven of 67 Drosophila exposed to a magnetic field at −8°C for 1 hour, survived, while all of the Drosophila in the control group died. It is well known that cold injuries cause mortality at much higher temperatures, typically around −5°C [14,15]. Our results suggest that the magnetic field reduced the cell damage induced by low temperatures. Recently, it was reported that either by direct effect on paramagnetic ROS (reactive oxygen species) or indirect changes to signaling pathways the magnetic field might affect the oxidative stress of reperfusion injury [16]. Moreover, electron paramagnetic resonance provides a non-invasive way to assay ROS build up during preservation if titration or the like are unappealing [17]. Therefore, ROS may be one of the important factors to explain the mechanisms of effects of magnetic filed on Drosophila.

In 1979, it was reported in the United States of America that leukemia frequently occurs in residents living near high voltage lines, and since then, experiments regarding the effect of magnetic fields on organisms have been strenuously advanced [18]. For example, in experiments using rats, it was reported that exposure to a 0.1 mT magnetic field promoted the onset of mammary gland tumors and ornithine decarboxylase activation [19]. Moreover, it is reported that when a 200 to 400 mT magnetic field was exposed to cells originating from rat pheochromocytoma (PC12-VG), change was observed in intracellular signaling, such as an influx of ions via the cell membrane [20]. Meanwhile, Bassett et al. [21] reported that a 0.2 mT magnetic field was effective for healing bone fractures in human and Oshibuchi [22] reported that a 10 mT magnetic field was effective for healing wounds in mice. However, many reports disproving these effects of magnetic fields have also been published and no uniform view has yet been reached [23]. Regarding the reasons thereof, the fact that no biological indicator showing high sensitivity in the magnetic field has yet been identified. Moreover, in experiments using magnetic fields, it is difficult to separate the effect of the magnetic field itself and the stimulation and/or heating effect on the excitable tissue, so problems with the experiments themselves have also been suggested [23]. In this experiment, a 0.5 mT magnetic field did not affect the transition of the temperature of distilled water or saline. Indeed it remains unclear whether the Drosophila are frozen or not under these conditions, but it is certain that a very weak magnetic field affects the life or death of such Drosophila under supercooled conditions. The present study involved a single condition of magnetic field (30 Hz, 0.5 mT). In the future, we plan to study the effects of other frequencies and intensities of magnetic fields on organisms under supercooled conditions.

Recently, Cho et al. [24] developed magnetic nanoparticles that turn on apoptosis cell signaling using a magnetic field in a remote and non-invasive manner. They also demonstrated that the

Table 1. The survival rate of the *Drosophila* at 24 hours after the preservation in the supercooled condition with or without the magnetic field.

	Time						
Temperature	**1h**	**6h**	**12h**	**24h**	**48h**	**72h**	**96h**
0°C	57/57 (60/60)	60/60 (39/39)	63/72 (54/60)	3/60 (54/64)	15/63 (3/63)	13/60 (0/47)	0/48 (0/36)
−2°C	45/45 (48/48)	44/44 (45/45)	62/68 (12/73)	25/54 (0/49)	0/43 (0/42)	–	–
−4°C	39/39 (44/46)	63/63 (44/64)	57/64 (3/60)	15/45 (0/57)	0/49 (0/39)	–	–
−6°C	68/68 (28/44)	0/62 (0/52)	–	–	–	–	–
−8°C	57/67 (0/66)	0/55 (0/50)	–	–	–	–	–
−10°C	0/49 (0/48)	–	–	–	–	–	–

The parentheses indicate the condition without the magnetic filed.

observed magnetic switch operates on the micrometre scale and it can also be applied to an in vivo system where apoptotic morphological changes of Zebrafish can be successfully induced. These techniques may therefore help to elucidate the mechanism of a magnetic field on Drosophila under supercooled conditions.

Acknowledgments

The authors wish to thank Ms. Miyuki Kitaoka, Mr. Shinichi Kawata, Ms. Akiko Abo and Ms. Yuki Ogawa for their excellent secretarial assistance. Additionally, the authors wish to thank Mr. Hiroyuki Iizuka and Mr. Kazuhiro Fujii for useful advice regarding the technical assistance.

Author Contributions

Conceived and designed the experiments: M. Naito S. Hirai M. Matsushita. Performed the experiments: M. Naito S. Hirai HT M. Matsushita. Analyzed the data: M. Naito S. Hirai HT M. Matsushita M. Mihara NH S. Hayashi MI. Contributed reagents/materials/analysis tools: MN S. Hirai S. Hayashi M. Matsushita MI. Wrote the paper: MN S. Hirai. Discussed the results and commented on the manuscript: MN S. Hirai M. Mihara HT NH S. Hayashi M. Matsushita MI.

References

1. Takahashi T, Kakita A, Takahashi Y, Yokoyama K, Sakamoto I, et al. (2001) Preservation of rat livers by supercooling under high pressure. Transplant Proc 33: 916–919.
2. Monzen K, Hosoda T, Hayashi D, Imai Y, Okawa Y, et al. (2005) The use of a supercooling refrigerator improves the preservation of organ grafts. Biochem Biophys Res Commun 18: 534–539.
3. Okamoto T, Nakamura T, Zhang J, Aoyama A, Chen F, et al. (2008) Successful sub-zero non-freezing preservation of rat lungs at −2 degrees C utilizing a new supercooling technology. J Heart Lung Transplant. 27: 1150–1157.
4. Nakagawa T, Mihara M, Noguchi S, Fujii K, Owada T, et al. (2012) Development of Pathology Specimen preparation Method by Supercooling Cryopreservation under Magnetic Field. ACSC (in press).
5. Iwasaka M, Onishi M, Kurita S, Owada N (2010) Effects of pulsed magnetic fields on light scattering property in the freezing process of aqueous solutions. J Appl Phys 107: 09B314.
6. Woods EJ, Benson JD, Agca Y, Critser JK (2004) Fundamental cryobiology of reproductive cells and tissues. Cryobiology 48: 146–156.
7. Sputtek A (2007) Cryopreservation of red blood cells and platelets. Methods Mol Biol 368: 283–301.
8. Gavish Z, Ben-Haim M, Arav A (2008) Cryopreservation of whole murine and porcine livers. Rejuvenation Res 11: 765–772.
9. Silber SJ (2012) Ovary cryopreservation and transplantation for fertility preservation. Mol Hum Reprod 18: 59–67.
10. Delhaye C, Mahmoudi M, Waksman R (2012) Hypothermia therapy: neurological and cardiac benefits. J Am Coll Cardiol 59: 197–210.
11. Storey K (1990) Life in a frozen state: adaptive strategies for natural freeze tolerance in amphibians and reptiles. Am J Physiol 258: 559–568.
12. Feeney R, Yeh Y (1978) Antifreeze proteins from fish bloods. Adv Protein Chem 32: 191–282.
13. Boyer BB, Barnes BM (1999) Molecular and Metabolic Aspects of Mammalian Hibernation. Bioscience 49: 713–724.
14. Strachan LA, Tarnowski-Garner HE, Marshall KE, Sinclair BJ (2011) The evolution of cold tolerance in Drosophila larvae. Physiol Biochem Zool 84: 43–53.
15. Koštál V, Korbelová J, Rozsypal J, Zahradníčková H, Cimlová J, et al. (2011) Long-term cold acclimation extends survival time at 0°C and modifies the metabolomic profiles of the larvae of the fruit fly Drosophila melanogaster. PLoS One 6: e25025.
16. Belova NA, Potselueva MM, Skrebnitskaia LK, Znobishcheva AV, Lednev VV (2010) Effects of weak magnetic fields on the production of reactive oxygen species in peritoneal neutrophils in mice. Biofizika 55: 657–663.
17. Valavanidis A, Fiotakis K, Bakeas E, Vlahogianni T (2005) Electron paramagnetic resonance study of the generation of reactive oxygen species catalysed by transition metals and quinoid redox cycling by inhalable ambient particulate matter. Redox Rep 10: 37–51.
18. Wertheimer N, Leeper E (1979) Electrical wiring configurations and childhood cancer. Am J Epidemiol 109: 273–284.
19. Mevissen M, Häussler M, Lerchl A, Löscher W (1998) Acceleration of mammary tumorigenesis by exposure of 7,12-dimethylbenz[a]anthracene-treated female rats in a 50-Hz, 100-microT magnetic field: replication study. J Toxicol Environ Health A. 53: 401–418.
20. Ohtsu S, Miyakoshi J, Tsukada T, Hiraoka M, Abe M, et al. (1995) Enhancement of beta-galactosidase gene expression in rat pheochromocytoma cells by exposure to extremely low frequency magnetic fields. Biochem Biophys Res Commun. 212: 104–109.
21. Bassett CA, Mitchell SN, Gaston SR (1981) Treatment of ununited tibial diaphyseal fractures with pulsing electromagnetic fields. J Bone Joint Surg Am 63: 511–523.
22. Oshibuchi H (1975) The experimental studies on the biological effect of the alternating magnetic fields; (1) Obsevation on the effect to the wound healing and the general condition. Nippon Geka Gakkai Zassi 76; 432–445 Japanese.
23. McCann J, Dietrich F, Rafferty C (1998) The genotoxic potential of electric and magnetic fields: an update. Mutat Res. 411: 45–86.
24. Cho MH, Lee EJ, Son M, Lee JH, Yoo D, et al. (2012) A magnetic switch for the control of cell death signaling in in vitro and in vivo systems. Nat Mater. (in press).

9

Methods for Cryopreservation of Guinea Fowl Sperm

Éva Váradi*, Barbara Végi, Krisztina Liptói, Judit Barna

Institute for Small Animal Research and Co-ordination Centre for Gene Conservation, Gödöllő, Hungary

Abstract

Conservation of indigenous poultry species is an important part of the new Hungarian agricultural strategy. Semen cryopreservation is the most practical method for the long term storage of poultry genetic material. The objective was to compare four protocols for cryopreservation of guinea fowl sperm (slow and fast programmable, freezing in nitrogen vapor, and pellet) and three cryoprotectants (10% ethylene glycol, 6% dimethyl-formamide and 6% dimethyl-acetamide). The efficiency of the methods was examined by *in vitro* tests (subjective motility scoring, sperm concentration, morphological and live/dead sperm analysis with eosin-aniline staining). Thereafter, the two most promising methods were tested by artificial insemination of frozen-thawed semen (3 times a week for 3 weeks using 300 million spermatozoa/hen), followed by candling of incubated eggs, assessment of fertilization, embryonic death, and hatching rate. The survival rate of live, intact spermatozoa was greatest ($p \leq 0.05$) in pellet method and the slow programmable protocol (with 10% ethylene glycol) (28.6 and 23.5%). The two best protocols (based on *in vitro* assessment of post-thaw semen quality) were subsequently tested *in vivo* with artificial insemination. The pellet method yielded a 64% fertility rate compared to slow protocol with only 30% fertility. Regardless, both freezing protocols significantly increased embryonic deaths compared to the control group (16,7; 9,1 and 8,3%, respectively). During the 3-week *in vivo* trial, fertility increased and early embryonic death decreased over time. According to the results the guinea fowl sperm could tolerate the fast freezing in pellet better than the slower freezing rates and resulted acceptable fertility rate.

Editor: Joel R. Drevet, Clermont-Ferrand Univ., France

Funding: The research was supported by the National Innovation Office (TET_09_FR_ANR_BIO-CryoBird) (www.nih.gov.hu). The funders had no role in study design, data collection and analysis, decision to publish, or preparation of the manuscript.

Competing Interests: The authors have declared that no competing interests exist.

* E-mail: varadievi@hotmail.com

Introduction

Biodiversity and gene conservation are of broad interest all over the world. Although much of the focus is on endangered species, these issues are also important for domestic livestock. With intense animal breeding and better, more homogenous environmental conditions, genetic diversity of domestic species has decreased rapidly [1]. Intensive poultry breeding pushed the indigenous breeds into the background disposing of valuable genes. According to the DAD-IS (Domestic Animal Diversity Information System), more than 50% of poultry species are in the endangered category [2]. Consequently, there was an urgent need to create gene- and databanks [3]. It is well known that for safe gene conservation, both *in vivo* and *in vitro* gene banks are necessary. Presently, indigenous poultry genetic materials are stored *in vitro* in only four national gene banks (France, Netherlands, North-America and Spain) [4], [5], [6]. In addition to *in vivo* management, *in vitro* conservation is a strategic tool to secure genetic diversity, considering the risk of epidemic diseases [7].

Semen cryopreservation is presently the most practical method for long-term storage of genetic material in poultry species [8], [9], [10]. The efficiency of sperm cryopreservation is variable among poultry species [11]. The freezing tolerance of guinea fowl spermatozoa seems to be the lowest among the various poultry species, presumably due to its low membrane fluidity, which was attributed to low concentrations of cholesterol. Moreover, the quality of guinea fowl semen is also lower than that most other poultry species [10].

In the apparently sole study of cryopreservation of guinea fowl spermatozoa, an intermediate freezing rate (15°C/min) and 6% dimethyl-formamide (DMF) was used; 37% sperm survival and 20% fertility were achieved [12]. *Schramm and Hübner* [13] successfully cryopreserved duck spermatozoa using a slow freezing protocol with ethylene glycol (EG). Freezing spermatozoa in nitrogen vapor has been widely used for fish [14] as well as for geese [15].

Sperm cryopreservation using slower freezing methods and pellet form using dimethyl-acetamide (DMA) looks also promising. This simple cryoconservation technique does not require complicated equipment, making it appropriate for widespread applications [16]. These methods have achieved high fertility rates for various poultry species [17], [18], although this approach has apparently not been reported for guinea fowl.

In the present study four freezing protocols with various cooling rates (slow and fast programmable method, freezing in nitrogen vapor, pellet method) and cryoprotectants (EG, DMF, DMA) were tested for cryopreservation of guinea fowl spermatozoa.

Materials and Methods

Ethics Statement

Keeping of animals and animal welfare prescriptions were according to the Hungarian Animal Protection Law (1998. XXVIII). Institute for Small Animal Research and Co-ordination Centre for Gene Conservation provides a permission for experimental animal research from the National Food Chain

Table 1. Experimental design of freezing protocols.

Protocols	Slow programmable		Fast programmable		Nitrogen vapor		Pellet
Type of cryo container	Cryo vial						
Diluents	Lake's						Tselutin's
Dilution rate	1:3						1:1
Equilibration time	25 min at 3°C		5 min at 5°C		5 min at 5°C		5 min at 2°C
CPs	10%EG	6%DMF	10%EG	6%DMF	10%EG	6%DMF	6%DMA
Freezing rate	−1°C/min to −30°C		−15°C/min to −30°C		4 cm above LN		directly into
	−30°C/min to −60°C		−30°C/min to −60°C		(120°C for 3 min)		LN

Safety Office, Animal Health and Animal Welfare Directorate, Budapest (permission number: 1/1512/49/15/2/2012). All applied experimental methods were permitted the own Institutional Ethical Review Board (No.7/2011) of the Institute for Small Animal Research and Co-ordination Centre for Gene Conservation.

Animals

Thirty rural Hungarian guinea fowl males (1 year old) were kept in individual cages, exposed to a 16L:8D photoperiod (natural and artificial light) and given *ad libitum* access to a commercial poultry ration and to water.

Semen Collection and Qualification

Males were selected firstly according to their tractability and response of sperm collection. Following a training period, assessment of semen quality was done to selecting the best sperm donors. For a two-month interval, semen was collected twice a week by dorso-abdominal massage [19]. During the sperm freezing period, *in vitro* assessment of semen quality was done before freezing and after thawing (fresh and frozen/thawed semen). Following sperm volume measurement, motility was determined by subjective scoring (0–5) by a well-trained person, while sperm concentration was estimated with a spectrophotometer (Accucell IMV Technologies, L'Aigle, France). Sperm morphology normal, abnormal and live/dead cell ratio was examined with aniline blue-eosin staining. This procedure is based on eosin (Certistain, 115935 Eosin Y, Merck Ltd., Budapest, Hungary) that is excluded by live, intact cell membranes but dead sperm membranes permit the eosin to stain the sperm. The aniline blue (415049 Sigma-Aldrich Ltd., Budapest, Hungary) provides a clear blue background in the smear to enhance the contrast of the white, unstained "live" sperm or the pinkish, stained "dead" sperm. Slides were evaluated microscopically (Zeiss, Axioscope; Carl Zeiss Microscopy GmbH., Göttingen, Germany) using an oil immersion objective and 1200× magnification. A total of 200 spermatozoa were assessed per slide.

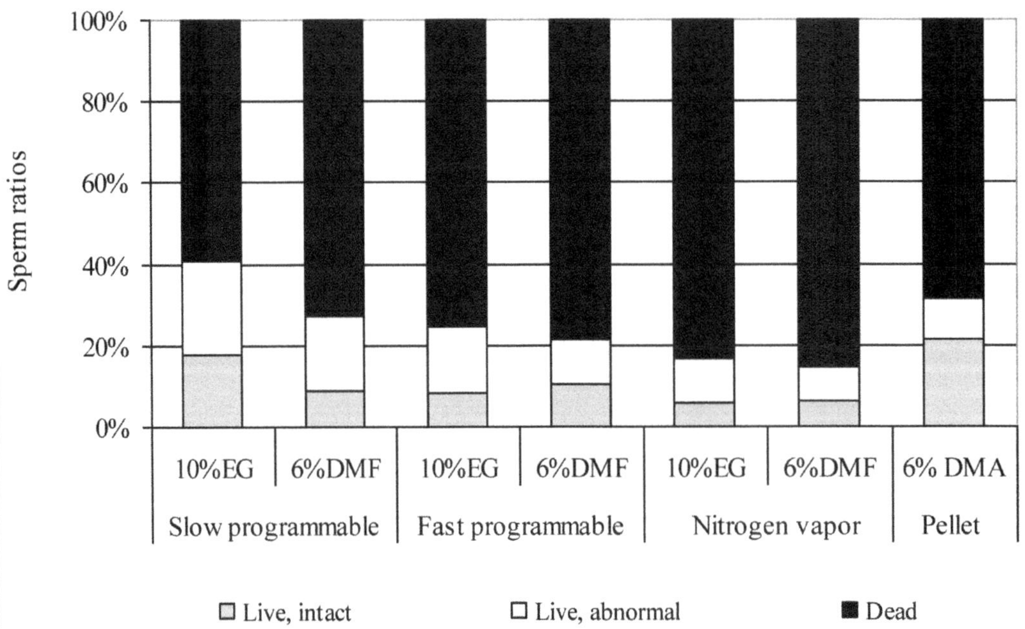

Figure 1. Quality of frozen/thawed spermatozoa in the different protocols.

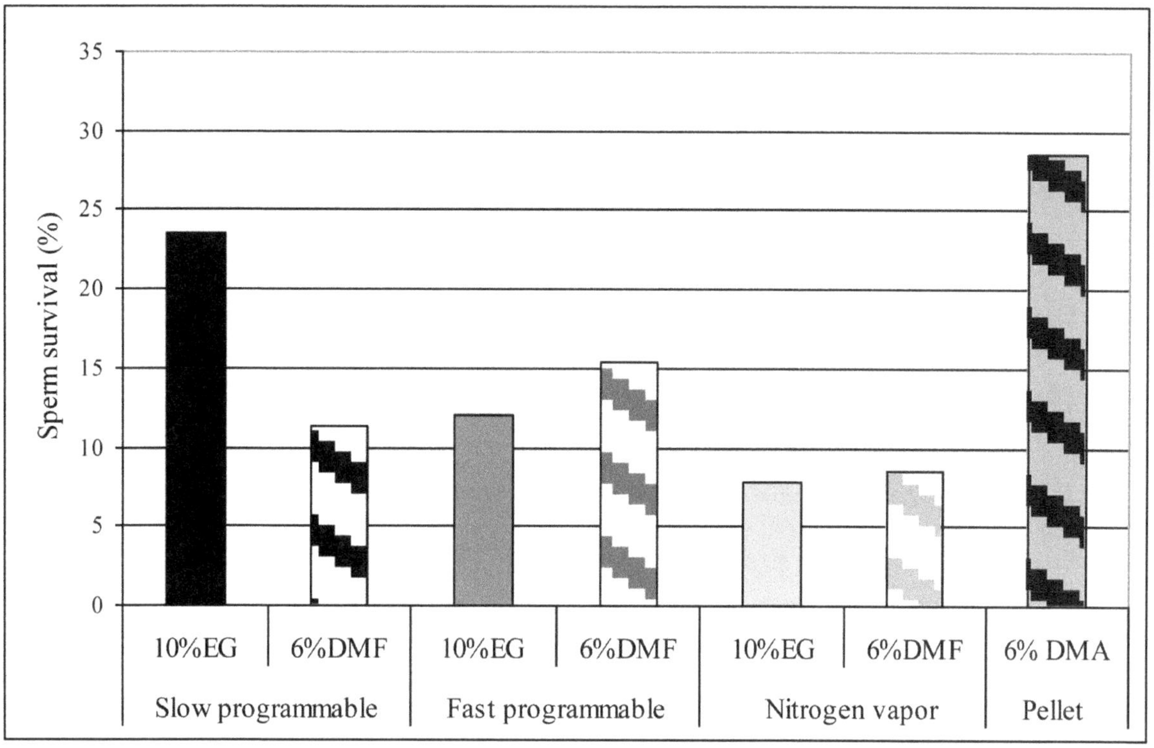

Figure 2. Survival rate of live, intact spermatozoa after various freezing methods.

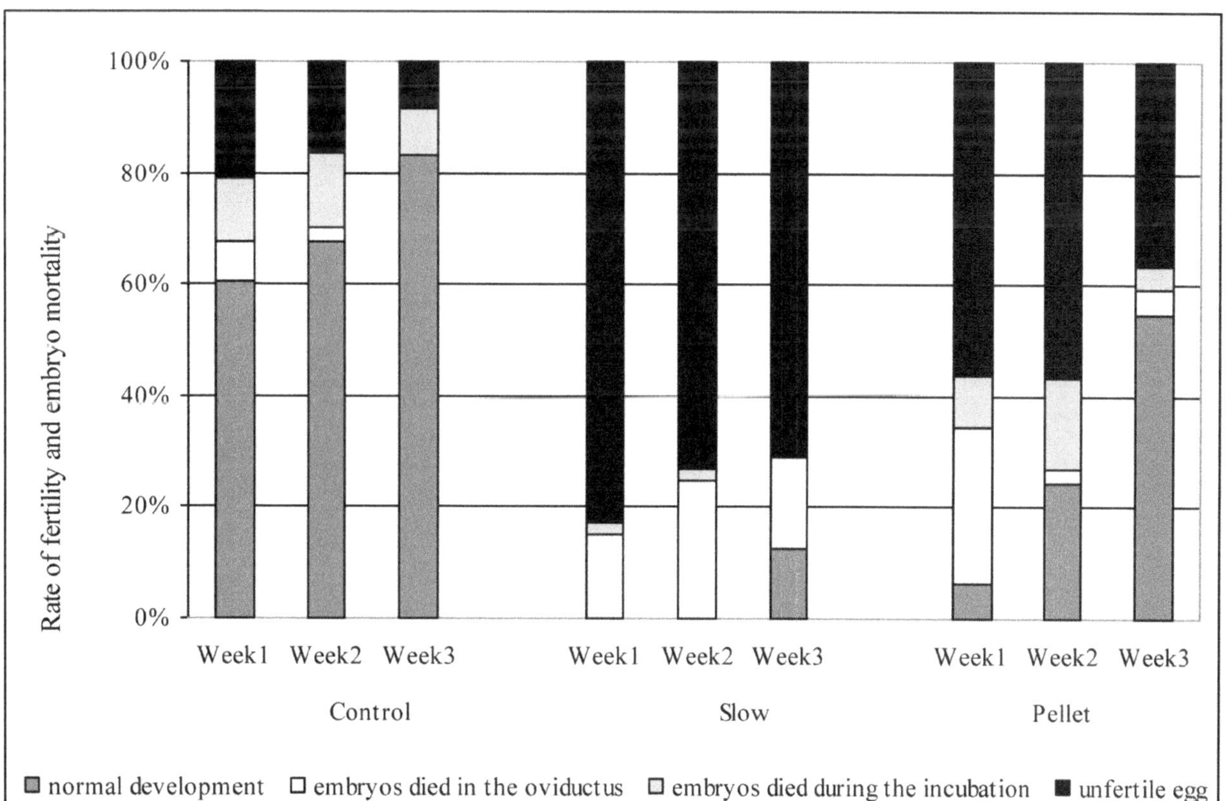

Figure 3. Fertility and embryo mortality in the three experimental groups following the artificial inseminations.

Freezing Protocols

In the study, four freezing protocols with various cooling rates (slow and fast programmable method, freezing in nitrogen vapor, pellet method) and three cryoprotectants (CPs) were tested: 10% ethylene glycol (EG; 09660-1-01-65, Reanal Laborvegyszer Ldt. Budapest, Hungary), 6% dimethyl-formamide (DMF; 444926, Carlo Erba Reagenti, Arese, Italy) and 6% dimethyl-acetamide (DMA; Sigma-Aldrich Ltd, Budapest, Hungary) *(Table 1)*.

Slow Programmable Protocol

After sperm collection, pooled semen was diluted 1:3 with Lake's extender [20] at room temperature. Diluted semen was divided in two equal parts: one sample contained 10% EG, the other 6% DMF as cryoprotectants. Aliquots (200 μL) of the samples were measured into cryo vials and placed into a programmable freezing machine (Planer KRYO10, Planer Products Ltd., Middlesex, U.K). Cooling and freezing was carried out in three consecutive steps, as described [12] with some modifications: Cooling was started at 20°C, with a −3°C/min freezing rate until 3°C, followed by 25 min equilibration. Then, a −1°C/min freezing rate was applied from 3°C to −30°C, than −30°C/min rate to −60°C. Finally cryo vials were placed into liquid nitrogen (LN). Samples were frozen for at least 2 weeks. Thawing was carried out at 5°C for 30 min in a cooler cabin (Brillant EURO-1-E, Ausseer Kälte und Edelstahltechnik GmbH, Kainisch, Austria).

Fast Programmable Protocol

Pretreatment of the semen was similar to the previous protocol, except the equilibration period, which lasted 5 minutes at 5°C for both 10% EG or 6% DMF. Aliquots (200 μL) from the samples were put into the cryo vials and placed in the freezing machine. Freezing started at 5°C with an intermediate freezing rate (−15°C/min) until −30°C, than samples were cooled (−30°C/min freezing rate) to −60°C. Cryo vials were placed into liquid nitrogen (LN). Thawing was done out at 5°C for 20 min in a cooler cabin (as described above).

Freezing in Nitrogen Vapor

Dilution and precooling of the semen were exactly the same as in fast programmable protocol. However, the cryo vials were put onto a vial rack which was placed in a plastic box at 4 cm above the liquid nitrogen. Thawing was carried out at 38°C in an incubator for 3 min (NB-201Q, N-Biotek Inc., Korea).

Pellet Method

After sperm collection, pooled semen was diluted 1:1 with Tselutin's extender [17] at room temperature. Cooling was 20 min at 2°C, whereas in the last 2 minutes 6% DMA was added to the sample. Treated semen (25 μL) was directly dropped into the liquid nitrogen using a mechanical pipette (Eppendorf Multipette plus, Eppendorf Austria GmbH, Wien, Austria). Pellets were collected with forceps and put in cryo vials filled with LN, which were then stored in a container containing LN. Thus, inside the vials the pellets were also in LN. Thawing was done at 70°C for 10 sec in a self-fabricated automatic warming device. The content of the vials was poured into the preheated metal funnel of the device and the thawed sperm was collected into a glass vessel.

Artificial Insemination

Thirty female guinea fowls were placed in individual cages (conditions were similar to those used for the males). The two best freezing methods (based on *in vitro* assessment of frozen-thawed sperm) were tested with artificial insemination (as an *in vivo* qualification method). Ten females (controls) were inseminated with fresh, diluted semen, whereas the remaining 20 were inseminated with frozen/thawed semen from slow and pellet method (10 birds per protocol). Artificial insemination was done within 10 minutes after thawing thrice weekly for 3 weeks using 250–300 million spermatozoa/hen/AI. The fertility of 300 incubated eggs was checked by candling at the 10th day of incubation. The rate of the true fertility (normal and dead embryos) was determined. Embryos that had died in the oviduct were detected by propidium iodide staining of the germinal disc. Egg yolks were separated from the albumen and placed into 0.9% NaCl solution. After visual fertility examination germinal discs seeming to be infertile were removed from the vitelline membrane, put into 0.9% NaCl solution and stained on a slide with propidium iodide (PI; P4170, Sigma-Aldrich Ltd, Budapest, Hungary). In case of fertile egg, propidium iodide (DNA-specific fluorescent dye of red color) stains the nucleus of the dividing embryonic cells which appear in lighting points [21].

Statistical Analyses

For statistical analysis of the data *Mann-Whitney test* and *Kruskal-Wallis test* were used (Statistica, Version 7.0, StatSoft Hungary Ltd., Budapest, Hungary).

Results

In vitro Examinations

The quality of spermatozoa after freezing and thawing is shown (Fig. 1). The best live sperm ratio (41.1%) was in the slow protocol with 10% EG; however, more than the half of these sperm were morphologically abnormal. With the same slow cooling rate using 6% DMF there were only 27.1% live sperm ($p \leq 0.05$). The same CPs with a faster cooling rate resulted in only 24.5 (for 10% EG) and 21.7% (for 6% DMF) live spermatozoa after thawing. The worst live cell ratios (16.8 and 14.8%) were in samples originating from nitrogen vapor, whereas the pellet method also resulted in high live sperm ratio (31.4%) and the highest live, intact cell ratio (21.4%).

The best ($p \leq 0.05$) survival rate of live, intact spermatozoa was in the pellet method and slow programmable protocol with 10% EG with 28.6 and 23.5%, respectively. The efficiency of the fast programmable protocol was rather low in both cases (12 and 15.5%), slow protocol with 6% DMF (11.4%) whereas freezing in nitrogen vapor produced the lowest survival rate (7.9 and 8.6%), irrespective of the type of CPs (Fig. 2).

Artificial Insemination

The slow protocol with 10% EG and pellet method yielded the best post-thaw sperm quality (based on *in vitro* assessments). Therefore, spermatozoa cryopreserved by these two methods were used for artificial insemination (*in vivo* test). To the end of the 3-week insemination period, fertility rates were 91.7, 29.1 and 63.6%, and rate of normal embryos were 83, 12, and 55% in fresh samples (control), slow and pellet method protocols, respectively. In sperm cryopreserved by the slow freezing, only dead embryos were detected in the first two weeks in the eggs. The rate of early embryonic deaths (embryo died in the oviduct) was quite high in the first week of insemination period (fresh 7, slow 15, pellet method 20%), but by the third week it had decreased significantly except for the slow protocol' group (fresh 0, pellet method 4.5%). The rate of the embryos that died during incubation decreased in all experimental groups to the end of the examined period (Fig. 3).

Discussion

Based on *in vitro* assessment, 6% DMF was a less effective cryoprotectant when used with a slow freezing rate, whereas 10% EG resulted in acceptable survival. It was noteworthy that the pellet method was the most promising for both sperm survival and fertility rate. This method has already been successfully adapted to turkey sperm freezing [22] with 42% sperm survival. Unfortunately, fertility was not reported. The pellet method produced higher fertility rate in the present study than that reported earlier in guinea fowl [12] using DMF with moderate freezing rate (64 vs 20%, respectively). The slow protocol with 10% EG had a good survival rate of live, intact spermatozoa, but fertility was very low. Consequently, using only *in vitro* tests to assess the effectiveness of sperm cryopreservation methods can be very misleading. Notable, there were higher rates of morphologically abnormal sperm in slower protocols than in the faster ones (nitrogen vapour or pellet method). The significantly highest rate of abnormal spermatozoa was detected in slow protocol using EG. Regarding specific defects, damage to the midpiece was the most common abnormality in all protocols (data not shown). Based on the present experiment, cryoprotectant had a lesser impact on sperm morphology than the rate of freezing. To our knowledge, there are no data available regarding sperm abnormalities of frozen thawed guinea fowl semen. However analysed the abnormal sperm ratio in frozen-thawed gander spermatozoa, the rate of midpiece injuries, namely "bent neck", were the most prevalent [23], similar to findings with emu spermatozoa [24].

When semen was preserved by the *slow* protocol, only 18–30% of eggs were fertilized; moreover 80–90% of them contained early dead embryos. According to earlier reports frozen–thawed avian spermatozoa are associated with an increased incidence of early embryonic mortality. Although only a single spermatozoon is required to activate an ovum, in the case of avian species, polyspermic fertilization occurs and an optimal number of sperm number are necessary for good fertility. The role of the other spermatozoa penetrated into the ovum is not fully understood yet. Previous studies reported that both too low [25], [26] and high [27] sperm concentration increase the chance of embryo mortality. Presumably, in the case of the present slow protocol there were not enough live, morphologically normal spermatozoa for the complete embryo development. Since the effect of freezing protocols was compared in the study, approximately similar amounts of sperm were used for inseminations. Certainly, the higher the sperm concentration, the higher the rate of intact spermatozoa and therefore the fertility. However, it is known that there is a limit in insemination doses since significant part of sperm is rejected from the cloaca if the doses are higher than 200 μL.

Furthermore, perhaps morphologically normal spermatozoa contained cryopreservation-induced damage to nuclear structures, including DNA [28]. In humans, DNA-damaged spermatozoa fertilized ova, but the rate of early embryonic death increased [29]. During storage of equine sperm, there is also a measurable increase in DNA damage as detected by the comet assay with both cooled and frozen storage [30], [31].

In conclusion, the pellet method was the most effective way for conservation of guinea fowl spermatozoa with a high rate of fertility. There are no data available about the long term duration of sperm in pellet form. *Tselutin et al.,* [17] prefer slow freezing protocols to pellet method for special long term gene conservation purposes in the case of domestic fowl. Further investigations are in progress to check the effects of storage conditions in the case of pellet form.

Author Contributions

Conceived and designed the experiments: ÉV BV JB. Performed the experiments: ÉV BV JB. Analyzed the data: ÉV BV. Contributed reagents/materials/analysis tools: ÉV BV KL. Wrote the paper: ÉV JB.

References

1. Tisdell C (2003) Socioeconomic causes of loss of animal genetic diversity: analysis and assesment. Ecol Econ 45: 365–376.
2. Hoffmann I (2005) Research and investment in poultry genetic resources-challenges and options for sustainable use. Worlds Poult Sci 61: 57–70.
3. Boa-Amponsem K, Scherf B, Hoffmann I (2004) Utilisation and conservation of poultry genetic resources: FAO Initiatives. World Poultry Congress, June 8–12, Istanbul, CD Proceedings.
4. Blackburn HD (2006) The national animal germplasm program: challenges and opportunities for poultry genetic resources. Poult Sci 85: 210–215.
5. Woelders H, Zuidberg CA, Hiemstra SJ (2006) Animal genetic resources conservation in The Netherlands and Europe: poultry perspective. Poult Sci 85: 216–222.
6. Santiago-Moreno J, Castaño C, Toledano-Díaz A, Coloma MA, López-Sebastián A et al. (2011) Semen cryopreservation for the creation of a Spanish poultry breeds cryobank: Optimalization of freezing rate and equilibration time. Poult Sci 90: 2047–2053.
7. Blesbois E, Seigneurin F, Grasseau I, Limouzin C, Besnard J et al. (2007) Semen Cryopreservation for Ex Situ Management of Genetic Diversity in Chicken: Creation of the French Avian Cryobank. Poult Sci 86: 555–564.
8. Gee GF (1995) Artificial insemination and cryopreservation of semen from nondomestic birds. In: Bakst M R – Wishart GJ (ed.) Proc. 1st Symposium on the Artificial Insemination in Poultry. 1995. Maryland. USA. 262–280.
9. Reedy SE, Leibo SP, Clark ME, Etches RJ (1995) Beyond Semen Freezing. In: Bakst MR – Wishart GJ (ed.) Proc. 1st Symposium on the Artificial Insemination in Poultry. 1995. Maryland. USA. 229–250.
10. Blesbois E, Grasseau I, Seigneurin F (2005) Membrane fluidity and the ability to survive cryopreservation in domestic bird spermatozoa. Reproduction. 129: 371–378.
11. Blesbois E, Dubos F, Grasseau I, Richard MM, Roman Y et al. (2006) Recherche d'indicateurs d'aptitude a la congelation de la semence chez les oiseaux et mise au point de la cryopreservation du sperme de jars landais. Les actes du BRG, 6: 415–431.
12. Seigneurin F, Blesbois E (2005) The first method of cryopreservation of guinea fowl semen. Proc. of the French Poultry Research Days. 2005. Brittary, France. 23.
13. Schramm GP, Hübner R (1989) Konservierung von Geflügelsperma. Arch Tierz 32: 51–61.
14. Horváth Á, Miskolczi E, Urbányi B (2003) Cryopreservation of common carp sperm. Aquat Living Resour 16: 457–460.
15. Barna J, Végi B, Váradi É, Liptói K (2010) Comparative study on cryopreservation procedures of gander sperm. Proc. XIII European Poultry Conference, Tours, France, 23–27 August 2010. Worlds Poult Sci J 66: 508.
16. Saint Jalme M, Lecoq R, Seigneurin F, Blesbois E, Plouzeau E (2003) Cryoconservation of semen from endangered pheasants: the first step towards a cryobank for endangered avian species. Theriogenology 59: 875–888.
17. Tselutin K, Narubina L, Mavrodina D, Tur B (1995) Cryopreservation of poultry semen. Br Poult Sci 36: 805–811.
18. Tselutin K, Seigneurin F, Blesbois E (1999) Comparison of cryoprotectants and methods of cryopreservation of fowl spermatozoa. Poult Sci 78: 586–590.
19. Burrows WH, Quinn JP (1937) The collection of spermatozoa of domestic fowl and turkey. Poult Sci 16: 19–24.
20. Lake PE (1986) The history and future of the cryopreservation of avian germ plasma. Poult Sci 65: 1–15.
21. Liptói K, Varga Á, Hidas A, Barna J (2004) Detection of the rate of true fertility in duck breeds by the combination of two in vitro methods. Acta Vet Hung 52: 227–233.
22. Iaffaldano N, Romagnoli L, Manchisi A, Rosato MP (2011) Cryopreservation of turkey semen by the pellet method: Effects of variables such as the extender, cryoprotectants concentration, cooling time and warming temperature on sperm quality determined through principal components analysis. Theriogenology 76: 794–801.
23. Lukaszewicz E (2002) An effective method for freezing White Italian gander semen. Theriogenology 58: 19–27.
24. Sood S, Malecki IA, Tawang A, Martin GB (2011) Response of spermatozoa from the emu (Dromaius novaehollandiae) to rapid cooling, hyperosmotic conditions and dimethylacetamide (DMA). Anim Reprod Sci 129: 89–95.
25. Bramwell RK (2002) Fertility and embryonic mortality in breeders. Avian Advice 4: 1–4.
26. Végi B, Varga Á, Szőke Zs, Liptói K, Várkonyi E, et al. (2005) Relationship Between Insemination of Frozen Sperm and Early Embryonic Death in

Domestic Fowl. Proc. 2nd Combined Workshop of Fundamental Physiology of the European Working Group of Physiology and Perinatal Development in Poultry. September 23–25. 2005. Berlin, 55.

27. Van Krey HP, Ogasawara FX, Lorenz FW (1966) Distribution of spermatozoa in the oviduct and fertility in domestic birds. J Reprod Fert 11: 257–262.
28. Watson PF (2000) The causes of reduced fertility with cryopreserved semen. Anim Reprod Sci 60–61: 481–492.
29. Morris ID, Ilott S, Dixon L, Brison DR (2002) The spectrum of DNA damage in human sperm assessed by single cell gel electrophoresis (Comet assay) and its relationship to fertilization and embryo development. Hum Reprod 17: 990–998.
30. Linfor J, Meyers SA (2002) Assessing DNA integrity of cooled and frozen-thawed equine sperm using single cell gel electrophoresis. J Androl 23: 107–113.
31. Baumber J, Ball BA, Linfor JJ, Meyers SA (2003) Reactive oxygen species and cryopreservation promote DNA fragmentation in equine spermatozoa. J Androl 24: 621–628.

10

Role of Baseline Antral Follicle Count and Anti-Mullerian Hormone in Prediction of Cumulative Live Birth in the First In Vitro Fertilisation Cycle: A Retrospective Cohort Analysis

Hang Wun Raymond Li*, Vivian Chi Yan Lee, Estella Yee Lan Lau, William Shu Biu Yeung, Pak Chung Ho, Ernest Hung Yu Ng

Department of Obstetrics and Gynaecology, Centre of Assisted Reproduction and Embryology, The University of Hong Kong, Queen Mary Hospital, Hong Kong

Abstract

Objective: This retrospective study determined for the first time the role of baseline antral follicle count (AFC) and serum anti-Mullerian hormone (AMH) level in the first in-vitro fertilisation (IVF) cycle in predicting cumulative live birth from one stimulation cycle.

Methods: We studied 1,156 women (median age 35 years) undergoing the first IVF cycle. Baseline AFC and AMH level on the day before ovarian stimulation were analysed. The main outcome measure was cumulative live birth in the fresh plus all the frozen embryo transfers after the same stimulation cycle.

Results: Serum AMH was significantly correlated with AFC. Both AMH and AFC showed significant correlation with age and ovarian response in the stimulated cycle and total number of transferrable embryos. Baseline AFC and serum AMH were significantly higher in subjects attaining a live birth than those who did not in the fresh stimulated cycle, as well as those attaining cumulative live birth. There was a significant trend of higher cumulative live birth rate in women with higher AMH or AFC. However, logistic regression revealed that both AMH and AFC were not significant predictors of cumulative live birth after adjusting for age and number of embryos available for transfer. Considering only one single predictor, the areas under the ROC curves for AMH (0.646, 95% CI 0.616–0.675) and age (0.648, 95% CI 0.618–0.677) were slightly higher than that for AFC (0.617, 95% CI 0.587–0.647) in predicting cumulative live birth. However, a model combining AMH (with or without AFC) and age of the women only classified an addition of less than 2% of subjects correctly compared to the model with age alone.

Conclusion: Baseline AFC and serum AMH have only modest predictive performance on the occurrence of cumulative live birth, and may not give additional value on top of the women's age.

Editor: Cornelis B Lambalk, VU University Medical Center, The Netherlands

Funding: The study was supported by Small Project Funding (200607176062), The University of Hong Kong. The funders had no role in study design, data collection and analysis, decision to publish, or preparation of the manuscript.

Competing Interests: The authors have declared that no competing interests exist.

* E-mail: raymondli@hku.hk

Introduction

In assisted reproduction programmes, a number of parameters known as ovarian reserve markers, such as serum follicle stimulating hormone (FSH) concentration, antral follicle count (AFC) and serum anti-Mullerian hormone (AMH) concentration, are widely used to predict ovarian responses to gonadotrophin stimulation during in-vitro fertilization (IVF) treatment. These markers may help to decide on the initial dose and regimen of stimulation [1,2]. Among the commonly used ovarian reserve markers, AFC and AMH provide the best performance in predicting both poor and excessive ovarian response [3,4,5]. However, most reports consistently showed a poor predictive value of these markers on pregnancy rate in the fresh IVF cycle [4,6].

AMH, also known as Mullerian-inhibiting substance, is a dimeric glycoprotein that belongs to the transforming growth factor-beta family. It is involved in the regression of the Mullerian ducts during development of the male fetus [7]. In the adult female, AMH is exclusively produced by granulosa cells of preantral and small antral follicles, and has been shown to correlate excellently with the primordial follicle pool [8]. Hence it would serve as an ovarian reserve marker. Since AMH performs equally well, if not better, than AFC in predicting ovarian response and that it is both operator- and menstrual cycle-independent, there has been a growing trend to adopt AMH assay as the first-line ovarian reserve test [2,9].

In modern-day assisted reproduction programmes, embryo cryopreservation has become an integral component. It allows the storage and subsequent usage of surplus good quality embryos in frozen-thawed embryo transfer (FET) cycles. Hence, to evaluate the outcome of an IVF cycle, it would be more logical to consider the cumulative live birth rate from the fresh and all FET cycles combined, instead of merely looking at the single fresh cycle outcome.

There have been limited data on the role of these ovarian reserve markers in predicting cumulative pregnancy or live birth rates in IVF programmes. A study looked at the use of AMH on Day 6 of ovarian stimulation in predicting cumulative ongoing pregnancy outcome [10]. Yet AMH levels can be altered after commencement of ovarian stimulation [11,12,13], and it may be of more practical value to study AMH level before commencement of stimulation, and to look at live birth as the outcome which would be what patients are ultimately concerned about.

Therefore, we conducted this retrospective analysis to evaluate the role of baseline AFC and AMH in the first stimulated IVF cycle in predicting the cumulative live birth from the stimulated cycle and all subsequent FET cycles derived from that stimulated cycle.

Materials and Methods

Subject selection

We reviewed all first IVF cycles carried out between January 2007 and December 2009 at the Centre of Assisted Reproduction and Embryology, The University of Hong Kong – Queen Mary Hospital, Hong Kong. Cycles carried out for pre-implantation genetic diagnosis or those using donor oocytes were excluded from this analysis. Clinical details of all treatment cycles were prospectively entered into a computerized database, which were retrieved for analysis.

Ethics statement

An ethics approval was obtained from the Institutional Review Board of the University of Hong Kong/Hospital Authority Hong Kong West Cluster for this retrospective study to be carried out using existing patient data in an anonymous manner without requiring written consent from individual patients. The project did not involve any additional intervention or modification from the standard treatment.

Stimulation cycle, AFC and AMH determination

All patients attended the clinic for blood test and pelvic ultrasound examination at the beginning of the IVF treatment cycle before commencing ovarian stimulation. Transvaginal scanning was performed using a 5 MHz vaginal probe (Voluson 730®, GE Healthcare, Wisconsin, USA) to determine AFC (2–9 mm) in both ovaries. Blood was taken for routine oestradiol (E2) assay, and residual serum samples were stored at −20°C. For the current study, these archived serum samples were retrieved and assayed for AMH using the AMH Gen II ELISA kit (Beckman Coulter, Webster, TX, USA; catalogue number A79765). The assay has a sensitivity of 0.08 ng/ml (0.6 pmol/l), and intra- and inter-assay coefficients of variation of less than 5.4 and 5.6%.

All subjects were treated either with the long GnRH agonist protocol or the GnRH antagonist protocol for pituitary down-regulation, with the latter mostly for selected patients with poor ovarian reserve. In the long GnRH agonist protocol, the women received buserelin (Suprecur®, Hoechst, Frankfurt, Germany) nasal spray 150 µg four times a day starting from the mid-luteal phase of the cycle preceding the treatment cycle, and received human menopausal gonadotrophin (HMG) or recombinant FSH for ovarian stimulation after return of a period. In the GnRH antagonist protocol, after confirming a basal serum oestradiol level, ovarian stimulation was started. The women received ganirelix or cetrorelix 250 µg daily starting from the sixth day of stimulation. The initial dose of stimulation was determined according to the baseline AFC (AFC≥15: 150 IU per day; AFC between 6 and 14: 300 IU for the first two days followed by 150 IU daily; AFC ≤5: 300–450 IU for the first two days followed by 225 IU daily). Human chorionic gonadotrophin (hCG) (Pregnyl® 5000 or 10000 units or Ovidrel® 250 µg) was given when the mean diameter of the leading follicle reached 18 mm and there were at least 3 follicles reaching a mean diameter of 16 mm or above, followed by transvaginal ultrasound-guided oocyte retrieval 36 hours later. Fertilisation was carried out in-vitro either by conventional insemination or intracytoplasmic sperm injection (ICSI) depending on semen parameters.

Subjects were allowed to have replacement of at most two embryos two days post-fertilisation. Embryo transfer (ET) was performed under transabdominal ultrasound guidance using a soft catheter (Sydney IVF Embryo Transfer Catheter®, Cook, Indiana, USA). Fresh embryo transfer would be cancelled and all the embryos with good quality were cryopreserved on day 2 after the

Table 1. Correlation of AMH and AFC with ovarian stimulation parameters in fresh stimulation cycle.

Parameter	Serum AMH		AFC	
	Correlation coefficient	p-value	Correlation coefficient	p-value
Age of women	−0.321	<0.001*	−0.335	<0.001*
Body mass index	0.001	0. 969	0.035	0.238
Duration of stimulation	−0.232	<0.001*	−0.224	<0.001*
Total dose of gonadotrophin	−0.590	<0.001*	−0.683	<0.001*
No of follicles >= 16 mm	0.548	<0.001*	0.449	<0.001*
Peak serum oestradiol	0.529	<0.001*	0.410	<0.001*
No. of oocytes retrieved	0.629	<0.001*	0.527	<0.001*
No. of transferrable embryos	0.438	<0.001*	0.364	<0.001*
No. of top quality embryos#	0.341	<0.001*	0.321	<0.001*

*statistically significant (p<0.005).
#defined as number of blastomeres ≥4 with grading 1 or 2 (data available in 593 subjects only).

Table 2. Comparison of demographic and clinical characteristics of subjects with or without achieving live birth in the fresh IVF cycle.

Parameter	Subjects achieving a live birth in fresh IVF cycle (n = 383)	Subjects not achieving a live birth in fresh IVF cycle (n = 643)	P value
Age (years)	34.0 (32.0–37.0)	36.0 (34.0–38.0)	<0.001*
Body mass index (kg/m^2)	21.0 (19.5–23.2)	21.2 (19.6–23.1)	0.927
Serum AMH (ng/ml)	3.27 (1.75–5.65)	2.64 (1.32–4.69)	<0.001*
Antral follicle count	10.0 (6–15)	9.0 (5–13)	0.001*
Cause of subfertility	236 (61.6%)	371 (57.7%)	0.056
Male factor	49 (12.8%)	98 (15.2%)	(Fisher's Exact test)
Tuboperitonal factor	19 (5.0%)	25 (3.9%)	
Endometriosis	6 (1.6%)	1 (0.2%)	
Anovulation	19 (5.0%)	41 (6.3%)	
Unexplained	54 (14.0%)	107 (16.7%)	
Mixed factors			
No. of embryos replaced	22 (5.7%)	116 (18.0%)	<0.001 (Fisher's
One	361 (94.3%)	527 (82.0%)	Exact test)
Two			

Continuous data are expressed as median (interquartile range).
*statistically significant.
Mann-Whitney U test unless otherwise stated.

retrieval if the subject had symptoms suggestive of ovarian hyperstimulation syndrome (OHSS) or serum E2 concentration on the day of hCG injection was >20,000 pmol/L.

Cryopreservation and frozen-thawed embryo transfer (FET)

Embryos were graded based on the number/regularity of blastomeres and the degree of fragmentation [14]. Surplus embryos of grades 1 to 4 were frozen on the day of ET, whereas embryos with poor quality (Grades 5 and 6) were discarded. Cryopreservation was performed by a slow freezing protocol using a programmable freezer (Planer Products Ltd.; Sunbury-On-Thames, UK) in straws of two. The details of the freezing and thawing protocols were reported previously [15]. The frozen embryos were thawed on the morning of FET. Embryos were discarded if more than 50% of the original blastomeres were lysed or degenerated upon thawing. Frozen-thawed embryos were transferred in natural cycles in ovulatory subjects. LH surge was determined by serial blood tests, and FET was performed on the third day after LH surge. For anovulatory subjects, FET was carried out in either clomiphene-induced or hormone replacement cycles. A maximum of two frozen embryos were allowed to be transferred in any one FET cycle.

Pregnancy outcome

A urine pregnancy test was done 16 days after embryo transfer. Pregnant women were offered an ultrasound examination 10–14 days later to confirm intrauterine pregnancy and the number of gestational sacs present. Pregnancy outcome was traced from all pregnant women by postal questionnaire or by phone.

Statistical analysis

The primary outcome measure was the cumulative live birth in the fresh and all FET cycles combined following the same index stimulation cycle. The age of the women used in analysis referred to the time of starting ovarian stimulation. Non-normally

Table 3. Binary logistic regression analysis of factors for prediction of (a) a live birth in the fresh IVF cycle and (b) a cumulative live birth from the fresh plus all frozen-thawed transfer cycles combined after the same index stimulation cycle.

(a) Factors to predict a live birth in the fresh IVF cycle	B	Exp(B), 95% CI	P-value
AFC	−0.002	0.998, 0.972–1.025	0.880
AMH	0.020	1.020, 0.966–1.077	0.470
Age	−0.122	0.885, 0.849–0.923	<0.001*
Number of embryos transferred	1.236	3.441, 2.124–5.577	<0.001*
(b) Factors to predict a cumulative live birth	**B**	**Exp(B), 95% CI**	**P-value**
AFC	−0.004	0.996, 0.968–1.025	0.791
AMH	0.003	1.003, 0.995–1.012	0.441
Age	−0.129	0.879, 0.841–0.918	<0.001*
Total number of transferrable embryos	0.259	1.295, 1.228–1.366	<0.001*

*statistically significant.

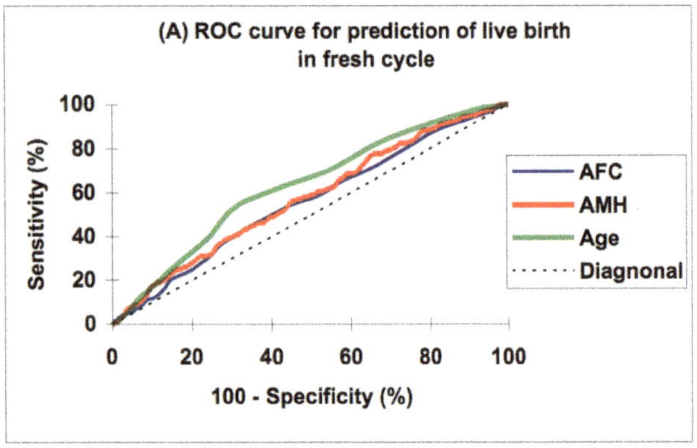

	Area under the curve	SE	95% CI
AFC	0.559	0.0184	0.528 to 0.589
AMH	0.575	0.0183	0.545 to 0.606
Age	0.629	0.0176	0.598 to 0.658

The area under the curves for AFC and AMH were significantly lower than that for age (p<0.05)

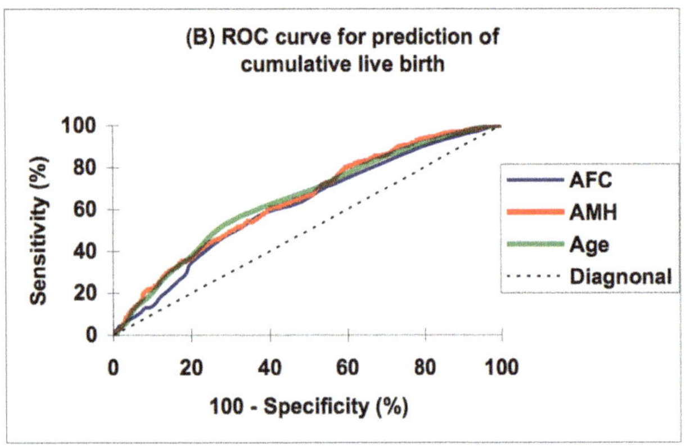

	Area under the curve	SE	95% CI
AFC	0.617	0.0174	0.587 to 0.647
AMH	0.646	0.0170	0.616 to 0.675
Age	0.648	0.0170	0.618 to 0.677

The area under the curve for AMH was significantly higher than that for AFC (p<0.05)

Figure 1. Receiver operator characteristic (ROC) curve analysis. Panal (A) shows the ROC curves of antral follicle count (AFC), serum AMH concentration and age of the women in predicting live birth from the fresh stimulation cycle, while Panal (B) shows those in predicting cumulative live birth from fresh and all frozen embryo transfers after the same index stimulation cycle.

distributed continuous variables were expressed as median (interquartile range) unless otherwise stated. Continuous and categorical variables were compared between groups using Mann-Whitney test and Fisher's Exact test respectively. Logistic regression analysis was used to examine factors predicting cumulative live birth, and receiver-operator characteristic (ROC) curves were compared for different ovarian reserve markers in predicting cumulative live birth. Statistical analysis was carried out using the Statistical Program for Social Sciences (SPSS Inc., Version 15.0, Chicago, U.S.A.) and MedCalc (Version 12, Belgium). The two-tailed value of P <0.05 was considered statistically significant.

Results

Of the 1,156 subjects who underwent the first IVF cycle during the study period, 1,050 (90.8%) were treated on the long GnRH agonist protocol, and the remaining 106 (9.2%) on the GnRH antagonist protocol. The age and body mass index (BMI) [median (interquartile range)] of our subjects were 35 (33–38) years and 21.2 (19.6–23.1) kg/m^2 respectively. The cause of subfertility included male factor (687; 59.4%), tubo-peritoneal factor (159; 13.8%), endometriosis (54; 4.7%), anovulation (8; 0.7%), unexplained (62; 5.4%) and mixed causes (186; 16.0%).

Correlation of AFC and AMH with other demographic and clinical characteristics

There was a significant correlation between serum AMH and AFC (R = 0.780, p<0.001). As shown in Table I, both serum AMH and AFC were significantly correlated with the women's age, duration of stimulation, total dose of gonadotrophin used, number of mature-sized follicles (≥16 mm), peak serum E2 concentration, number of oocytes retrieved and transferrable embryos obtained (for fresh transfer and/or freezing) and number of top quality embryos (p<0.001 for all). There was no significant correlation between serum AMH or AFC with BMI.

AFC, AMH and live birth in the fresh cycle

There were 1,026 subjects (88.8%) who had fresh ET. Among them, 510 (49.7%) had a positive pregnancy test in the fresh cycle, and 383 (37.3%) attained live birth. Among those who were pregnant, there was no significant difference (p>0.05) in AFC and serum AMH between those who miscarried and those who had live birth. The clinical and demographic characteristics of subjects with or without a live birth in the fresh cycle are compared in Table II. Subjects who attained a live birth were significantly younger (median 34.0 vs 36.0 years), had significantly higher AFC (median 10.0 vs 9.0) and serum AMH [median 3.27 ng/ml (23.3 pmol/l) vs 2.64 ng/ml (18.8 pmol/l)], and more of them had double embryo transfer (94.3% vs 82.0%). There were no statistically significant differences in BMI between subjects who had live birth in the fresh cycle or not, nor in the distribution of the cause of subfertility.

Binary logistic regression using the enter method was used to analyse the prediction on live birth in the fresh IVF cycle by the women's age, AFC, serum AMH and number of embryos replaced. The women's age and the number of embryos replaced, but not AFC or serum AMH, were the significant factors which independently predicted the likelihood of a live birth in the fresh cycle (Table IIIa).

Figure 1A shows the ROC curve analysis of AFC and serum AMH in predicting a live birth in the fresh cycle. The area under the curve (AUC) for AFC and serum AMH were 0.559 (95% CI 0.528–0.589) and 0.575 (95% CI 0.545–0.606) respectively, which were both inferior (p<0.05, Delong analysis) to that for age of the women (0.629, 95% CI 0.598–0.658).

AFC, AMH and cumulative live birth

Among the subjects included in this study, 1,038 (89.8%) had completed replacement of all available embryos and with a live birth outcome available. Of them, 558 (53.8%) attained live birth cumulatively from the fresh plus all FET cycles after the same index IVF cycle.

Table IV depicts the demographic and clinical characteristics of subjects who had a cumulative live birth and those who did not. Subjects who attained a cumulative live birth were significantly younger (median 34.0 vs 36.0 years), had significantly higher AFC (median 11.0 vs 8.0) and serum AMH [median 3.7 ng/ml (26.4 pmol/l) vs 2.3 ng/ml (16.4 pmol/l), and had significantly higher number of transferrable embryos (median 5 vs 2). The minimum values of AFC and serum AMH in subjects who attained a cumulative live birth were 0 and 0.15 ng/ml (1.1 pmol/

Table 4. Comparison of demographic and clinical characteristics of subjects with or without achieving a cumulative live birth from the fresh plus all the frozen-thawed embryo transfer cycles combined after the same index stimulation cycle.

Parameter	Subjects achieving a cumulative live birth (n = 558)	Subjects not achieving a cumulative live birth (n = 480)	P value
Age (years)	34 (32–37)	36 (34–38)	<0.001*
Body mass index (kg/m^2	21.1 (19.5–23.1)	21.2 (19.7–23.2)	0.426
Serum AMH (ng/ml)	3.7 (1.9–6.1)	2.3 (1.1 6–4.2)	<0.001*
Antral follicle count	11 (7–16)	8 (5–12)	<0.001*
Cause of subfertility Male factor Tuboperitonal factor Endometriosis Anovulation Unexplained Mixed factors	348 (62.4%) 69 (12.4%) 23 (4.1%) 7 (1.3%) 31 (5.6%) 80 (14.2%)	274 (57.1%) 77 (16.0%) 22 (4.6%) 0 (0%) 27 (5.6%) 80 (16.7%)	0.048* (Fisher's Exact test)
Total number of transferrable embryos	5 (3–8)	2 (2–4)	<0.001*

Continuous data are expressed as median (interquartile range).
*statistically significant.
Mann-Whitney U test unless otherwise stated.

l) respectively (vs 0 and <0.08 ng/ml (<0.6 pmol/l) respectively in those without a cumulative live birth).

Figure 2 shows the cumulative live birth rate in subjects categorised according to baseline AMH and AFC. There was an overall significant trend of a higher cumulative live birth rate in subjects with higher serum AMH levels or AFC compared to those at the lower end ($p<0.001$ for both AMH and AFC, χ^2 test for trend).

When the women's age, AFC, serum AMH and total number of transferrable embryos derived from the index IVF cycle were entered into the binary logistic regression model using the enter method, the women's age and total number of transferrable embryos, but not AFC or AMH, were significant independent factors in predicting a cumulative live birth (Table IIIb).

We further analysed the predictive performance of AFC and AMH in additional to age of the women versus age alone on cumulative live birth. (Table V). The addition of AMH, but not AFC, to age provided additional predictive value compared to using age alone. However, although statistically significant ($p<0.05$, Delong analysis), the model combining the women's age and AMH, with or without AFC, only respectively classified an additional 1.7% or 1.9% of subjects correctly compared to the model with age alone.

Figure 1B shows ROC curve analysis of AFC, serum AMH and age in predicting a cumulative live birth. The AUC for serum AMH and age were 0.646 (95% CI 0.616–0.675) and 0.648 (95% CI 0.618–0.677) respectively, which were both significantly higher ($p<0.05$) than that for AFC (0.617, 95% CI 0.587–0.647).

Discussion

In assisted reproduction programmes, ovarian reserve markers such as FSH, AFC, and more recently AMH, have been widely used for predicting ovarian response and prognostic counseling. Recent systematic reviews have suggested that AFC and AMH were the best predictors of excessive and suboptimal ovarian response and both had comparable performance in this regard [4,5]. AMH has the additional benefits of being not operator-dependent and having negligible inter-cycle variations [16], and hence has gained increasing favour in its clinical applications [2,9]. However, the majority of reported studies consistently demonstrated that both AFC and AMH had poor performance in predicting pregnancy outcome in IVF cycles [4,6].

Nonetheless, most of the reported studies evaluated the pregnancy or live birth outcome in the fresh IVF cycle only. In modern assisted reproduction programmes, cryopreservation of surplus embryos with subsequent transfers in thawed cycles has become an integral part. It has been estimated that in programmes which incorporate embryo cryopreservation, up to 42% of all conceptions could be derived from FET [17]. As there is a recent trend in reducing the number of embryos replaced to reduce the risk of multiple pregnancy, the number of live births from the stimulated cycle only is not a good indicator of outcome of treatment. Hence, we feel that it would be more meaningful to study the cumulative live birth per stimulated cycle as the treatment outcome. This would take into account the outcomes in the transfer of fresh as well as frozen embryos derived from the same index stimulation cycle, the latter of which should not be ignored. There was one study which reported that serum AMH on day 6 of stimulation was significantly higher in subjects attaining cumulative ongoing pregnancy than those who did not [10]. However, it has been reported that serum AMH is suppressed during ovarian stimulation compared to baseline level under both GnRH agonist and antagonist protocols [11,12,13]. Practically, measurement of the baseline AMH level before stimulation would be of more relevance for either prognostic prediction or individualization of gonadotrophin dosing. To our knowledge, the current study is the first to report on the role of baseline serum AMH and AFC in the index stimulation cycle in predicting cumulative live birth outcome so derived in a large cohort.

Our results demonstrated that women who attained a cumulative live birth had significantly higher serum AMH and AFC at baseline before ovarian stimulation. However, the predictive performance of both parameters on the absolute occurrence of cumulative live birth was only modest and not better than age as demonstrated in the ROC curves. This is obviously comprehensible by observing the trend of cumulative live birth rate over serial increase in serum AMH or AFC as shown in Figure 2. There was a gradual increase in cumulative live birth rate in women with higher AMH or AFC, and yet it varied over a continuum with no clear cut-off where the cumulative live birth rate could show an

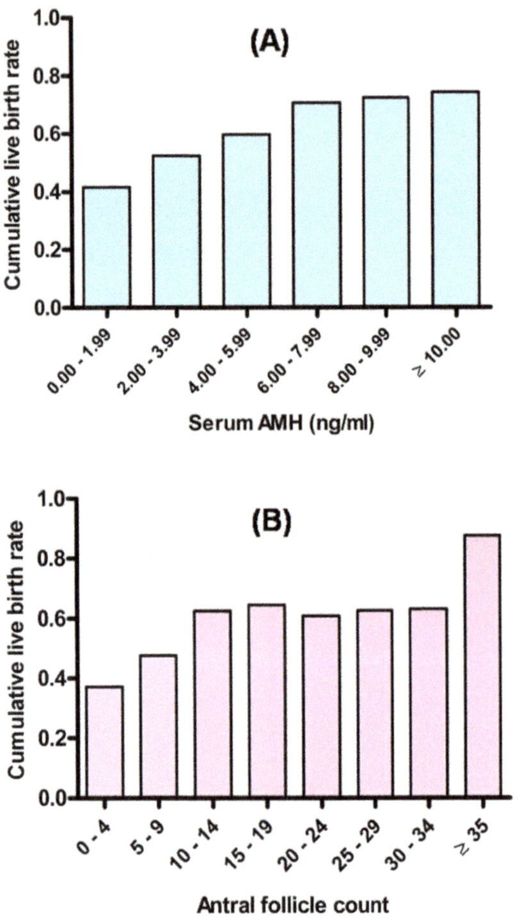

Figure 2. Cumulative live birth rate stratified according to serum AMH and antral follicle count (AFC). Panal (A) shows the association between serum AMH level and cumulative live birth rate from the fresh plus all frozen embryo transfer cycles combined, whereas Panal (B) shows that between AFC and cumulative live birth rate. There was a significant trend of increase across the bars from the lower to the higher ends of AMH and antral follicle count (p<0.001 for both, χ^2 test for trend).

abrupt change. Hence, it would be impossible for any set value of AMH or AFC to predict the all-or-none occurrence of a cumulative live birth.

Hypothetically, women with higher AMH or AFC have better ovarian reserve and hence higher oocyte yield and a larger potential pool of good quality embryos to be selected for transfer. This was supported by our findings that subjects who attained cumulative live birth had higher number of transferable embryos, and that the latter was an independent factor predicting cumulative live birth in logistic regression analysis. This is in line with a recent study which suggested that the total number of good-quality embryos and total number of transferable embryos were the two most important predictors of cumulative clinical pregnancy, particularly in women aged below 40 years [18]. Analysis of our data showed that the cumulative live birth rate increased with increasing number transferrable embryos up till 15, after which it plateaued (data not shown). The same has been recently reported on a large cohort of 400,135 IVF treatment cycles, in which the live birth rate in the fresh IVF cycle correlated positively with the number of retrieved oocytes up till 15, after which it plateaued and actually declined after going beyond 20

Table 5. Comparison of AMH, AFC in addition to age and age only in predicting cumulative live birth by logistic regression analysis.

Model	Predictors	OR (95% CI)	Classification rate	#AUC (95% CI)
A	Age	0.86 (0.81–0.92)	63.6%	[a]0.69 (0.64–0.74)
	AMH	1.02 (0.98–1.07)		
	AFC	1.06 (0.97–1.15)		
B	Age	0.86 (0.80–0.92)	63.8%	[b]0.69 (0.63–0.74)
	AMH	1.09 (1.02–1.16)		
C	Age	0.86 (0.81–0.92)	61.7%	[c]0.68 (0.63–0.73)
	AFC	1.04 (1.01–1.07)		
D	Age	0.84 (0.79–0.90)	61.9%	0.66 (0.60–0.71)

#The AUCs were derived from the logistic regression models.
[a]$p = 0.026$ compared with AUC for model A and AUC for model D.
[b]$p = 0.018$ compared with AUC for model B and AUC for model D.
[c]$p = 0.083$ compared with AUC for model C and AUC for model D.

[19], though the study did not take into account the FETs. Therefore, at the higher end of embryo yield, further increase in embryo number does not necessarily mean better live birth outcome because the occurrence of live birth is subjected to influence by other factors as well, such as embryo quality and endometrial receptivity.

With regard to endometrial receptivity, women with high AFC and/or serum AMH levels are at increased risk of ovarian hyperstimulation, and the resultant higher peak E2 levels could have detrimental effects on endometrial receptivity [20,21]. This may offset the potentially favourable implication of higher ovarian response in the fresh stimulated cycle. Embryo quality is another major issue, which seemingly cannot be predicted by either AMH or AFC, as illustrated by previously reported studies which revealed no consistent correlation between AMH and embryo morphology nor aneuploidy rate [22,23,24]. Hence all these parameters do not seem to have good prediction on the qualitative aspect of embryo competence. Probably there is no single factor which can precisely predict cumulative live birth outcome, which is actually determined by the complicated interplay of multiple factors. Indeed, among our subjects who had serum AMH <0.5 ng/ml (<3.6 pmol/l), a cumulative live birth rate of 27% could still be attained, and in those with AFC ≤1, again a cumulative live birth rate of 22% was observed (data not shown). Moreover, a cumulative live birth did occur with a minimum baseline AMH level of 0.15 ng/ml (1.1 pmol/l) and AFC of zero.

We also analysed the value of adding AFC and AMH on top of the women's age in the prediction model for cumulative live birth. Although statistically significant, the combined model with the women's age and AMH (with or without AFC) classified an addition of less than 2% of subjects correctly compared to the model with age alone. This very modest improvement in the prediction may not be very meaningful clinically.

Based on our results, we would recommend that it is not worthwhile to use any of the ovarian reserve markers including serum AMH or AFC as the basis to exclude subjects from attempting assisted reproduction treatments, and when used for prognostic counselling the women should understand the limitations and imprecision of these indices in anticipating the treatment outcome.

Inherent to the retrospective design, our results might have been subjected to verification bias, as the gonadotrophin dosing regimen has been preselected based on AFC in our protocol. We have repeated the analysis on the subgroup of subjects stimulated with an average daily dose of 225 IU or more, which presumably gave maximal stimuation. There was no significant difference in the areas under the ROC curve for AMH, AFC and age in predicting cumulative live birth (data not shown).

The strength of our study was the inclusion of a large cohort of subjects over a wide age range from 22 to 45 years with diversified ovarian reserve to enhance generalisability. Although some of the subjects had not completed replacement of all available frozen embryos and were hence excluded from the cumulative live birth analysis, they did not differ significantly from the included subjects in sociodemographic parameters.

In conclusion, our results confirmed a good correlation of baseline serum AMH and AFC with ovarian response parameters in IVF treatment. Baseline AFC and serum AMH were significantly higher in subjects attaining cumulative live birth, and there was a significant trend of increase in cumulative live birth rate in women with higher AMH or AFC. However, the predictive role of AFC and AMH on absolute occurrence of cumulative live birth was only modest and may not give much additional value on top of the women's age.

Acknowledgments

The authors would like to thank Ms. Benancy P.C. Wong for performing the AMH assay, Mr. Tak-Ming Cheung, Ms. Sharon Lee and Ms. Emily So for data management, and Ms. Joyce Yuen for blood sample retrieval. We also thank Mr. Chi-Tao Ng, Medical Statistician of the Clinical Trial Centre, The University of Hong Kong, for advice on data analysis.

Author Contributions

Conceived and designed the experiments: HWRL PCH EHYN. Performed the experiments: HWRL VCYL EYLL WSBY. Analyzed the data: HWRL VCYL EHYN. Wrote the paper: HWRL PCH EHYN. Laboratory work: EYLL WSBY.

References

1. Nelson SM, Yates RW, Fleming R. (2007) Serum anti-Mullerian hormone and FSH: prediction of live birth and extremes of response in stimulated cycles-implications for individualization of therapy. Hum Reprod 22(9): 2414–2421.
2. Nelson SM, Anderson RA, Broekmans FJ, Raine-Fenning N, Fleming R, et al. (2012) Anti-Mullerian hormone: clairvoyance or crystal clear? Hum Reprod 27(3): 631–636.
3. Broekmans FJ, Kwee J, Hendriks DJ, Mol BW, Lambalk CB. (2006) A systematic review of tests predicting ovarian reserve and IVF outcome. Hum Reprod Update 12: 685–718.
4. Broer SL, Mol BWJ, Hendriks D, Broekmans FJM. (2009) The role of antimullerian hormone in prediction of outcome after IVF: comparison with the antral follicle count. Fertil Steril 91(3):705–714.
5. Broer SL, Dolleman M, Opmeer BC, Fauser BC, Mol BW, et al. (2011) AMH and AFC as predictors of excessive response in controlled ovarian hyperstimulation: a meta-analysis. Hum Reprod Update 17(1):46–54.
6. Broer SL, Mol B, Dolleman M, Fauser BC, Broekmans FJ. (2010) The role of anti-Mullerian hormone assessment in assisted reproductive technology outcome. Curr Opin Obstet Gynecol 22: 193–201.
7. Lee MM, Donahoe PK, Hasegawa T, Silverman B, Crist GB, et al. (1996) Mullerian inhibiting substance in humans: normal levels from infancy to adulthood. J Clin Endocrinol Metab 81: 571–576.
8. Hensen KR, Hodnett GM, Knowlton N, Craig LB. (2011) Correlation of ovarian reserve tests with histologically determined primordial follicle number. Fertil Steril 95(1):170–175.
9. Anderson RA, Nelson SM, Wallace WHB. (2012) Measuring anti-Mullerian hormone for the assessment of ovarian reserve: when and for whom is it indicated? Maturitas 71(1): 28–33.
10. Kini S, Li HWR, Morrell D, Pickering S, Thong KJ. (2010) Anti-mullerian hormone and cumulative pregnancy outcome in in-vitro fertilization. J Assist Reprod Genet 27(8): 449–456.
11. Fanchin R, Schonauer LM, Righini C, Frydman N, Frydman R, et al. (2003) Serum anti-Mullerian hormone dynamics during controlled ovarian hyperstimulation. Hum Reprod 18:328–332.
12. La Marca A, Malmusi S, Giulini S, Tamaro LF, Orvieto R, et al. (2004) Anti-Mullerian hormone plasma levels in spontaneous menstrual cycle and during treatment with FSH to induce ovulation. Hum Reprod 19:2738–2741.
13. Lee JR, Kim SH, Kim SM, Jee BC, Ku SY, et al. (2010) Anti-Mullerian hormone dynamics during controlled ovarian hyperstimulation and optimal timing of measurement for outcome prediction. Hum Reprod 25: 2597–2604.
14. Veeck LL (1999) An atlas of human gametes and conceptuses. New York: The Parthenon Publishing Group.
15. Yeung WSB, Li RHW, Cheung TM, Ng EHY, Lau EYL, et al. (2009) Frozen-thawed embryo transfer cycles. Hong Kong Med J 15(6):420–426.
16. Van Disseldorp J, Lambalk CB, Kwee J, Looman CW, Eijkemans MJ, et al (2010) Comparison of inter- and intra-cycle variability of anti-Mullerian hormone and antral follicle counts. Hum Reprod 25: 221–227.
17. Borini A, Cattoli M, Bulletti C, Goticchio G. (2008) Clinical efficiency of oocyte and embryo cryopreservation. Ann N Y Acad Sci 1127:49–58.
18. Cai QF, Wan F, Huang R, Zhang HW. (2011) Factors predicting the cumulative outcome of IVF/ICSI treatment: a multivariable analysis of 2450 patients. Hum Reprod 26(9): 2532–2540.
19. Sunkara SK, Rittenberg V, Raine-Fenning N, Bhattacharya S, Zamora J, et al (2011) Association between the number of eggs and live birth in IVF treatment: an analysis of 400 135 treatment cycles. Hum Reprod 26(7):1768–1774.
20. Mitwally MF, Bhakoo HS, Crickard K, Sullivan MW, Batt RE, et al. (2006) Estradiol production during controlled ovarian hyperstimulation correlates with treatment outcome in women undergoing in vitro fertilization-embryo transfer. Fertil Steril 86:588–596.
21. Ng EHY. (2009) Adverse effects of excessive ovarian response on the pregnancy rate of in vitro fertilization treatment. Gynecol Endocrinol 25(1):2–7.
22. Lie Fong S, Baart EB, Martini E, Schipper I, Visser JA, et al. (2008) Themmen AP, de Jong FH, Fauser BJ, Laven JS. Anti-Mullerian hormone: a marker for oocyte quantity, oocyte quality and embryo quality? Reprod Biomed Online 16(5):664–670.
23. Plante BJ, Beamon C, Schmitt CL, Moldenhauer JS, Steiner AZ. (2010) Maternal antimullerian hormone levels do not predict fetal aneuploidy. J Assist Reprod Genet 27:409–414.
24. Kline JK, Kinney AM, Levin B, Kelly AC, Ferin M, et al. (2011) Trisomic pregnancy and elevated FSH: implications for the oocyte pool hypothesis. Hum Reprod 26(6):1537–1550.

11

Prevention of Acute Kidney Injury by Tauroursodeoxycholic Acid in Rat and Cell Culture Models

Sandeep Gupta[1*§], Shunan Li[1], Md. Joynal Abedin[1], Kajohnsak Noppakun[1], Lawrence Wang[1], Tarundeep Kaur[1], Behzad Najafian[2], Cecília M. P. Rodrigues[3], Clifford J. Steer[1,4*§]

1 Department of Medicine, University of Minnesota Medical School, Minneapolis, Minnesota, United States of America, **2** Department of Laboratory Medicine and Pathology, University of Minnesota Medical School, Minneapolis, Minnesota, United States of America, **3** Research Institute for Medicines and Pharmaceutical Sciences (iMed.UL), Faculty of Pharmacy, University of Lisbon, Lisbon, Portugal, **4** Department of Genetics, Cell Biology and Development, University of Minnesota Medical School, Minneapolis, Minnesota, United States of America

Abstract

Background: Acute kidney injury (AKI) has grave short- and long-term consequences. Often the onset of AKI is predictable, such as following surgery that compromises blood flow to the kidney. Even in such situations, present therapies cannot prevent AKI. As apoptosis is a major form of cell death following AKI, we determined the efficacy and mechanisms of action of tauroursodeoxycholic acid (TUDCA), a molecule with potent anti-apoptotic and pro-survival properties, in prevention of AKI in rat and cell culture models. TUDCA is particularly attractive from a translational standpoint, as it has a proven safety record in animals and humans.

Methodology/Principal Findings: We chose an ischemia-reperfusion model in rats to simulate AKI in native kidneys, and a human kidney cell culture model to simulate AKI associated with cryopreservation in transplanted kidneys. TUDCA significantly ameliorated AKI in the test models due to inhibition of the mitochondrial pathway of apoptosis and upregulation of survival pathways.

Conclusions: This study sets the stage for testing TUDCA in future clinical trials for prevention of AKI, an area that needs urgent attention due to lack of effective therapies.

Editor: Demetrios Vavvas, Massachusetts Eye & Ear Infirmary, Harvard Medical School, United States of America

Funding: This work was supported in part by National Institutes of Health grant R21 DK078706 to Sandeep Gupta. No additional external funding was received for this study. The funders had no role in study design, data collection and analysis, decision to publish, or preparation of the manuscript.

Competing Interests: The authors have declared that no competing interests exist.

* E-mail: gupta024@gmail.com (SG); steer001@umn.edu (CJS)

§ These authors contributed equally to this work.

Introduction

Kidneys are acutely injured when deprived of nutrients and exposed to nephrotoxins. Acute kidney injury (AKI) has reached epidemic proportions and has grave short- and long-term consequences on patient health and cost of care [1]. Even kidneys that regain normal function following AKI have persistent maladaptive alterations that may result in a higher incidence of hypertension and chronic kidney disease [2–5]. Even in situations where the onset of AKI is predictable, such as perioperative kidney injury, none of the current therapies can prevent AKI. Thus, there is a critical need to develop therapies for the prevention of AKI.

Following acute kidney injury, cells die either immediately by necrosis or over hours to days by apoptosis, or programmed cell death. Cells under stress resist death by upregulating survival pathways. AKI can be prevented under experimental conditions by upregulating survival pathways by pro-survival molecules such as Survivin [6] or by ischemic preconditioning [7,8]. Similarly, anti-apoptotic molecules have been shown to prevent AKI in animal models [9,10]. However, these experimental approaches are limited in their translational potential by toxicity. Therefore, an ideal therapy for prevention of AKI should be nontoxic, pro-survival, and anti-apoptotic.

The liver may provide clues for developing such a therapy for AKI. Liver cells are exposed to toxic compounds and have well-developed cytoprotective mechanisms. Of the known mechanisms, protection by ursodeoxycholic acid (UDCA) and its taurine conjugate, tauroursodeoxycholic acid (TUDCA), has been well studied. U/TUDCA prevent cell death by stabilizing the cell membranes, inhibiting apoptosis, and upregulating survival pathways [11–15]. Furthermore, protection by U/TUDCA extends beyond liver to other cells in the body [16–18]. For example, hibernating animals such as black bears have high blood levels of UDCA, which prevents cell death under low nutrient conditions encountered during long periods of hibernation [19,20]. In contrast, humans have very low blood levels of UDCA. Black bear bile has been used in traditional Chinese medicine for more than 3000 years; and western medicine is increasingly recognizing the therapeutic value of U/TUDCA. U/TUDCA have been used effectively for treating human liver diseases [21–26], usually at dosages up to 20 mg/kg/day orally for long periods. In experimental models of acute injury such as myocardial infarction [16], stroke [17,18], and spinal cord injury [27–29], dosages as high as 500 mg/kg/day were administered intraperitonially or intravenously as single or short period injections. Furthermore, several studies have shown U/TUDCA to be safe for animal [16,18,29,30] and human applications [22,31], making them attractive molecules from a translational standpoint.

AKI is often predictable in clinical situations such as following surgery; exposure to nephrotoxic medications; and donor nephrectomy during cryopreservation. However, none of the current therapies can prevent AKI. Our vision in planning these studies was to develop a therapy with high translational potential that can be administered for prevention of AKI. Thus in this study we tested our hypothesis that TUDCA can prevent AKI. We chose TUDCA over UDCA because of its higher solubility at physiological pH, a characteristic that permits rapid parenteral administration in high doses and avoids precipitation during cryopreservation of donor kidneys. Accordingly, in this study we determined the efficacy and mechanisms of action of TUDCA in a rat model of AKI and a human kidney cell culture model of cryopreservation injury.

Results

In vivo Experiments

Functional protection. Rats were given 400 mg/kg/day of TUDCA or equal volume of vehicle from three days before until five days following the induction of AKI. Renal function was determined by daily measurements of blood urea levels. Rats in the TUDCA group had significantly less elevation in blood urea levels on days 1 ($p<0.001$) and 2 ($p<0.01$) following AKI as compared to those in the vehicle group (Figure 1). Although on days 3–5 blood urea was lower in the TUDCA-treated rats, the difference was not statistically significant. Interestingly, the blood urea continued to decline in the TUDCA-treated group until the day of euthanasia (day 5), while it stabilized above baseline in the vehicle-treated group.

Structural protection. Kidneys were harvested five days following the induction of ischemic AKI. Many proximal tubules from the deep cortex (Figure 2A, a) had significant injury in vehicle-treated rats; in contrast, TUDCA-treated rats had minimal injury (Figure 2A, b). There was significantly less injury in the superficial ($p<0.05$) and deep ($p<0.001$) cortex in TUDCA-treated rats as compared to the vehicle-treated rats (Figure 2B). There were no differences in the medulla or papilla (data not shown).

Protection against apoptosis. As TUDCA is a potent anti-apoptotic molecule, we quantified cells undergoing apoptosis in kidney sections by the transferase mediated deoxyuridine triphosphate (dUTP)-digoxigenin nick-end labeling (TUNEL) assay (Figure 3A and B). There were significantly less TUNEL-positive cells in the superficial and deep cortices ($p<0.05$) and outer strip of the outer medulla ($p<0.001$) in the TUDCA-treated rats as compared to the vehicle-treated rats (Figure 3B). The apoptotic cells were exclusively limited to the proximal tubules, which were identified by morphology in the periodic acid Schiff stain (PAS) stained sections.

Apoptosis pathway analysis. We determined activation of the mitochondrial, death-receptor, and endoplasmic reticulum (ER)-stress pathways of apoptosis by Western blot analysis for active caspase-9, caspase-8, and caspase-12, respectively. Activation of caspase-9 was significantly inhibited by TUDCA ($p<0.01$) (Figure 4); and the results were confirmed by densitometry using β-actin as the loading control. Interestingly, there was no significant difference in the activation of caspase-8 and caspase-12 between the TUDCA- and vehicle-treated rats.

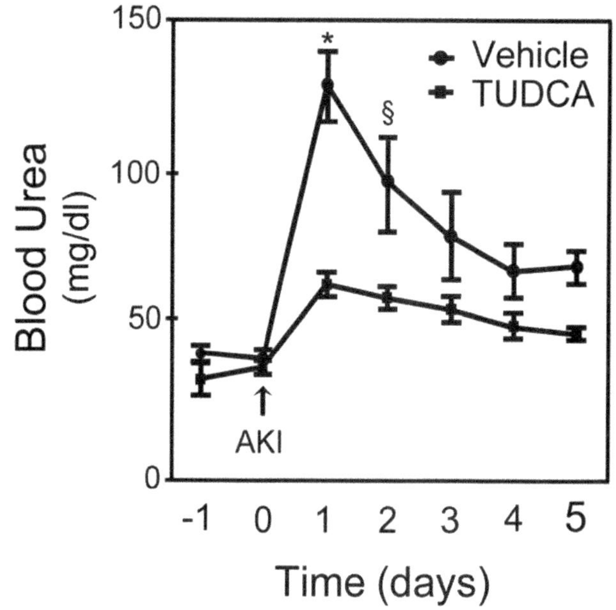

Figure 1. Functional protection against AKI by TUDCA. Daily administration of 400 mg/kg of TUDCA protected against ischemic AKI. Renal function was determined by daily measurement of blood urea. Rats that received TUDCA (squares) had significantly less elevation in blood urea levels on day 1 and day 2 following ischemia-reperfusion injury as compared to rats that received vehicle (circles) ($p<0.01$). Similarly, on days 3–5, blood urea levels continued to remain lower in the TUDCA-treated rats; however, the difference was not statistically significant. Results are expressed as mean ± standard deviation of a least 3 different animals in each group. *$p<0.001$ and §$p<0.01$ from day 0.

Survival pathway analysis. Extracellular regulated kinase 1/2 (ERK1/2), c-Jun N-terminal kinase (JNK), and p38, which are the key components of mitogen-activated protein kinase (MAPK) survival pathway, were analyzed by Western blot. There was increased activation of ERK1/2 in the TUDCA group, although without statistical significance, when compared with controls (Figure 5). No difference was detected in the amount of p-JNK and p-p38 between the two groups.

Cell Culture Experiments

Toxicity studies of TUDCA. We treated human renal proximal tubular epithelial cells (RPTE) cells with different concentrations of TUDCA, and performed cytotoxicity and viability assays. TUDCA in concentrations from 15 to 600 μM did not cause cytotoxicity; only 1200 μM of TUDCA was cytotoxic ($p<0.05$) (Figure 6A). None of the tested concentrations of TUDCA decreased cell viability (Figure 6B). We chose concentrations up to 600 μM of TUDCA for subsequent experiments.

Protection against apoptosis. We determined activation of the final common pathway of apoptosis by caspase-3 activity assay. Cryopreservation injury significantly activated caspase-3 in RPTE cells ($p<0.01$), which was significantly inhibited by 150–600 μM of TUDCA in a dose-dependent fashion ($p<0.05$) (Figure 7A).

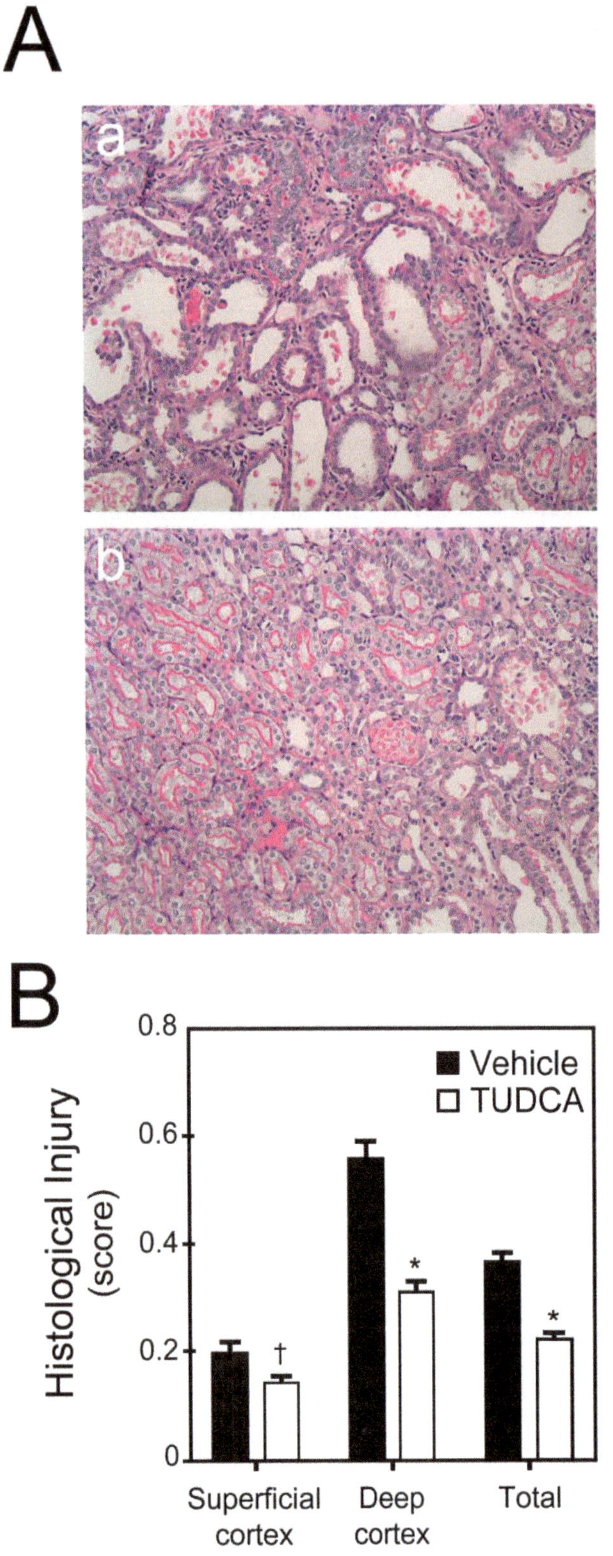

Figure 2. Protection against AKI-induced histological damage by TUDCA. A, Representative PAS stained images from deep cortex from animals that received vehicle control (a) or TUDCA (b). B, Animals that received TUDCA as compared to controls, showed significantly less damage in the deep cortex where the S3 segment is located. Results are expressed as mean ± standard deviation of a least 3 different animals in each group. *$p<0.001$ and †$p<0.05$ from vehicle-injected controls.

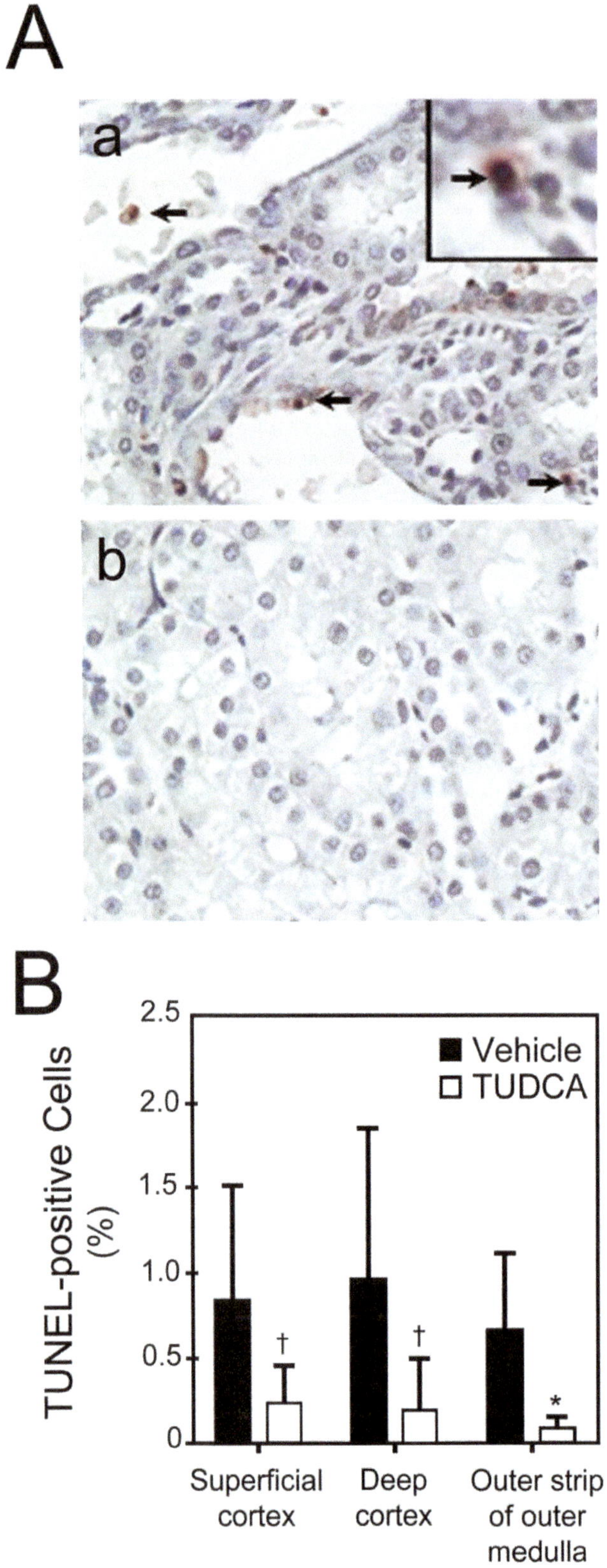

Figure 3. Protection against AKI-induced apoptosis by TUDCA. A, Representative images from cortico-medullary junction from vehicle (a; control) and TUDCA-treated (b) groups. Brown staining and arrows identify TUNEL-positive cells. B, There were significantly less TUNEL-positive cells in the TUDCA-treated group as compared to the vehicle-treated (control) group in the cortex (*p<0.05*) and outer strip of the outer medulla ($p<0.001$). Results are expressed as mean ± standard

deviation of a least 3 different animals in each group. *$p<0.001$ and †$p<0.05$ from vehicle-injected controls.

Activation of the mitochondrial pathway of apoptosis was determined via caspase-9 activity assay (Figure 7B). Cryopreservation injury consistently activated caspase-9 in RPTE cells ($p<0.01$), which was significantly inhibited by 100 and 150 μM of TUDCA ($p<0.05$). Next, we determined the activation of death receptor and ER-stress pathways of apoptosis by performing Western blot analysis for caspase-8 and procaspase-12 (Figure 7C). There was no activation of procaspase-8 and procaspase-12 in our model of cryopreservation injury, and TUDCA had no effect on the activation. Thus, TUDCA protected against cryopreservation injury by inhibiting the mitochondrial pathway of apoptosis.

Survival pathway analysis. We performed Western blot analysis for active forms of the MAPKs (ERK, JNK, and p38) to determine activation of survival pathways. Treatment with 100 and 150 μM of TUDCA activated ERK1/2 (Figure 8; $p<0.05$ and $p<0.01$, respectively); however, there was no effect of TUDCA on JNK or p38 (data not shown). Thus, upregulation of ERK1/2 by TUDCA contributed to protection against cryopreservation injury.

Discussion

This is the first study to determine the protective properties of TUDCA against AKI and sets the stage for developing relevant therapies for clinical use. We chose two models of AKI to simulate commonly encountered clinical scenarios: 1) the warm ischemia reperfusion model of AKI in rats recapitulates clinical AKI in the native kidney due to poor perfusion; and 2) the cellular model of cryopreservation injury recapitulates cryopreservation-associated AKI in the donor kidney. We preferred a cellular model to in vivo models of cryopreservation injury to prevent systemic and donor factors such as immunity, inflammation, and donor age from confounding the results.

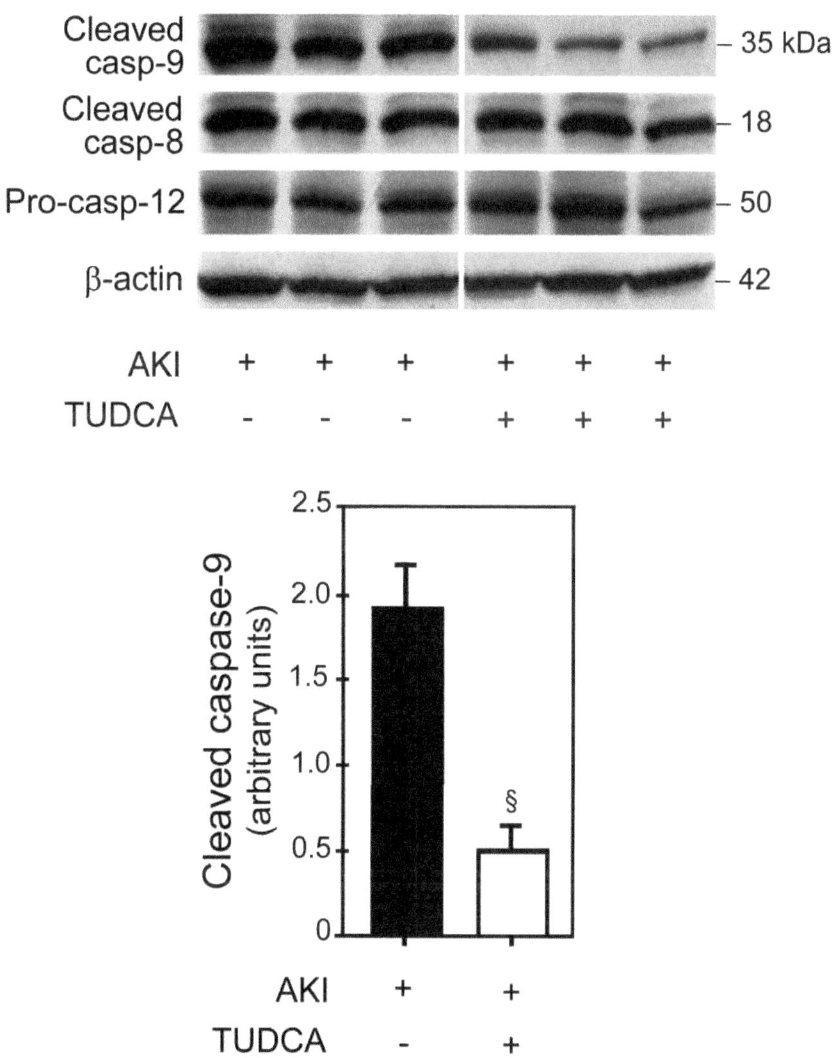

Figure 4. Apoptosis pathway analysis. TUDCA treatment significantly blocked activation of caspase-9 following AKI as compared to vehicle treatment (rats 1, 2, and 3) (top panel). There was no difference in the activation of caspase-8 and caspase-12 between the TUDCA- and vehicle-treated rats. Densitometry analysis of cleaved caspase-9 normalized for β-actin (lower panel). When densitometry results for caspase-9 were compared between the TUDCA- and vehicle-treated groups, there was significantly less ($p<0.01$) activation of caspase-9 in the TUDCA group. Results are expressed as mean ± standard deviation of a least 3 different animals in each group. §$p<0.01$ from vehicle-injected controls.

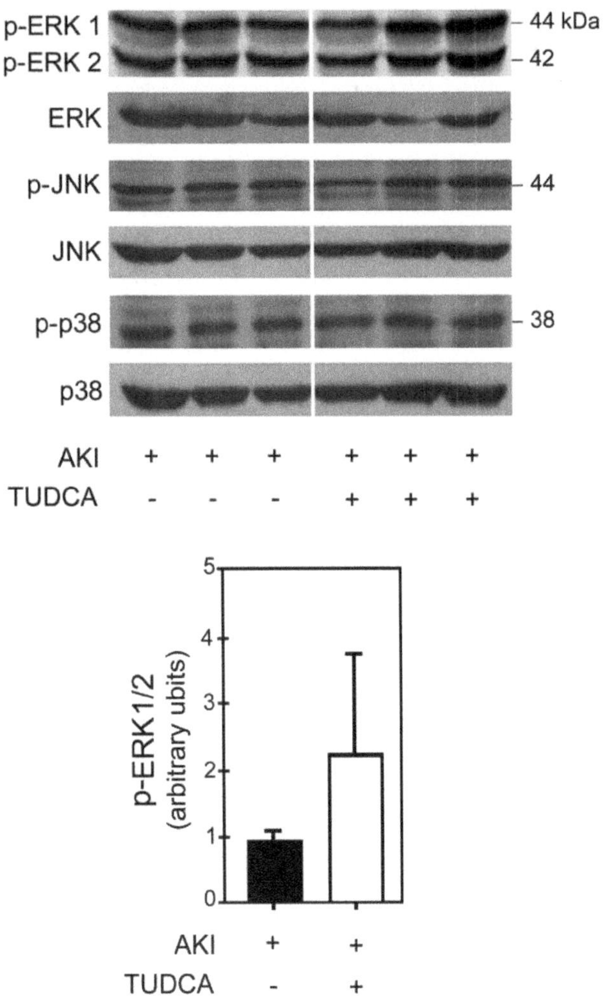

Figure 5. Survival pathway analysis. TUDCA treatment upregulated ERK1/2 following ischemia-reperfusion injury to the kidney in rats (top panel); however, there was no statistical significant difefrence as compared to the vehicle-treated rats (top panel). There was no difference in JNK and p38 proteins between the TUDCA and vehicle groups. Densitometry analysis of ERK1/2 in each rat in the TUDCA group were compared with those in the vehicle group (lower panel); the difference was not significant ($p = 0.29$). Results are expressed as mean ± standard deviation of a least 3 different animals in each group.

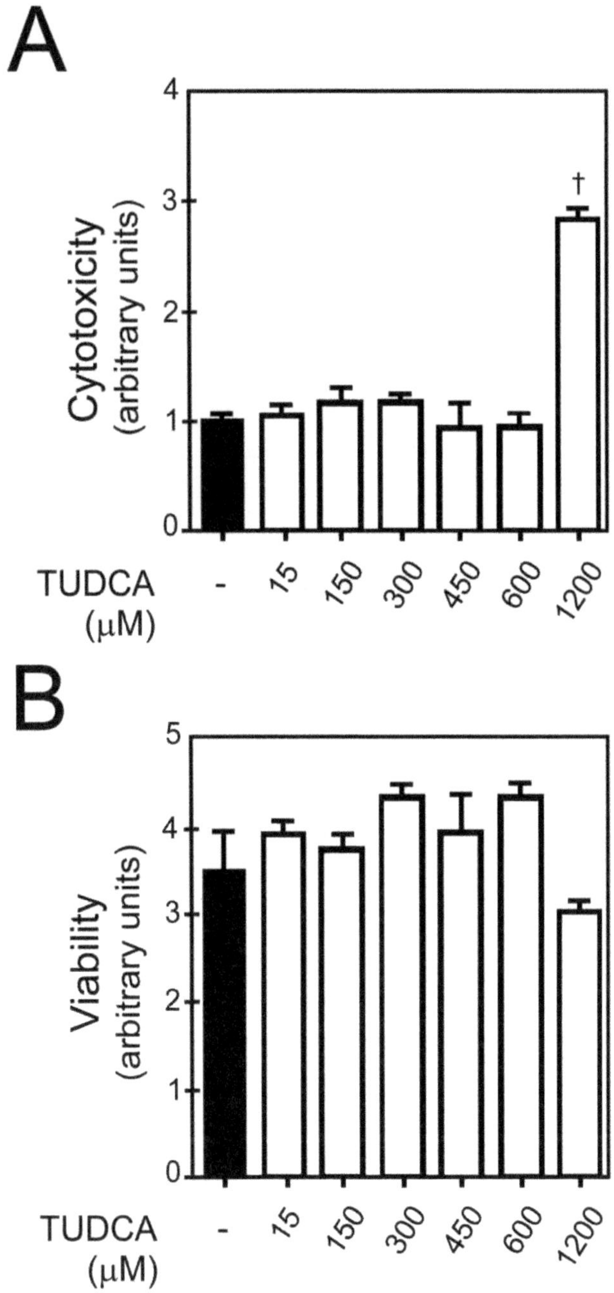

Figure 6. Cytotoxicity studies of TUDCA in primary human RPTE cells. Cells were treated with either vehicle (control) or 15 to 1200 μM of TUDCA for 24 hours. A, TUDCA was not cytotoxic in concentrations from 15 to 600 μM. Significant cytotoxicity was seen only with 1200 μM of TUDCA, as compared to the vehicle ($p<0.05$). B, TUDCA did not decrease cell viability in all the tested concentrations from 15 μM to 1200 μM. Results are expressed as mean ± standard deviation. All experiments were performed in triplicate. †$p<0.05$ from vehicle-treated control.

Warm ischemia reperfusion injury for 45 minutes at 37°C produced a very consistent injury in the rat model. The peak rise in blood urea occurred 24 hours following bilateral renal artery clamping in both the vehicle- and TUDCA-treated rats. As compared to rats treated with vehicle, rats treated with TUDCA had significantly less peak elevation in blood urea following AKI. Blood urea values subsequently declined until day 4, when they plateaued in the vehicle group, while continued to improve in the TUDCA group until the day of euthanasia. Thus, TUDCA reduced the severity of AKI without postponing its onset.

The functional protection against AKI by TUDCA was supported by less severe histological injury seen in kidneys of TUDCA-treated rats. The deep cortex, where the S3 segment of the proximal tubule is located and which sustains maximum injury following ischemia-reperfusion, was better preserved in the TUDCA group. Furthermore, like several published studies in other organs, TUDCA provided protection against apoptosis following AKI. There were significantly less TUNEL-positive cells in the superficial and deep cortices and in the outer strip of the outer medulla in the TUDCA-treated rats as compared to the vehicle-treated rats. Based on the morphological appearance, the TUNEL-positive cells were present exclusively in the proximal tubules. We did not find apoptotic cells in the distal tubules following AKI as reported by other investigators [32,33]. Thus,

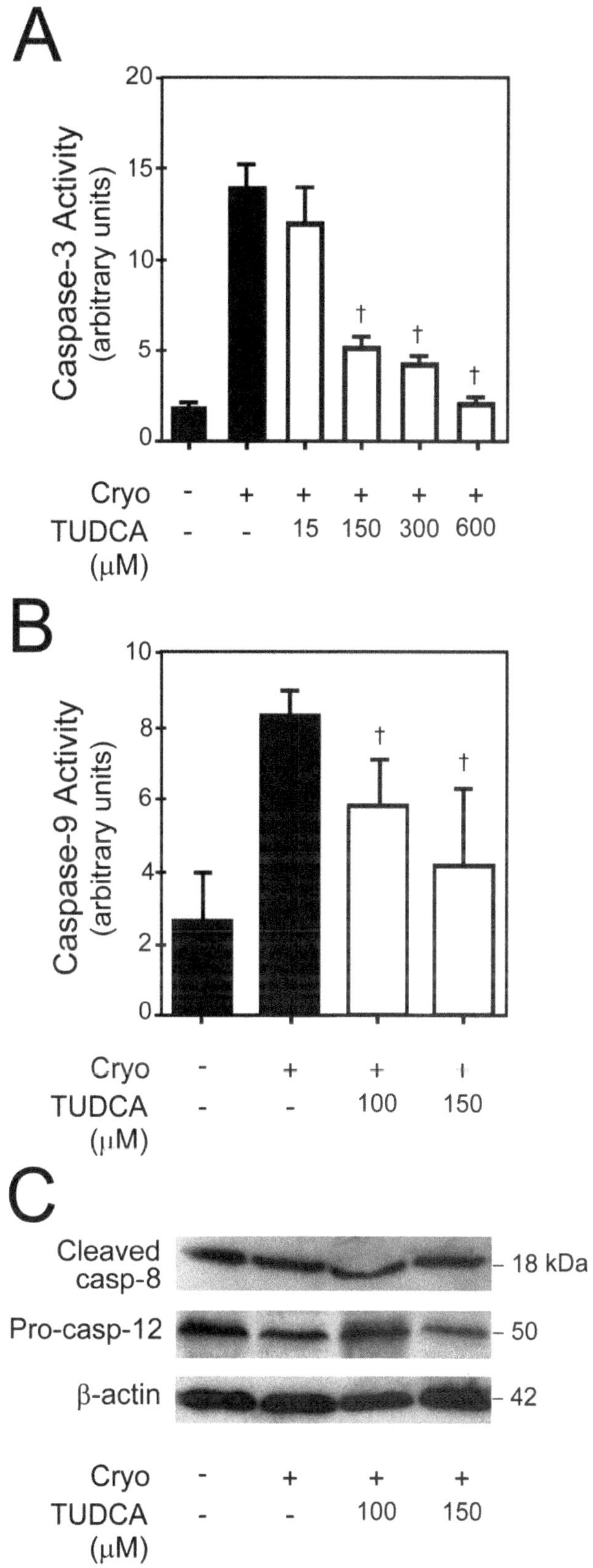

Figure 7. Caspase activation following cryoinjury and treatment with TUDCA in primary human RPTE cells. A, Caspase-3 activity following cryoinjury in RPTE cells treated with either vehicle or different concentrations of TUDCA. Caspase-3 activity in cryoinjured RPTE cells was compared with that in uninjured RPTE cells. There was signficant activation of caspase-3 following cryoinjury ($p<0.05$), which was significantly inhibited by 150–600 μM of TUDCA in a dose-dependent fashion. 15 μM of TUDCA did not inhibit activation of caspase-3 following cryoinjury. B, Caspase-9 activity following cryoinjury in RPTE cells treated with either vehicle or different concentrations of TUDCA. There was statistically significant increased caspase-9 activity in cryoinjured cells as compared to uninjured cells ($p<0.05$). Both 100 and 150 μM of TUDCA significantly decreased caspase-9 activity. C, Caspase-8 and caspase-12 analysis following cryoinjury in RPTE cells treated with either vehicle or different concentrations of TUDCA. There was no difference in the amount of caspase-8 and procaspase-12 between the uninjured and cryoinjured cells treated with vehicle or TUDCA. Results are expressed as mean ± standard deviation. All experiments were performed in triplicate. †$p<0.05$ from vehicle-treated control. Cryo, cryoinjury.

these results were consistent with the published studies on the protection by TUDCA against acute injury in animal models of stroke and spinal cord injury [18,29]. It would also be interesting to test the protective effect of TUDCA in non-hemodynamic injury-mediated AKI.

Activation of caspase-9, which represents the mitochondrial pathway of apoptosis, was significantly inhibited by TUDCA following AKI. TUDCA has been shown to inhibit the mitochondrial pathway of apoptosis in primary hepatocytes, neurons, and in animal models of ischemic injury such as stroke [15,17,18,34]. Similar to the mitochondrial pathway of apoptosis, the ER-stress pathway of apoptosis plays an important role in the pathogenesis of glomerular, tubular, and interstitial kidney diseases [35]. In particular, following ischemic kidney injury, investigators have shown activation of the ER-stress pathway of apoptosis and

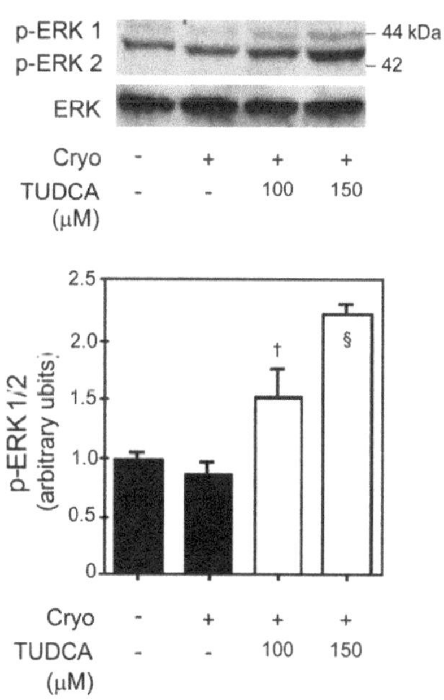

Figure 8. Survival pathway analysis following cryoinjury and treatment with TUDCA in primary human RPTE cells. Phosphorylated ERK1/2 protein in cryoinjured cells that were treated with either vehicle (control) or 100 μM or 150 μM TUDCA as compared to uninjured cells (top panel). There was no difference in the amount of phosphorylated ERK1/2 between uninjured and cryoinjured cells treated with vehicle. Densitometry analysis of phosphorylated ERK1/2 (lower panel). Results are expressed as mean ± standard deviation. All experiments were performed in triplicate. §$p<0.01$ and †$p<0.05$ from vehicle-treated control. Cryo, cryoinjury.

protection by its inhibitors [36,37]. In contrast to the mitochondrial pathway, in our model, TUDCA did not have any effect on the ER-stress and death receptor pathways of apoptosis. This is somewhat surprising in light of recent studies demonstrating the ability of TUDCA to reduce ER-stress induced caspase-12 activation [13,38–41].

MAPKs constitute important survival pathways in response to several stimuli in mammals, which include JNK, p38, and ERK. Activation of JNK and p38 has been shown to facilitate cell death, while that of ERK1/2 to promote cell survival following acute injury [42]. Furthermore, ischemic preconditioning of kidneys, which is protective against AKI, acts through activation of ERK1/2 [43]. TUDCA has been shown to inhibit apoptosis in hepatocytes and following acute stroke in rats by activating survival pathways [15,17]. In our model of AKI, TUDCA increased ERK1/2 levels in two out of three rats. TUDCA had no effect on JNK and p38 pathways. Although this finding is encouraging, it needs further confirmation before one can conclude that activation of ERK1/2 by TUDCA following AKI plays a protective role.

Similar to native kidneys, AKI in the donor kidney is a significant clinical problem. Cessation of blood supply following harvesting results in AKI. To minimize this risk, donor kidneys are currently cryopreserved in specialized solutions such as University of Wisconsin solution. Although a major advancement in the field, the current cryopreservation techniques still results in significant graft injury [44]. Cryopreservation injury to the donor kidney leads to increased incidence of delayed graft function, acute and chronic rejection, and poor short- and long-term graft outcome [45–49]. Furthermore, the current cryopreservation time in the USA has remained long at approximately 21 hours. Therefore, there is a pressing need to improve the current cryopreservation techniques. The current studies were performed to set the stage for developing improved cryopreservation solutions for clinical use.

Kidney cells die by necrosis during cryopreservation and by apoptosis during warm reperfusion following transplantation [50]. There is activation of apoptosis pathways during cryopreservation [51], and survival pathways play an important role in resisting cell death [52–54]. Furthermore, TUDCA has been shown to reduce structural injury to human livers following cryopreservation; however, in this study, the mechanism of protection was not studied [22].

In the current study, we investigated the efficacy and mechanisms of action of TUDCA against cryopreservation injury. TUDCA was not cytotoxic to RPTE cells in concentrations up to 600 μM and it did not decrease cell viability in all the tested concentrations. Although TUDCA has been shown to be safe in concentrations ranging from 100 nM up to 5 mM in cell culture experiments, most studies have studied 100 to 500 μM of TUDCA [34,38–40,55]. Our chosen concentrations of 100 and 150 μM of TUDCA were similar to those used by other investigators.

Caspase-3 is activated following cryopreservation of cells and caspase inhibitor provides protection [56]. Similar to the published literature, caspase-3 was consistently activated following cryopreservation injury to the RPTE cells, and was inhibited by TUDCA in a dose-dependent fashion. Similarly, there was activation of mitochondrial pathway of apoptosis in our model of cryopreservation injury, and 100 and 150 μM of TUDCA provided protection [57–60]. This is an advancement of the previously known anti-apoptotic properties of TUDCA in models of warm ischemia-reperfusion injury [16,61]. Unlike the mitochondrial pathway of apoptosis, there was no activation of the ER-stress and death receptor pathway of apoptosis following cryopreservation injury. Further, ERK1/2 survival pathways were activated by both 100 and 150 μM of TUDCA following cryopreservation injury to the RPTE cells, similarly to upregulation of ERK1/2 seen in two of three rats in the in vivo model of AKI. This is the first report of activation of ERK1/2 pathways by TUDCA following cryopreservation injury. Much like in the *in vivo* model, there was no effect of TUDCA on JNK and p38 pathways following cryopreservation injury.

In conclusion, TUDCA was protective in the rat model of ischemia-reperfusion induced AKI and cellular model of cryopreservation injury. It provided protection in the tested models of AKI by inhibiting the mitochondrial pathway of apoptosis and upregulating ERK1/2 survival pathways. Results of this study and a proven safety profile of TUDCA in humans will open the door for conducting human feasibility studies in patients with AKI, an important area of investigation that currently lacks effective therapy. We anticipate administration of TUDCA prior to precipitating events will prevent AKI in humans.

Materials and Methods

Ethics Statement

All animal work was approved by the Institutional Animal Care and Use Committee at the University of Minnesota (Minneapolis, MN) and received humane care according to the criteria outlined in the "Guide for the Care and Use of Laboratory Animals" prepared by the National Academy of Sciences and published by the National Institutes of Health (NIH publication 86-23; revised 1985).

Materials

Normal human RPTE cells were purchased from Lonza Inc. (Walkersville, MD). The University of Wisconsin solution was obtained from BTL solutions LLC (Columbia, SC). TUDCA and Cell Lytic MT Mammalian Tissue Lysis/Extraction Reagent were purchased from Sigma (St Louis, MO). Protease Inhibitor Cocktail and Bicinchoninic Acid Protein Assay kits were purchased from Pierce Biotechnology Inc (Rockford, IL). 12% Tris-HCl sodium dodecyl sulfate-polyacrylamide gel electrophoresis (SDS-PAGE) pre-cast gel was obtained from Bio-Rad Laboratories Inc (Hercules, CA). Amersham Hybond ECL nitrocellulose membrane, Amersham hyperfilm and peroxidase-labeled anti-mouse/rabbit IgG were purchased from GE Healthcare (Waukesha, WI). Anti-phospho-ERK1/2 antibody was purchased from New England BioLabs (Boston, MA); anti-phospho-p38 and anti-phospho-JNK from Santa Cruz Biotechnology Inc. (Santa Cruz, CA); anti-caspase-8 and anti-caspase-12 from BioVision Inc. (Mountainview, CA); anti-caspase-9 antibody from Enzo Life Sciences (Plymouth Meeting, PA); and anti-mouse β-actin from Calbiochem (Spring Valley,CA). SuperSignal West Femto Maximum Sensitivity Substrate Kit was from Thermo Fisher Scientific (Waltham, MA). MultiTox-Glo Multiplex Cytotoxicity Assay and Caspase-9 Glo Assay kits were purchased from Promega Corp. (Madison, WI). Apo-One® Homogeneous Caspase-3/7 Assay kit and Caspase-9 Assay kit were purchased from Promega (Madison, WI). Male Sprague Dawley rats were purchased from Harlan Laboratories (Indianapolis, IN). QuantiChromTM Urea Assay kit was purchased from BioAssay Systems (Hayward, CA), APO-DIRECT™ kit was purchased from BD Pharmingen (San Diego,CA). Reflex Clips Applier and Reflex 9 mm Clips were obtained from World Precision Instruments, Inc (Sarasota, FL). 4-0 absorbable sutures were purchased from Ethicon, Johnson and Johnson (Somerville, NJ).

In vivo Experiments

Rat model of AKI. All experiments were performed in accordance with the Institutional Animal Care and Use Committee. Six to eight week old Sprague-Dawley rats were anaesthetized by isoflurane gas, a midline abdominal incision was made, and bilateral renal pedicles were clamped for 45 minutes maintaining body temperature at 37°C. After removing the clamps, abdomen was closed in two layers by using 4-0 absorbable sutures and Reflex 9 mm clips. Blood samples were obtained daily by tail vein puncture. Blood urea levels were measured by the improved Jung method using the QuantiChromTM Urea Assay Kit as per manufacturer's protocol. Kidneys were harvested five days following surgery.

Administration of TUDCA. 200 mg/ml of TUDCA (Sigma) was dissolved in phosphate buffered saline at pH 7.5. 400 mg/kg of TUDCA or equal volume of vehicle was administered to rats by daily intraperitoneal injection from three days prior to surgery (day 3) to five days (day 5) following surgery. The TUDCA dose was based on previous studies [17,18] and its solubility at physiological pH.

Histology and TUNEL assay. 4% paraformaldehyde-fixed, paraffin-embedded 5 μm kidney sections were stained with PAS stain using standard methods. Histological examination was performed by a renal pathologist in a blinded fashion. Histological injury was scored based on the percentage of tubular cell necrosis, dilation, and cell detachment as per published protocol [62]. In brief, the following criteria were used: 0, no abnormality; 1+, changes affecting less than 25% of sample; 2+, changes affecting 25%–50%; 3+, changes affecting 50%–75%; and 4+, changes affecting >75% of the sample. Apoptotic cells were detected by TUNEL assay using APO-DIRECT^{TM} kit as per manufacturer's protocol. The average injury score and percentage of TUNEL-positive cells from the renal cortex (superficial and deep), medulla (outer and inner), and papilla were counted and calculated separately and averaged to obtain the total score. At least 10 fields (under X 200 magnification) were reviewed at each location.

Cell Culture Experiments

Cell culture. RPTE cells were grown in Renal Epithelial Cell Basal Medium (REBM) with full supplements at 37°C in 5% CO_2 incubator as per supplier's instructions. RPTE cells were able to proliferate for 6–8 passages under the culture conditions.

Cryopreservation injury. We have utilized a published cell culture model of cryopreservation injury [51,63]. In brief, RPTE cells were grown to 80% confluence in the complete medium containing TUDCA or vehicle. The complete medium was then replaced with UW solution containing TUDCA or vehicle. The culture plates were subsequently incubated in a temperature-regulated refrigerator at 4°C for 48 hours. To simulate warm reperfusion phase of kidney transplantation, UW solution was replaced with complete medium containing TUDCA or vehicle, and cells were cultured for an additional 24 hours at 37°C. We used 100 or 150 μM TUDCA for these experiments.

Viability and cytotoxicity assays. Cytotoxicity and viability were determined using MultiTox-Glo Multiplex Cytotoxicity Assay kit as per manufacturer's protocols. RPTE cells were seeded in 96-well culture plates at a density of 1.2×10^4 cells per well. Subsequently, TUDCA was added to the wells to achieve final concentrations of 15, 150, 300, 450, 600, and 1200 μM. Following 24 hours of culture, to determine viability, 50 μL of GF-AFC Reagent was added to each well. The plates were gently shaken and incubated at 37°C for 30 minutes in the dark. The cell viability was determined by measuring fluorescence at 400 nmEx/505 nmEm. Subsequently, to determine cytotoxicity, 50 μL of AAF-Glo Reagent was added to each well. The plates were shaken gently and incubated at room temperature for 15 minutes in the dark. Cytotoxicity was determined by measuring luminescence as per manufacturer's protocol.

Other Methods

Protein extraction. Frozen kidney tissue was ground in liquid nitrogen using a pestle and mortar. One ml of Tissue Protein Extraction Reagent with 1× protease and phosphatase inhibitor was added to the ground kidney tissue (per 30 mg of tissue) or RPTE cells (per 1.2×10^6 cells). The lysate was incubated at 4°C for 10 minutes with vigorous shaking and subsequently centrifuged at 4°C for 10 minutes at 13,000 g. The resultant supernatant was immediately frozen in liquid nitrogen and stored at −80°C until further analysis. The amount of protein present in the solution was quantified by the Bicinchoninic Acid Protein Assay kit as per manufacturer's protocol.

Western blot analysis. 20 μg of protein was mixed with SDS-PAGE sample buffer, boiled for 5 minutes, electrophoresed on a 12% Tris-HCl SDS-PAGE gel for 1 hour at 200 V, and electroblotted onto nitrocellulose membrane. The membrane was blocked for 1 hour at room temperature in 1× Tris-buffered saline with 0.1% Tween-20 at pH 7.4 containing 5% dry milk powder. To detect p-ERK1/2, p-JNK, p-p38, caspase-8, caspase-9, and caspase-12, the membranes were incubated overnight with respective primary antibodies at 4°C and appropriate secondary antibodies at room temperature for one hour. The immunoblot was detected using SuperSignal West Femto Maximum Sensitivity Substrate Kit. To verify equal loading of proteins, the membrane was stripped at room temperature with 1× Tris-buffered saline at pH 2.5 for 30 minutes and re-probed with the mouse anti-β-actin antibody and corresponding secondary antibody.

Caspase-3 activity assay. The caspase-3 activity in kidney and RPTE extracts was quantified using Apo-One® Homogeneous Caspase-3/7 Assay kit as per the manufacturer's protocol. In brief, 100 μg of protein in 100 μL of lysis buffer was added to 100 μL of the assay buffer containing non-fluorescent caspase-3 substrate, bis-N-CBZL-aspartyl-L-glutamyl-L-valyl-L-aspartic acid amide (Z-DEVD-R110). The mixture was then incubated for 1 hour at 30°C during which Z-DEVD-R110 was converted into a fluorescent substrate by the active caspase-3 enzyme. Fluorescence was measured by Spectramax M12 fluorescent plate reader (Molecular Devices) using wavelengths of 485 nmEx/520 nmEm. The fluorescent signal was expressed in relative fluorescent units.

Caspase-9 assay. The caspase-9 activity was measured in RPTE cells using Caspase-9 Assay kit. RPTE cells were suspended in 50 μL of chilled Cell Lysis Buffer and incubated on ice for 10 minutes. Subsequently, 50 μL of 2× Reaction Buffer and 5 μL of 1 mM LEHD-AFC substrate was added to each samples and the mixture was incubated at 37°C for 2 hours. The caspase-9 activity was quantified by measuring luminescence.

Statistical Analysis

Data were expressed as mean and standard deviation unless otherwise stated. The differences between normally distributed data were analyzed by independent Student's T-test. Nonparametric unpaired Mann-Whitney test was used if the data was not normally distributed. Multiple group comparisons were performed using ANOVA with post-test according to Bonferroni. A p value of less than 0.05 was considered statistically significant.

Author Contributions

Conceived and designed the experiments: SG CJS. Performed the experiments: SG SL MJA KN LW BN. Analyzed the data: SG SL MJA KN LW BN. Wrote the paper: SG SL. Read and edited the manuscript: TK CMPR CJS.

References

1. Soni SS, Ronco C, Katz N, Cruz DN (2009) Early diagnosis of acute kidney injury: the promise of novel biomarkers. Blood Purif 28: 165–174.
2. Ishani A, Xue JL, Himmelfarb J, Eggers PW, Kimmel PL, et al. (2009) Acute kidney injury increases risk of ESRD among elderly. J Am Soc Nephrol 20: 223–228.
3. Finkenstaedt JT, Merrill JP (1956) Renal function after recovery from acute renal failure. N Engl J Med 254: 1023–1026.
4. Lowe KG (1952) The late prognosis in acute tubular necrosis; an interim follow-up report on 14 patients. Lancet 1: 1086–1088.
5. Venkatachalam MA, Griffin KA, Lan R, Geng H, Saikumar P, et al. (2010) Acute kidney injury: a springboard for progression in chronic kidney disease. Am J Physiol Renal Physiol 298: F1078–1094.
6. Kindt N, Menzebach A, Van de Wouwer M, Betz I, De Vriese A, et al. (2008) Protective role of the inhibitor of apoptosis protein, survivin, in toxin-induced acute renal failure. FASEB J 22: 510–521.
7. Hausenloy DJ, Yellon DM (2006) Survival kinases in ischemic preconditioning and postconditioning. Cardiovasc Res 70: 240–253.
8. Park KM (2001) Prevention of kidney ischemia/reperfusion-induced functional injury and jnk, p38, and MAPK kinase activation by remote ischemic pretreatment. J Biol Chem 276: 11870–11876.
9. Chien CT, Shyue SK, Lai MK (2007) Bcl-xL augmentation potentially reduces ischemia/reperfusion induced proximal and distal tubular apoptosis and autophagy. Transplantation 84: 1183–1190.
10. Ortiz A, Justo P, Catalan MP, Sanz AB, Lorz C, et al. (2002) Apoptotic cell death in renal injury: The rationale for intervention. Curr Drug Targets Immune Endocr Metabol Disord 2: 181–192.
11. Rodrigues CM, Ma X, Linehan-Stieers C, Fan G, Kren BT, et al. (1999) Ursodeoxycholic acid prevents cytochrome c release in apoptosis by inhibiting mitochondrial membrane depolarization and channel formation. Cell Death Differ 6: 842–854.
12. Rodrigues CM, Fan G, Ma X, Kren BT, Steer CJ (1998) A novel role for ursodeoxycholic acid in inhibiting apoptosis by modulating mitochondrial membrane perturbation. J Clin Invest 101: 2790–2799.
13. Xie Q, Khaoustov VI, Chung CC, Sohn J, Krishnan B, et al. (2002) Effect of tauroursodeoxycholic acid on endoplasmic reticulum stress-induced caspase-12 activation. Hepatology 36: 592–601.
14. Azzaroli F, Mehal W, Soroka CJ, Wang L, Lee J, et al. (2002) Ursodeoxycholic acid diminishes fas-ligand-induced apoptosis in mouse hepatocytes. Hepatology 36: 49–54.
15. Schoemaker MH, Conde de la Rosa L, Buist-Homan M, Vrenken TE, Havinga R, et al. (2004) Tauroursodeoxycholic acid protects rat hepatocytes from bile acid-induced apoptosis via activation of survival pathways. Hepatology 39: 1563–1573.
16. Rivard AL, Steer CJ, Kren BT, Rodrigues CM, Castro RE, et al. (2007) Administration of tauroursodeoxycholic acid (TUDCA) reduces apoptosis following myocardial infarction in rat. Am J Chin Med 35: 279–295.
17. Rodrigues CM, Sola S, Nan Z, Castro RE, Ribeiro PS, et al. (2003) Tauroursodeoxycholic acid reduces apoptosis and protects against neurological injury after acute hemorrhagic stroke in rats. Proc Natl Acad Sci U S A 100: 6087–6092.
18. Rodrigues CM, Spellman SR, Sola S, Grande AW, Linehan-Stieers C, et al. (2002) Neuroprotection by a bile acid in an acute stroke model in the rat. J Cereb Blood Flow Metab 22: 463–471.
19. Hagey LR, Crombie DL, Espinosa E, Carey MC, Igimi H, et al. (1993) Ursodeoxycholic acid in the Ursidae: Biliary bile acids of bears, pandas, and related carnivores. J Lipid Res 34: 1911–1917.
20. Sola S, Garshelis DL, Amaral JD, Noyce KV, Coy PL, et al. (2006) Plasma levels of ursodeoxycholic acid in black bears, Ursus Americanus: Seasonal changes. Comp Biochem Physiol C Toxicol Pharmacol 143: 204–208.
21. Arat M, Idilman R, Soydan EA, Soykan I, Erden E, et al. (2005) Ursodeoxycholic acid treatment in isolated chronic graft-vs-host disease of the liver. Clin Transplant 19: 798–803.
22. Falasca L, Tisone G, Palmieri G, Anselmo A, Di Paolo D, et al. (2001) Protective role of tauroursodeoxycholate during harvesting and cold storage of human liver: A pilot study in transplant recipients. Transplantation 71: 1268–1276.
23. Festi D, Montagnani M, Azzaroli F, Lodato F, Mazzella G, et al. (2007) Clinical efficacy and effectiveness of ursodeoxycholic acid in cholestatic liver diseases. Curr Clin Pharmacol 2: 155–177.
24. Kuiper EM, Hansen BE, de Vries RA, den Ouden-Muller JW, van Ditzhuijsen TJ, et al. (2009) Improved prognosis of patients with primary biliary cirrhosis that have a biochemical response to ursodeoxycholic acid. Gastroenterology 136: 1281–1287.
25. Omata M, Yoshida H, Toyota J, Tomita E, Nishiguchi S, et al. (2007) A large-scale, multicentre, double-blind trial of ursodeoxycholic acid in patients with chronic hepatitis C. Gut 56: 1747–1753.
26. Paumgartner G, Beuers U (2002) Ursodeoxycholic acid in cholestatic liver disease: Mechanisms of action and therapeutic use revisited. Hepatology 36: 525–531.
27. Sola S, Aranha MM, Steer CJ, Rodrigues CM (2007) Game and players: Mitochondrial apoptosis and the therapeutic potential of ursodeoxycholic acid. Curr Issues Mol Biol 9: 123–138.
28. Amaral JD, Viana RJS, Ramalho RM, Steer CJ, Rodrigues CMP (2009) Bile acids: Regulation of apoptosis by ursodeoxycholic acid. J Lipid Res 50: 1721–1734.
29. Colak A, Kelten B, Sagmanligil A, Akdemir O, Karaoglan A, et al. (2008) Tauroursodeoxycholic acid and secondary damage after spinal cord injury in rats. J Clin Neurosci 15: 665–671.
30. Duan WM, Rodrigues CM, Zhao LR, Steer CJ, Low WC (2002) Tauroursodeoxycholic acid improves the survival and function of nigral transplants in a rat model of parkinson's disease. Cell Transplant 11: 195–205.
31. Kars M, Yang L, Gregor MF, Mohammed BS, Pietka TA, et al. (2010) Tauroursodeoxycholic acid may improve liver and muscle but not adipose tissue insulin sensitivity in obese men and women. Diabetes 59: 1899–1905.
32. Hall AM, Unwin RJ, Parker N, Duchen MR (2009) Multiphoton imaging reveals differences in mitochondrial function between nephron segments. J Am Soc Nephrol 20: 1293–1302.
33. Beeri R, Symon Z, Brezis M, Ben-Sasson SA, Baehr PH, et al. (1995) Rapid DNA fragmentation from hypoxia along the thick ascending limb of rat kidneys. Kidney Int 47: 1806–1810.
34. Rodrigues CM, Stieers CL, Keene CD, Ma X, Kren BT, et al. (2000) Tauroursodeoxycholic acid partially prevents apoptosis induced by 3-nitropropionic acid: Evidence for a mitochondrial pathway independent of the permeability transition. J Neurochem 75: 2368–2379.
35. Inagi R (2009) Endoplasmic reticulum stress in the kidney as a novel mediator of kidney injury. Nephron Exp Nephrol 112: e1–9.
36. Bailly-Maitre B, Fondevila C, Kaldas F, Droin N, Luciano F, et al. (2006) Cytoprotective gene bi-1 is required for intrinsic protection from endoplasmic reticulum stress and ischemia-reperfusion injury. Proc Natl Acad Sci U S A 103: 2809–2814.
37. Bando Y, Tsukamoto Y, Katayama T, Ozawa K, Kitao Y, et al. (2004) Orp150/hsp12a protects renal tubular epithelium from ischemia-induced cell death. FASEB J 18: 1401–1403.
38. Ozcan U, Yilmaz E, Ozcan L, Furuhashi M, Vaillancourt E, et al. (2006) Chemical chaperones reduce ER stress and restore glucose homeostasis in a mouse model of type 2 diabetes. Science 313: 1137–1140.
39. Lee YY, Hong SH, Lee YJ, Chung SS, Jung HS, et al. (2010) Tauroursodeoxycholate (TUDCA), chemical chaperone, enhances function of islets by reducing ER stress. Biochem Biophys Res Commun 397: 735–739.
40. Chen Y, Liu CP, Xu KF, Mao XD, Lu YB, et al. (2008) Effect of taurine-conjugated ursodeoxycholic acid on endoplasmic reticulum stress and apoptosis induced by advanced glycation end products in cultured mouse podocytes. Am J Nephrol 28: 1014–1022.
41. Malo A, Krüger B, Seyhun E, Schäfer C, Hoffmann RT, et al. (2010) Tauroursodeoxycholic acid reduces endoplasmic reticulum stress, trypsin activation, and acinar cell apoptosis while increasing secretion in rat pancreatic acini. Am J Physiol Gastrointest Liver Physiol 299: G877–886.
42. Xia Z, Dickens M, Raingeaud J, Davis RJ, Greenberg ME (1995) Opposing effects of ERK and JNK-p38 MAP kinases on apoptosis. Science 270: 1326–1331.
43. Park KM, Chen A, Bonventre JV (2001) Prevention of kidney ischemia/reperfusion-induced functional injury and JNK, p38, and MAPK kinase activation by remote ischemic pretreatment. J Biol Chem 276: 11870–11876.
44. Quiroga I, McShane P, Koo DD, Gray D, Friend PJ, et al. (2006) Major effects of delayed graft function and cold ischaemia time on renal allograft survival. Nephrol Dial Transplant 21: 1689—1696.
45. Bronzatto EJ, da Silva Quadros KR, Santos RL, Alves-Filho G, Mazzali M (2009) Delayed graft function in renal transplant recipients: Risk factors and impact on 1-year graft function: A single center analysis. Transplant Proc 41: 849–851.
46. Jushinskis J, Trushkov S, Bicans J, Suhorukov V, Shevelev V, et al. (2009) Risk factors for the development of delayed graft function in deceased donor renal transplants. Transplant Proc 41: 746–748.
47. Mikhalski D, Wissing KM, Ghisdal L, Broeders N, Touly M, et al. (2008) Cold ischemia is a major determinant of acute rejection and renal graft survival in the modern era of immunosuppression. Transplantation 85: S3–9.
48. Ojo AO, Wolfe RA, Held PJ, Port FK, Schmouder RL (1997) Delayed graft function: Risk factors and implications for renal allograft survival. Transplantation 63: 968–974.
49. Salahudeen AK, Haider N, May W (2004) Cold ischemia and the reduced long-term survival of cadaveric renal allografts. Kidney Int 65: 713–718.
50. Burns AT, Davies DR, McLaren AJ, Cerundolo L, Morris PJ, et al. (1998) Apoptosis in ischemia/reperfusion injury of human renal allografts. Transplantation 66: 872–876.
51. Salahudeen AK, Huang H, Joshi M, Moore NA, Jenkins JK (2003) Involvement of the mitochondrial pathway in cold storage and rewarming-associated apoptosis of human renal proximal tubular cells. Am J Transplant 3: 273–280.
52. Omori K, Valiente L, Orr C, Rawson J, Ferreri K, et al. (2007) Improvement of human islet cryopreservation by a p38 MAPK inhibitor. Am J Transplant 7: 1224–1232.
53. Omori K, Valiente L, Orr C, Rawson J, Ferreri K, et al. (2005) Inhibition of p38 mitogen-activated protein kinase protects human islets from cryoinjury and improves the yield, viability, and quality of frozen-thawed islets. Transplant Proc 37: 3422–3423.

54. Yoshinari D, Takeyoshi I, Kobayashi M, Koyama T, Iijima K, et al. (2001) Effects of a p38 mitogen-activated protein kinase inhibitor as an additive to University of Wisconsin solution on reperfusion injury in liver transplantation. Transplantation 72: 22–27.
55. Malo A, Kruger B, Seyhun E, Schafer C, Hoffmann RT, et al. (2010) Tauroursodeoxycholic acid reduces endoplasmic reticulum stress, trypsin activation, and acinar cell apoptosis while increasing secretion in rat pancreatic acini. Am J Physiol 299: G877–886.
56. Stroh C, Cassens U, Samraj AK, Sibrowski W, Schulze-Osthoff K, et al. (2002) The role of caspases in cryoinjury: Caspase inhibition strongly improves the recovery of cryopreserved hematopoietic and other cells. FASEB J 16: 1651–1653.
57. Anderson CD, Belous A, Pierce J, Nicoud IB, Knox C, et al. (2004) Mitochondrial calcium uptake regulates cold preservation-induced Bax translocation and early reperfusion apoptosis. Am J Transplant 4: 352–362.
58. Huang J, Nakamura K, Ito Y, Uzuka T, Morikawa M, et al. (2005) Bcl-xL gene transfer inhibits Bax translocation and prolongs cardiac cold preservation time in rats. Circulation 112: 76–83.
59. Kuznetsov AV, Schneeberger S, Seiler R, Brandacher G, Mark W, et al. (2004) Mitochondrial defects and heterogeneous cytochrome c release after cardiac cold ischemia and reperfusion. Am J Physiol Heart Circ Physiol 286: H1633–1641.
60. Yagi T, Hardin JA, Valenzuela YM, Miyoshi H, Gores GJ, et al. (2001) Caspase inhibition reduces apoptotic death of cryopreserved porcine hepatocytes. Hepatology 33: 1432–1440.
61. Ramalho RM, Viana RJ, Castro RE, Steer CJ, Low WC, et al. (2008) Apoptosis in transgenic mice expressing the p301l mutated form of human tau. Mol Med 14: 309–317.
62. Gupta S, Li S, Abedin MJ, Wang L, Schneider E, et al. (2010) Effect of notch activation on the regenerative response to acute renal failure. Am J Physiol Renal Physiol 298: F209–215.
63. Salahudeen AK, Joshi M, Jenkins JK (2001) Apoptosis versus necrosis during cold storage and rewarming of human renal proximal tubular cells. Transplantation 72: 798–804.

Survival, Growth and Reproduction of Cryopreserved Larvae from a Marine Invertebrate, the Pacific Oyster (*Crassostrea gigas*)

Marc Suquet[1]*, Catherine Labbé[2], Sophie Puyo[3], Christian Mingant[1], Benjamin Quittet[3], Myrina Boulais[1], Isabelle Queau[1], Dominique Ratiskol[1], Blandine Diss[4], Pierrick Haffray[3]

1 Ifremer, UMR 6539, PFOM Department, Station Expérimentale d'Argenton, Argenton, France, **2** INRA, UR 1037, LPGP, Rennes, France, **3** SYSAAF, LPGP, Rennes, France, **4** Satmar, Barfleur, France

Abstract

This study is the first demonstration of successful post-thawing development to reproduction stage of diploid cryopreserved larvae in an aquatic invertebrate. Survival, growth and reproductive performances were studied in juvenile and adult Pacific oysters grown from cryopreserved embryos. Cryopreservation was performed at three early stages: trochophore (13±2 hours post fertilization: hpf), early D-larvae (24±2 hpf) and late D-larvae (43±2 hpf). From the beginning (88 days) at the end of the ongrowing phase (195 days), no mortality was recorded and mean body weights did not differ between the thawed oysters and the control. At the end of the growing-out phase (982 days), survival of the oysters cryopreserved at 13±2 hpf and at 43±2 hpf was significantly higher ($P<0.001$) than those of the control (non cryopreserved larvae). Only the batches cryopreserved at 24±2 hpf showed lower survival than the control. Reproductive integrity of the mature oysters, formely cryopreserved at 13±2 hpf and 24±2 hpf, was estimated by the sperm movement and the larval development of their offspring in 13 crosses gamete pools (five males and five females in each pool). In all but two crosses out of 13 tested ($P<0.001$), development rates of the offspring were not significantly different between frozen and unfrozen parents. In all, the growth and reproductive performances of oysters formerly cryopreserved at larval stages are close to those of controls. Furthermore, these performances did not differ between the three initial larval stages of cryopreservation. The utility of larvae cryopreservation is discussed and compared with the cryopreservation of gametes as a technique for selection programs and shellfish cryobanking.

Editor: Tilmann Harder, University of New South Wales, Australia

Funding: The present research was supported by the national project CRECHE (Ofimer 136/08/C) and CRYOAQUA (GIS IBISA) and by the European Union (FEP 30906-2009). The funders had no role in study design, data collection and analysis, decision to publish, or preparation of the manuscript.

Competing Interests: The private hatchery, Satmar, declares its affiliation in the competing interests section of the online manuscript form, along with any other relevant declarations relating to employment, consultancy, patents, products in development or marketed products.

* E-mail: marc.suquet@ifremer.fr

Introduction

Shellfish farming worldwide is mostly based on wild spat. The development of hatcheries combined with the recent adoption of genetic technologies such as selection or polyploidization (see for review [1], [2]) has created new genotypes with higher genetic value than wild stocks. The mid- and long-term cryopreservation of these new genetic resources has become of prime interest for research, genetic improvement (selection, hybridization or polyploidization) and for restoration programs for endangered species or populations. Obtaining optimized rearing potential for thawed embryos or larvae would result in an extension of the type of material that can be preserved in cryobanks [3].

Since the pioneering work carried out by Renard [4], most publications on the cryopreservation of shellfish embryos or larvae have been focused on the definition of protocols (cooling rate, type and concentration of cryoprotectant, embryo concentration in straws, etc…) and they have rarely reported good survival of thawed individuals or good performances after their settlement. In blue mussel (*Mytilus galloprovincialis*), the survival of 21-day old larvae after thawing was 12.5% of the control group [5]. Only 2.8% of the frozen/thawed Greenshell mussel (*Perna canaliculus*) trochophores survived up to competent pediveligers and their size was lower than that observed on unfrozen larvae [6].

In Pacific oysters, only one thawed larva succeeded in settling after 29 days of rearing [7]. During the growing-out phase, the survival of 850 four month old Eastern oyster (*Crassostrea virginica*), produced from thawed trochophore larvae, was not different from the control group [8]. However, this work did not provide information on later stages of development that could have validated larval cryopreservation as an effective tool for the maintenance of genetic resources. The recent improvement in survival reported from Pacific oyster larvae frozen after the trochophore stage [9] provided an opportunity to compare performances of one full generation in order to assess the feasibility of larval cryopreservation for genetic resources management purposes.

The aim of the present work was to describe the long-term effects of cryopreservation on the subsequent performances of Pacific oysters. The survival and growth rate of oysters which were cryopreserved at larval stages were assessed during the ongrowing and growing-out phases. Furthermore, the reproductive performances of these oysters were estimated up to the D-larval stage of their progenies.

Materials and Methods

Obtaining thawed larvae

In June 2009, oyster embryos were obtained from crosses made at the Ifremer experimental hatchery in Argenton (Northern Brittany, France; no specific permissions were required): three females were stripped during the natural spawning period. For each female 500 000 oocytes were fertilized in 300 ml seawater with a pool of sperm collected from three males (1.3 10^6 spermatozoa ml^{-1}) [10]. For each female, the embryos were incubated in 5L beakers (30 000 embryos L^{-1}, 20°C). They were then pooled and batches were cryopreserved at 13±2 hours post fertilization: hpf (trochophore stage), 24±2 hpf (early D-larval stage) or 43±2 hpf (late D-larval stage), according to a protocol previously published [11] modified as follows: briefly, the larvae at the required stage were filtered on 20 µm mesh and diluted at a 1:1 volume ratio in cryoprotectant (10% ethylene glycol with 1% PVP and 200 mM trehalose in bi-distilled water, final concentration). The larvae were frozen in 0.5 ml straws (23 000 larvae per straw, 18 straws per larval stage), using a Kryo 10 (Planer, Sunbury, U.K.) The freezing curve was: -1°C min^{-1} from 0 to -10°C, hold for 5 min at -10°C and then, -0.3°C min^{-1} from -10 to -35°C. Finally, the straws were plunged into liquid nitrogen. In September 2009, three months after cryopreservation, the straws were thawed in a water bath (37°C, 10 s) and the larvae were reared in a 5 L flow through culture system for 19 days (24°C), according to a method previously developed by Rico-Villa et al. [12]. A control group was created using a pool of oocytes and spermatozoa collected from three females and three males. At the end of the larval rearing phase which lasted 21 days, the survival was: 28.9±25.8% in the fresh control, 0.1±0.0% in the 13±2 hpf batch, 0.9±0.7% in the 24±2 hpf batch and 0.4±0.1% in the 43±2 hpf batch. Spat (juvenile oysters) were then filtered on 180 µm mesh, transferred to mesh bottomed sieve tanks (45×20×6 cm, mesh size: 150 µm) and maintained in a race way (flow rate: 12 L h^{-1} $tank^{-1}$, seawater 24°C). They were fed daily with a mixture of two micro-algae (*Isochrysis galbana* and *Chaetoceros calcitrans*: at 65% and 35% respectively, maintaining a volume of 1 500 $µm^3$ algae $µl^{-1}$ seawater) for a 2 month period (November-December 2009).

Ongrowing phase (this phase refers to 1 to 4 g oysters and aged of 88 to 195 days: the age is given without taking into account the period during which larvae were maintained in liquid nitrogen)

After filtration on a 4 mm mesh, spat were maintained in the same mesh bottom sieve tanks and reared according to the conditions described above. Spat number in each tank and the number of tanks were adapted to the number of spat available in each group, this number being related to the initial spat survival for each treatment (control: 280 spat/3 tanks; oysters produced from larvae cryopreserved at 13±2 hpf: 110 spat/1 tank, 24±2 hpf: 290 spat/3 tanks and 43±2 hpf: 245 spat/1 tank). Spat were counted and weighed individually (n = 30 per batch) at the end of the 4-month period (January to April 2010). The survival was calculated from the number of spat at the beginning of this ongrowing phase.

Growing-out phase (4–80 g, 196–982 days)

The spat were then transferred to the natural site of Aber Benoit (North Brittany, France) where water temperature ranges from 9 to 17°C (The owner of the site gave permissions to conduct the field study and this study did not involve endangered or protected species). Spat were maintained in oyster bags (90×45×6 cm, mesh size 1.5 cm), spat number and number of bags were the same as those of the tanks in the ongrowing-phase. Oyster survival and mean weight (n = 30 per batch) were assessed at the end of the 26 month rearing period (April 2010-June 2012).

Reproduction phase (>80 g, >982 days)

In June 2012, adult oysters were transferred from the Aber Benoit rearing site to the Ifremer experimental hatchery in Argenton, during the natural spawning period and their reproductive performances were assessed. Five groups of parents were created using five males and five females in each case: one group was the control made with the oysters which had never been frozen, two groups were composed of oysters formerly cryopreserved at 13±2 hpf (13.1 and 13.2) and the two last groups contained oysters produced from larvae cryopreserved at 24±2 hpf (24.1 and 24.2).

Gametes from each group were individually collected and pooled according to previously published techniques [10]. Therefore, each group produced one pool of sperm (from the five males in the group), and one pool of eggs (from the five females in the group). Sperm movement characteristics were assessed for each of the five pools of five males (13.1, 13.2, 24.1, 24.2 and control). Sperm motility was estimated using a two-step dilution procedure: firstly, 10 µL sperm (sperm concentration: 10^9 to 10^{10} sperm ml^{-1}) were diluted in 90 µL SW (seawater, salinity 35‰ +5 g L^{-1} BSA, 19°C), secondly, 2 µL of this first suspension was diluted again in 100 µL SW (final dilution 1:500). Sperm samples of 7 µL were transferred to a Thomas cell and the movement characteristics were observed under a phase contrast microscope (Olympus BX51, ×20 magnification), connected to a camera (Qicam Fast 1394). The percentage of motile spermatozoa and their VAP (Velocity of the Average Path) were assessed using a CASA plug-in developed for the Image J software [13]. Calibration settings of the software were as follows, frame rate: 25 frame sec^{-1}; sperm size range: 0.9 to 7.5 µm; minimum VAP for motile sperm: 10 µm sec^{-1}; minimum track length: 6 frames; minimum number of sperm observed: 30.

The fertilization capacity of the four sperm pools (13.1, 13.2, 24.1, 24.2) was then estimated in triplicate (50 000 oocytes for each pool; fertilization volume 100 ml) [10]. Firstly, each sperm pool was used to fertilize a pool of oocytes collected from the control group (four crosses). Secondly, the sperm was used to fertilize the oocytes of its respective female groups (four crosses). Because sperm quality was being studied, a limiting sperm to oocyte ratio of 300, corresponding to 150 000 sperm ml^{-1}, was used. Furthermore, oocyte quality was estimated using the four oocyte pools (13.1, 13.2, 24.1, 24.2) and a pool of fresh sperm collected from the control group (four crosses). The pools of oocytes were fertilized using the method described above (except that a non limiting sperm to oocyte ratio of 800 was used to study oocyte quality, corresponding to 400 000 spermatoza ml^{-1}). Then, a control (male control x female control) was added. For both sperm and oocyte quality studies, the number of D-larvae was counted 48 hours post fertilization (3×100 µL samples) for the 13 crosses carried out. The D-larval survival was calculated as a percentage of the oocyte number used for fertilization.

Statistical analysis

Data are presented as mean ± standard deviation. Percentages were arcsin square root transformed before analysis. Means were compared using one way ANOVA. When differences were significant ($P<0.05$), a Tukey a posteriori test was used for mean comparison.

Results

At the end of the ongrowing phase, no mortality was recorded in the four different oyster batches (control, oysters produced from larvae cryopreserved at 13±2 hpf, 24±2 hpf and 43±2 hpf). Furthermore, the mean weights of the four oyster batches were not significantly different (Figure 1).

At the end of the growing-out phase, the survival of two cryopreserved batches (13±2 hpf and 43±2 hpf) was significantly higher than the control (Figure 2). Only the 24±2 hpf had survival lower than the fresh control ($P<0.001$). The mean weight of the cryopreserved batches was either higher (13±2 hpf and 24±2 hpf; $P<0.05$) or similar (43±2 hpf) to that of the control.

During the reproductive phase, the percentage of motile spermatozoa observed for pool 13.2 (cryopreservation 13±2 hpf of the gamete pool n°2) was significantly higher ($P<0.001$) than that observed for the four other pools (control, 13.1, 24.1 and 24.2; Figure 3). A percentage of motile spermatozoa higher than 70% was recorded for four of the pools, but not for pool 24.2. This observation highlights the good quality of the sperm produced in four pools out of five. The VAP observed for the pool 24.2 was significantly higher ($P<0.001$) than that of the control. Furthermore, the VAP observed for the pools 13.2 and 24.1 was lower than that recorded for the control. In all, the VAP recorded for three (24.2, 13.1 and 13.2) out of the four sperm pools from thawed oysters was never below 70% of the VAP value recorded for the control.

Larges variations in the D-larval survival (1.5 to 49.6%; Figure 4) were observed depending on the crossing but they were not significantly different from the control (male control x female control), except for the progenies issued from control x 13.2 and 13.2×13.2 ($P<0.001$).

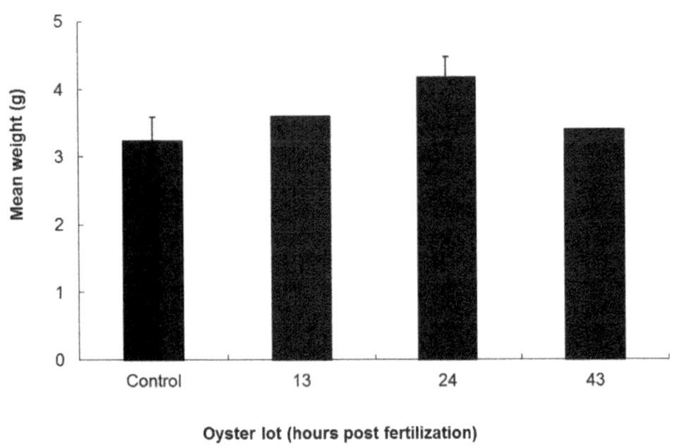

Figure 1. Effect of larvae cryopreservation at different stages post fertilization on the subsequent growth rate of oysters at the end of the ongrowing phase (195 days; spat number in each tank/number of tanks, control: 280/3, 13±2 hpf batch: 110/1, 24±2 hpf batch: 290/3, 43±2 hpf batch: 245/1).

Discussion

This study reports survival and growth to reproduction at 2.7 years of age of thawed Pacific oyster larvae previously cryopreserved at three different larval stages. This is the first report of the successful development of cryopreserved individuals to reproduction in molluscs and the second report of viable progenies produced from cryopreserved parents in aquatic invertebrates. Indeed a study was already published on the rotifer, *Brachionus plicatilis* [14]. In this species, the thawed larvae started to lay eggs 2 to 3 days after thawing. Therefore, any damage caused by cryopreservation may not have had time to be expressed compared with oysters in the present work, as these reproduced more than 2 years after. Moreover, gamete quality was not evaluated in *B. plicatilis* because it reproduces by parthenogenesis, thereby producing amictic eggs. Similar positive results had already been reported in other invertebrates such as in insects (see review in [15]). However, major differences exist between such studies and the present model: permeabilization of the insect egg shell, deshydratation and vitrification are not required in Pacific oyster. The simpler cryopreservation procedure used in oyster is associated with the eco-physiology of marine mollusc larvae which do not have a thick external cuticle that could limit osmotic exchanges and cryoprotectant efficacy.

In the present study, no marked differences were observed in the ongrowing, growing-out and reproductive capacities of oysters produced from cryopreserved larvae compared with those of oysters that had been grown directly from fresh larvae with no cryopreservation step. These results confirm a pioneering study carried out by Paniagua-Chavez and co-workers [8], assessing the effects of cryopreservation at the trochophore stage on the subsequent development of Eastern oysters until the age of four months after thawing: the survival of larvae produced from thawed embryos was not different from the control. Furthermore, our work compares the effect of cryopreservation at three different stages of development. No marked differences were found in the rearing performances between these treatments and the control group. This indicates that after the initial dramatic mortalities observed during the larval development, the surviving larvae yielded similar post-settlement rearing performances to the ones that have never been frozen. The protocol of the present study differs from that used by Paniagua-Chavez et al. [8] in many respects (equilibration phase, cooling rate, cryoprotectant and thawing temperature). Because of similar conclusions at least for the immature stage, it can be suggested that there is only a limited interaction between the freezing protocol and the performances of the thawed larvae after their settlement. However, this hypothesis must be confirmed after the improvement of initial larvae survival observed just after thawing.

Some limitations of the study include the long-term protocol used. Firstly, oyster density maintained in tanks during the ongrowing phase and in bags during the growing-out phase differed between batches. Oyster density of each batch depended on the survival observed during the previous rearing phase. However, these densities are considered as non limiting for oyster growth [16]. Furthermore at the end of the ongrowing phase, the low survival observed for the 24±2 hpf batch which had the highest oyster density was moderated by its high growth rate assessed during the same rearing phase and its high subsequent reproductive performances. Secondly, parental effects on rearing performances of spat produced from thawed larvae were not studied in the present work. Such effects could influence a number of different rearing parameters: the survival of D-larvae produced from thawed oocytes collected from eight Pacific oyster females

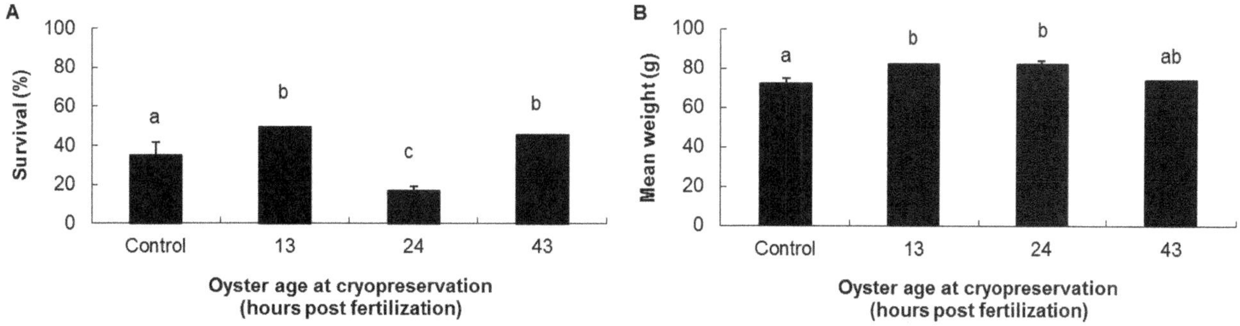

Figure 2. Effect of larvae cryopreservation at different stages post fertilization on the subsequent performances of oysters at the end of the growing-out phase (982 days). A: survival, B: mean weight observed at the end of the phase (spat number in each bag and number of bags are the same as those reported for Figure 1; different letters indicate significantly different results).

ranged from 0.1 to 30.1% [11]. However, the present 32 month protocol would be too difficult to conduct because it would require keeping separated progenies from different families. As a consequence, a pool of females and a pool of males were used to obtain the larvae used for cryopreservation. This choice is similar to previous studies aiming to describe the long-term rearing performances of molluscs produced from thawed larvae [5,6].

The high rearing performances of surviving thawed larvae may suggest an absence of genome alterations on the individuals that succeeded in settling, allowing subsequent development of these larvae. DNA alteration of thawed embryos has mostly been investigated in mammalian species [17] but rarely in invertebrates. Several instances of DNA damage were identified in the ragworm (*Nereis virens*) [18] and in insects [15], the most life-threatening of these is thought to be damages to the gut epithelium in ragworm. DNA integrity has been more greatly studied in cryopreserved fish gametes and several techniques to assess DNA integrity have been developed, including Tunel (Terminal deoxynucleotidyl transferase-nick-end-labelling), SCSA (Sperm Chromatin Structure Assay) and Comet assays (single gell electrophoresis) [19]. Inter-species differences have been found: cryopreservation caused damage to rainbow trout (*Onchorhynchus mykiss*) sperm compared with fresh sperm, but not to sperm of gilthead sea bream (*Sparus aurata*) [20]. In molluscs, the percentage of Pacific oyster spermatozoa with damaged DNA significantly increased after cryopreservation [21] but the consequences of such damage for the F_1 progenies were not estimated. Because a principal application of cryopreservation is to maintain the genetic variability of domesticated and wild animal species, the long-term development capacities of thawed larvae must be studied.

The discussion sections of scientific papers on embryo cryopreservation are often generally limited to scientific or technical questions, without addressing the relative benefits offered by larvae cryopreservation compared to gamete cryopreservation. It was shown that the development capacities of surviving D-larvae up to spat stage was not altered by oocyte cryopreservation [11]. Together with the success of sperm cryopreservation, and with the results showed in our study on larvae, it can be inferred that both haploid and diploid genomes of Pacific oyster can now be cryopreserved. The freezing of haploid spermatozoa or oocytes or of diploid spermatozoa collected from tetraploid males may be useful for controlled crosses using parents presenting desired characteristics, thus supporting the production of original genotypes for genetic selection and of triploids. The rationale of larvae cryopreservation is different since it cannot create new genetic variation *per se* but preserves diploid family genomes already created by controlled crosses. The cryopreservation of larvae has several other interests for shellfish aquaculture. First, it secures investments in selection programs as the families created can be conserved frozen. This could allow breeder to return to the best families of earlier-made crosses for selection purposes (with a loss of a one generation interval for sib selection). Second, it could allow a rapid restoration of genetic variation from previous generations if genetic drift or inbreeding depression are observed. Third, it could allow inter-generation estimation of genetic progress, reducing costs and risks associated with maintenance of

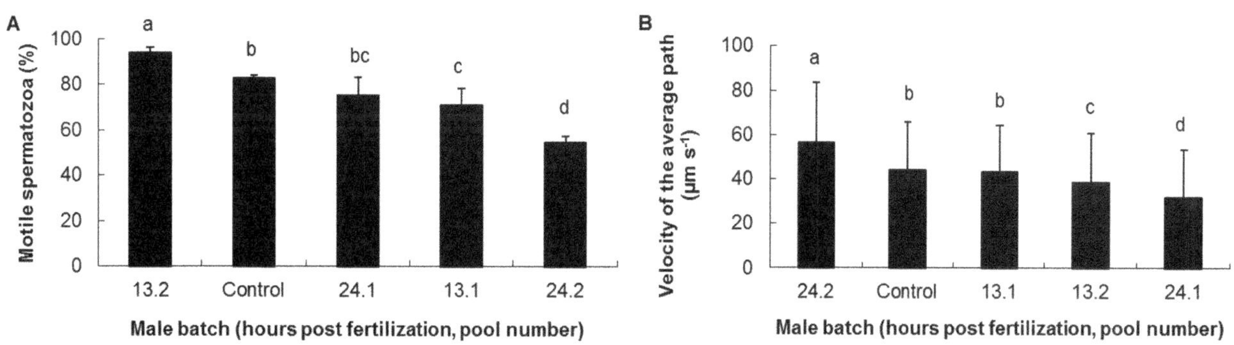

Figure 3. Effect of larvae cryopreservation at different stages post fertilization on the subsequent performances of oysters during the reproductive phase (>982 days). A: percentage of motile spermatozoa, B: Velocity of the Average Path (n>30 spermatozoa scored for each male gamete pool; n = 5 males contributed to each pool; different letters indicate significantly different results).

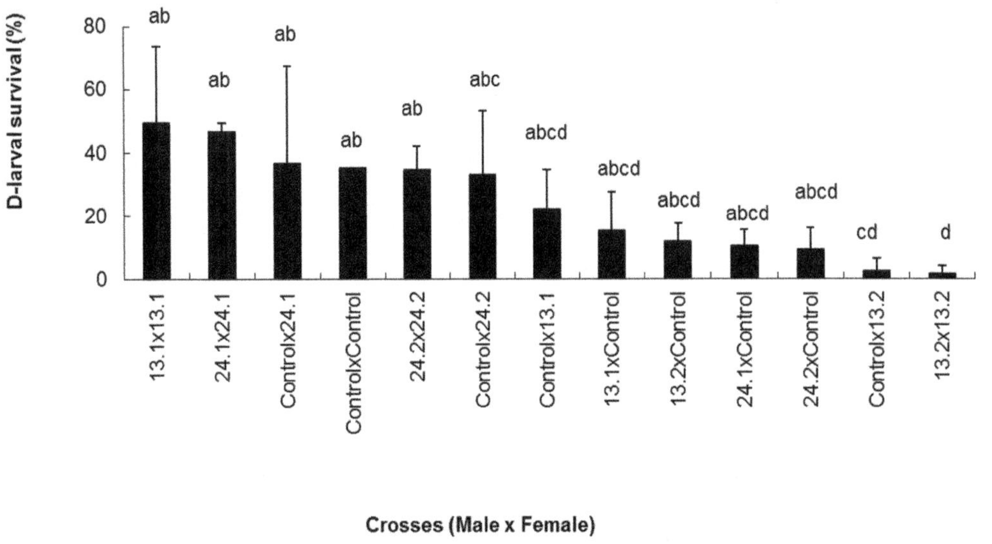

Figure 4. Effect of larvae cryopreservation at different stages post fertilization on the subsequent D-larval survival in progenies. (n = 5 oysters contributed to each gamete pool; different letters indicate to significantly different results).

live control lines that are never really equivalent to a true control because of genetic drift across generations, non-intentional selection or domestication. More generally, the cryopreservation of diploid progenies may also provide suitable means for diffusing genetic progress and genetic exchange of material worldwide in using only sanitary approved genotypes ready to be used as future broodstock. The application of a cryopreservation procedure developed for diploid larvae may also facilitate maintenance and genetic improvement of tetraploid oyster stocks by cryopreservation of tetraploid larvae [2].

Conclusion

In the present work, We demonstrated that the subsequent growing-out and reproductive performances of surviving thawed oysters are similar to those observed for unfrozen ones. From thawing up to the end of the growing out phase (982 days), oyster survival observed was respectively 0.05% (13±2 hpf), 0.15% (24±2 hpf), 0.18% (43±2 hpf) and 10.06% (control). Because of the very high fecundity observed in Pacific oyster (10 to 50×10^6 oocytes per female), larvae cryopreservation allows the production of 11,000 to 42,000 good quality adult oysters for each female, depending on oyster development stage at cryopreservation. These yields can be largely increased by the further optimisation of the cryopreservation protocols designed for oyster larvae.

Pacific oyster is the first aquatic farmed species for which sperm, oocyte and larval genomes can be stored in this way. Freezing larvae will help to preserve the genomes of wild oyster populations and of selected oyster families. Furthermore, thawed oysters may be used to evaluate the genetic progress recorded between successive generations. Such advances in embryo cryopreservation will contribute to the establishment of oyster cryobanks and maintenance of the genetic diversity of domestic and wild stocks of other mollusc species.

Acknowledgments

Many thanks to H. McCombie-Boudry for corrections of the English.

Author Contributions

Conceived and designed the experiments: MS CL SP PH. Performed the experiments: MS CL SP CM BQ MB IQ DR BD PH. Analyzed the data: MS CL SP PH. Contributed reagents/materials/analysis tools: CM BQ MB IQ DR BD. Wrote the paper: MS CL PH.

References

1. Boudry P, Dégremont L, Haffray P (2008) The genetic basis of summer mortality in Pacific oyster spat and potential for improving survival by selective breeding in France. In: Samain JF, McCombie H, editors. Summer mortality of Pacific oyster - The Morest Project. Versailles, France: Quae Editions. pp.153–196.
2. Piferrer F, Beaumont A, Falguière JC, Colombo L, Flajshans M, et al. (2009) The use of induced polyploidy in the aquaculture of fish and shellfish for performance improvement and genetic containment. Aquaculture 293: 125–156.
3. Hiemstra SJ, van der Lende T, Woelders H (2006) The potential of cryopreservation and reproductive technologies for animal genetic resources conservation strategies. In: Ruane J, Sonnino A, editors. The role of biotechnology in exploring and protecting agricultural genetic resources. Rome: : FAO. pp. 45–59.
4. Renard P (1991) Cooling and freezing tolerances in embryos of the Pacific oyster, *Crassostrea gigas*: methanol and sucrose effects. Aquaculture 92: 43–57.
5. Wang H, Li X, Wang M, Clarke S, Gluis M, et al. (2011) Effects of larval cryopreservation on subsequent development of the blue mussels, *Mytilus galloprovincialis* Lamarck. Aquaculture Res 42: 1816–1823.
6. Paredes E, Adams SL, Tervit HR, Smith J, McGowan LT, et al. (2012) Cryopreservation of Greenshell mussel (*Perna canaliculus*) trochophore larvae. Cryobiol 65: 256–262.
7. Usuki H, Hamaguchi M, Ishioka H (2002) Effects of developmental stage, seawater concentration and rearing temperature on cryopreservation of Pacific oyster *Crassostrea gigas* larvae. Fish Sci 68: 757–762.
8. Paniagua-Chavez CG, Buchanan JT, Supan JE, Tiersch TR (2000) Cryopreservation of sperm and larvae of the Eastern oyster. In: Tiersch TR, Mazik PM, editors. Cryopreservation in aquatic species. Baton rouge, Louisiana: WAS. pp. 230–239.
9. Suquet M, Le Mercier A, Rimond F, Mingant C, Haffray P, et al. (2012) Setting tools for the early assessment of the quality of thawed Pacific oyster (*Crassostrea gigas*) D-larvae. Theriogenol 78: 462–467.
10. Song YP, Suquet M, Quéau I, Lebrun L (2009) Setting of a procedure for experimental fertilisation of Pacific oyster (*Crassostrea gigas*) oocytes. Aquaculture 287: 311–314.
11. Tervit HR, Adams SL, Roberts RD, McGowan LT, Pugh PA, et al. (2005) Successful cryopreservation of Pacific oyster (*Crassostrea gigas*) oocytes. Cryobiol 51: 142–151.

12. Rico-Villa B, Woerther P, Mingant C, Lepiver D, Pouvreau S, et al. (2008) A flow- through rearing system for ecophysiological studies of Pacific oyster, *Crassostrea gigas*, larvae. Aquaculture 282: 54–60
13. Wilson-Leedy JG, Ingermann RL (2007) Development of a novel CASA system based on open source software for characterization of zebrafish sperm motility parameters. Theriogenol 67: 661–672.
14. Toledo JD, Kurokura H (1990) Cryopreservation of the euryhaline rotifer *Brachionus plicatilis* embryos. Aquaculture 91: 385–394.
15. Leopold RA, Rinehart JP (2010) A template for insect cryopreservation. In: Denlinger DL, Lee RE, editors. Low temperature biology of insects. Cambridge: Cambridge University Press. pp. 325–341.
16. Honkoop PJC, Bayne BL (2002) Stocking density and growth of the Pacific oyster (*Crassostrea gigas*) and the Sydney rock oyster (*Saccostrea glomerata*) in Port Stephens, Australia. Aquaculture 213: 171–186
17. Smith GD, Silva CASE (2004) Developmental consequences of cryopreservation of mammalian oocytes and embryos. Reprod Bioled Online 9, 171–178. Available: http://ac.els-cdn.com/S1472648310621268/1-s2.0-S1472648310621268-main.pdf?_tid=92971c4c-7f41-11e2-874f-00000aab0f6b&acdnat=1361793218_93bf401c2c411ee833e5454b804763fe. Accessed 25 January 2013.
18. Wang WB, Olive JW (2000) Effects of cryopreservation on the ultrastructural anatomy of 3-segment *Nereis virens* larvae (Polycheta). Inv Biol 119: 309–318.
19. Cabrita E, Sarasquete C, Martinez-Paramo S, Robles V, Beirao J, et al. (2010) Cryopreservation of fish sperm: applications and perspectives. J Appl Ichthyol 26: 623–635.
20. Cabrita E, Robles V, Rebordinos L, Sarasquete C, Herraez MP (2005) Evaluation of DNA damage in rainbow trout (*Onchorhynchus mykiss*) and giltead sea bream (*Sparus aurata*) cryopreserved sperm. Cryobiol 50: 144–153.
21. Gwo JC, Wu CY, Chang WS, Cheng HY (2003) Evaluation of damage in Pacific oyster (*Crassostrea gigas*) spermatozoa before and after cryopreservation using comet assay. Cryoletters 24: 171–180.

13

High-Throughput Cryopreservation of *In Vivo*-Derived Swine Embryos

Lee D. Spate, Clifton N. Murphy, Randall S. Prather*

Division of Animal Sciences, University of Missouri, Columbia, Missouri, United States of America

Abstract

Cryopreservation of swine embryos is inefficient. Our goal was to develop a non-invasive method for "relatively" high-throughput cryopreservation of in vivo-produced swine embryos. Since removal of the lipid droplets within early swine embryos improves cryosurvival we wanted to apply a technique of high osmolality treatment followed by centrifugation that was first developed for in vitro-produced swine embryos to in vivo-produced swine embryos. The first aim was to determine how sensitive the in vivo-produced zygote and 2-cell stage embryo was to various high osmolality conditions for a short duration. Culture for 6, 12 or 18 min at 300, 400 or 500 milliosmoles (mOsm) had no detectable affect on the resulting blastocyst stage embryos (number of inner cell mass nuclei, trophectoderm nuclei, total number of nuclei, ratio of the trophectoderm to inner cell mass nuclei or percent blastocyst). However there was an effect of gilt on each of these parameters. For the second aim we focused on 300 mOsm for 6 min, 400 mOsm for 12 min, 500 mOsm for 12 min, and 500 mOsm for 18 min. The embryos were centrifuged for the duration of high osmolality treatment, then cultured to the blastocyst stage and vitrified. After vitrification and thawing the 500 mOsm for 18 min had the highest percent re-expansion with no difference in the total number of nuclei. While requiring a different base culture medium than in vitro-produced embryos, in vivo-derived embryos also survive cryopreservation without damage to their zona pellucida.

Editor: Jason Glenn Knott, Michigan State University, United States of America

Funding: Funding was provided by the National Institutes of Health R21 RR25879 and U42 OD011140. The funders had no role in study design, data collection and analysis, decision to publish, or preparation of the manuscript.

Competing Interests: The University of Missouri has filed for patent protection for the procedures described in the document. The patent application citation is: Li, R., L. Spate, C.N. Murphy, R.S. Prather. "High throughput cryopreservation of swine embryos", Provisional Patent filed August 29, 2008. US61/190,515. This patent will not inhibit the authors' ability to provide all the details of how the invention works.

* E-mail: PratherR@Missouri.Edu

Introduction

Cryopreservation of early mammalian embryos provides prospects for the preservation of germplasm as well as the movement of genetics nationally and internationally. Unfortunately the swine embryo has been more resistant to cryopreservation than the embryos of many mammals. The first significant advance toward the successful cryopreservation of swine embryos was based on the observation that they are very sensitive to hypothermic conditions and that removal of intracellular lipids (delipation) alleviates this sensitivity [1–4]. Alternatives for physically removing the lipids at an early stage include culturing to the blastocyst stage and then removing the lipids [5], destabilizing the cytoskeleton [6], altering the vitrification conditions [7–9], or chemically removing the lipids before cryopreservation [10]. While the chemical lipid removal appears attractive, there have yet to be any offspring produced by this method. Other cumbersome (slow-cooling) cryopreservation methods have been developed, but there is tremendous batch to batch variation [11].

The current methods for cryopreservation of swine embryos are time consuming and troublesome. One of the most successful methods of cryopreservation of swine embryos is to centrifuge the cells to polarize the lipids and then use a micromanipulator to remove the lipids from the embryo [4,12]. Interestingly, removal of the lipids not only permits the embryos to survive low temperatures [13], but in some cases they actually develop to the blastocyst stage at a higher rate [14,15]. Unfortunately this current procedure results in a break in the zona pellucida. The International Embryo Transfer Society guidelines state that the zona pellucida cannot be compromised if the health status is to be maintained [16,17]. This is especially important for embryos that are to be frozen and transported to another facility, as swine are generally raised in specific pathogen-free environments and the movement of pathogens can have a devastating effect on the recipient herd.

While hyperosmotic treatments and centrifugation have been developed for lipid removal that does not damage the zona pellucida, these techniques have been only reported for in vitro-produced (IVP) embryos [18]. The limited information available on IVP swine oocytes suggests that they are very sensitive to osmotic stresses [19]. These authors showed that any deviation away from 290 milliosmolal (mOsm) resulted in a dramatic reduction in development after fertilization. The IVP blastocyst stage embryo is less sensitive to osmotic stress. When sucrose was used adjust the osmolality as long as the osmolality did not go beyond 170% of isotonic, i.e. ~600 mOsm, then the embryos survived the osmotic insult [20].

Our goal was to develop a non-invasive method for "relatively" high-throughput cryopreservation of in vivo-produced swine embryos. Our first aim was to confirm that the amount of perivitteline space in the zona pellucida can be increased so that more complete separation of the lipids and the zygote cytoplasm can be achieved by centrifugation. The first aim was to determine how sensitive the in vivo-produced embryo is to various high osmolality conditions for a short duration. The second aim was to use the added perivitteline space, caused by high osmolality treatment, to aid in separating the lipids within the zona pellucida (via centrifugation) prior to cryopreservation and to determine their viability.

Results

The experimental treatments for the first aim were designed to determine if high osmolality for a short duration will negatively affect development of zygotes and 2-cell stage embryos to the blastocyst stage. The specific treatments were: 1. 6 min at 300 mOsm, 2. 6 min at 400 mOsm, 3. 6 min at 500 mOsm, 4. 12 min at 400 mOsm, 5. 18 min at 400 mOsm, 6. 12 min at 500 mOsm, and 7. 18 min at 500 mOsm. These treatments were chosen based on the successful treatment of IVP embryos. As predicted we observed no treatment effects ($p>0.15$) on the number of inner cell mass (ICM) nuclei, trophectoderm (TE) nuclei, total number of nuclei, the ratio of the number of TE/ICM nuclei, or the percent blastocyst (Table 1). The power of the test ranged from 0.16 to 0.77. However, it was quite interesting to observe an effect due to gilt on every parameter (Table 1). While the power is not extremely high, we did have enough confidence in our results to move forward with the more important test of the ability of these embryos to survive the centrifugation and cryopreservation. In conclusion we found that there was no negative effect of osmotic stress on development of the embryos.

Experiments addressing the second aim were designed to reduce the number of treatment groups to maximize viability, and decrease the number of embryos that we would need to complete the project. Based on the previous data the treatments for this experiment were as follows: 1) 300 mOsm centrifuged for 6 min, 2) 400 mOsm centrifuged for 12 min, 3) 500 mOsm centrifuged for 12 min, or 4) 500 mOsm centrifuged for 18 min. Each of these treatments successfully disconnected the membrane-bound lipids from the remainder of the embryo (see Figure 1 for an example). It was expected that the ability to break this connection would be directly proportional to their subsequent development after cryopreservation. The percent of treated embryos that developed to the blastocyst stage was recorded. Blastocysts were then vitrified in open pulled straws (OPS) [18]. The embryos were later thawed in TCM 199 using a 2-step sucrose rehydration method [18]. The embryos were cultured for an additional 18 hours in PZM3 [21], assessed for blastocoel re-expansion and were processed to count total cell number. The percent blastocyst was higher in the 400 mOsm for 12 min and 500 mOsm for 18 min, than for 500 mOsm for 12 min (Fig. 2). After cryopreserving and thawing the embryos and culturing an additional 18 hours the re-expansion rates were higher from the 500 mOsm 18 min group as compared to the 300 mOsm group, but the total number of nuclei was not different (Fig. 2).

Discussion

Swine oocytes and embryos have large, abundant lipid droplets in the early stage swine embryo and gradually decline in size and abundance as the embryo advances to and beyond the blastocyst stage [22,23]. Coincident with reduced lipid content is an increased freezability of late stage embryos. Late stage embryos (hatched blastocyst) with low lipid content and smaller size of lipid droplets survive cryopreservation better than early stage embryos (2-cell, 4-cell and 8-cell) [24–26]. Interestingly, the large lipid droplets in the early stage embryos are easier to remove by centrifugation than the smaller droplets in the later stage embryos. While there is very little in the literature about the exact composition of the lipids in the swine oocyte; one study shows that 74 ng of fatty acids could be extracted from a single swine oocyte [27], and that this was significantly higher (3X) than even sheep and cattle oocytes (whose cytoplasm is also opaque).

Successful cryopreservation of IVP embryos has been reported after mechanical delipation through centrifugation and micromanipulation [14,28]. However mechanical delipation substantially increases the potential of pathogen transmission because of the damage inflicted upon the zona pellucida during micromanipulation. It is also labor–intensive and time-consuming. Another option is to use hatched blastocysts as they are more cold tolerant than are zona enclosed blastocysts [29]. However the use of hatched blastocysts creates problems with maintaining the embryo's barrier to diseases. International shipment of embryos requires that the zona pellucida remain intact. Development of a practical non-invasive means of lipid removal for cryopreservation of swine embryos is still necessary.

Removal of the lipids after centrifugation appears to be the most promising technique. However, after centrifugation of the swine oocyte or embryo with an intact zona pellucida the polarized lipid droplets remain connected with the cytoplasm of the oocyte or blastomere of the embryo via a bridge-like structure [13]. If the lipids are not immediately removed, or the embryo frozen, the polarized lipid droplets redistribute into the oocyte or blastomere during subsequent culture. Thus embryos need to be cryopreserved immediately after centrifugation in order to prevent lipid redistribution prior to cryopreservation [30]; however this is not practical since the lipid removal is best achieved during the early cleavage stages, and the embryos survive the best if they have developed to the blastocyst stage. If the perivitelline space is enlarged, the bridge-like structure may more easily break after

Table 1. In vitro development of in vivo-derived swine embryos after hyperosmotic treatment and centrifugation (the number of gilts providing embryos for this experiment was 21).

	Mean ± SEM (N)	Treatment Effect (p Value)	Power of Test	Gilt Effect (p Value)
# ICM Nuclei	7.83±0.38 (45)	0.60	0.35	<0.00
# TE Nuclei	23.20±1.00 45)	0.36	0.77	0.01
Total # Nuclei	32.89±0.87 (113)	0.15	0.37	0.01
Ratio TE/ICM	3.28±0.21 (45)	0.94	0.16	0.06
Percent Blastocyst	73.0±8.1 (135)	0.50	0.33	<0.01

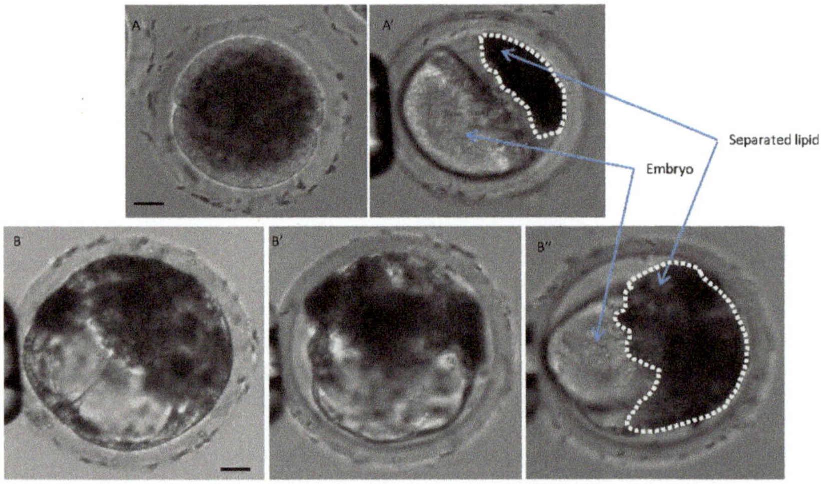

Figure 1. Separation of lipids in early pig embryos by centrifugation and hypertonicity. Two views of the same 2-cell stage embryo (A, A') after exposure to a hypertonic solution and centrifugation. A 2-cell stage embryo that was previously subjected to a hypertonic solution and centrifugation followed by culture to the blastocyst stage (B, B', B''). The same embryo is held at different angles. In panel B' and B'' the embryo has been placed into another hypertonic solution to shrink the cells so that the lipids can be better visualized. The white dashed line encircles the extruded lipids. Scale bar = 25 microns.

centrifugation and the lipid droplets will not redistribute into the cytoplasm of the oocyte or the blastomere of the embryo, but will stay within the intact zona pellucida. At least two methods might make the perivitelline space larger. One is to swell the zona pellucida through partial enzymatic digestion (such as trypsin or pronase, etc). Another option is to condense the volume of the oocyte or embryo by high osmolality treatment as described here. Our treatment appeared to successfully break this connection between the embryo and the extruded lipids (Fig. 1). However based on this experiment alone, since it is difficult to visualize any lipids moving back to the embryo it cannot be concluded that both high osmolality and centrifugation are necessary for subsequent cryosurvival. But since centrifugation alone results in the lipids returning to the embryo and the embryo not surviving cryopreservation and since the high osmolality treatment occurs days before cryopreservation, it is likely that both are required for survival. Indeed, we propose that the higher centrifugation and hypertonic treatment work together to improve the lipid separation, and the higher cryosurvival is a direct indication of the success of lipid separation. Since this describes the development of a high throughput method of cryopreservation dozens of embryos can be treated at single time and thus avoids the necessity of

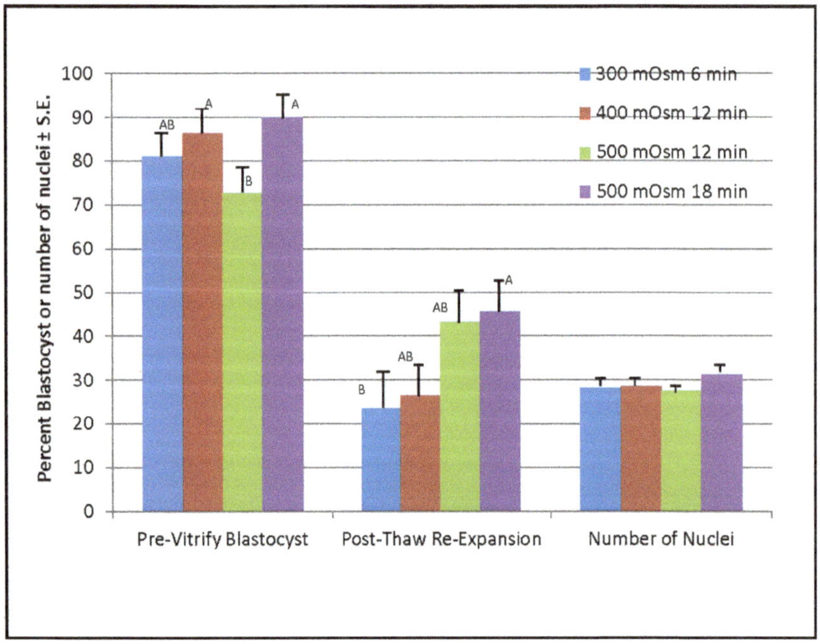

Figure 2. Blastocyst development and number of nuclei in embryos after various osmolality treatments. ($^{AB}P<0.05$; N = 160 for blastocyst development, and 113 for number of nuclei; 23 different gilts provided embryos for this experiment.).

micromanipulating each individual embryo to assure complete removal of the lipids.

Based on the above observations various groups have treated IVP swine embryos with trypsin to swell the zona pellucida [31,32], or with high osmolality to shrink the embryo followed by centrifugation to separate the lipids. The high osmolality treated embryos were cultured to the blastocyst stage and subjected to cryopreservation in an OPS system. Live piglets resulted from both trypsin and high osmolality treated IVP embryos [18]. In another report IVP embryos were vitrified in a closed straw system and also produced live piglets [33]. The results presented here are somewhat different from those previously reported for IVP, in that 400 mOsm for 6 minutes of centrifugation was sufficient to completely separate the lipids and result in cryosurvival [18]. Similarly, the conditions for somatic cell nuclear transfer embryos was also different, e.g. the osmolality did not need to be as high for complete lipid separation [18].

Our original plan was to use the same media for in vivo-produced embryos that worked for generation of our preliminary data on our IVP embryos TCM199: [18]). Unfortunately, placing the in vivo-produced zygotes and 2-cell stage embryos in TCM199 for even a few min resulted in the embryos blocking at the 4-cell stage (unbpublished). This was quite surprising to us. While we knew that long term culture in TCM199 resulted in the embryos blocking [34], since the IVP embryos developed quite well after exposure to TCM199 we anticipated no problems with what we thought would be hardier in vivo-produced embryos. Nevertheless they blocked at the 4-cell stage. Thus we changed the base medium to TL-Hepes for centrifugation, and then the embryos developed to the blastocyst stage. The TL-Hepes was then used throughout the experiments.

There were significant differences due to the gilt from which the embryos were collected. Unfortunately since most of the collections were single animals on any given day, this variation is confounded with day. Similar differences due to gilt have been reported by us [35] and others [36] and indicate the tremendous variation due to different animals providing the embryos.

Treating in vivo-derived embryos with high osmolality and centrifugation to polarize intercellular lipid is compatible with development to blastocyst as seen with the development data. The rate of re-expansion of the blastocyst illustrates the survivability post-thawing of these embryos.

Materials and Methods

All chemicals were purchased from Sigma (St. Louis, MO) unless otherwise indicated.

Gilt Treatment and Embryo Collection

All experiments with animals were approved by the University of Missouri Institutional Animal Care and Use Committee (#6523). Surgery was performed under anesthesia, and analgesics were administered during the surgery. Pre-pubertal gilts were monitored for signs of estrous activity. When estrus was detected (day 0) the gilts were artificially inseminated 8 hours later. On Day 2 the gilt was anesthetized and oviducts flushed [37] with TL-Hepes [38] and embryos recovered. They were cultured in PZM3 (NaCl 108.0 mmol/L, KCl 10.0 mmol/L, KH_2PO_4 0.35 mmol/L, $MgSO_4*7H_2O$ 0.4 mmol/L, $NaHCO_3$ 25.07 mmol/L, Na-pyruvate 0.2 mmol/L, Ca $(Lactate)_2$ * 5 H_2O 2.0 mmol/L, Glutamine 1.0 mmol/L, Hypotaurine 5.0 mmol/L, BME amino acid solution 20 ml/L, MEM amino acid solution 10 ml/L, Gentamicin 0.01 mg/mL, bovine serum albumin (BSA) 3 mg/mL, Osmolarity 288±2, pH 7.3±0.2) in 90% N_2, 5% O_2 and 5% CO_2 at 38.5°C until treatment assignment. Embryos were randomly assigned to the various treatments (see Statistical Analysis below).

High Osmolality Treatments

For the first aim zygotes and 2-cell stage embryos were collected and divided into 3 different osmolality treatment groups (300, 400 or 500 mOsm; achieved by adding sucrose) for 6 min and centrifuged for 0, 12, or 18 min. The durations were chosen to simulate what happens during the cryopreservation procedures as described above, i.e. they were all equilibrated for 6 min and then centrifuged for 6, 12 or 18 min. Then embryos were washed 3 times in PZM3 and cultured for 6 days in PZM3. This experiment determined how sensitive the in vivo-produced embryos are to high osmolality treatments induced by sucrose.

Based on the results from the first aim the following treatments were chosen for the second aim, for preparation prior to vitrification: 1) 300 mOsm centrifuged for 6 min, 2) 400 mOsm centrifuged for 12 min, 3) 500 mOsm centrifuged for 12 min, or 4) 500 mOsm centrifuged for 18 min. The embryos were cultured to the blastocyst stage and vitrified.

Vitrification

Vitrification of embryos was carried out by using the OPS method. The OPS straws were purchased from LEC Instruments P/L and Minitube. All solutions used during vitrification were prepared with holding medium (Oocyte manipulation medium with 20% fetal calf serum instead of 3 mg/mL BSA). Blastocysts were placed in equilibration solution (10% ethylene glycol, 10% dimethyl sulfoxide [DMSO]) for 2 min, followed by exposure to vitrification solution (20% ethylene glycol, 20% DMSO). Embryos were loaded into an OPS and immediately plunged into liquid nitrogen. All processes before plunging into nitrogen were conducted on a 37°C warm stage. The time from exposure to the vitrification solution to plunging was 25–30 sec.

For embryo warming, cryopreserved embryos were warmed by removing the straw from liquid nitrogen for a few seconds and immersing the end of the OPS into 0.3 M sucrose for 6 min at 37°C, then transferring them to 0.2 M sucrose for 6 min, and then holding medium for 6 min. The warmed embryos were cultured for 18 hours in PZM3 with 10% FBS to check re-expansion, and nuclear number.

Statistical Analysis

Embryos from a single gilt were selected at random and assigned to the treatments. Since often the number of embryos from a single gilt did not match the number of treatments additional embryos were assigned to other treatments in order. So, for example, if there are seven treatments and 10 embryos are collected, then seven treatments received 1 embryo and three treatments received 2 embryos. The next gilt provided embryos first to the last four treatments that only had a single embryo, and then the remainder were distributed to the other treatments. In contrast, the same strategy was applied if fewer than seven embryos were collected. It should be noted that in some cases 2- and 4-cell stage embryos were collected. In this case the 4-cell stage embryos were not used. Similarly, some oocytes were unfertilized; these also were not used. This strategy permitted a robust analysis of the treatment effects as well as providing an opportunity to detect gilt effects. All data were analyzed by using the PROC GLM commands in SAS 9.1 (Cary, NC, USA). The main effects of the number of nuclei in the blastocysts were tested as appropriate in each experiment. Gilt was also included in the model to remove variation due to different animals. Percentage data was arcsine transformed prior to analysis. Differences between treatment groups (time by osmolality) were assessed via post-hoc pairwise comparisons of least squares mean values. Significance was assigned at P-values <0.05.

Acknowledgments

The authors would like to acknowledge the efforts of Lonnie and Jason Dowell at the Swine Research Complex at the University of Missouri. The University of Missouri has filed for patent protection for the procedures described in this manuscript.

Author Contributions

Conceived and designed the experiments: RSP LDS CNM. Performed the experiments: LDS CNM. Analyzed the data: RSP LDS CNM. Contributed reagents/materials/analysis tools: RSP LDS CNM. Wrote the paper: RSP LDS CNM.

References

1. Polge C, Wilmut I, Rowson LEA (1977) The low temperature preservation of cow, sheep, and pig embryos. Cryobiology 11: 560.
2. Wilmut I (1972) The low temperature preservation of mammalina embryos. Journal of Reproduction and Fertility 31: 513–514.
3. Dobrinsky JR (1996) Cellular Approach to Cryopreservation of Embryos. Theriogenology 45: 17–26.
4. Nagashima H, Kashiwazaki N, Ashman RJ, Grupen CG, Seamark RF, et al. (1994) Removal of Cytoplasmic Lipid Enhances the Tolerance of Porcine Embryos to Chilling. Biology of Reproduction 51: 618–622.
5. Beebe LFS, Cameron RDA, Blackshaw AW, Keates HL (2005) Changes to porcine blastocyst vitrification methods and improved litter size after transfer. Theriogenology 64: 879–890.
6. Dobrinsky JR, Pursel VG, Long CR, Johnson LA (2000) Birth of piglets after transfer of embryos cryopreserved by cytoskeletal stabilization and vitrification. Biology of Reproduction 62: 564–570.
7. Berthelot F, Martinat-Botte F, Perreau C, Terqui M (2001) Birth of piglets after OPS vitrification and transfer of compacted morula stage embryos with intact zona pellucida. Reproduction, Nutrition, Development 41: 267–272.
8. Beebe LFS, Cameron RDA, Blackshaw AW, Higgins A, Nottle MB (2002) Piglets born from centrifuged and vitrified early and peri-hatching blastocysts. Theriogenology 57: 2155–2165.
9. Misumi K, Suzuki M, Sato S, Saito N (2003) Successful production of piglets derived from vitrified morulae and early blastocysts using a microdroplet method. Theriogenology 60: 253–260.
10. Men HS, Agca Y, Riley LK, Critser JK (2006) Improved survival of vitrified porcine embryos after partial delipation through chemically stimulated lipolysis and inhibition of apoptosis. Theriogenology 66: 2008–2016.
11. Fujino Y, Kikuchi K, Nakamura Y, Kobayashi H, Yonemura I, et al. (2006) Batchwise assessment of porcine embryos for cryotolerance. Theriogenology 67: 413–422.
12. Nagashima H, Kashiwazaki N, Ashman RJ, Grupen CG, Nottle MB (1995) Cryopreservation of porcine embryos. Nature 374: 416.
13. Nagashima H, Cameron RD, Kuwayama M, Young M, Beebe L, et al. (1999) Survival of porcine delipated oocytes and embryos after cryopreservation by freezing or vitrification. Journal of Reproduction and Development 45: 167–176.
14. Li R, Lai L, Wax D, Hao Y, Murphy CN, et al. (2006) Cloned transgenic swine via in vitro production and cryopreservation. Biology of Reproduction 75: 226–230.
15. Lai LX, Kang JX, Li RF, Wang JD, Witt WT, et al. (2006) Generation of cloned transgenic pigs rich in omega-3 fatty acids. Nature Biotechnology 24: 435–436.
16. Wrathall AE, Sutmoller P (1998) Potential of embryo transfer to control transmission of disease. In: Stringfellow DA, Seidel SM, editors. Manual of the International Embryo Transfer Society. 3rd ed. Savoy, IL: International Embryo Transfer Society. 17–44.
17. Stringfellow DA (1998) Recommendations for the sanitary handling of in-vivo-derived embryos. In: Stringfellow DA, Seidel SM, editors. Manual of the International Embryo Transfer Society. 3rd ed. Savoy, IL: International Embryo Transfer Society. 79–84.
18. Li R, Murphy CN, Spate L, Wax D, Isom C, et al. (2009) Production of piglets after cryopreservation of embryos using a centrifugation-based method for delipation without micromanipulation. Biol Reprod 80: 563–571.
19. Mullen SF, Rosenbaum M, Critser JK (2007) The effect of osmotic stress on the cell volume, metaphase II spindle and developmental potential of in vitro matured porcine oocytes. Cryobiology 54: 281–289.
20. Men HS, Agca Y, Mullen SF, Critser ES, Critser JK (2005) Osmotic tolerance of in vitro produced porcine blastocysts assessed by their morphological integrity and cellular actin filament organization. Cryobiology 51: 119–129.
21. Yoshioka K, Suzuki C, Tanaka A, Anas IMK, Iwamura S (2002) Birth of piglets derived from porcine zygotes cultured in a chemically defined medium. Biology of Reproduction 66: 112–119.
22. Norberg HS (1973) Ultrastructural aspects of the preattached pig embryo: cleavage and early blastocyst stage. Z Anat Entwicklungsgesch 143: 95–114.
23. Kikuchi K, Ekwall H, Tienthai P, Kawai Y, Noguchi J, et al. (2002) Morphological features of lipid droplet transition during porcine oocyte fertilisation and early embryonic development to blastocyst in vivo and in vitro. Zygote 10: 355–366.
24. Nagashima H, Kato Y, Yamakawa H, Matsumoto T, Ogwa S (1989) Changes in freezing tolerance of pig blastocysts in peri-hatching stages. Jpn J Animal Reproduction 35: 130–134.
25. Nagashima H, Yamakawa H, Niemann H (1992) Freezability of porcine blastocysts at different peri-hatching stages. Theriogenology 37: 839–850.
26. Li R, Hosoe M, Shioya Y, Bou S (2002) The preliminary research on freezing viability of bovine in vitro fertilized embryos. Chinese J Scientia Agricultura Sinica 35: 1125–1129.
27. McEvoy TG, Coull GD, Broadbent PJ, Hutchinson JSM, Speake BK (2000) Fatty acid composition of lipids in immature cattle, pig and sheep oocytes with intact zona pellucida. Journal of Reproduction & Fertility 118: 163–170.
28. Nagashima H, Hiruma K, Saito H, Tomii R, Ueno S, et al. (2007) Production of live piglets following cryopreservation of embryos derived from in vitro-matured oocytes. Biology of Reproduction 76: 900–905.
29. Dobrinsky JR (2002) Advancements in cryopreservation of domestic animal embryos. Theriogenology 57: 285–302.
30. Cameron RDA, Beebe LFS, Blackshaw AW, Keates HL (2004) Farrowing rates and litter size following transfer of vitrified porcine embryos into a commercial swine herd. Theriogenology 61: 1533–1543.
31. Esaki R, Ueda H, Kurome M, Hirakawa K, Tomii R, et al. (2004) Cryopreservation of porcine embryos derived from in vitro-matured oocytes. Biology of Reproduction 71: 432–437.
32. Du Y, Zhang Y, Li J, Kragh PM, Kuwayama M, et al. (2007) Simplified cryopreservation of porcine cloned blastocysts. Cryobiology 54: 181–187.
33. Men H, Zhao C, Si W, Murphy CN, Spate L, et al. (2011) Birth of piglets from in vitro-produced, zona-intact porcine embryos vitrified in a closed system. Theriogenology 76: 280–289.
34. Hagen DR, Prather RS, Sims MM, First NL (1991) Development of one-cell porcine embryos to the blastocyst stage in simple media. J Anim Sci 69: 1147–1150.
35. Prather RS, Boice ML, Gibson J, Hoffman KE, Parry TW (1995) In vitro development of embryos from Sinclair miniature pigs: a preliminary report. Theriogenology 43: 1001–1007.
36. Petters RM, Johnson BH, Reed ML, Archibong AE (1990) Glucose, glutamine and inorganic phosphate in early development of the pig embryo in vitro. J Reprod Fertil 89: 269–275.
37. Machaty Z, Day BN, Prather RS (1998) Development of Early Porcine Embryos in Vitro and in Vivo. Biology of Reproduction 59: 451–455.
38. Im G-S, Lai L, Liu Z, Hao Y, Prather RS (2004) In vitro development of preimplantation porcine nuclear transfer embryos cultured in different media and gas atmospheres. Theriogenology 61: 1125–1135.

Mechanical Compliance and Immunological Compatibility of Fixative-Free Decellularized/ Cryopreserved Human Pericardium

Maria Cristina Vinci[1], Giulio Tessitore[2], Laura Castiglioni[3], Francesca Prandi[1], Monica Soncini[4], Rosaria Santoro[1], Filippo Consolo[4], Francesca Colazzo[3], Barbara Micheli[5], Luigi Sironi[3], Gianluca Polvani[2], Maurizio Pesce[1]*

1 Laboratorio di Ingegneria Tissutale Cardiovascolare, Centro Cardiologico Monzino-IRCCS, Milan, Italy, **2** Dipartimento di Scienze Cardiovascolari, Università degli Studi di Milano, Milan, Italy, **3** Dipartimento di Farmacologia, Università di Milano, Milan, Italy, **4** Dipartimento di Bioingegneria, Politecnico di Milano, Milan, Italy, **5** Banca dei Tessuti Cardiovascolari della Regione Lombardia, Milan, Italy

Abstract

Background: The pericardial tissue is commonly used to produce *bio*-prosthetic cardiac valves and patches in cardiac surgery. The procedures adopted to prepare this tissue consist in treatment with aldehydes, which do not prevent post-graft tissue calcification due to incomplete xeno-antigens removal. The adoption of fixative-free decellularization protocols has been therefore suggested to overcome this limitation. Although promising, the decellularized pericardium has not yet used in clinics, due to the absence of proofs indicating that the decellularization and cryopreservation procedures can effectively preserve the mechanical properties and the immunologic compatibility of the tissue.

Principal Findings: The aim of the present work was to validate a procedure to prepare decellularized/cryopreserved human pericardium which may be implemented into cardiovascular homograft tissue Banks. The method employed to decellularize the tissue completely removed the cells without affecting ECM structure; furthermore, uniaxial tensile loading tests revealed an equivalent resistance of the decellularized tissue to strain, before and after the cryopreservation, in comparison with the fresh tissue. Finally, immunological compatibility, showed a minimized host immune cells invasion and low levels of systemic inflammation, as assessed by tissue transplantation into immune-competent mice.

Conclusions: Our results indicate, for the first time, that fixative-free decellularized pericardium from cadaveric tissue donors can be banked according to Tissue Repository-approved procedures without compromising its mechanical properties and immunological tolerance. This tissue can be therefore treated as a safe homograft for cardiac surgery.

Editor: Hani A. Awad, University of Rochester, United States of America

Funding: The present work has been financed by Ricerca Corrente, Centro Cardiologico Monzino. The funders had no role in study design, data collection and analysis, decision to publish, or preparation of the manuscript.

* E-mail: maurizio.pesce@ccfm.it

Introduction

Pericardium is employed in cardiac surgery to repair congenital heart defects or to perform valve reconstruction. In addition, pericardium of animal origin (pig, cow) is the elective material to fabricate bio-prosthetic valves to be employed in surgical replacement of insufficient and/or stenotic valves. Indeed, thanks to its abundant collagen bundles and elastic fibers composition, pericardium has a relatively high resistance to mechanical stress, comparable to that of the native heart valves leaflets.

The technology commonly used to produce heart valves leaflets from animal-derived pericardium consists in treating the tissue with low concentration aldehydes (e.g. glutaraldehyde, GA), necessary to create chemical bonds among the extracellular matrix (ECM) components, thereby further increasing the tissue mechanical resistance, and to prevent acute host immune rejection [1]. On the other hand, aldehyde treatment has also drawbacks regarding the tissue long-term durability. In fact, clinical data from long term follow up of patients receiving pericardium-made bio-prosthetic valves, has clearly indicated a severe impact of structural valve deterioration (SVD) and calcification, due to permanence of fixative remnants having potent cytotoxic effects. In addition, GA fails in removing animal-specific antigens such as the (α1, 3)-Gal epitope [2–6], which elicits chronic rejection of the implanted prostheses. Importantly, the impact of SVD is highly correlated to the age of recipients, with younger patients being un-favored - and needing redo strategies - as early as 10–15 years following the first surgery in a significant number of cases [7,8].

In order to overcome these limitations, basic research has actively pursued fixation-free protocols (i.e., avoiding the use of aldehydes) aimed at enhancing pericardial tissue durability. In particular, novel decellularization methods have been proposed, based on treatment with hypotonic "decellularization" buffers, containing detergents and DNA/RNA digesting enzymes to eliminate cell debris and the remaining nucleic material from the tissue. These strategies have been employed in the production of both animal [9] and human-derived [10,11] pericardium with relevant implications for clinical use. Moreover, they have been proposed for the generation of tissue engineered heart valves (TEHVs) [12].

Despite a very active research in this field, animal-derived and human-derived decellularized pericardium have not been yet introduced in the routine clinical practice. This also depends on the absence of convincing evidences that this tissue may be prepared and cryopreserved according to Tissue Repository-compliant procedures, without compromising its mechanical properties and immunological tolerance. In the present study, we aimed at filling this technological gap by setting up a fixative-free decellularization process followed by tissue cryopreservation procedure of human pericardial tissue obtained by cadaveric tissue donors. We validated our protocol with specific histological and biomechanical analysis of the tissue, which revealed, *i)* a preserved structure of the collagen bundles and elastin fibers, and *ii)* the maintenance of suitable mechanical properties. Remarkably, the immunological compatibility of decellularized/cryopreserved pericardium was proven by subcutaneous transplantation for up to 60 days into immune-competent mice. From these results, we conclude the feasibility of a Tissue Repository approved procedure for the decellularization and cryopreservation of pericardial samples from tissue donors that may be released as a novel homograft tissue.

Results

Maintenance of tissue integrity and complete genetic material removal by fixative-free decellularization procedure

Histological analysis (Fig. 1A, 1B and 1C) was used as a first line of inspection to qualitatively assess, *i)* the efficiency of cells removal after decellularization, *ii)* the degree of the tissue histo-architecture preservation, and *iii)* the content and distribution of the collagen bundles and elastic fibers in the extracellular matrix (ECM). In detail, Masson's staining of decellularized (DE) or decellularized/cryopreserved (DE/CR) pericardium transversal histological sections showed a complete removal of cells. The treatment was not found to cause major deterioration of the tissue, as witnessed by preservation of collagen bundles and elastic fibers. In addition, DE or DE/CR samples did not appear swollen compared with fresh tissue. To verify the removal of cells, staining with Hoechst 33258 of the histological sections was also performed [10], followed by fluorescence microscopy analysis. As shown in Figure 2A, this showed an efficient removal of cells, with no evidence of cellular or nuclear residues observed in the sections. Absence of remaining genetic material from DE/CR samples was confirmed by Real Time-PCR on DNA extracted from three pericardium samples, using primer couples specific for the *hMYO-D*, *hGAPDH* and *hNFκB* genes promoter/coding sequences (Figure 2B, Table 1).

Mechanical integrity of human pericardium after cryopreservation procedure

Uniaxial tensile loading (UTL) tests were carried out on fresh, DE and N_2 vapors-cryopreserved DE pericardium (DE/CR) samples to quantitatively characterize the biomechanical properties of the tissue. In particular, we used UTL tests to reveal any potential alteration of the biomechanical characteristics of the

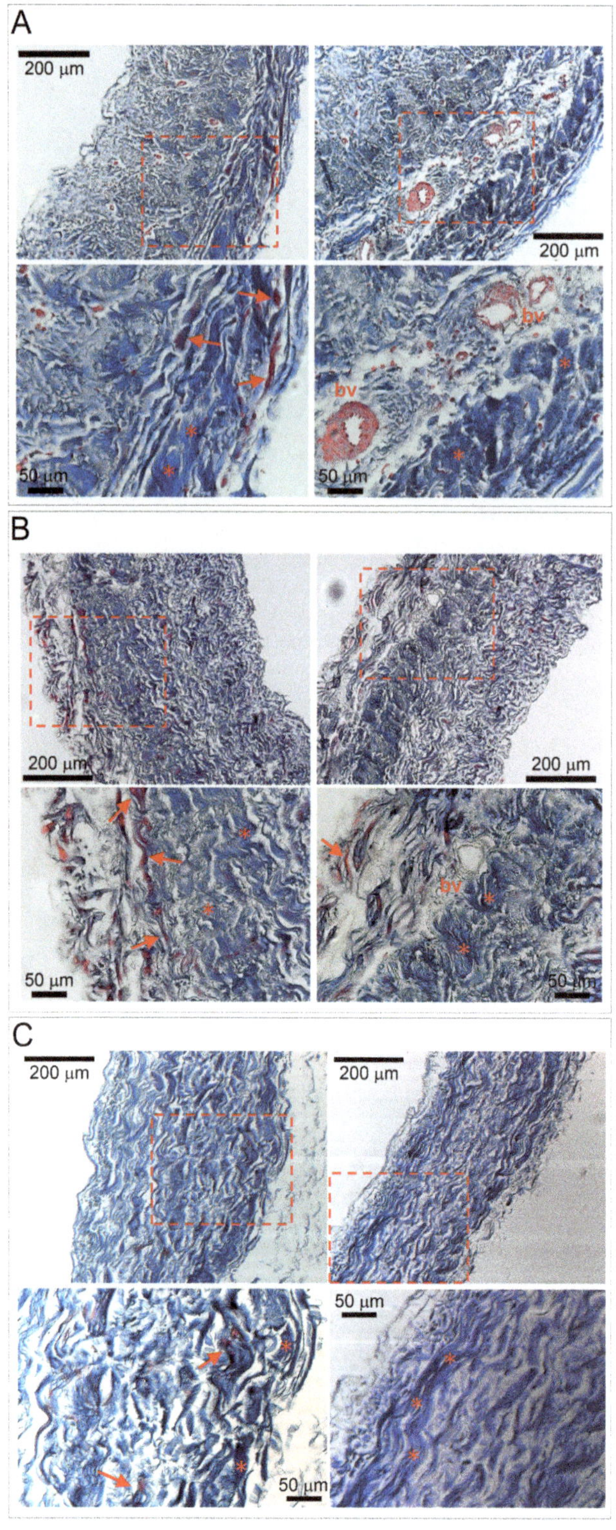

Figure 1. Histological appearance and decellularization of pericardial samples. (A, B, C) Masson's trichrome staining of the pericardial tissue before (A), after decellularization (B) and after decellularization/cryopreservation (C). Upper pictures in each panel show low magnification of the tissue, while lower pictures show magnification of the boxed areas. In these pictures it is evident that decellularization did not affect the structure of collagen bundles (*) and of elastic fibers (arrows). Bv: blood vessels.

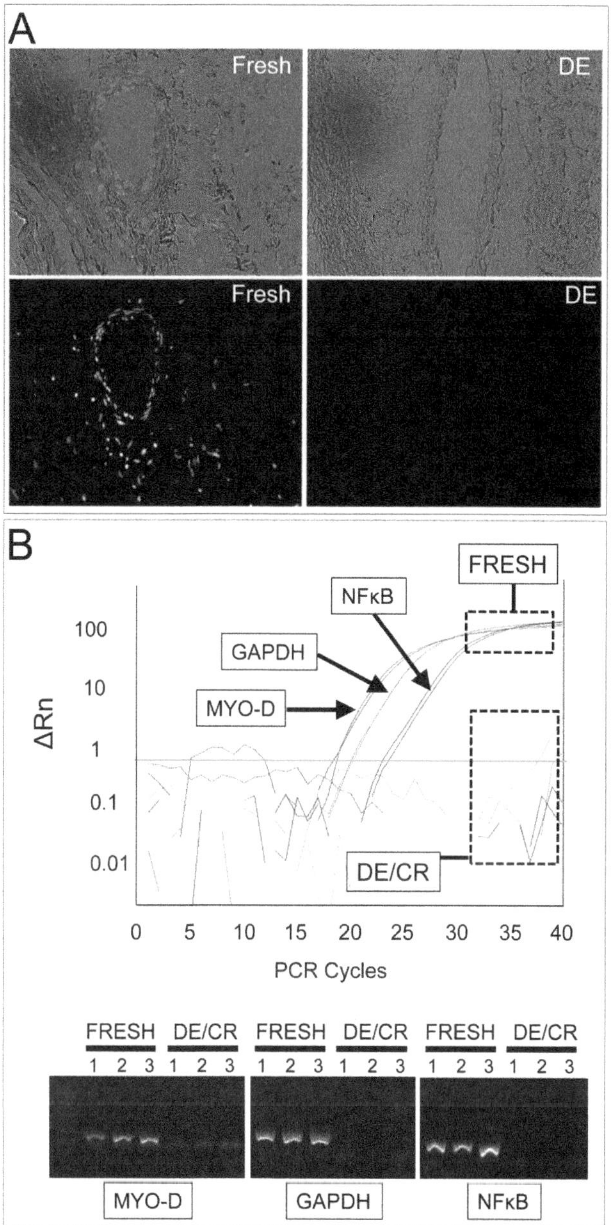

Figure 2. Efficiency of the decellularization process. (A) Hoechst 33258 staining was used to histologically assess the performance of the decellularization process. The upper panels show the overlapped bright field and Hoechst dye channels of the fresh and DE tissues without counterstaining, while lower panels show the dark field image of the Hoechst dye channel only in the same microscopic field. The lack of any signal in DE samples indicates a complete removal of double strand DNA from the tissue. (B) Q-PCR on DNA extracted from fresh and DE/CR samples shows complete removal of DNA by decellularization procedure using primers specific for the *hMYO-D*, *hGAPDH* and *hNFκB* genes promoter/coding sequences. The upper panel shows a representative amplification plot from one of the three FRESH samples with each primers pair, while the curves under the green bar are underthreshold signals generated by the corresponding DE/CR sample. The lower panel shows an agarose gel run of the PCR amplification products; as expected, no amplification bands were obtained by none of the DE/CR samples.

Table 1. CT values (mean±SE) obtained by real time PCR amplification of DNA extracted from FRESH and DE/CR samples.

	CT (mean±SE)	
GENE	**FRESH**	**DE/CR**
MYO-D	19,25±0,54	0
GAPDH	26,82±2,83	0
NFκB	18,68±0,53	0

Each amplification was performed in duplicate using 3 independent FRESH and DE/CR samples.

tissue induced by the cryopreservation procedure. Figure 3A–D shows the preparation steps of pericardial specimens for UTL tests, while 3E represents three typical stress/strain curves obtained from fresh, DE and DE/CR specimens strained to rupture. The stress-strain behavior for each specimen of the three groups was analyzed by means of six parameters [13], including: the elastic modulus (i.e., the stress-strain curve slope) at low (E_{low}) and high strain (E_{high}), representative of the tissue resistance due to the contribution of the elastin and collagen fibers composing the ECM, respectively; the tensile stress (σ_{trans}) and strain (ε_{trans}) values at the transition between the elastin and collagen stress-strain curve slope; the maximum tensile stress (σ_{max}) and strain (ε_{max}) characterizing the failure phase of the tissue sample. The comparison of the six parameters in fresh *vs.* DE and DE/CR groups did not show statistically significant differences (Fig. 3F), suggesting that the procedure adopted to obtain the decellularized pericardium not modify the tissue resistance to mechanical strain. This result suggests that the integrity and the arrangement of ECM components were maintained in DE and DE/CR pericardium samples, ensuring a biomechanical performance similar to that of native tissue.

Immunological compatibility of decellularized pericardium, before and after cryopreservation

The immunological compatibility of human- and animal-derived pericardium has been previously tested by transplantation in animals, where the reaction of the host against the graft, and the graft calcification were assessed mainly by *ex post* histological analysis of inflammatory/immune cells invasion [11,14–16]. To assess whether the decellularization and the decellularization/cryopreservation procedure of the human pericardium adopted in the present study led to changes in the tissue immune-tolerance, we adopted a similar strategy. Fresh, DE and DE/CR pericardium fragments were transplanted in subcutaneous position into immune-competent mice for 30 and 60 days followed by histology, immunohistochemistry and analysis of inflammatory-competent cells in peripheral circulation. As shown in Figure 4A, the histological appearance of DE and DE/CR samples was strikingly different from that of the fresh tissue. In fact, high levels of round cells with macrophage/lymphocyte morphology, forming deep layers of granulation tissue were observed in fresh pericardium samples; by contrast an overall lower number of cells with a predominant spindle-shape morphology and elongated nuclei, was observed in in DE and DE/CR samples. Of note, foci of donor-derived inflammatory cells infiltration and tissue matrix reabsorption were never observed in any of the DE and DE/CR samples explanted from mice at both time points, indicating a high - but not different - degree of compatibility of DE and DE/CR *vs.* fresh

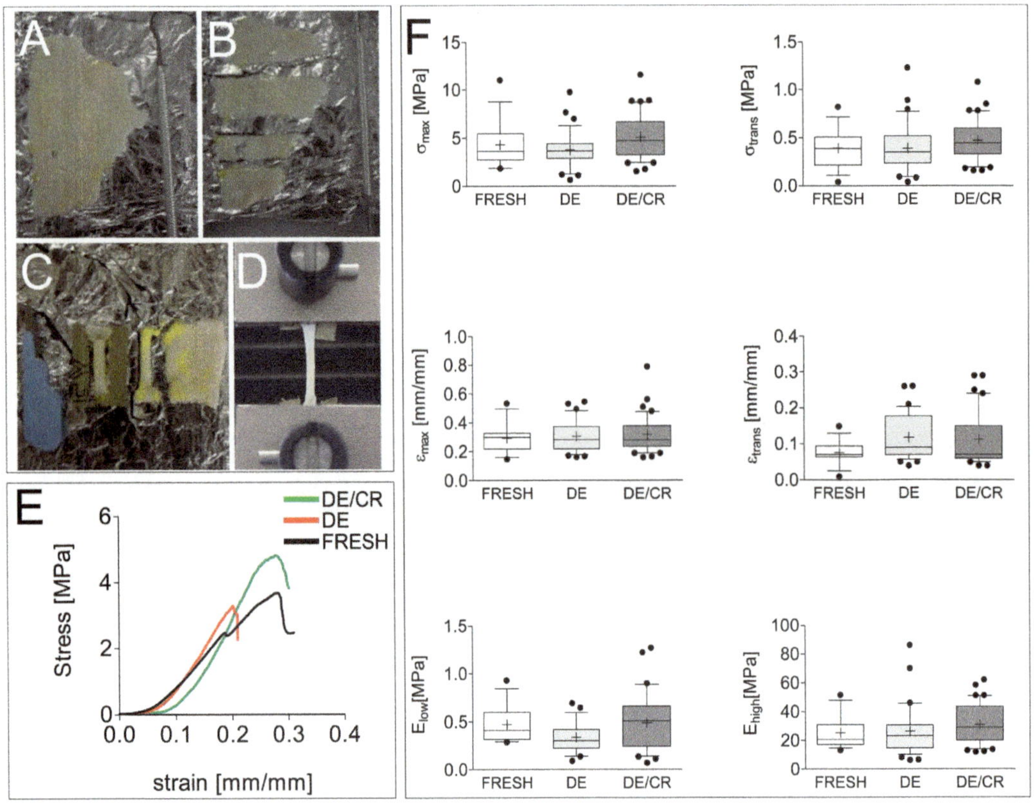

Figure 3. Uniaxial mechanical testing of fresh, DE and DE/CR pericardial samples. (A–D) Preparation steps of pericardial specimens; intact pericardium (A), strips of pericardium (B), dog-shaped specimens obtained from the strips (C), specimen mounted into the clamps of the uniaxial test machine (D). (E) Representative curves showing the stress/strain relationship of fresh, DE and DE/CR samples. (F) Graphical representation of the six mechanical parameters obtained for fresh, DE and DE/CR samples: elastic modulus at low (E_{low}) and high (E_{high}) strain values, transition stress (σ_{trans}) and strain (ε_{trans}), maximum tensile stress (σ_{max}) and strain (ε_{max}). Results are graphically represented with inter-quartile range box plots and whiskers indicating 10th and 90th percentile, plus outlier values (●). Continuous lines in box plots indicate median values of each data set, while crosses indicate mean of the same data. None of these values in DE or DE/CR samples were significantly different from fresh tissue values as verified by Kruskall-Wallis test (n = 15, fresh samples; n = 36, DE samples; n = 41, DE/CR samples; pericardial tissues from 3 independent cadaveric donors).

pericardium. To confirm the overall lower infiltration by host cells, an automatic counting of cellular nuclei was performed in fixed area images (0.15 μm^2) of hematoxyilin/eosin stained sections of fresh, DE and DE/CR samples. Results of this survey confirmed an overall lower presence of cells in DE and DE/CR compared with fresh pericardial samples (Fig. 4B).

We and others have found that patients receiving allograft valve transplantation show increased serum levels of anti HLA-1 antibodies [17] and of circulating cytotoxic and helper T-lymphocytes (CTLs, HTLs) [18,19].

Since an elevation of inflammatory markers and immune cell circulation has been indicated as one of the main causes of SVD, a time course analysis of circulating T-lymphocytes was performed in blood samples obtained from mice receiving fresh, DE and DE/CR pericardium at 15, 30, 45 and 60 days after implantation. As a reference for the dynamics of the T-cell mediated rejection of pericardial tissue, the relative ratio of the $CD4^+/CD8^+$ lymphocytes was calculated after appropriate recognition of total circulating T-lymphocytes stained with the *pan* T-cell marker CD3. As shown in Figure 5, a statistically significant decrease of this ratio, caused by a relative increase of circulating $CD8^+$ cells, was observed at all times in mice receiving fresh pericardial samples. By contrast, a constant $CD4^+/CD8^+$ cell ratio around or above 3 was observed in mice implanted with DE or DE/CR, suggesting absence of a strong T-cell-mediated immune response. Finally, no statistically significant difference between DE and DE/CR receiving mice was observed, suggesting that cryopreservation did not alter the immunological compatibility of human decellularized pericardium.

The host immune response has a striking impact on *in vivo* durability of GA-fixed and fixative-free *bio*-prosthetic valve grafts, due to chronic inflammatory process [20] and calcification [21], which leads to progressive SVD and decay of mechanical performance. To reveal the inflammatory cellular species invading the implanted pericardium specimens, an immunohistochemistry staining of fresh, DE and DE/CR samples was performed using anti mouse CD3 and CD11b antibodies, which recognized host-derived lymphocytes and macrophages, respectively (Fig. 6A). The results showed the presence of an elevated number of $CD3^+$ and $CD11b^+$ cells in fresh pericardial samples explanted at 30 and 60 days; these numbers were significantly higher than in DE or DE/CR explanted pericardial tissue (Fig. 6B). Interestingly, a lower T cells number in fresh pericardium was observed at 60 compared with 30 days, suggesting an attenuation of the host T-cell-mediated immune rejection at longer times post-implantation. A similar decrease in macrophages ($CD11b^+$ cells) content was not observed, showing a sustained innate immunity response.

Finally, to assess whether calcification secondary to inflammatory/immune response occurred in the transplanted pericardial specimens, Von Kossa staining was performed on histologic

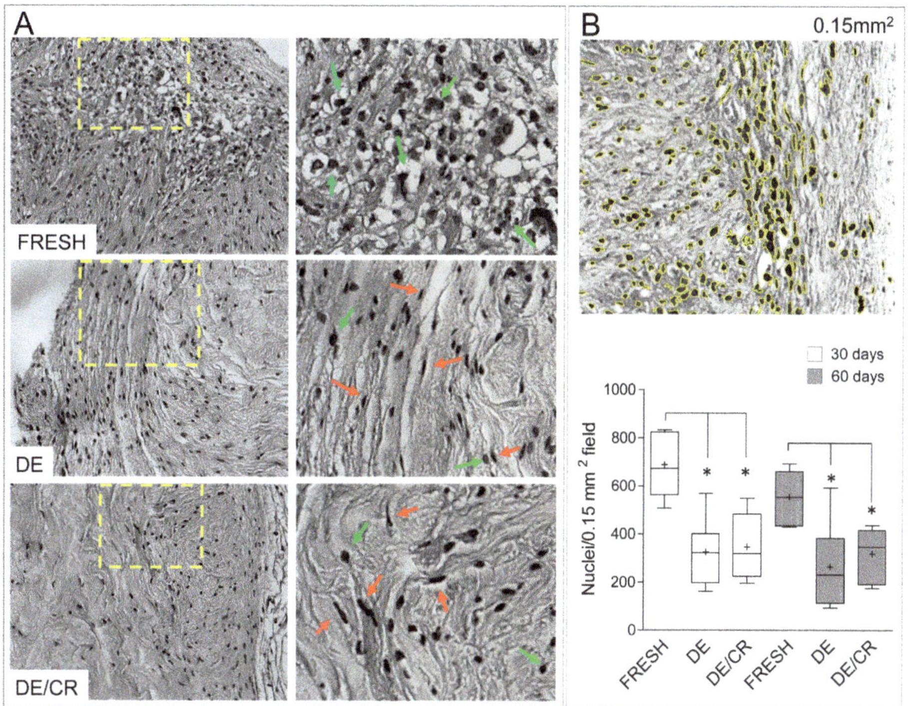

Figure 4. Gross morphology and host cell invasion quantitative analysis at 30 and 60 days following fresh, DE and DE/CR tissues implantation into imunocompetent mice. (A) Left panels show low magnification of fresh, DE and DE/CR pericardium after recovery from the subcutaneous position at 60 days. It is evident an overall higher number of cells in fresh *vs.* DE and DE/CR samples. Foci of inflammation with signs of tissue absorption were often observed in fresh samples, while these were never found in both DE and DE/CR tissues, indicating lower rejection. Panels on the right show high magnification of the yellow-boxed areas in left panels. It is evident that in fresh samples, infiltration was mainly due to cells with round nuclei, resembling inflammatory cells (green arrows), while the majority of infiltrating cells in DE and DE/CR pericardium specimens had elongated nuclei, resembling fibroblasts (red arrows). (B) Quantification of infiltrating cells was performed by computer assisted nuclei counting after appropriate filtering of the B/W images. Picture at the top is a 0.15 mm^2 micrograph where nuclei were automatically recognized and contoured in yellow by Image-J software; box-plot in the bottom indicates the results of nuclei counting at 30 and 60 days in fresh, DE and DE/CR samples. Results are graphically represented with inter-quartile range box plots and whiskers indicating 10th and 90th percentile. Continuous lines in box plots indicate median values of each data set, while crosses indicate mean of the same data. * indicates $P<0.05$ for statistical comparison of fresh *vs.* DE and DE/CR samples at both time points by Kruskall-Wallis test with Dunn's post-*hoc* analysis ($5\leq n\leq 8$).

sections of tissue explants at 60 days post surgery. As shown in Figure S1, no staining was observed in DE or DE/CR explanted samples; small calcium deposits were occasionally found in fresh pericardium specimens recovered after 60 days, showing that inflammatory response was not associated to major pericardium calcification in any of the conditions considered in the present study.

Discussion

Devising novel procedures to produce replacement allograft or bio-prosthetic valve conduits offers significant expectations to ameliorate the clinical outcome of the current heart valves disease treatments. In fact, the durability of these devices is still unacceptably limited, with an elevated burden for the patients' quality of life. As an example, studies reporting the clinical outcome of patients receiving bio-prosthetic valve implants clearly indicated a progressive SVD due to host inflammatory response and immune rejection. Strikingly, the clinical data indicated a higher propensity of younger subjects susceptible to develop implant failure more rapidly than older patients and thus requiring re-operation [7,8]. Depending on the type of the employed prostheses, different causes are at the basis of the insufficient durability. For example, the employment of non-decellularized/ cryopreserved valve homografts obtained from cadaveric tissue donors does not resolve the problem of the allogenic (man to man) immune response [20], while the employment of the so called "biological" valves - valve conduits of animal origin (porcine valves) or bio-prosthetic valves fabricated with leaflet mimicking biological tissues (typically pig and cow pericardium) - has its major limitation in the use of GA-based fixation protocols, which leave residual free aldehyde groups in collagen abundant tissues [22], form chemical bonds between ECM components, and fail in removing xeno-antigens such as the (α1, 3)-Gal epitope [5,6], recognized as "non-self" by the human species. Although advanced protocols to reduce the overall cytotoxicity of GA-fixed tissues have been devised [14,16,23–25], novel procedures to prepare leaflet and pericardial tissue patches preventing the use of aldehydes have been also explored. These procedures are based on decellularization of the tissue through osmotic lysis followed by exposure to ionic (e.g. SDS) and/or non-ionic (e.g. Triton-X 100) detergents to efficiently remove cell remnants. The apparent advantage of these procedures over the GA-based fixation is the lack of aldehyde residues, the absence of chemical bonds between ECM components, and the removal of xeno-antigens. On the other hand, whether decellularized tissues have or not an improved immunologic compatibility and an optimal mechanical performance is still matter of debate requiring further investigation [4,9,26].

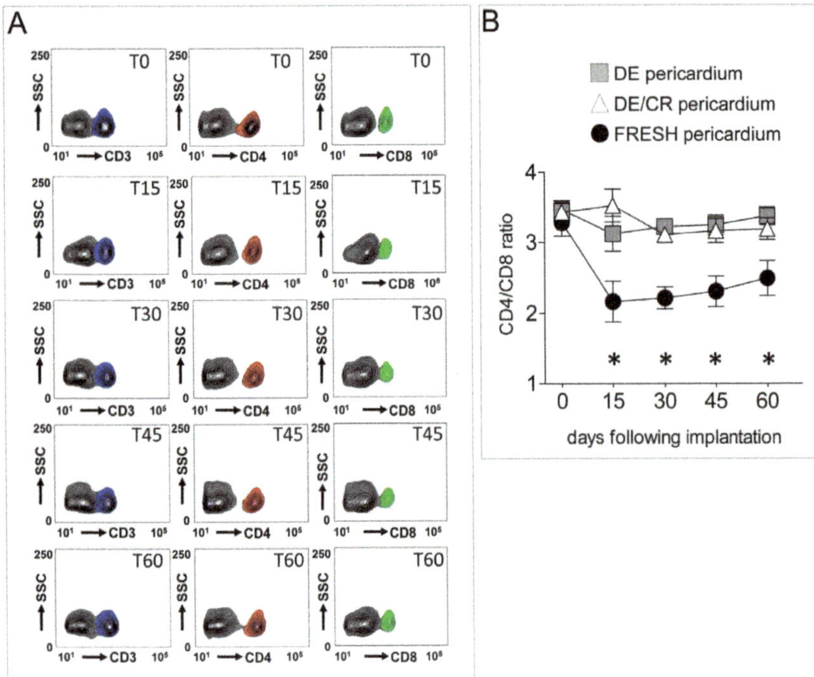

Figure 5. Dynamics of CD3$^+$, CD4$^+$ and CD8$^+$ cells mobilization in mice implanted with fresh, DE and DE/CR samples. (A) Representative flow cytometry detection of CD3$^+$ (left contour plots column), CD4$^+$ (center contour plots column) and CD8$^+$ (right contour plots column) circulating cells in a mouse receiving a non decellularized (FRESH) pericardial sample at 0, 15, 30, 45 and 60 days. CD3$^+$, CD4$^+$ and CD8$^+$ cells are indicated, respectively by blue, red and green contours on the right of the main PBMNCs population (grey contours), established by staining with isotype control antibodies. (B) CD4$^+$/CD8$^+$ lymphocytes ratio was determined at all the time points and plotted. Mice receiving fresh pericardium showed a reduced ratio suggesting increase in the number of circulating T-lymphocytes implicated in T cell-mediated tissue rejection. * indicate a statistical difference ($P<0.05$, by two-ways ANOVA with Bonferroni post-*hoc* test, n = 5) between the CD4/CD8 ratio in mice receiving fresh *vs.* those implanted with DE and DE/CR tissues.

To our knowledge, human-derived pericardium decellularized with an osmotic/detergent-based procedure has not yet clinically employed. In fact, preservation of mechanical properties and *in vivo* compatibility of this tissue were previously assessed [10,11]; however, the absence of data related to the maintenance of the tissue material properties after a cryopreservation storage period does not allow conclude whether this tissue may be eventually routinely prepared, banked and distributed by officially recognized homograft Repositories. This particular issue is not trivial. In fact, similar to the production process of cells for cell-based therapy according to the Good Manufacturing Practice [27], the homografts approval for clinical employment requires crucial validation steps to demonstrate the biological and the functional integrity of the tissue. In this regard, the UTL tests performed in the present study (Figure 3) showed that none of the parameters describing the mechanical properties of the human pericardium were modified by the decellularization or the decellularization/cryopreservation procedure. Thus, our data extend the observations by Mirsadraee et al. [10] and underline the relevance of fixation-free decellularization procedures to prepare human pericardium with intact mechanical properties for future clinical employments.

The result of subcutaneous implantation into immune-competent mice was performed taking as a reference the human fresh pericardium. This was done to assess the efficiency of the decellularization protocol at removing cellular material representing a source of allogenic tissue rejection, or promoting secondary calcification. As shown in Figure 4, the overall invasion by host cells was significantly lower in mice receiving DE and DE/CR tissue samples; moreover, in both cases, the number of infiltrating cells was significantly lower than in mice implanted with fresh pericardium. In addition, DE and DE/CR samples showed no signs of reabsorption and formation of granulomas. This finding is in line with the lower numbers of CD3$^+$ cells and CD11b$^+$ cells in DE and DE/CR compared with the fresh pericardium (Fig. 6) [11]. To further explore the role of systemic inflammation in mice implanted with human pericardial samples, the ratio of circulating CD4$^+$/CD8$^+$ lymphocytes, a parameter that is commonly taken as a reference to monitor rejection in solid organ transplantation [28] and mouse models [29], was also determined. These tests (Fig. 5) revealed a striking increase in the relative amount of circulating CD8$^+$ cells as early as at 15 days post-implantation in mice receiving the fresh pericardium, suggesting a rapid initiation of the rejection process. By contrast, DE and DE/CR pericardial implants were never found to cause an imbalance in lymphocytes ratio for the whole duration of the experiment. Although we did not specifically investigate the subclasses of CD8$^+$ circulating cells, this observation appears particularly relevant in the view of homograft transplantation of the DE/CR tissue. In fact, GA-fixed pericardial implants are known to cause an increase of circulating cytotoxic [18] or helper [19] T cells and production of anti HLA-I antibodies, which correlate with graft failure [17]. Thus, the fixative-free decellularization/cryopreservation method used in the present study may be helpful to reduce the inflammatory and cell-mediated immune response in recipients, improving the long-term durability of the implants.

In summary, thanks to unaltered mechanical properties, the potentially reduced cytoxicity and the lowered immunogenicity,

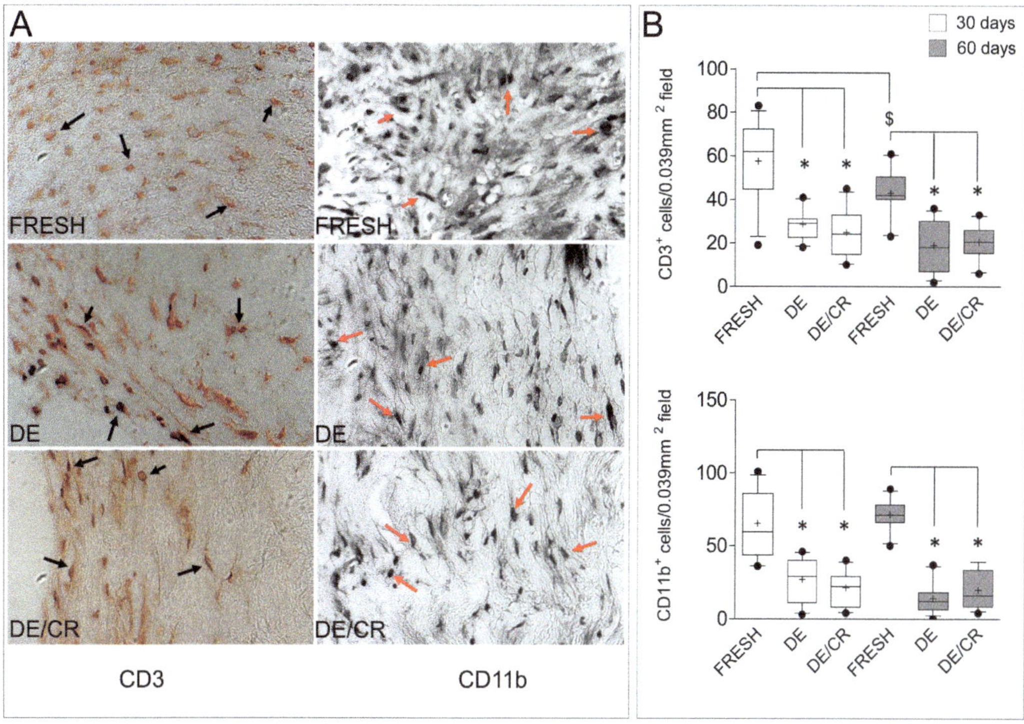

Figure 6. Infiltration of the implanted pericardial tissue by CD3$^+$ lymphocytes and CD11b$^+$ macrophages. (A) Representative images of fresh, DE and DE/CR pericardium tissue sections recovered by a mouse at 60 days and stained with CD3− and CD11b-specific antibodies. Black arrows and red arrows indicate, respectively CD3$^+$ and CD11b$^+$ cells present in the tissue. Note the higher number of cells in fresh pericardial specimens compared to DE or DE/CR samples. (B) Quantification of CD3$^+$ (upper plot) and CD11b$^+$ cells (lower plot) by software-assisted manual counting (Image J) into fixed (0.039 mm^2) tissue sections areas. At both time points the number of CD3$^+$ and CD11b$^+$ cells was significantly higher in mice receiving fresh vs. DE or DE/CR pericardium (* indicate $P<0.05$ by 1-way ANOVA with Newman-Keuls post-*hoc*; n = 12). Interestingly, the number of lymphocytes was significantly lower at 60 days compared with 30 days in mice receiving fresh pericardium ($ indicates P<0.05 by Mann-Whitney rank sum test; n = 12).

we propose the decellularized human pericardium produced with the protocol described here as a novel and safe homograft tissue to be clinically employed. In addition, since the adoption of fixative-free decellularization strategies is recommendable for production of biological-derived material for recellularization with living cells and derivation of TEHVs [16,25,30–33], the procedure described here may also be useful for devising novel living implants, with higher degree of immunological compatibility and lower potential for SVD, with significant benefits for the patients.

Materials and Methods

Ethics statement

Human pericardial samples were obtained as discharge material during heart dissection in the routine processing of valve homografts preparation at the Lombardy Cardiovascular Tissue Repository, the official Lombardy's Facility for collection of cadaveric cardiovascular tissues and cardiovascular homograft preparation and storage (www.cardiologicomonzino.it/Clinica/ChirurgiaCardiovascolare); authorization number: DGR VII/12848 (April 28, 2003 – Official Lombardy Bullettin). Samples from a total of six cadaveric tissue donors were used in the present study. Collection of hearts from these donors was performed after signature of an approved informed consent by donor relatives.

The procedures concerning animal care, surgery, and euthanasia were carried out in accordance to the "Guide for the Care and Use of Laboratory Animals" and the Helsinki declaration. They were further compliant with European directives and guidelines (Legislative Decree September 19, 1994, n. 626; 89/391/CEE, 89/654/CEE, 89/655/CEE, 89/656/CEE, 90/269/CEE, 90/270/CEE, 90/394/CEE, 90/679/CEE). The procedures used in the present study, were further approved and authorized by the Italian Ministry of Health and the University of Milan Ethical Committee (approval number:1242003-A 13/10/2003).

Decellularization procedure and histology

In the present study, a decellularization procedure similar to that developed by Mirsadree et al. [10,11] was adopted. After an accurate surgical elimination of fat, pericardial tissue was first washed for 90 min in phosphate-buffered saline (PBS) containing protease inhibitors (aprotinin, 10 KIU/ml, Trasylol, Bayer, Germany; 0.1% w/v EDTA, BDH, Lutterworth, United Kingdom) under continuous agitation, and with buffer changing every 30 min. To achieve cells removal from the tissue, the tissue was treated with hypotonic buffer (10 mM Tris-HCl; pH 8.0) for 16 h under continuous agitation at 4°C in the presence of protease inhibitors, followed by incubation for 24 h in 0.1% (w/v) sodium dodecylsulphate (SDS) in hypotonic buffer at room temperature under continuous agitation. To remove nucleic acid material, the samples were washed in sterile PBS three times for 30 min under agitation and incubated for 3 h at 37°C in a reaction buffer containing 50 U/mL deoxyribonuclease I from bovine pancreas [DNase, Sigma- Aldrich, Germany] and 1 U/mL ribonuclease A from bovine pancreas [RNase, USB] in 10 mM Tris-HCl (pH 7.5; Sigma-Aldrich) under gentle agitation on a gyrorocker. Finally, samples were washed in PBS for three periods of 30 min under agitation. To check the completeness of cell removal, pericardial tissue was fixed immediately after the end of the decellularization procedure and embedded into paraffin. Tissue sections (4–6 µm) were cut and stained with ematoxylin/eosin, Masson's trichrome solution, Von Kossa staining solution, or with Hoechst 333258 nuclear dye.

DNA extraction and Q-PCR methods

Total DNA was isolated with TriPure Isolation Reagent (Roche Applied Science) from 100 mg of fresh and DE pericardium samples. Gycogen was added to samples before precipitation to see pellets. Quantitative real-time PCR (qRT-PCR) amplifications were carried out for MYOD, GAPDH and NFkB gene promoters using Power SYBR Green PCR Master Mix (Applied Biosystems) on a 7900 Fast Real-Time PCR System (Applied Biosystems). Primer sequences were as follows:

hMYOD promoter: sense-5′-CCTCTTTCGGTCCCT-CTTTC-3′, antisense-5′-ATGGGTAGAGCGGCTGTAGA-3′;

hGAPDH: sense-5′-TACTAGCGGTTTTACGGGCG-3′, antisense-5′ -TCGAACAGGAGGAGCAGAGAGCGA-3′;

hNFκB promoter: sense-5′-CAGCCGATGAGAGCCGG-CAG-3′, antisense-5′-CCGGACCTCCCAGCCTGACA-3′.

Cryopreservation and thawing

Pericardium samples were frozen and thawed according to the procedures used by Lombardia Cardiovascular Tissue Bank. Briefly, samples were frozen in a double sterile bag containing freezing solution (RPMI 1640 plus 10% of DMSO) by the use of liquid nitrogen ultra-low-temperature cooler (Planer - SOL S.p.A., Milan, Italy) and stored in N_2 vapors in temperature range between −140°C and −185°C. The samples were thawed inside the sterile bag and washed three times in fresh RPMI 1640 [34,35] immediately before use.

Mechanical testing

Uniaxial tensile loading (UTL) was performed on native (fresh), decellularized (DE) and decellularized/cryo-preserved (DE/CR) pericardial samples. Before the test, the samples were maintained immersed in saline solution at 4°C. Dog-bone shaped specimens of 4 mm width and 20 mm height were dissected; in detail, 15, 36 and 41 specimens were retrieved and tested from the fresh, DE and the DE/CR groups, respectively. No preferential orientations were considered when cutting the specimens, assuming pericardium isotropy [10,36]. The thickness of each specimen was measured by means of a micrometer (0.05 mm of scale) in 3 different positions along the specimen height and the mean value (t_{av}) was used for mechanical parameters calculation. The specimens were then mounted onto the clamps of the testing machine; a sandpaper frame was constructed around the specimen to facilitate uniform gripping and to avoid sample slippage [37]. Tissue specimens were preloaded up to 0.01 N and subjected to a number of preconditioning loading-unloading cycles (ranging from 6 to 8 cycles), at 15% maximum strain with an elongation rate of 10 mm/min, until the loading-unloading curves were almost superimposed. After tissue preconditioning the specimen was preloaded up to 0.01N, the specimen initial length (l_0) was measured, and UTL test conducted until specimen failure, at a constant velocity of 10 mm/min. The samples were maintained immersed in PBS at 4°C until the UTL tests begun. All the tests were performed at room temperature and the specimens were maintained hydrated with PBS for the whole test duration.

The load (F) and displacement (l) data were acquired from the load cell and the stroke of the cross-head of the testing machine and the engineering stress (σ) and strain (ε) were calculated for each data point, according to equation 1 and 2:

$$\sigma = F/A \tag{1}$$

$$\varepsilon = (l - l_0)/l_0 \tag{2}$$

where A is the cross-sectional area of the specimen at zero strain, expressed as (eq. 3):

$$A = wxt_{av} \tag{3}$$

with, w, t_{av} and l_0 representing the specimen width, average thickness, and length at zero strain, respectively.

The stress strain behavior was described by means of 6 parameters according to Korossis et al. [13]: elastic modulus at low (E_{low}) and high (E_{high}) strain values, transition stress (σ_{trans}) and strain (ε_{trans}), maximum tensile stress (σ_{max}) and strain (ε_{max}). The biomechanical parameters calculated for verifying pericardium mechanical properties are based on engineering stress and strain values assuming that the cross-sectional area reduction is negligible during specimen deformation. Even if the real calculated stress, considering the cross-sectional reduction of the specimens may be more accurate, its measurement requires the use of optical-based systems, equipment not commonly integrated in the standard UTL apparatuses. On the other hand, most of the published UTL data consider the calculation of engineering stress and strain [10,35,36,37].

Surgical procedures, peripheral blood collection, pericardium recovery and image analysis

Subcutaneous implantation of pericardium fragments (10×10 mm) was performed in male CD1 mice (n = 54; 9 mice/group) (Charles River, Calco Italy) of about 30gr weight. Before implantation, mice were anesthetized with a ketamine (100 mg/Kg) - xylazine (20 mg/kg) solution injected intraperitoneally; at the same time, blood sampling was performed. Peripheral blood was also collected in 5% EDTA containing tubes at 15, 30, 45 and 60 days after implantation following mild anesthesia with isoflurane (1% in 100% oxygen, v/v). Pericardial tissue specimens recovered at 30 and 60 days after implantation were placed into 4% buffered formalin immediately after recovery from the animals and processed for conventional histological analysis (see before) and immunohistochemistry. Tissue sections were observed under light/fluorescence microscopes (Axiovert/Akioskop) equipped with image analysis software (Zeiss, Cell Quest). Images were acquired at a fixed magnification and analyzed using Image-J software (http://rsbweb.nih.gov/ij), after appropriate grayscale filtering, for automatic/manual nuclei/cells counting.

Immunohistochemistry

For immunohistochemistry, dewaxed and re-hydrated sections (5 µm thick) were incubated with 5% bovine serum albumin (BSA) in PBS to block nonspecific binding. The sections were then immunostained with rabbit mAbs recognizing mouse CD3 (Abcam, Cambridge, UK, ab5690) or CD11b (Abcam, Cambridge, UK, ab75476), followed by biotinylated goat anti-rabbit IgG, using an avidin–horseradish peroxidase–biotin complex method (Vector Laboratories, Burlingam, CA) and DAB or NovaRed substrate (Vector Laboratories). A manual counting protocol of $CD3^+$ and $CD11b^+$ cells was performed by using Image-J software.

Flow cytometric evaluation of circulating monocytes

Before staining with antibodies, blood was treated with ammonium chloride to lyse red blood cells. Lymphocyte population ($CD3^+$ cells) and $CD8^+$ cells were identified by staining with allophycocyanin-conjugated anti-mouse CD3 (Ab553066), and fluorescein-isothiocyanate-conjugated anti-mouse CD8 (Ab553031), while the percentage of $CD4^+$ population was evaluated by the use of phycoerythrin-conjugated anti-mouse CD4 antibody (Ab553653). All antibodies were from BD Biosciences, Milan, Italy. Cells were analyzed using a FACSCalibur flow cytometer (BD Biosciences, Milan, Italy). Data on $CD4^+/CD8^+$ cells ratio were derived by acquisition of at least 10^4 events with appropriate logical gating of the $CD3^+$ cellular population.

Statistical analyses

Statistical comparisons were performed with Graph-Pad 5 (Prism) statistical software. All data were analyzed with Shapiro-Wilk normality test. In case of normal data distribution, statistical comparisons were performed using parametric tests - 1-way/2ways ANOVA with Newman-Keuls/Bonferroni post-hoc analysis. In case of non-normal data distribution, the statistical methods used to compare data were the Kruskall-Wallis test with Dunn's multiple comparison post-hoc analysis or the Mann-Whitney rank sum test. The methods adopted to perform statistical comparisons are indicated in each figure legend.

Supporting Information

Figure S1 Von Kossa staining of fresh, DE and DE/CR pericardium specimens recovered from mice at 60 days following implantation. Except for few small calcium deposits (arrows) in fresh samples, calcification was not observed.

Acknowledgments

The Authors are very grateful to Dr. Eleonora Penza, Dr. Samer Kassem and Prof. Marco Agrifoglio (II Division Cardiac Surgery, Centro Cardiologico Monzino-IRCCS) for their continuous support during the execution of the present work.

Author Contributions

Conceived and designed the experiments: MCV MS GP MP LS. Performed the experiments: MCV GT LC FP MS RS F. Colazzo F. Consolo BM. Analyzed the data: MCV MS MP. Contributed reagents/materials/analysis tools: BM F. Colazzo MS MCV FP RS GT. Wrote the paper: MP.

References

1. Carpentier A, Lemaigre G, Robert L, Carpentier S, Dubost C (1969) Biological factors affecting long-term results of valvular heterografts. J Thorac Cardiovasc Surg 58: 467–483.
2. Grabenwoger M, Sider J, Fitzal F, Zelenka C, Windberger U, et al. (1996) Impact of glutaraldehyde on calcification of pericardial bio-prosthetic heart valve material. Ann Thorac Surg 62: 772–777.
3. Vincentelli A, Latremouille C, Zegdi R, Shen M, Lajos PS, et al. (1998) Does glutaraldehyde induce calcification of bio-prosthetic tissues? Ann Thorac Surg 66: S255–258.
4. Bloch O, Golde P, Dohmen PM, Posner S, Konertz W, et al. (2011) Immune response in patients receiving a bio-prosthetic heart valve: lack of response with decellularized valves. Tissue Eng Part A 17: 2399–2405.
5. Galili U (2005) The [alpha]-gal epitope and the anti-Gal antibody in xenotransplantation and in cancer immunotherapy. Immunol Cell Biol 83: 674–686.
6. Konakci KZ, Bohle B, Blumer R, Hoetzenecker W, Roth G, et al. (2005) Alpha-Gal on bioprostheses: xenograft immune response in cardiac surgery. Eur J Clin Invest 35: 17–23.
7. McClure RS, Narayanasamy N, Wiegerinck E, Lipsitz S, Maloney A, et al. (2010) Late Outcomes for Aortic Valve Replacement With the Carpentier-Edwards Pericardial Bioprosthesis: Up to 17-Year Follow-Up in 1,000 Patients. The Annals of Thoracic Surgery 89: 1410–1416.
8. Puvimanasinghe JPA, Takkenberg JJM, Eijkemans MJC, Steyerberg EW, van Herwerden LA, et al. (2005) Prognosis After Aortic Valve Replacement With the Carpentier-Edwards Pericardial Valve: Use of Microsimulation. The Annals of Thoracic Surgery 80: 825–831.
9. Mendoza-Novelo B, Avila EE, Cauich-Rodríguez JV, Jorge-Herrero E, Rojo FJ, et al. (2010) Decellularization of pericardial tissue and its impact on tensile viscoelasticity and glycosaminoglycan content. Acta Biomaterialia 7: 1241–1248.
10. Mirsadraee S, Wilcox HE, Korossis SA, Kearney JN, Watterson KG, et al. (2006) Development and characterization of an acellular human pericardial matrix for tissue engineering. Tissue Eng 12: 763–773.
11. Mirsadraee S, Wilcox HE, Watterson KG, Kearney JN, Hunt J, et al. (2007) Biocompatibility of acellular human pericardium. J Surg Res 143: 407–414.
12. Dainese L, Guarino A, Burba I, Esposito G, Pompilio G, et al. (2012) Heart valve engineering: decellularized aortic homograft seeded with human cardiac stromal cells. J Heart Valve Dis 21: 125–134.
13. Korossis SA, Booth C, Wilcox HE, Watterson KG, Kearney JN, et al. (2002) Tissue engineering of cardiac valve prostheses II: biomechanical characterization of decellularized porcine aortic heart valves. J Heart Valve Dis 11: 463–471.
14. Jorge-Herrero E, Fonseca C, Barge AP, Turnay J, Olmo N, et al. (2010) Biocompatibility and calcification of bovine pericardium employed for the construction of cardiac bioprostheses treated with different chemical crosslink methods. Artif Organs 34: E168–176.
15. Sinha P, Zurakowski D, Susheel Kumar TK, He D, Rossi C, et al. (2012) Effects of glutaraldehyde concentration, pretreatment time, and type of tissue (porcine versus bovine) on postimplantation calcification. The Journal of Thoracic and Cardiovascular Surgery 143: 224–227.
16. Maizato MJS, Pires MD, Canzian M, Higa OZ, Pitombo RNM, et al. (2008) Histological Evaluation of Biocompatibility of Lyophilized Bovine Pericardium Implanted Subcutaneously in Rats. Artificial Organs 32: 268–271.
17. Pompilio G, Polvani G, Piccolo G, Guarino A, Nocco A, et al. (2004) Six-year monitoring of the donor-specific immune response to cryopreserved aortic allograft valves: Implications with valve dysfunction. The Annals of Thoracic Surgery 78: 557–563.
18. Oei FBS, Welters MJP, Knoop CJ, Vaessen LMB, Stegmann APA, et al. (2000) Circulating donor-specific cytotoxic T lymphocytes with high avidity for donor human leukocyte antigens in pediatric and adult cardiac allograft valved conduit recipients. European Journal of Cardio-Thoracic Surgery 18: 466–472.
19. Welters MJP, Oei FBS, Vaessen LMB, Stegmann APA, Bogers AJJC, et al. (2001) Increased numbers of circulating donor-specific T helper lymphocytes after human heart valve transplantation. Clinical & Experimental Immunology 124: 353–358.
20. Koolbergen DR, Hazekamp MG, de Heer E, Bruggemans EF, Huysmans HA, et al. (2002) The pathology of fresh and cryopreserved homograft heart valves: An analysis of forty explanted homograft valves. The Journal of Thoracic and Cardiovascular Surgery 124: 689–697.
21. Manji RA, Zhu LF, Nijjar NK, Rayner DC, Korbutt GS, et al. (2006) Glutaraldehyde-Fixed Bio-prosthetic Heart Valve Conduits Calcify and Fail From Xenograft Rejection. Circulation 114: 318–327.
22. Levy RJ, Schoen FJ, Sherman FS, Nichols J, Hawley MA, et al. (1986) Calcification of subcutaneously implanted type I collagen sponges. Effects of formaldehyde and glutaraldehyde pretreatments. Am J Pathol 122: 71–82.
23. Guldner NW, Bastian F, Weigel Gn, Zimmermann Hr, Maleika M, et al. (2012) Nanocoating with titanium reduces iC3b- and granulocyte-activating immune response against glutaraldehyde-fixed bovine pericardium: A new technique to improve biologic heart valve prosthesis durability? The Journal of Thoracic and Cardiovascular Surgery 143: 1152–1159.
24. Guldner NW, Jasmund I, Zimmermann Hr, Heinlein M, Girndt B, et al. (2009) Detoxification and Endothelialization of Glutaraldehyde-Fixed Bovine Pericardium With Titanium Coating. Circulation 119: 1653–1660.
25. Rodas ACD, Polak R, Hara PH, Lee EI, Pitombo RNM, et al. (2011) Cytotoxicity and Endothelial Cell Adhesion of Lyophilized and Irradiated Bovine Pericardium Modified With Silk Fibroin and Chitosan. Artificial Organs 35: 502–507.
26. Rieder E, Seebacher G, Kasimir M-T, Eichmair E, Winter B, et al. (2005) Tissue Engineering of Heart Valves: decellularized porcine and human valve scaffolds differ importantly in residual potential to attract monocytic cells. Circulation 111: 2792–2797.
27. Gaipa G, Tilenni M, Straino S, Burba I, Zaccagnini G, et al. (2010) GMP-based CD133+ cells isolation maintains progenitor angiogenic properties and enhances standardization in cardiovascular cell therapy. J Cell Mol Med 14: 1619–1634.

28. Creemers P, Brink J, Wainwright H, Moore K, Shephard E, et al. (2002) Evaluation of peripheral blood CD4 and CD8 lymphocyte subsets, CD69 expression and histologic rejection grade as diagnostic markers for the presence of cardiac allograft rejection. Transplant Immunology 10: 285–292.
29. Chen CH, Yeh YC, Wu GJ, Huang YH, Lai WF, et al. (2010) Tracking the rejection and survival of mouse ovarian iso- and allografts in vivo with bioluminescent imaging. Reproduction 140: 105–112.
30. Vesely I (2005) Heart Valve Tissue Engineering. Circulation Research 97: 743–755.
31. Wong ML, Leach JK, Athanasiou KA, Griffiths LG (2012) The role of protein solubilization in antigen removal from xenogeneic tissue for heart valve tissue engineering. Biomaterials 32: 8129–8138.
32. Dong X, Wei X, Yi W, Gu C, Kang X, et al. (2009) RGD-modified acellular bovine pericardium as a bio-prosthetic scaffold for tissue engineering. J Mater Sci Mater Med.
33. Tedder ME, Liao J, Weed B, Stabler C, Zhang H, et al. (2009) Stabilized collagen scaffolds for heart valve tissue engineering. Tissue Eng Part A 15: 1257–1268.
34. Barili F, Dainese L, Cheema FH, Dell'Antonio G, Topkara VK, et al. (2007) Rates of cycling cells in cryopreserved valvular homograft: a preliminary study. Artif Organs 31: 152–154.
35. Dainese L, Barili F, Topkara VK, Cheema FH, Formato M, et al. (2006) Effect of cryopreservation techniques on aortic valve glycosaminoglycans. Artif Organs 30: 259–264.
36. Lee JM, Boughner DR (1985) Mechanical properties of human pericardium. Differences in viscoelastic response when compared with canine pericardium. Circulation Research 57: 475–481.
37. Dahms SE, Piechota HJ, Dahiya R, Lue TF, Tanagho EA (1998) Composition and biomechanical properties of the bladder acellular matrix graft: comparative analysis in rat, pig and human. Br J Urol 82: 411–419.

15

Laser-Assisted *In Vitro* Fertilization Facilitates Fertilization of Vitrified-Warmed C57BL/6 Mouse Oocytes with Fresh and Frozen-Thawed Spermatozoa, Producing Live Pups

Stephanie E. Woods, Peimin Qi, Elizabeth Rosalia, Tony Chavarria, Allan Discua, John Mkandawire, James G. Fox, Alexis García*

Transgenic Core Facility, Division of Comparative Medicine, Massachusetts Institute of Technology, Cambridge, Massachusetts, United States of America

Abstract

The utility of cryopreserved mouse gametes for reproduction of transgenic mice depends on development of assisted reproductive technologies, including vitrification of unfertilized mouse oocytes. Due to hardening of the zona pellucida, spermatozoa are often unable to penetrate vitrified-warmed (V-W) oocytes. Laser-assisted *in vitro* fertilization (LAIVF) facilitates fertilization by allowing easier penetration of spermatozoa through a perforation in the zona. We investigated the efficiency of V-W C57BL/6NTac oocytes drilled by the XYClone laser, compared to fresh oocytes. By using DAP213 for cryoprotection, 83% (1,470/1,762) of vitrified oocytes were recovered after warming and 78% were viable. Four groups were evaluated for two-cell embryo and live offspring efficiency: 1) LAIVF using V-W oocytes, 2) LAIVF using fresh oocytes, 3) conventional IVF using V-W oocytes and 4) conventional IVF using fresh oocytes. First, the groups were tested using fresh C57BL/6NTac spermatozoa (74% motile, 15 million/ml). LAIVF markedly improved the two-cell embryo efficiency using both V-W (76%, 229/298) and fresh oocytes (69%, 135/197), compared to conventional IVF (7%, 12/182; 6%, 14/235, respectively). Then, frozen-thawed C57BL/6NTac spermatozoa (35% motile, 15 million/ml) were used and LAIVF was again found to enhance fertilization efficiency, with two-cell embryo rates of 87% (298/343) using V-W oocytes ($P<0.05$, compared to fresh spermatozoa), and 73% (195/266) using fresh oocytes. Conventional IVF with frozen-thawed spermatozoa using V-W (6%, 10/168) and fresh (5%, 15/323) oocytes produced few two-cell embryos. Although live offspring efficiency following embryo transfer was greater with conventional IVF (35%, 18/51; LAIVF: 6%, 50/784), advantage was seen with LAIVF in live offspring obtained from total oocytes (5%, 50/1,010; conventional IVF: 2%, 18/908). Our results demonstrated that zona-drilled V-W mouse oocytes can be used for IVF procedures using both fresh and frozen-thawed spermatozoa, producing live pups. The ability to cryopreserve mouse gametes for LAIVF may facilitate management of large-scale transgenic mouse production facilities.

Editor: Stefan Schlatt, University Hospital of Münster, Germany

Funding: SEW was supported by a National Institutes of Health (NIH) National Research Service Award (T32-RR070036; http://grants.nih.gov/training/nrsa.htm). PQ, ER, AD, JGF, AG and in part TC were supported by departmental funding. JM and in part TC were supported by a NIH National Cancer Institute Program Project (P01CA10451; http://grants.nih.gov/grants/guide/pa-files/PAR-09-025.html). The funders had no role in study design, data collection and analysis, decision to publish, or preparation of the manuscript.

Competing Interests: The authors have declared that no competing interests exist.

* E-mail: agarcia@mit.edu

Introduction

Transgenic mice are used broadly in biomedical research and the utility of cryopreserved mouse gametes for reproduction depends on development of assisted reproductive technologies, including vitrification of unfertilized mouse oocytes. In contrast to another cryopreservation technique called 'slow-freezing', vitrification prevents physiological damage caused by the formation of ice crystals both within and outside the cells, and reduces chilling damage by allowing for more rapid freezing [1,2]. This technique is established for cryopreservation of mammalian embryos, but is a developing technology for oocytes [1,3,4]. The recent development of a simple, efficient and general-purpose protocol for vitrification of mouse oocytes using DAP213 (2 M DMSO, 1 M acetamide, 3 M propylene glycol) has made possible further applications of vitrified-warmed (V-W) oocytes in both large- and small- scale transgenic mouse production facilities [5,6].

Spermatozoa are often unable to penetrate V-W mouse oocytes due to hardening of the zona pellucida (ZP) following premature release of cortical granules [7–10]. Previous studies have evaluated intracytoplasmic sperm injection (ICSI) [9], partial ZP digestion [7] and piezo-actuated zona drilling [8] or incision [10] for easier penetration, and found fertilization advantage. One study also examined laser-assisted zona drilling prior to vitrification of mouse oocytes and demonstrated improved post-warming *in vitro* fertilization (IVF) using spermatozoa from a subfertile transgenic mouse [11].

Laser-assisted zona drilling offers promise for assisted reproduction in both humans and mice [11–17]. A recent study using laser-assisted IVF (LAIVF) to derive subfertile genetically-modified

(GM) mouse lines achieved fertilization rates 4 to 10 times that of conventional IVF [12]. Further, fertilization rates have been shown to be higher with LAIVF in mice with poor quality sperm [11,13,16–19].

To test whether zona-drilled V-W mouse oocytes can be used for IVF procedures using both fresh and frozen-thawed spermatozoa, we investigated the fertilization efficiency (% two-cell embryos) and % live offspring of V-W C57BL/6NTac oocytes zona-drilled by the XYClone laser, compared to fresh oocytes. Our study suggests that laser-assisted zona drilling after vitrification of mouse oocytes for IVF may facilitate the production of transgenic mice.

Methods

Ethics Statement

Mice were housed in an Association for Assessment and Accreditation of Laboratory Animal Care International-accredited facility. This study was carried out in strict accordance with the recommendations in the Guide for the Care and Use of Laboratory Animals. All animal use was approved by the M.I.T. Committee on Animal Care (Animal Welfare Assurance #: A-3125-01).

Mice

Four-month-old C57BL/6NTac proven breeder males were used for collection of spermatozoa, and 4–5-week-old C57BL/6NTac superovulated females were used for oocyte collection. Two-cell embryo transfer recipients were 0.5 dpc (days post coitum) pseudopregnant Crl:CD1(ICR) females (2–4 months old). Vasectomized male mice [Crl:CD1(ICR); 2–8 months old] were used to induce pseudopregnancies. Macroenvironmental housing conditions included a 14:10 and 12:12 light/dark cycle for oocyte donors and two-cell embryo recipients, respectively, with temperature maintenance at 68±2°F.

Media

All media used for spermatozoa collection and evaluation, oocyte collection and warming, and conventional and laser-assisted IVF were equilibrated and warmed prior to use. An incubator (Forma Scientific, Inc., Marietta, OH) set at 37°C with 5% CO_2 was used throughout the experiment for pre-warming and incubation. Potassium simplex optimized medium (KSOMaa) and human tubal fluid medium (HTF) were purchased from Zenith Biotech (Guilford, CT); bovine serum albumin (Sigma-Aldrich, Saint Louis, MO) was added to the media at 4 mg/ml. FHM medium, a modified KSOM medium using HEPES instead of bicarbonate as the buffer, and M2 medium were both purchased from Millipore (Billerica, MA).

Spermatozoa Collection, Cryopreservation and Analysis

For collection of fresh spermatozoa after CO_2 euthanasia of the donor, the epididymides and vasa deferentia were removed and placed into a 200 μl drop of HTF medium. Multiple tears in the epididymides and vasa deferentia were made using an 18-gauge needle under light microscopy, and spermatozoa were left to swim out into the medium during a 10-minute incubation period at 37°C. Spermatozoa collected for cryopreservation was obtained similarly from a donor, but into a 200 μl drop of cryoprotectant agent (CPA; 3% milk, 18% raffinose, 488 mM L-glutamine). Ten μl of spermatozoa suspension was loaded into each 0.25 ml plastic straw. Straws were slowly cooled in liquid nitrogen vapor (about −150°C) for 10–15 minutes, then plunged into liquid nitrogen for storage; spermatozoa were kept frozen for at least 1 week prior to use.

One person was responsible for evaluating concentration and motility of both fresh and frozen-thawed (pre- and post-freeze) spermatozoa. Spermatozoa concentration of suspension was determined by adding 4 μl of the spermatozoa in HTF medium or CPA into 96 μl of sterile water (dilution factor of 25). Ten μl of homogenized, diluted spermatozoa were loaded under the coverslip on each side of an improved Neubauer hemocytometer (Bright-Line, American Optical, Buffalo, NY.) and placed into a pre-wetted incubation chamber for 20 minutes prior to analysis. Spermatozoa were counted in the 5 designated squares under light microscopy at 100x magnification and the count averaged between the two sides, with concentration per ml calculated as dilution factor *count in 5 squares *0.05×10^6 [20].

To determine motility of spermatozoa, 4 μl of the spermatozoa suspended in HTF medium or CPA was added to 96 μl of HTF medium and incubated at 37°C with 5% CO_2 for 10 minutes. Following incubation, 10 μl of spermatozoa uniformly dispersed in HTF medium was pipetted onto a microscope slide and covered with a coverslip. At 100x magnification, the first 100 spermatozoa manually visualized were counted to establish % progressively motile, % non-progressively motile and % immotile. Based on visual observation, progressive motility was defined as spermatozoa swimming in a forward manner [21].

Oocyte Collection

Pregnant mare serum gonadotropin (National Hormone and Peptide Program, Torrance, CA) and human chorionic gonadotropin (hCG; Sigma-Aldrich, Saint Louis, MO) were administered intraperitoneally (5 IU per mouse) 48 hours apart to superovulate oocyte donors. Oviducts were collected from females euthanized by cervical dislocation starting at 13 hours after hCG administration; approximately 20–30 oocytes are collected per donor. Cumulus oocyte complexes (COCs) were released from the oviduct ampullas into HTF (for conventional IVF) or FHM (for LAIVF) medium containing 0.1% hyaluronidase (HA). The dish of COCs with medium containing HA was incubated for 5 minutes at 37°C to remove cumulus cells from oocytes. Cumulus-free oocytes were then collected and washed 3 times with HTF or FHM medium.

Vitrification and Warming of Oocytes

For oocyte vitrification, a modified Nakagata et al. protocol was followed [5,6]. Cumulus-free oocytes were equilibrated in a drop of FHM medium containing 1 mM DMSO (DMSO/FHM) at room temperature, and loaded into a cryogenic vial to then sit at 4°C for 3–5 minutes; each vial contained approximately 50 oocytes in 5 μl DMSO/FHM. Ninety microliters of 4°C cryoprotectant DAP213 (2 M DMSO, 1 M acetamide, 3 M propylene glycol) was added to the vial and the lid was loosely closed. The vial was then immediately submerged in liquid nitrogen and stored until use, for at least 1 week. The vitrification procedure took less than 10 minutes.

On the morning of IVF, vials were periodically removed from liquid nitrogen; within 30 seconds, 0.9 ml of pre-warmed (37°C) FHM medium containing 0.25 M sucrose (sucrose/FHM) was added and mixed by gently pipetting up-and-down. An additional 0.5 ml of sucrose/FHM medium was used to rinse each vial to maximize oocyte recovery. Medium containing oocytes was placed in a 35 mm dish for evaluation. Survived V-W oocytes were washed 2 times with pre-equilibrated HTF or FHM medium and used for conventional IVF or LAIVF, as described below.

Conventional and Laser-assisted IVF

On the morning of IVF, the straws of frozen spermatozoa were thawed in a 37°C water bath for 2 minutes. The ends of the straws were cut and the contents of one straw (10 µl; approximately 3 million spermatozoa) were purged into a 200 µl IVF drop of HTF medium (15 million spermatozoa/ml). For IVF using fresh spermatozoa, 10 µl of gently mixed spermatozoa suspension (approximately 3 million spermatozoa) was added to each 200 µl IVF drop of HTF medium (15 million spermatozoa/ml). Frozen-thawed and fresh spermatozoa were allowed to pre-incubate in the IVF drop for at least 20 minutes prior to the addition of oocytes.

For conventional IVF, approximately 100 survived V-W oocytes or freshly harvested cumulus-free oocytes were incubated for 4–6 hours in a 200 µl IVF drop of HTF medium containing capacitated spermatozoa. Following fertilization, viable oocytes were washed twice to remove spermatozoa and cultured overnight with KSOMaa medium.

For LAIVF, one ZP perforation per fresh or V-W oocyte in FHM medium was made at room temperature on a microscope (Nikon Diaphot 200) equipped with the XYClone laser (Hamilton Thorne Biosciences, Beverly, MA) set at 100% power and 300 µs; the associated computer software was used to monitor the power and view the image. ZP perforation was performed on approximately 100 oocytes at a time and took approximately 10–15 minutes per dish. Perforated oocytes were washed twice in pre-equilibrated HTF medium and moved into an IVF drop containing spermatozoa, as described above.

Embryo Transfer Surgery and Evaluation

The morning following IVF, all cells were counted and assessed for viability; the total number of cells, numbers of non-viable and one-cells, and number of two-cell embryos were recorded. Two-cell embryos were surgically transferred into the oviducts (14 two-cell embryos per oviduct) of the embryo transfer surgery recipients. Pseudopregnancies of recipients were induced following mating with vasectomized males 0.5 days prior to surgery; a recipient was selected based on the observation of a copulatory plug the morning of surgery.

To evaluate *in vivo* development, all recipients of embryos obtained through conventional IVF were sacrificed at embryo day 17 (E17); resorption and implantation sites were counted and the numbers of non-viable and viable E17 embryos were recorded. Approximately half of the recipients that received embryos from LAIVF were also sacrificed and evaluated as described above. The remaining embryo transfer surgery recipients were pair-housed by group and allowed to deliver pups naturally. The number of live pups delivered per recipient pair was determined, and pups were maintained for at least 2 weeks after birth.

Statistics

Percent two-cell embryos (fertilization efficiency), viable E17 embryos, live pups and live offspring per group are presented as means ± standard error of means (SEM). Percent two-cell embryos were calculated as the number of two-cell embryos/ number of total cells *100. Percent viable E17 embryos or % live pups were calculated as the number of viable E17 embryos or number of live pups/two-cell embryos transferred *100. Percent live offspring was calculated as the number of viable E17 embryos and live pups/total number of two-cell embryos transferred. All percentage data were subjected to arcsine transformation prior to statistical analysis. Data were analyzed by Student's t-tests, where $P<0.05$ was considered statistically significant and 2–7 replicates were included in each group compared. Statistical analysis was performed using STATA/IC 12.1 for Mac (StataCorp, College Station, TX) and Prism Version 5.0 (GraphPad Software, La Jolla, CA).

Results

Vitrification of Mouse Oocytes Resulted in Satisfactory Recovery and Survival

In total, 1,762 oocytes were vitrified, and 83% (n = 1,470) were successfully recovered from the cryogenic vials post-warming. Of those V-W oocytes recovered, 78% (n = 1,151) were viable and proceeded to be used for IVF. There were no discernible differences microscopically between survived V-W and freshly harvested oocytes before and after zona perforation, and upon generation of two-cell embryos (Figure 1).

Cryopreservation of Mouse Spermatozoa Caused Decreased Motility, but Improved Generation of Two-cell Embryos Using Zona-drilled V-W Mouse Oocytes

The concentrations of spermatozoa in suspension obtained from each of the two donors were similar (fresh: 275×10^6/ml; frozen-thawed, pre-freeze: 298×10^6/ml; Table 1). Further, fresh spermatozoa from both donors had comparable motility (progressive+ non-progressive motility; fresh: 74%; frozen-thawed, pre-freeze: 70%; Table 1). After freezing, the concentration of spermatozoa remained commensurate (278×10^6/ml), but motility was decreased (35%) (Table 1).

Using LAIVF, V-W oocytes with frozen-thawed spermatozoa produced significantly more two-cell embryos (87.1±2.7%; Table 2) than with fresh spermatozoa (76.2±2.5%; Table 2) ($P<0.05$; Figure 2), but results were comparable when fresh oocytes were used. There were no significant differences in two-cell embryo efficiency between frozen-thawed and fresh spermatozoa using conventional IVF (Table 2; Figure 2).

LAIVF Improved % Two-cell Embryos Using both V-W and Fresh Mouse Oocytes

In all groups compared, two-cell embryo efficiency was greater when using LAIVF than conventional IVF, with both frozen-thawed and fresh spermatozoa. For instance when frozen-thawed spermatozoa were used, LAIVF with V-W oocytes generated 87.1±2.7% two-cell embryos, while conventional IVF resulted in 5.7±1.0% two-cell embryos ($P<0.001$; Table 2; Figure 2). Likewise, when fresh oocytes were used, 73.1±3.4% two-cell embryos formed with LAIVF, compared to 4.7±1.2% using conventional techniques ($P<0.01$; Table 2; Figure 2).

Zona-drilled V-W Mouse Oocytes Effectively Generated Two-cell Embryos

There were no statistically significant differences in using V-W versus fresh oocytes in generation of two-cell embryos. However, % two-cell embryos were higher with zona-drilled V-W oocytes compared to fresh oocytes, using both frozen-thawed (V-W oocytes: 87.1±2.7%; fresh oocytes: 73.1±3.4%; $P=0.06$) and fresh spermatozoa (V-W oocytes: 76.2±2.5%; fresh oocytes: 68.5±1.8%; $P=0.12$) (Table 2; Figure 2).

LAIVF with V-W Mouse Oocytes can be Used to Produce Live Pups

Normal live offspring were obtained from all groups except from conventional IVF using V-W oocytes and frozen-thawed spermatozoa, due to presumed failed induction of pseudopreg-

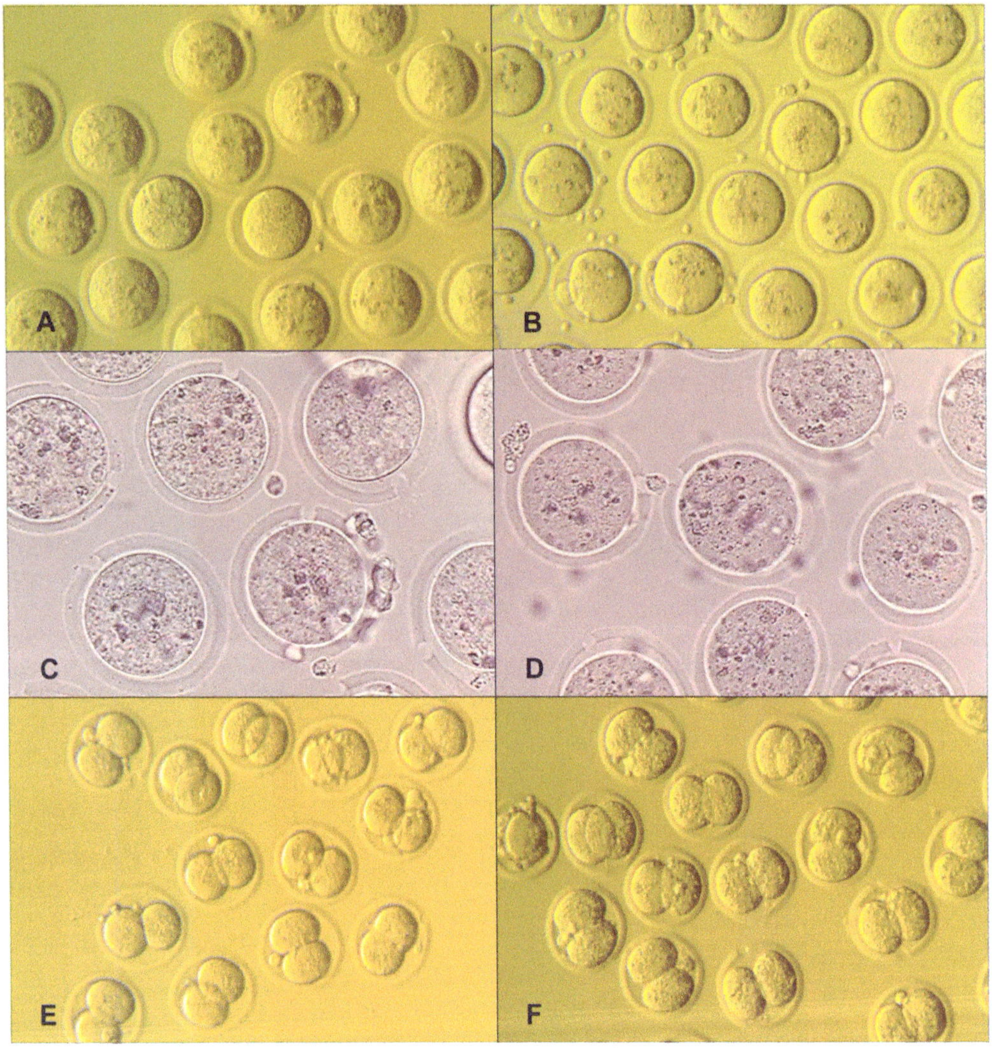

Figure 1. Microscopic images of V-W and fresh oocytes, and two-cell embryos generated by LAIVF. A: Zona-intact vitrified-warmed (V-W) oocytes, 200×. **B:** Zona-intact fresh oocytes, 200×. **C:** Zona-drilled V-W oocytes, 400×. **D:** Zona-drilled fresh oocytes, 400×. **E:** Two-cell embryos from laser-assisted IVF (LAIVF) using V-W oocytes, 200×. **F:** Two-cell embryos from LAIVF using fresh oocytes, 200×.

nancy of the embryo transfer surgery recipient. Sixteen-day-old pups from LAIVF using V-W oocytes and fresh spermatozoa are pictured in Figure 3. While no statistically significant differences were obtained between live offspring efficiencies of zona-drilled V-W and fresh oocytes, % live offspring from fresh oocytes was higher than from V-W oocytes, using both frozen-thawed (V-W

Table 1. Fresh and frozen-thawed spermatozoa suspension concentration and motility from 4-month-old C57BL/6NTac proven breeders.

			Motility (%)		
Spermatozoa		**[a]Concentration ($\times 10^6$/ml)**	**Progressive**	**Non-Progressive**	**Immotile**
Fresh		275	28%	46%	26%
Frozen-Thawed	Pre-Freeze	298	47%	23%	30%
	Post-Freeze	278	21%	14%	65%

[a]Concentration of spermatozoa suspension per ml used for IVF was determined via a hemocytometer under light microscopy at 100× magnification, and calculated as dilution factor *count in 5 squares *0.05×10^6, where the dilution factor was 25 [20]. Ten µl of spermatozoa suspension (approximately 3 million spermatozoa) was added to each 200 µl IVF drop of HTF medium (15 million spermatozoa/ml). At 100× magnification, the first 100 spermatozoa manually visualized were counted to establish % progressively motile, % non-progressively motile and % immotile. Based on visual observation, progressive motility was defined as spermatozoa swimming in a forward manner [21].

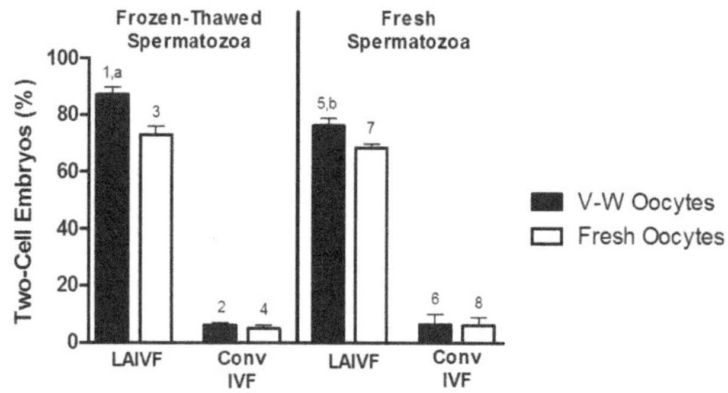

Figure 2. Percent two-cell embryos following LAIVF and conventional IVF using V-W and fresh oocytes. Key denotes vitrified-warmed (V-W) and fresh oocytes by black and white bars, respectively. Bars presented on the left represent data from IVF using frozen-thawed spermatozoa, and bars on the right represent data using fresh spermatozoa. Data are presented as means ± SEM. Different superscripts denote significant difference ($P<0.05$), where $^{a\text{-}b}$ denotes difference between frozen-thawed and fresh spermatozoa, and $^{1\text{-}2,\ 3\text{-}4,\ 5\text{-}6,\ 7\text{-}8}$ between laser-assisted IVF (LAIVF) and conventional IVF (Conv IVF).

oocytes: 3.8±1.6%; fresh oocytes: 9.9±3.3%; P = 0.09) and fresh spermatozoa (V-W oocytes: 4.5±1.6%; fresh oocytes: 8.9±1.8%; P = 0.15) (Table 3; Figure 4).

Statistical analysis on live offspring efficiency was not performed on groups where conventional IVF was used because too few two-cell embryos were generated to have more than one replicate per group. Overall, live offspring efficiency appeared much greater when using conventional IVF than with LAIVF: conventional IVF using V-W oocytes and fresh spermatozoa resulted in 66.7% live offspring, whereas with LAIVF, 4.5±1.6% live offspring were obtained (Table 3; Figure 4).

Discussion

This study demonstrated that zona-drilled V-W mouse oocytes can be used for IVF procedures using both fresh and frozen-thawed spermatozoa to produce live pups. Mouse oocytes were successfully cryopreserved and recovered, and demonstrated satisfactory survival rates using DAP213 as the cryoprotectant introduced by Nakagata et al. [5,6]. To reduce unintentional loss through adherence of V-W oocytes to the cryovial, recovery may be improved by using plastic straws or open pulled straws [8] with other cryoprotectants, although the cryovial may be preferred for temperature stability and safe transport [22]. Recent studies have investigated survival and development of cryopreserved oocytes, resulting in significant progress [1,5,6,23]. For instance, one study examined a vitrification protocol of oocytes from C57BL/6J mice, and found that vitrifying COCs produced more two-cell embryos and blastocysts than cumulus-free oocytes, and vitrified COCs developed to term at a rate equivalent to the use of fresh COCs [1]. Despite the loss of oocytes to incomplete recovery and survival following vitrification, zona-drilled V-W oocytes trended toward producing more two-cell embryos than fresh oocytes. This could possibly be due to preferential selection for healthy gametes through cryopreservation; Sakamoto et al. documented that frozen-thawed oocytes were highly tolerant to damage by injection during ICSI [9]. Because two-cell embryo and live offspring efficiency are comparable between V-W and fresh oocytes, advantage can be gained using V-W oocytes in overall convenience and in reducing after-hours labor costs associated

Table 2. Two-cell embryo efficiency of laser-assisted IVF (LAIVF) and conventional IVF (Conv IVF) using vitrified-warmed (V-W) and fresh oocytes.

Spermatozoa	IVF	Oocytes	# Total Cells	# Non-Viable Oocytes	# One-Cells	# Two-Cell Embryos	a% Two-Cell Embryos, Mean ± SEM
Frozen-Thawed	LAIVF	V-W	343	17	28	298	87.1±2.7%
		Fresh	266	15	56	195	73.1±3.4%
	Conv IVF	V-W	168	3	155	10	5.7±1.0%
		Fresh	323	15	293	15	4.7±1.2%
Fresh	LAIVF	V-W	298	11	58	229	76.2±2.5%
		Fresh	197	9	53	135	68.5±1.8%
	Conv IVF	V-W	182	5	165	12	6.8±3.6%
		Fresh	235	4	217	14	5.7±2.9%

[a]Mean ± SEM of % two-cell embryos was calculated from % two-cell embryos (# two-cell embryos/# total cells *100) of each replicate; 2–4 replicates were included in each group.

Figure 3. Sixteen-day-old pups from laser-assisted IVF using vitrified-warmed oocytes and fresh spermatozoa.

with hormone-priming and per diem rates for extended housing of oocyte donors [5].

In recent years advancements have been made in improving the efficiency of conventional IVF for both inbred and GM mice, yet imprecision and the need for technical refinement still exist. A protocol used for training in LAIVF by the European Mouse Mutant Archive reports a mean of ~25% two-cell embryos using conventional techniques and fresh C57BL/6NTac mouse gametes [24], while another study found ~50% to ~80% two-cell embryos, depending on the IVF medium used [25]. In this study, we obtained 6% two-cell embryos using fresh, but cumulus-free, oocytes and spermatozoa with conventional IVF. We suspect that optimizing our technique would not only improve the percentage of two-cell embryos generated, but also show recognizable differences between using cryopreserved versus fresh gametes. One possible improvement to conventional IVF would be to maintain cumulus cell presence during culture, as this has been demonstrated to improve *in vitro* embryo development [26,27] and fertilization efficiency [28–31]. One study found that the presence of an intact cumulus layer on oocytes increased the *in vitro* fertilizing ability of fresh capacitated mouse spermatozoa, with < 10% of cumulus-free oocytes from GM female mice inseminated, compared to almost 40% two-cell embryos with COCs [28] Because cumulus cells must be removed for rapid and effective

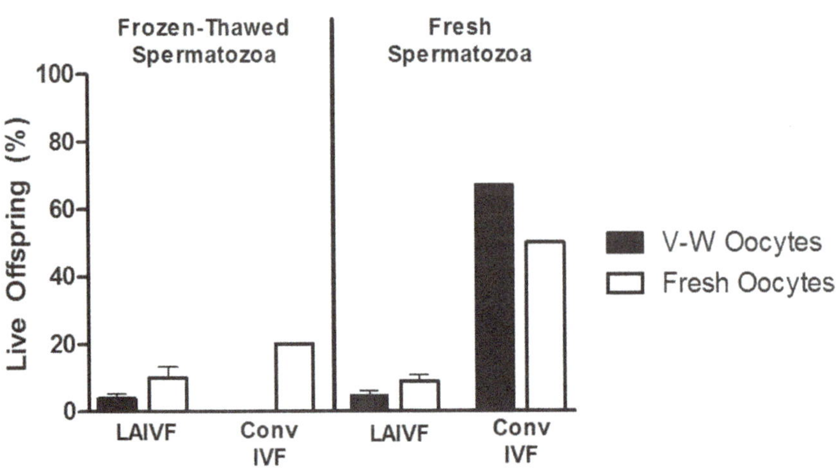

Figure 4. Percent live offspring following LAIVF and conventional IVF using V-W and fresh oocytes. Key denotes vitrified-warmed (V-W) and fresh oocytes by black and white bars, respectively. Bars presented on the left represent data from laser-assisted IVF (LAIVF) and conventional IVF (Conv IVF) using frozen-thawed spermatozoa, and bars on the right represent data using fresh spermatozoa. Data are presented as means ± SEM.

Table 3. Live offspring efficiency of laser-assisted IVF (LAIVF) and conventional IVF (Conv IVF) using vitrified-warmed (V-W) and fresh oocytes.

			Viable E17 Embryo Efficiency						Live Pup Efficiency			
Spermatozoa	IVF	Oocytes	# Two-Cell Embryos Transf	# Resorp Sites	# Implant Sites	# Non-Viable E17 Embryos	# Viable E17 Embryos	[a]% Viable E17 Embryos, Mean ± SEM	# Two-Cell Embryos Transf	[b]# Live Pups	[a]% Live Pups, Mean ± SEM	[c]% Live Offspring, Mean ± SEM
Frozen-Thawed	LAIVF	V-W	112	9	1	0	2	1.8±1.8%	168	11	6.6±2.1%	3.8±1.6%
		Fresh	56	6	0	0	6	10.8±7.2%	112	10	9.0±3.6%	9.9±3.3%
	Conv IVF	V-W	10	0	0	0	0	0.0%	NA			0.0%
		Fresh	15	2	0	0	3	20.0%	NA			20.0%
Fresh	LAIVF	V-W	112	27	1	0	5	4.5±2.2%	112	5	4.5±2.7%	4.5±1.6%
		Fresh	56	8	2	0	4	7.1±0.0%	56	7	12.5%	8.9±1.8%
	Conv IVF	V-W	12	3	0	0	8	66.7%	NA			66.7%
		Fresh	14	2	0	1	7	50.0%	NA			50.0%

Resorption is abbreviated as resorp, and implantation is abbreviated as implant.

[a]Means ± SEM of % viable E17 embryos and % live pups were calculated from % viable E17 embryos or % live pups [# viable E17 embryos or # live pups/# two-cell embryos transferred (transf) *100] of each replicate; 2–4 replicates were included in each group where SEM is indicated.

[b]One replicate included number of live pups obtained from 2 embryo transfer surgery recipients co-housed together post-operatively, with a combined total of 56 transferred two-cell embryos.

[c]Mean ± SEM of % live offspring was calculated from % viable E17 embryos and live pups (# viable E17 embryos and live pups/total # two-cell embryos transferred *100) of each replicate; 3–7 replicates were included in each group where SEM is indicated.

NA indicates data not available; too few two-cell embryos were obtained via conventional IVF to evaluate live pup efficiency.

Table 4. Summary of two-cell embryo and live offspring efficiencies of current and prior studies evaluating various techniques to improve penetration of the zona pellucida (ZP) of vitrified-warmed (V-W) mouse oocytes by spermatozoa.

Technique [Reference]	Spermatozoa	[a]Donor Mice	[b]% Two-Cell Embryos	[c]% Live Offspring
LAIVF, drilling post-vit (current study)	Frozen-Thawed	C57BL/6NTac	87.1% (298/343)	3.8% (13/280)
	Fresh		76.2% (229/298)	4.5% (10/224)
Piezo-actuated zona drilling [8]	Frozen-Thawed	Kunming	NA	NA
	Fresh		85.5% (112/131)	8.1% (12/149)
Intracytoplasmic sperm injection [9]	Frozen-Thawed	C57BL/6J Jcl	72.3% (136/188)	42.0% (68/162)
	Fresh		70.3% (78/111)	NA
Piezo-actuated zona incision [10]	Frozen-Thawed	Kunming	69.9% (137/196)	10.0% (17/170)
	Fresh		NA	NA
LAIVF, drilling pre-vit [11]	Frozen-Thawed	Oocytes: C57BL/6; Spermatozoa: C57BL/6N-Tg(UCP/FAD2)U8	NA	NA
	Fresh		52.7% (48/91)	13.3% (6/45)
Partial ZP digestion [7]	Frozen-Thawed	Kunming	48.8% (61/125)	3.4% (6/174)
	Fresh		NA	NA

Laser-assisted IVF is abbreviated as LAIVF, and vitrification is abbreviated as vit.
[a]Kunming mice are outbred, and C57BL/6 mice are inbred.
[b]Percent two-cell embryos was calculated as # two-cell embryos/# total cells post-recovery and prior to treatment *100; data presented differently in the original manuscript were reconfigured to match the format of data presentation in the current study.
[c]Percent live offspring was calculated as # live offspring/# two-cell embryos transferred to all recipients (pregnant and nonpregnant) *100; data presented differently in the original manuscript were reconfigured to match the format of data presentation in the current study.
NA indicates data not available.

zona perforation, we decided to also remove cumulus cells for conventional IVF, to avoid this as a possible confounder. In an experiment performed under identical conditions, pairing instead COCs with fresh spermatozoa, we doubled the fertilization rate (14% two-cell embryos, n = 54/394; unpublished data), suggesting a disadvantage to using cumulus-free oocytes for fertilization in this study. In this same experiment, we also demonstrated improvement using fresh spermatozoa, compared to cryopreserved spermatozoa (2% two-cell embryos, n = 7/393) that are perhaps defective at penetrating the COC. Further, our use of a traditional fertilization medium (HTF) may also have contributed to lower fertilization rates obtained in this study, as HTF has been found to underperform in conventional IVF of C57BL/6 mice, compared to other media commonly used for inbred mouse strains, specifically modified Krebs-Ringer bicarbonate medium (TYH) and minimal essential medium (MEM) [25]. Other reasons the fertilization rates following conventional IVF in this study were suboptimal could be due to our selection of uniform spermatozoa samples for IVF rather than preference for the most motile, the incubator atmosphere, including a lack of nitrogen [32], and reactive oxygen species present in the fertilization milieu, perhaps from damaged frozen-thawed spermatozoa, that are known to inhibit fertilization [33]. Bath et al. found that with 4 inbred mouse strains, fertilization was increased significantly by adding reduced glutathione to the fertilization medium [33].

We found that LAIVF was significantly more efficient at producing two-cell embryos than conventional IVF using inbred mouse gametes. We recently performed a retrospective evaluation of 4 years of conventional and laser-assisted IVF data gathered from our in-house, fee-for-service operation (unpublished data). Using GM C57BL/6 mice, our fertilization success, recognizing differences in methodologies, for conventional IVF ranged from 3% to 33% two-cell embryos, depending on IVF application (mean: 12%, n = 459/3,786), generating live offspring from 7 of 45 GM lines. Whereas, LAIVF resulted in 42% two-cell embryos (n = 9,301/21,997), and was useful for obtaining live pups from 55 of 56 GM lines. Li et al. found similarly low fertilization rates using fresh spermatozoa from subfertile GM C57BL/6 mice with conventional IVF (<15% two-cell embryos), while rates using LAIVF ranged from 25% to 85% two-cell embryos [12].

As presented in Table 4, the greatest % two-cell embryos were obtained in our study with LAIVF using V-W oocytes and frozen-thawed spermatozoa (87.1%). Prior studies have evaluated various techniques to augment fertilization of V-W oocytes and demonstrated increased fertilization compared to non-treated V-W oocytes [7–11]. The use of oocytes zona-drilled after DAP213 vitrification in our study provided the highest % two-cell embryos reported to date using such techniques, to the best of our knowledge. While 100% of oocytes that survive ICSI with frozen-thawed spermatozoa develop to form two-cell embryos, 78% of those injected do not survive, reducing the overall % two-cell embryos to 72.3% [9]. ZP digestion using acidic Tyrode's solution resulted in 48.8% fertilization with cryopreserved spermatozoa, compared to 22.1% without digestion [7]. Meng et al. reported a fertilization rate of 85.5% using piezo-actuated zona drilling and fresh outbred mouse spermatozoa following open pulled straw vitrification [8], while Wang et al. used zona incision to obtain 69.9% fertilization of cryopreserved oocytes with frozen-thawed spermatozoa [10]. Finally, zona drilling of oocytes prior to vitrification and subsequent pairing with fresh spermatozoa from a subfertile transgenic mouse produced a two-cell embryo rate of 52.7% [11]. Note that a recent study on a new IVF system using oocytes cryopreserved with DAP213 and specialty media alone (FERTIUP Mouse Sperm Preincubation Medium and CARD MEDIUM; KYUDO CO., LTD., Japan) reported two-cell embryo rates of 72–97% [5], showing that in prime conditions assistive techniques such as zona drilling may not be necessary to efficiently fertilize V-W oocytes.

Without a highly optimized and efficient conventional IVF, advantage was seen with using zona-drilled V-W and fresh oocytes for producing live offspring. In total, conventional IVF generated 18 live offspring from 908 oocytes (2.0%; 100% of two-cell embryos obtained were transferred), whereas LAIVF resulted in 50 live offspring using 1,010 oocytes (5.0%; 91.5% of two-cell embryos obtained were transferred). Even with optimal conventional techniques, LAIVF may still offer an advantage in circumstances where producing two-cell embryos is a challenge, such as when cryopreserved and/or GM/inbred mouse gametes are used. The % live offspring following two-cell embryo transfer documented in this study is consistent with that documented elsewhere following other treatments to improve fertilization of V-W oocytes (Table 4) [7,8,10,11]. There may be a slight benefit to using zona-drilled fresh oocytes compared to V-W oocytes in producing live offspring, but this trend was not statistically significant. Meng et al. found that the rate of development to term was lower with vitrified piezo-drilled oocytes (16.6%), than with vitrified intact (36.0%) and fresh (51.3%) oocytes [8]. The advances made in this study, in terms of successful use of cryopreserved gametes with LAIVF and increased % two-cell embryos, should have an even greater impact on the production of transgenic mice once live offspring efficiency is improved.

While we weren't able to statistically evaluate the difference between % live offspring using LAIVF and conventional IVF (and nor was this a primary focus of our study), our data suggests that more of the embryos transferred developed to produce live offspring using conventional techniques. This finding has also been demonstrated in other studies, where mouse embryos generated through laser-assisted zona drilling developed to the blastocyst stage *in vitro* at a rate similar to conventional IVF [13,17,18], but had significantly lower birth rates [11,12,17,34,35]. Although not ruled out here, neither parthenogenic activation, nor polyspermy appear to be problems of LAIVF with mouse gametes [11,13,16–18]. Liow et al. reported an absence of polyspermy in mouse oocytes with one laser-drilled hole in the zona paired with spermatozoa ranging in concentration from 50,000 to 2 million/ml, and only found 1% polyspermy when 2 holes were present with 1 and 2 million spermatozoa/ml [16]. While a high concentration of spermatozoa and a defect in the zona, as in our study, could both allow for more than one spermatozoon to penetrate an oocyte, zona hardening following oocyte vitrification, impaired motility of frozen-thawed spermatozoa and a precise, small hole in the zona as obtained by laser [16] could reduce the potential for polyspermy. Loss of blastomeres through the zona perforation site [10] or cytotoxic thermal damage from the laser [11,12,17,34,35] may rather be responsible for the low birth rates following LAIVF. While the true role that thermal damage plays on impaired development of embryos from LAIVF is disputed, one study found that addition of 0.25 M sucrose to the culture medium during zona drilling completely mitigated the lower birth rate [12]. Other studies have also evaluated offspring efficiency when using sucrose to osmotically shrink oocytes relative to the ZP [11,17,34,35].

Frozen-thawed spermatozoa were used in this study to provide a reference for how zona-drilled V-W oocytes would perform in IVF with cryopreserved spermatozoa, with extrapolation to other spermatozoa-ZP penetration defects. Cryopreservation reduced fertilization potential of spermatozoa by decreasing the percentage of motile spermatozoa, as previously reported [6,11,36,37]. The same concentration of spermatozoa was used for IVF using both fresh and frozen-thawed spermatozoa, and spermatozoa used were representative and not selective for those with good motility. The efficiency of frozen-thawed spermatozoa largely compared to that of fresh spermatozoa, but frozen-thawed spermatozoa paired with V-W oocytes for LAIVF did produce more two-cell embryos than fresh spermatozoa, suggesting again that cryopreservation may select for stronger gametes. This study demonstrates that when both male and female subfertility factors exist, LAIVF reliably ensures a high percentage of two-cell embryos and the production of healthy pups. This allows for oocyte and spermatozoa banking and shipment [5,11], and conserves genetic resources by eliminating waste of suboptimally preserved gametes.

The ability to cryopreserve mouse gametes should considerably facilitate management of large-scale transgenic production facilities, and may ultimately support more rapid and efficient recovery of mouse lines using LAIVF. This study also contributes to the overall improvement in use of V-W oocytes in all animals, including endangered species and livestock, and in clinical applications for human assisted reproduction.

Acknowledgments

We would like to thank Alexandre Viana for his invaluable technical assistance with mouse colony management, and preparing mice for oocyte collection and embryo transfer surgery.

Author Contributions

Conceived and designed the experiments: SEW PQ AG. Performed the experiments: SEW PQ ER TC AD JM. Analyzed the data: SEW. Wrote the paper: SEW. Critical review of the manuscript, including suggestion to expand comparison of current study to past related studies (Table 4): JGF.

References

1. Kohaya N, Fujiwara K, Ito J, Kashiwazaki N (2013) Generation of live offspring from vitrified mouse oocytes of C57BL/6J strain. PLoS One 8: e58063.
2. Kuwayama M (2007) Highly efficient vitrification for cryopreservation of human oocytes and embryos: the Cryotop method. Theriogenology 67: 73–80.
3. Whittingham DG, Leibo SP, Mazur P (1972) Survival of mouse embryos frozen to −196 degrees and −269 degrees C. Science 178: 411–414.
4. Rall WF, Fahy GM (1985) Ice-free cryopreservation of mouse embryos at −196 degrees C by vitrification. Nature 313: 573–575.
5. Nakagata N, Takeo T, Fukumoto K, Kondo T, Haruguchi Y, et al. (2013) Applications of cryopreserved unfertilized mouse oocytes for in vitro fertilization. Cryobiology 67: 188–192.
6. Nakao K, Nakagata N, Katsuki M (1997) Simple and efficient vitrification procedure for cryopreservation of mouse embryos. Exp Anim 46: 231–234.
7. Fan ZQ, Wang YP, Yan CL, Suo L, Zhu SE (2009) Positive effect of partial zona pellucida digestion on in vitro fertilization of mouse oocytes with cryopreserved spermatozoa. Lab Anim 43: 72–77.
8. Meng Q, Li X, Wu T, Dinnyes A, Zhu S (2007) Piezo-actuated zona-drilling improves the fertilisation of OPS vitrified mouse oocytes. Acta Vet Hung 55: 369–378.
9. Sakamoto W, Kaneko T, Nakagata N (2005) Use of frozen-thawed oocytes for efficient production of normal offspring from cryopreserved mouse spermatozoa showing low fertility. Comp Med 55: 136–139.
10. Wang YP, Zhao XM, Zhou GB, Hou YP, Fan ZQ, et al. (2008) Production of normal offspring from partially zona-incised vitrified mouse oocytes fertilized with cryopreserved spermatozoa using an optimized protocol. Cryo Letters 29: 111–119.
11. Anzai M, Nishiwaki M, Yanagi M, Nakashima T, Kaneko T, et al. (2006) Application of laser-assisted zona drilling to in vitro fertilization of cryopreserved mouse oocytes with spermatozoa from a subfertile transgenic mouse. J Reprod Dev 52: 601–606.
12. Li MW, Kinchen KL, Vallelunga JM, Young DL, Wright KD, et al. (2013) Safety, efficacy and efficiency of laser-assisted IVF in subfertile mutant mouse strains. Reproduction 145: 245–254.
13. el-Danasouri I, Westphal LM, Neev Y, Gebhardt J, Louie D, et al. (1993) Zona opening with 308 nm XeCl excimer laser improves fertilization by spermatozoa from long-term vasectomized mice. Hum Reprod 8: 464–466.
14. Laufer N, Palanker D, Shufaro Y, Safran A, Simon A, et al. (1993) The efficacy and safety of zona pellucida drilling by a 193-nm excimer laser. Fertil Steril 59: 889–895.

15. Antinori S, Versaci C, Fuhrberg P, Panci C, Caffa B, et al. (1994) Seventeen live births after the use of an erbium-yytrium aluminum garnet laser in the treatment of male factor infertility. Hum Reprod 9: 1891–1896.
16. Liow SL, Bongso A, Ng SC (1996) Fertilization, embryonic development and implantation of mouse oocytes with one or two laser-drilled holes in the zona, and inseminated at different sperm concentrations. Hum Reprod 11: 1273–1280.
17. Kaneko T, Fukumoto K, Haruguchi Y, Kondo T, Machida H, et al. (2009) Fertilization of C57BL/6 mouse sperm collected from cauda epididymides after preservation or transportation at 4 degrees C using laser-microdissected oocytes. Cryobiology 59: 59–62.
18. Enginsu ME, Schutze K, Bellanca S, Pensis M, Campo R, et al. (1995) Micromanipulation of mouse gametes with laser microbeam and optical tweezers. Hum Reprod 10: 1761–1764.
19. Germond M, Nocera D, Senn A, Rink K, Delacretaz G, et al. (1996) Improved fertilization and implantation rates after non-touch zona pellucida microdrilling of mouse oocytes with a 1.48 microm diode laser beam. Hum Reprod 11: 1043–1048.
20. The Jackson Laboratory, Cryopreservation Course, Fall 2005.
21. Borg CL, Wolski KM, Gibbs GM, O'Bryan MK (2010) Phenotyping male infertility in the mouse: how to get the most out of a 'non-performer'. Hum Reprod Update 16: 205–224.
22. Mochida K, Hasegawa A, Li MW, Fray MD, Kito S, et al. (2013) High osmolality vitrification: a new method for the simple and temperature-permissive cryopreservation of mouse embryos. PLoS One 8: e49316.
23. Agca Y (2000) Cryopreservation of oocyte and ovarian tissue. ILAR J 41: 207–220.
24. European Mutant Mouse Archive, Laser-Assisted IVF Protocol. https://dev.infrafrontier.eu/sites/infrafrontier.eu/files/upload/public/pdf/Knowledgebase/Laser-assisted-IVF-protocol_Feb-2013.pdf.
25. Liu L, Nutter LM, Law N, McKerlie C (2009) Sperm freezing and in vitro fertilization in three substrains of C57BL/6 mice. J Am Assoc Lab Anim Sci 48: 39–43.
26. Omar Farouk FN, Vlad M (2008) In vitro development of mouse pronuclear embryos to blastocysts in sequential media with and without co-culture of autologous cumulus cells. J Reprod Dev 54: 385–390.
27. Omar Farouk FN, Stott D, Vlad M (2011) Mouse embryo co-culture with autologous cumulus cells and fetal development post-embryo transfer. Anim Sci J 82: 420–427.
28. Jin M, Fujiwara E, Kakiuchi Y, Okabe M, Satouh Y, et al. (2011) Most fertilizing mouse spermatozoa begin their acrosome reaction before contact with the zona pellucida during in vitro fertilization. Proc Natl Acad Sci U S A 108: 4892–4896.
29. Salustri A, Garlanda C, Hirsch E, De Acetis M, Maccagno A, et al. (2004) PTX3 plays a key role in the organization of the cumulus oophorus extracellular matrix and in in vivo fertilization. Development 131: 1577–1586.
30. Shimada M, Yanai Y, Okazaki T, Noma N, Kawashima I, et al. (2008) Hyaluronan fragments generated by sperm-secreted hyaluronidase stimulate cytokine/chemokine production via the TLR2 and TLR4 pathway in cumulus cells of ovulated COCs, which may enhance fertilization. Development 135: 2001–2011.
31. Tanii I, Aradate T, Matsuda K, Komiya A, Fuse H (2011) PACAP-mediated sperm-cumulus cell interaction promotes fertilization. Reproduction 141: 163–171.
32. The Jackson Laboratory, IVF Protocol.
33. Bath ML (2010) Inhibition of in vitro fertilizing capacity of cryopreserved mouse sperm by factors released by damaged sperm, and stimulation by glutathione. PLoS One 5: e9387.
34. Peters DD, Lepikhov K, Rodenacker K, Marschall S, Boersma A, et al. (2009) Effect of IVF and laser zona dissection on DNA methylation pattern of mouse zygotes. Mamm Genome 20: 664–673.
35. Peters DD, Marschall S, Mahabir E, Boersma A, Heinzmann U, et al. (2006) Risk assessment of mouse hepatitis virus infection via in vitro fertilization and embryo transfer by the use of zona-intact and laser-microdissected oocytes. Biol Reprod 74: 246–252.
36. Nakagata N (2000) Cryopreservation of mouse spermatozoa. Mamm Genome 11: 572–576.
37. Sztein JM, Farley JS, Young AF, Mobraaten LE (1997) Motility of cryopreserved mouse spermatozoa affected by temperature of collection and rate of thawing. Cryobiology 35: 46–52.

Scaling-Up of Dental Pulp Stem Cells Isolated from Multiple Niches

Nelson F. Lizier[1,2]*, Alexandre Kerkis[1], Cícera M. Gomes[1], Josimeri Hebling[3], Camila F. Oliveira[1], Arnold I. Caplan[4], Irina Kerkis[1,2]

1 Laboratory of Genetics, Butantan Institute, Sao Paulo, SP, Brazil, **2** Departament of Morphology of Federal University of Sao Paulo (UNIFESP), Sao Paulo, SP, Brazil, **3** Department of Orthodontics and Pediatric Dentistry of State University of Sao Paulo (UNESP), Araraquara, SP, Brazil, **4** Skeletal Research Center, Department of Biology of Case Western Reserve University, Cleveland, Ohio, United States of America

Abstract

Dental pulp (DP) can be extracted from child's primary teeth (deciduous), whose loss occurs spontaneously by about 5 to 12 years. Thus, DP presents an easy accessible source of stem cells without ethical concerns. Substantial quantities of stem cells of an excellent quality and at early (2–5) passages are necessary for clinical use, which currently is a problem for use of adult stem cells. Herein, DPs were cultured generating stem cells at least during six months through multiple mechanical transfers into a new culture dish every 3–4 days. We compared stem cells isolated from the same DP before (early population, EP) and six months after several mechanical transfers (late population, LP). No changes, in both EP and LP, were observed in morphology, expression of stem cells markers (nestin, vimentin, fibronectin, SH2, SH3 and Oct3/4), chondrogenic and myogenic differentiation potential, even after cryopreservation. Six hours after DP extraction and *in vitro* plating, rare 5-bromo-2′-deoxyuridine (BrdU) positive cells were observed in pulp central part. After 72 hours, BrdU positive cells increased in number and were found in DP periphery, thus originating a multicellular population of stem cells of high purity. Multiple stem cell niches were identified in different zones of DP, because abundant expression of nestin, vimentin and Oct3/4 proteins was observed, while STRO-1 protein localization was restricted to perivascular niche. Our finding is of importance for the future of stem cell therapies, providing scaling-up of stem cells at early passages with minimum risk of losing their "stemness".

Editor: Jan Pruszak, University of Freiburg, Germany

Funding: This work was supported by grant 2010/51051-6 from FAPESP (Sao Paulo Research Foundation). The funder had no role in study design, data collection and analysis, decision to publish, or preparation of the manuscript.

Competing Interests: The authors have declared that no competing interests exist.

* E-mail: nelson_lizier@yahoo.com.br

Introduction

Isolation of stem cells (SC) from human adult and deciduous teeth has been reported in the last decade [1,2]. In this short period of time, considerable progress has been achieved, in particular, with deciduous teeth stem cells (DTSC) [3]. It has been demonstrated that the use of different handling methods of dental pulp (DP) can lead to the isolation of SC populations with distinct properties. These DTSC populations are similar to mesenchymal stem cells (MSCs) or epithelial SCs or they are composed by a mixed population of both cell types [3]. We previously isolated a population of multipotent DTSCs, which were referred to as "immature" (Immature Dental Pulp Stem Cells, IDPSCs). Along with MSC markers, IDPSCs express embryonic stem (ES) cells markers (Oct3/4, Nanog and Sox2) and undergo spontaneous differentiation into a wide range of cell types *in vitro* [4]. These cells showed expressive capacity to contribute into multiple tissues in response to the cellular milieu during human/mouse pre-termed chimeras development [5]. After transplantation of IDPSCs into different adult animals, including mouse, rabbit and dog, neither immune rejection nor teratoma formation was observed [6,7,8]. IDPSCs and other dental stem/progenitor cells were recently used to obtain induced pluripotent SCs [9,10]. These cells demonstrated higher efficiency of reprogramming than fibroblasts, providing a model for the study of pediatric diseases and disorders. Taken together, these data strongly suggest that IDPSCs are a hopeful source for the future of SC therapies [11].

Recent SC research studies revealed a promising potential of MSCs to treat at least ten human diseases: heart disease, diabetes, Crohn's disease, deafness, autoimmune disorders, leukemia, cancers, sickle cell disease, amyotrophic lateral sclerosis and metabolic disorders. Since MSCs are present at low relative amounts in bone marrow and other adult tissues, significant *in vitro* expansion is necessary in order to generate sufficient quantities of these cells to treat human disease [12,13]. The expansion process itself induces senescence of MSCs and loss of their stemness as shown by a decline in proliferative and differentiation capacity [14,15]. In addition, prolonged culturing of MSCs increases the probability of genetic changes, which could affect their safe use in clinical trials and future therapies [16,17]. Therefore, studies which provide adequate production of SC of excellent quality at early passages derived from the same donor are of importance.

Our group was the first to use explant culture of DP to obtain DTSCs, which in combination with appropriate cell culture conditions, provides isolation of a relatively pure (not homogeneous) population of IDPSC [4]. Subsequent study demonstrated the advantages of DP explant culture for the differentiation and proliferation potentials of SC [18]. Herein, we aimed to establish a new method based on tissue explant culture and mechanical (non-enzymatic) transfer in order to obtain a long-term culture of DP providing substantial quantities of DTSCs without aberrant genetic and biologic changes. DP was maintained in culture following mechanical transfer during several months. We evaluated such characteristics as: morphology, expression of specific MSC-phenotypes and ES cell proteins and genes, karyotype, growth rate and differentiation ability of IDPSCs just after DP extraction (early population, EP) and after multiple DP transfer (late population, LP). Some of these parameters were evaluated after cryopreservation and with culturing IDPSCs in three distinct culture media. The used of antibody against BrdU incorporated in DP just after plating and three days after DP cultivation gave insight into the mechanism of IDPSCs generation by explant culture. Additionally, to distinguish SCs in DP, immunohistochemical staining against nestin, vimentin, Oct3/4 and STRO-1 has been performed.

Results

Long-term Culture of DP

Freshly extracted DP is a tissue, which contains large nerve trunks and blood vessels in the central region of the coronal and radicular pulp (Figure 1A). First outgrowing fibroblast-like cells appeared between three to four days after DP plating (Figure 1B). Long-term culture was performed by mechanical transfer of DP into new culture dish without using enzymatic treatment. After each transfer, DP produces large numbers of outgrowing cells approximately every three or four days, thus allowing constant production of SCs at passage zero (P0) (Figure 2). We obtained successful isolations with all samples (n = 10) of deciduous teeth. We performed multiple DP transfers during, at least, six months (LP of IDPSCs). Both, EP and LP of IDPSCs maintained their morphology (Figure 1C). Transmission electron microscopy revealed two types of IDPSCs morphology: ES-like cells with low cytoplasm-to-nucleus ratio, low cytoplasm density, which are poor of organelles (Figure 1D). IDPSCs of MSC-like cells have a high number of stretched out pseudopodes, which serve to explore substrate and more cytoplasm and organelles when compared with IDPSCs of ES-like cells (Figure 1D, E). Figure 1F documents that IDPSCs showed a relatively uniform population in respect of these two cell types. IDPSCs karyotype was confirmed here to be unchanged, suggesting that during culture, numerical and gross structural chromosomal abnormalities did not occur as shown by routine G-banding technique (Figure 1G). Further, *in vitro* expansion of IDPSCs was performed using enzymatic treatment and the passages was called P1 (Figure 2).

Immunophenotyping of EP and LP of IDPSCs

EP and LP were characterized using several markers, which were employed in our previous original study [4], such as SH2/CD105, SH3/CD73 (MSCs markers) and Oct3/4 (ES cells marker). Additionally, expression of MSCs markers such as vimentin, nestin, fibronectin has been evaluated in both EP and LP. The reason of choosing these markers will be explained in discussion section. Representative Figure 3 (A1–E1, A2–E2) showed that all MSC markers were expressed in both populations (EP and LP) and slightly declined in LP after six month of multiple DP transfer (Figure 3A2–E2). A percentage of IDPSCs, which showed positive immunostaining for these markers, was evaluated by flow cytometry and was 99.10% to EP and 96.60% to LP for SH2/CD105; 99.60% to EP and 98.40% to LP for SH3/CD73; 97.76% to EP and 94.56% to LP for nestin; 99.45% to EP and 95.60% to LP for vimentin; 97.10% to EP and 96.30% to LP for fibronectin (Figure 3A1–E1, A2–E2). Interestingly, that in this particular IDPSC population (which is not a rule) a very low percentage of Oct3/4 positive cells ~0.75% was observed in EP, which increased to ~10.03% in LP (Figure 3F1, F2).

Flow cytometry data have been confirmed by cells immunostaining using antibodies against the same MSC and ES cell markers. Their expression was observed in both EP and LP of IDPSCs. In Figure 3, the expression of MSC markers in LP is presented (Figure 3A3–E3). As expected, Oct-3/4 protein expression was observed in the cell nuclei (Figure 3F3). The expression of all these markers in LP was similar to EP (data not shown).

Culture Media Influence EP and LP Growth Rate and Gene Expression

Proliferative capacity of EP and LP of IDPSCs before and after cryopreservation was studied using three different culture media: DMEM/F12, DMEM-LG, and MEM-alpha. Starting from P2, non-cryopreserved cells were harvested following enzymatic dissociation and counted daily during 15 consecutive passages. IDPSCs cultured in DMEM/F12 and MEM-alpha medium, presented constant proliferative rate during initial passages, which achieved their peak growth = 5±2 (Figure 4A, Table 1). Based on growth curves presented in Figure 4, statistical analyses were performed considering the cell number from passage 3 to 7 (Table 1). Using the same parameters, proliferative rate of EP and LP cultivated in DMEM/F12 and MEM-alpha media were evaluated after thawing and showed similar proliferative potential, when compared with those before cryopreservation (Figure 4B, Table 1). Non-cryopreserved EP and LP of IDPSCs cultured in DMEM-LG presented spontaneous differentiation into osteogenic lineage (data not shown) and demonstrated rapid decrease of proliferative potential (Figure 4A, Table 1). Interestingly, EP and LP of IDPSCs, cultured in DMEM-LG after thawing, maintained their proliferative state (Figure 4B, Table 2). DMEM/F12 and MEM-alpha media did not induce any spontaneous differentiation in non-cryopreserved and cryopreserved EP and LP of IDPSCs.

The gene expression pattern of pluripotent ES cell and MSC markers were analyzed by RT-PCR in EP and LP after thawing. Both were cultured in different basal media (DMEM/F12, MEM-alpha, and DMEM-LG) for seven passages. Overall, both EP and LP showed similar expression pattern of vimentin, SH2/CD105 and SH3/CD73 (Figure 4C). Similar expression pattern was observed to fibronectin, nestin and Oct3/4, when IDPSCs were cultured in DMEM/F12 and DMEM-LG. However, it was distinct when cultivated in MEM-alpha, in which three of these genes (fibronectin, nestin and Oct3/4) did not show any expression (Figure 4C).

Chondrogenic and Myogenic Differentiation

Multipotential capacity of IDPSCs was reported elsewhere [4]. Therefore, only two differentiation assays were chosen to demonstrate their differentiation capacity.

At day 21 after induction of chondrogenic differentiation, IDPSCs demonstrated the formation of an extracellular cartilage matrix which was intensively stained by Massoǹs trichrome (Figure 5A, Inset). Toluidine blue staining was used to detect essential cartilage matrix proteins such as proteoglycans (Figure 5B). IDPSCs maintained in basal culture medium (control) did not form any cell pellet (data not shown). Additionally, chondrogenic differentiation was confirmed by the expression of COMP (Cartilage Oligomeric Matrix Protein) gene, which

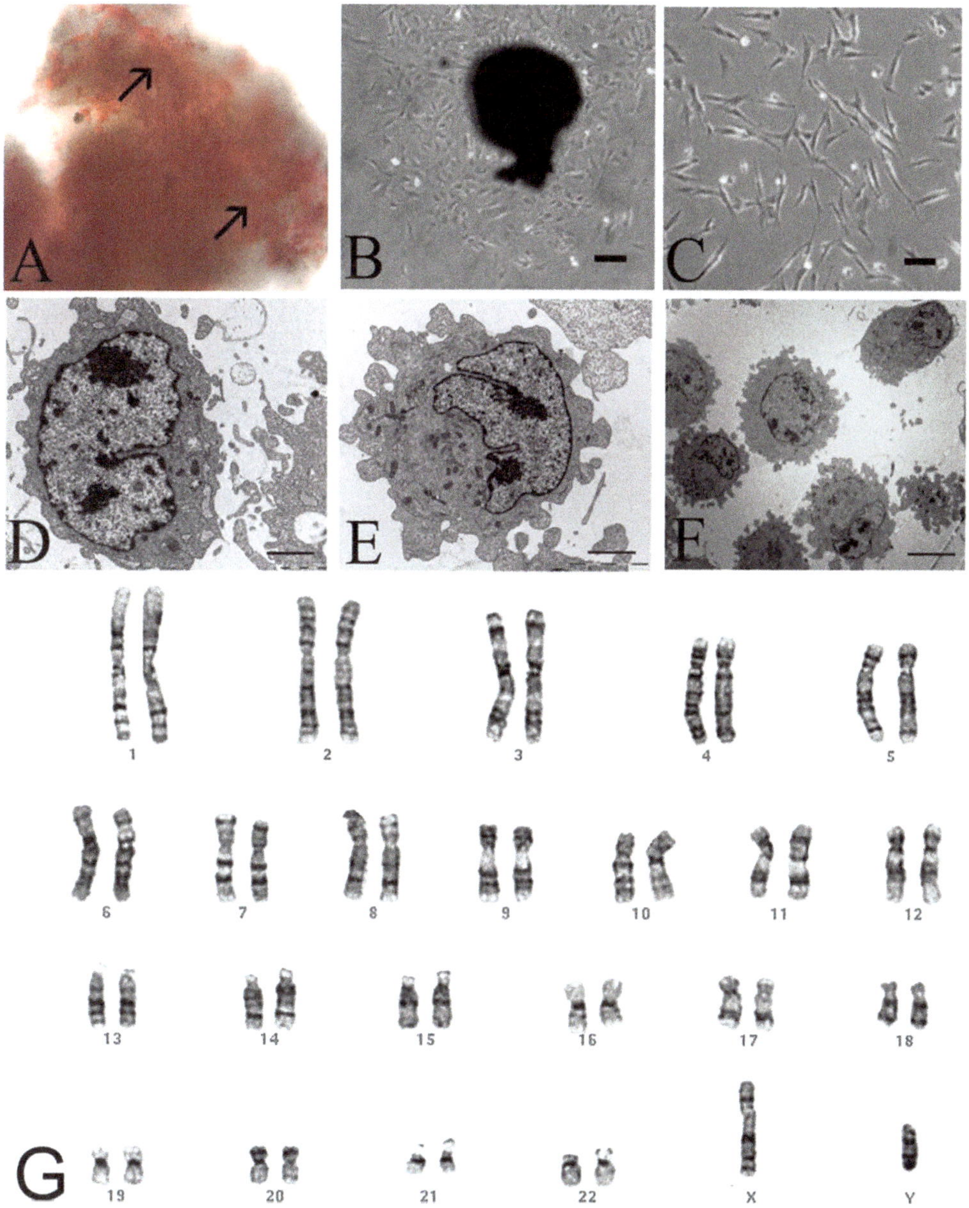

Figure 1. Dental pulp and IDPSCs. A) Highly vascularized (black arrows) DP just after extraction. **B**) Explant culture of DP with outgrowing IDPSCs. **C**) Culture of IDPSCs at 1st passage. **D**) IDPSCs showing ES-like cells morphology with a large nucleus. **E**) IDPSCs showing MSC-like morphology with several pseudopodes. **F**) IDPSCs showing uniform morphology resembling ES cells and MSCs. **G**) Karyotype of IDPSCs (LP): chromosomes in pairs, ordered by size and position did not reveal any numerical changes in chromosome number; G-banding analysis. A–C, G) Light Microscopy; D–F) Transmission Electron Microscopy; A = 20X, G = 63X; Scale bars: B = 20 μm; C, F = 10 μm; D, E = 3 μm.

encodes a pentameric non-collagenous matrix protein that is mainly expressed in articular cartilage. The expression of COMP was observed in both EP and LP of IDPSCs (Figure 5C). It is important to highlight that chondrogenic differentiation of IDPSCs was uniform even in the absence of TGF-β, which is known to be a strong inductor of chondrogenesis in bone marrow-derived MSCs [19,20].

Following myogenic differentiation, IDPSCs showed typical cells elongation and fusion leading to small myotubes formation at day 7 (Figure 5D). At day 21, this cell fusion was obvious and most of the cells formed small myofibers (Figure 5E). MyoD transcription factor, which is a master regulatory gene of skeletal muscle differentiation, as expected, was expressed in IDPSC-derived myoblasts in nucleus or in perinuclear space following immunostaing using anti-MyoD1 antibody (Figure 5F). These myoblasts further form myosacs and MyoD1 protein was observed in the cytoplasm of these more mature cells (Figure 5G). Titin is the third most abundant skeletal muscle filamentous protein that forms a separate myofilament system in both skeletal and cardiac muscle. It was expressed in IDPSC-derived muscle cells at more advanced stages of differentiation (Figure 5H). Some titin negative cells were also observed (Figure 5I).

Troponin I is a protein responsible for immobilizing the actin-tropomyosin complex in place. The expression of this protein was visualized in more mature myofibers-derived from IDPSCs (Figure 5J). Human specific anti-actinin and anti-myosin antibod-

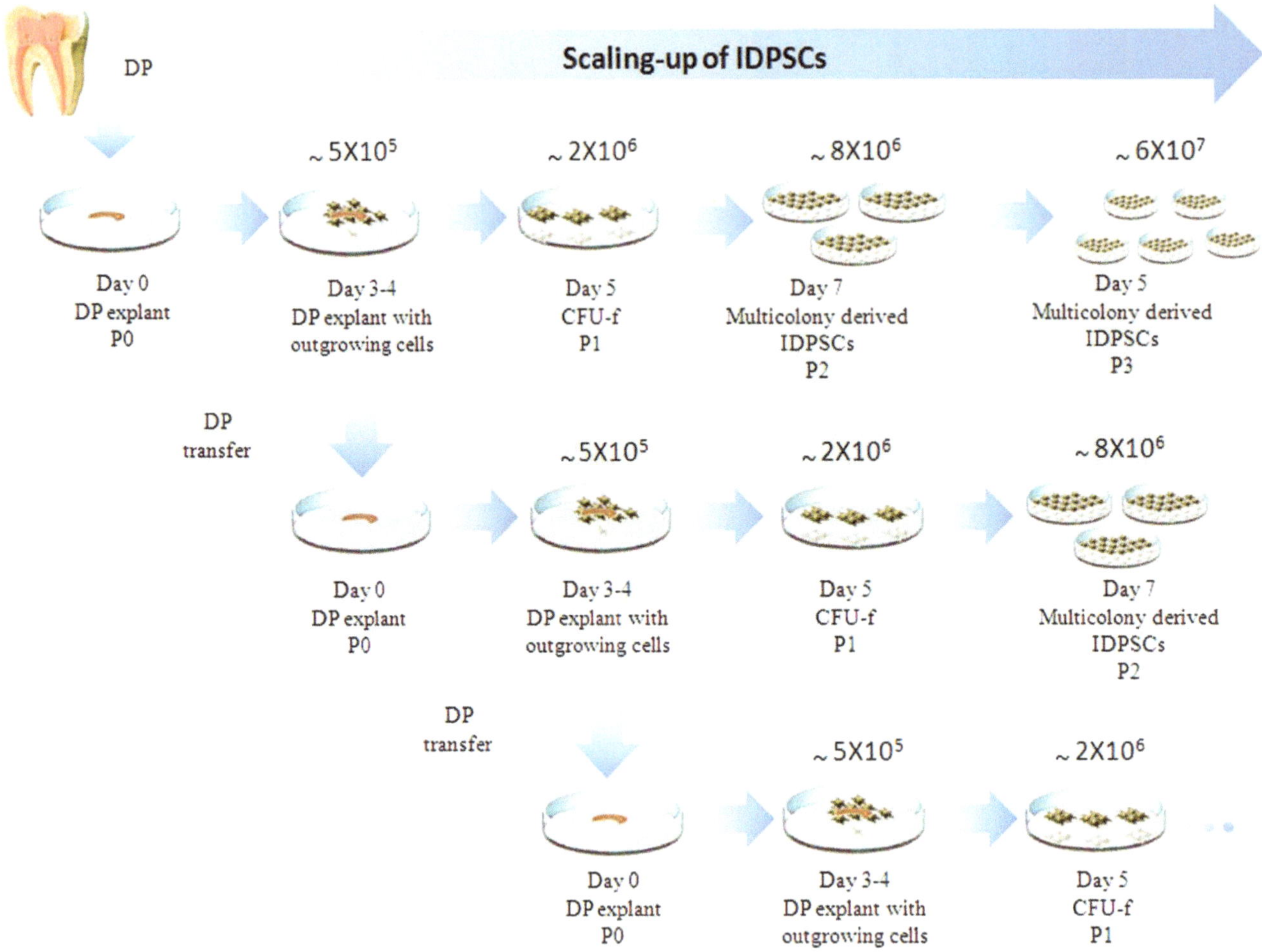

Figure 2. Scaling-up of IDPSCs. Horizontally, the process of DP *in vitro* plating (Day 0, P0) followed by DP adherence and cells outgrowth (Day 3–4). This process is followed by enzymatic treatment (P1) of the cells and formation of multiple colonies (CFU-f - Colony Forming Units-fibroblast). After 5 days, enzymatic treatment was performed to harvest multicolony-derived IDPSCs (P2) population. Next, *in vitro* expansion of IDPSCs (P3) has been performed. Upper numbers represent approximate quantity of harvested IDPSCs in each passage. Vertically, the same process is shown, albeit after multiple DP mechanical transfer.

ies reacted positively with differentiated IDPSCs (Figure 5K–N). Myosin positive immunostaining was observed in myofibers (Figure 5K) and also in differentiated small cells, which presented spot-like immunolabeling (Figure 5L). Singular binuclear differentiated IDPSCs were alpha-actinin positive (Figure 5M). This marker showed differential expression pattern within myosacs: some cells were strongly positive, while others presented only shadow-like immunostaining (Figure 5N). RT-PCR was used to verify the expression of MyoD1 and ACTB (Beta cytoskeletal actin) genes during IDPSCs myogenic differentiation. Both genes were found to be expressed in EP and LP (Figure 5O). EP and LP of IDPSCs showed similar chondrogenic and myogenic differentiation before and after cryopreservation. Control culture of IDPSCs did not present any signals of myogenic differentiation (data not shown).

Expression of MSCs and ES Cells Markers in DP

We observed that both EP and LP of IDPSCs cultured *in vitro* are rich in nestin and vimentin positive cells (Figure 3C1–C3). Therefore, we attempted to identify the exact niche of nestin, vimentin and Oct3/4 positive cells within DP using immunohistochemical assay (Figure 6A–Q). Nestin positive cells were found in all zones of DP: in cell rich zone (innermost pulp layer which contains fibroblasts and undifferentiated mesenchymal cells) (Figure 6A–C), in cell free zone, nestin expression was observed in both capillaries and nerve networks (Figure 6D–F); as well as in odontoblastic layer (outermost layer which contains odontoblasts and lies next to the predentin and mature dentin) (Figure 6G, H). Nestin positive cells in cell rich zone showed fibroblast-like as well as ES-like cell morphologies (Figure 6B, C). In cell free zone, nestin protein was found to be expressed in intermediate filaments in the cells from plexus of nerves (Figure 6D), as well as nestin positive cells were embedded in the wall of small capillaries (Figure 6E) and in adjacent regions of these capillaries (Figure 6F). In odontoblastic layer, several round ES cell-like and large columnar cells were also nestin positive (Figure 6G, H). STRO-1 antibody was used as control, once this marker was described to be specific for stem cells/pericytes from DP. The expression of this marker was mainly observed in small capillaries and middle size blood vessels (Figure 6J), as well as in plexus of nerves in the cell free zone (Figure 6K). We also verified localization of vimentin expressing cells in DP. As expected, vimentin positive cells were localized in capillaries and in innermost pulp layer, locals where nestin positive cells were also found (Figure 6L, M). Once low

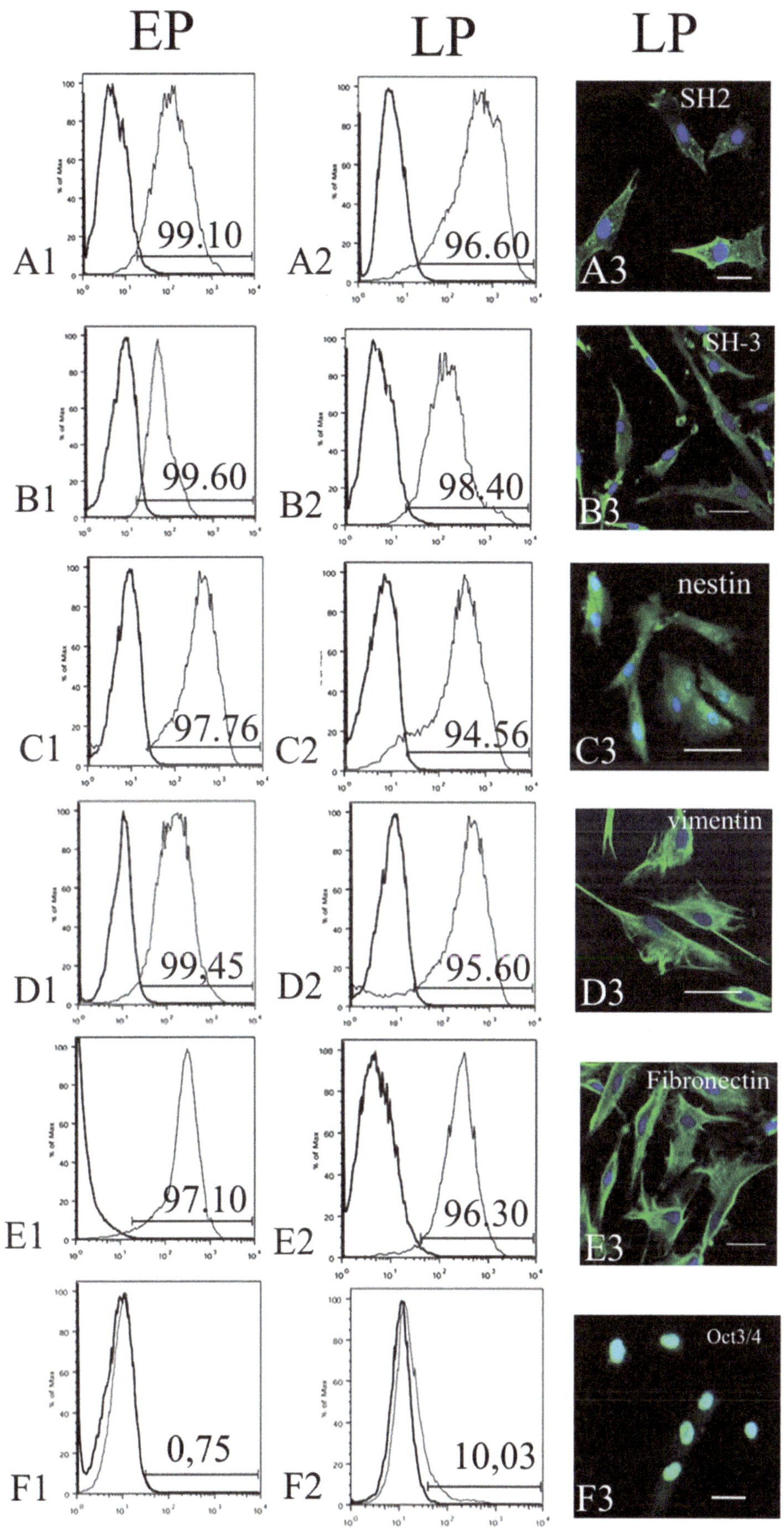
EP
LP
LP
A1
99.10
A2
96.60
SH2
A3
B1
99.60
B2
98.40
SH-3
B3
C1
97.76
C2
94.56
nestin
C3
D1
99,45
D2
95.60
vimentin
D3
E1
97.10
E2
96.30
Fibronectin
E3
F1
0,75
F2
10,03
Oct3/4
F3

Figure 3. Characterization of EP and LP of IDPSCs. A1–F1) Flow cytometry showing EP of IDPSCs, which highly expressed such markers as SH2/CD105 (A1); SH3/CD73 (B1); nestin (C1); vimentin (D1); fibronectin (E1). **F1)** Low expression of Oct3/4 in EP; **A2–F2)** Flow cytometry showing LP of IDPSCs, which expressed same markers as EP. **F2)** Higher expression of Oct3/4 in LP, than in F1. **A3–F3)** Immunofluorescence of LP of IDPSCs using same markers as in (A2–E2). **F3)** Nuclear localization of Oct3/4 can be observed. A3–F3) Epi-fluorescence, nuclei stained with DAPI (blue). Scale bars: A3, B3, E3, F3 = 5 μm; C3, D3 = 10 μm.

percentage of Oct3/4 positive cells, which increased with time of DP *in vitro* cultivation, was observed in IDPSCs, the expression of this protein in DP was also checked. Strong expression of Oct3/4 in the cells nuclei, localized in DP capillaries and in innermost pulp layer can be observed (Figure 6N–Q).

Localization of BrdU Positive Cells in DP during *in vitro* Cultivation

In order to understand continuous process of IDPSCs generation, DP was treated with BrdU just after extraction and *in vitro* plating (Figure 6R–T). After 6 h, only few anti-BrdU antibody positive cells were found in the central part of DP (Figure 6R). After 48 h, BrdU positive cells were observed in the periphery of DP (Figure 6S), while after 72 h, it seems, that BrdU positive cells increased in number and were also found in the periphery of DP in the apical part, close to IDPSCs outgrowing zone (Figure 6T). Morphological aspect of DP with and without enzymatic treatment (collagenase/dispase) was compared (Figure 6U, V). DP, without any treatment, maintains their integrity especially in the region where BrdU positive cells were observed (Figure 6U), while after enzymatic digestion this region was destroyed (Figure 6V).

Discussion

Stem cells reside in a quiescent state within all organs of organism in their special niche and they start to proliferate and to migrate when their niche experiences changes [21–24]. Thus, culture of adult SC niche may provide harvesting of SCs at high scale. We developed a method of long-term DP culture, which allowed harvesting of large quantities of IDPSCs (Figure 2). Mechanical transfer, before and after DP cryopreservation, can also be performed as long as it is necessary and in our experience, it may stop due to, e.g. occasional DP explant contamination. IDPSCs are uniform in respect of morphology (light microscopy and TEM analyzes) (Figure 1A–F) and karyotype of cells remained unchanged (Figure 1G). These cells express high percentage of SC markers such as SH2/CD105, SH3/CD73, nestin, vimentin, fibronectin and low percentage of Oct3/4 (Figure 3A1–F3) without losing their original properties [25].

To date, standardized protocol of SCs culture from DP is not available. Therefore, we directed our study to optimization culture medium conditions for scaling-up of IDPSCs. DMEM-LG, MEM-alpha and DMEM/F12 are culture media which are commonly used for the isolation and expansion of MSCs [26]. In the present work, we verified the effect of DMEM-LG, MEM-alpha and DMEM/F12 on proliferation rate and gene expression pattern of IDPSCs (Figure 4). These analyses indicated that MEM-alpha and DMEM/F12 were the most appropriate media for the isolation and long-term expansion of IDPSCs (Figure 4A, B, Table 2); distinct gene expression patterns were observed between these media (Figure 4C). DMEM-LG was not efficient for the isolation, but it was able to support long-term expansion of these cells after their cryopreservation (Figure 4A, B, Table 1). It seems that our present data (Figure 4, Table 1) are in contrast with previous observations, which showed IDPSCs exponential growth following multiple passages [4]. However, in the present study, cells were counted daily, while in previous study, passages were performed every 3–4 days. Therefore, enzymatic treatment used daily seems to have hampered the IDPSCs.

Analysis of differentiation potential toward chondrogenic and myogenic lineages evidenced high differentiation potential of LP and EP of IDPSCs (Figure 5A–O) comparable to those described previously [4]. IDPSCs showed similar chondrogenic and myogenic differentiation before and after cryopreservation (data not shown) as was reported previously for other SCs from DP [27,28]. Cryopreservation process preserves the proliferative and differentiation capacity of IDPSCs and, thus, allows the opportunity to bank these valuable DTSCs [29].

Recently, new populations of DTSCs were isolated and were shown to be distinct from DPSC (Dental Pulp Stem Cells from permanent teeth)/SHED (Stem Cells from Human Exfoliated Deciduous teeth) [1,2,4,30–35]. As reported in several original publications, DPSC/SHED are supposed to be pericytes, which are isolated from perivascular niche [36,37]. To delineate the anatomic localization of IDPSCs inside the pulp, we performed *in*

Table 1. Number of IDPSCs cultured in three different growth media and at different passages before and after cryopreservation.

Cell line	Growth media	Passage number													
		2	3	4	5	6	7	8	9	10	11	12	13	14	15
Before cryopreservation	MEM-alpha	100	165.5	364	431.5	476.5	394	378.5	321	266.5	189	176.5	147.5	119.5	90.5
	DMEM-LG	100	139.5	109.5	100	78	38.5	a	a	a	a	a	a	a	a
	DMEM/F12	100	125	262.5	302	333	305.5	273.5	218	193	161.5	149.5	124.5	110.5	89
After cryopreservation	MEM-alpha	100	153	329	433.5	345	325.25	245	191.5	175.5	172.5	166.5	112.5	64.6	60.5
	DMEM-LG	100	115.5	265	314	266	268.50	209	186.5	165.5	162	154.5	50	64.5	35.75
	DMEM/F12	100	99	251.5	296.5	250.5	232.5	191	180.5	166.5	142	128	117.5	59.5	57

[a]DMEM-LG did not support cell growth.

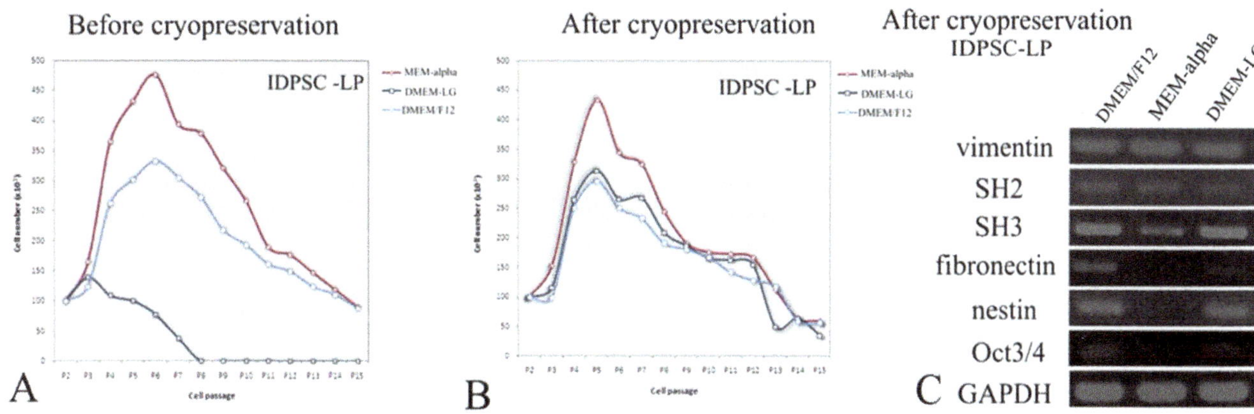

Figure 4. Proliferation rate and gene expression of IDPSCs after cultivation in three distinct culture media. A) Proliferation curve of LP before cryopreservation; **B**) Proliferation curve of LP after cryopreservation. **C**) Gene expression of LP after cryopreservation.

situ analysis using markers of MSCs and ES cells. Our study suggested that DP has multiple SC niches, which are localized in capillaries and nerve networks in cell free zone (Figure 6D–F); in innermost pulp layer in cell rich zone (Figure 6A–C) and in outermost layer, which contains odontoblasts (Figure 6G, H). All these niches are rich of nestin positive cells [38], which can present fibroblast-, epithelial- and odontoblast-like morphologies (Figure 6A–I). In accordance with our finding, recently, rare quiescent multipotent nestin positive MSCs were found in bone marrow in association with hematopoietic SCs and adrenergic nerve fibers. However, bone marrow derived colonies of MSCs cultured *in vitro* exhibit a low percentage (~4%) of such nestin positive cells [39]. Meanwhile, STRO-1 positive cells localization in DP, as expected, is restricted to endothelium (Figure 6J), albeit very weak positive immunostaining was observed in nervous plexus (Figure 6K). Vimentin showed distribution within DP similar to nestin, however, a smaller amount of cells positive for this markers can be observed (Figure 6L, M). Overall, these data suggest that IDPSCs constitute a mixed population of both MSCs and epithelial SCs, among which the pericytes (SHED) are also present. Surprisingly, multiple Oct3/4 positive cells were found in DP (Figure 6N–Q), showing the appropriate nuclear localization (Figure 6Q). After isolation, IDPSCs did not show as high as expected percentage of Oct3/4 positive cells. Previously, we succeeded to isolate IDPSCs population, which contained ~20% of these cells. We supposed that in the pulp, Oct3/4 positive cells are highly pluripotent and of epithelial type. When these cells start to migrate, they undergo epithelial-mesenchymal transition, which leads to a decrease or lost Oct3/4 expression. Thus, another method can be developed in order to isolate naïve Oct3/4 positive cells from their niche in DP.

DP is capable to produce long-term culture of SCs; however the process of such ability is unknown. BrdU incorporation into DP demonstrates that, as expected, only very rare BrdU positive cells were observed just after DP extraction and plating. Following further cultivation, such cells increased in number and are located in the periphery of DP. It is plausible to suggest, that the isolation of DP stimulates the mechanisms leading to SC proliferation and migration (Figure 6R–T). In contrast to previous original report [1], which used the method of DP enzymatic dissociation, we choose DP explant as a main method of SCs isolation [4]. Comparative morphological analysis demonstrate that prior SCs isolation enzymatic treatment is not recommended for DP which may destroy future "niche" of SCs (Figure 6U, V).

In conclusion, our method provides the isolation of a high purity SC population in substantial quantities. It is based on natural (intrinsic) mechanisms of SCs activation similar which occur during tissue trauma or injury, when in response to their damage, quiescent SCs are activated. Our method can be applied to isolate SCs from single and multiple niches from any type of adult tissues, such as bone marrow, adipose tissue, umbilical cord, muscles, skin and others. This protocol diminishes a probability of occurrence of spontaneous genomic mutations and eventual karyotype abnormalities, which may arise during multiple passages in SCs. This method is simple, does not requires long time DP preparation, guarantees sterility (DP can be transferred using sterile instruments or even pipette) and avoids any type of SCs selection, which is undesirable for future clinical applications.

Table 2. Cell number (x10^3) of IDPSCs (passage range P3–P7) cultured in different growth media before and after cryopreservation.

Cryopreservation	Growth media		
	MEM-alpha	**DMEM-LG**	**DMEM/F12**
Before	317.15±101.81[a]	245.80±75.70[a]	226.00±74.82[a]
After	366.30±119.88[a]	93.10±37.68[b]	265.60±82.52[a]

[a]Values are mean±standard deviation, n = 5. Means followed by the same letter are not statistically different (Tukey, p>0.05).

Materials and Methods

Human Dental Pulp Extraction and Cell Culture

The investigation was approved by the Ethical Committee of the Federal University of Sao Paulo (Protocol N° 0139/10). Human DP was extracted from ten deciduous teeth of ten healthy subjects (range 6–9 years) following previously established protocol [4]. All patients agreed to participate of this study as well as, their next of kin, carers or guardians on the behalf of the minors/children participants signed a written informed consent. Next, DP was gently rinsed in phosphate-buffered solution (PBS) (Invitrogen, Carlsbad, CA, USA), slightly dissected and placed into 35 mm plastic tissue culture dishes (Corning Inc., Corning, NY, USA). Tissue explants were cultured in Dulbecco's-modified Eagle's

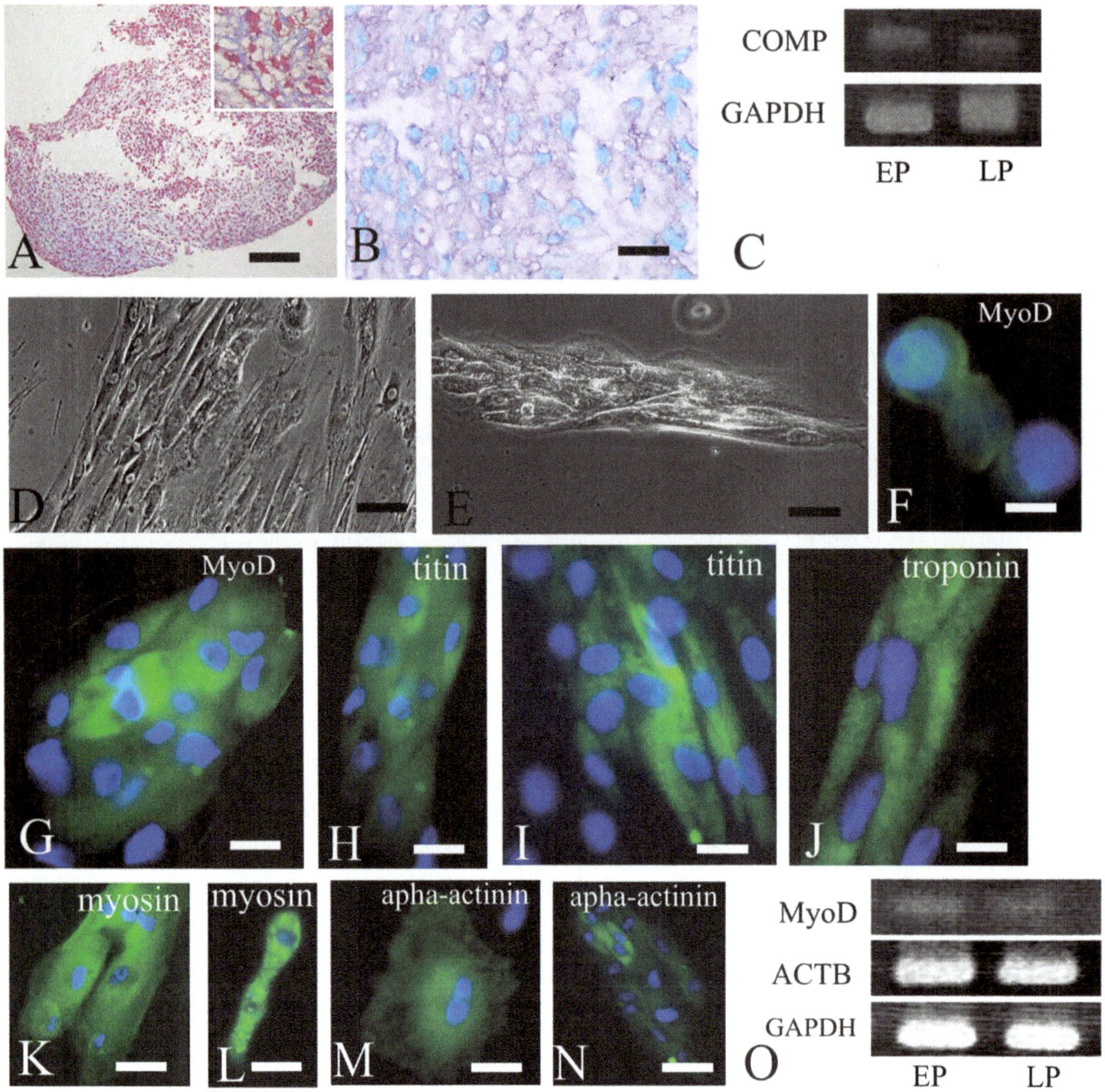

Figure 5. ***In vitro*** **differentiation potential of IDPSCs. A–C)** Chondrogenic differentiation. **A)** Pellet culture: collagen fibers intensively stained by Massońs thrichrome. Inset: same as in (A) high magnification. **B)** The proteoglycans presence was revealed by Toloudine blue staining. **C)** RT-PCR shows the expression of COMP gene in EP and LP of IDPSCs. Housekeeping gene GAPDH is used as control. **D–O)** Myogenic differentiation. **D, E)** Morphological aspect showing stages of muscle fibers formation. **F)** Nuclear expression of MyoD1 protein in LP of IDPSCs-derived myocyte-like cells. **G)** Myosac composed by MyoD1 positive cells. **H, I)** Titin protein expression in LP of IDPSCs-derived myotubes. **J)** Expression of troponin I in Z-bands of myofibers. **K)** Myosin protein expression. **L)** Very small, satellite-like cells, showing positive myosin immunostaining. **M)** Binuclear cell positive for alpha-actinin (spot-like labeling). **N)** Fused myotubes, which deferentially express alpha-actinin protein. **O)** RT-PCR shows the expression of MyoD1 and ACTB genes in EP and LP of IDPSCs. Housekeeping gene GAPDH is used as control. A, B, D, E) Light Microscopy; F-N) Epi-fluorescence, nuclei stained with DAPI (blue). Scale bars: A = 200 μm; B = 20 μm; D = 50 μm; E, N = 10 μm; F–M = 5 μm.

medium (DMEM)/Ham's F12 (DMEM/F12, Invitrogen Corporation - Carlsbad, CA, USA) supplemented with 15% fetal bovine serum (FBS, Hyclone, Logan, Utah, USA), 100 units/ml penicillin, 100 μg/ml streptomycin, 2 mM L-glutamine, and 2 mM nonessential amino acids (all from Invitrogen) in a 5% CO_2 humid atmosphere at 37°C. After a period of 3 or 4 days, fibroblast-like cells were generated from adherent explants. Explants were transferred to another Petri dish under the same culture conditions; this procedure was repeated several times (Figure 2). Fibroblast-like cells growing in monolayer were further washed twice with PBS and subjected to 0.5 g/L trypsin and 0.53 mmol/L Ethylenediamine tetra-acetic acid (EDTA) (Invitrogen) for 3 to 5 minutes at 37°C. Passage 1 was counted after the first enzymatic digestion. Trypsin action was inactivated by culture medium supplemented with 10% FBS and cells ($\sim 5 \times 10^5$) were placed into 25 cm^2 cell culture flask (Corning). This subculturing was performed each 3–4 days and the culture medium was changed daily. For cryopreservation, 90% FBS and 10% dimethylsulfoxide (DMSO) (Sigma, St. Louis, Mo., USA) were used as freezing medium. Frozen cells were maintained in sealed vials at −196°C.

Karyotype Analyses

Karyotyping of subconfluent EP and LP of IDPSCs cultured in DMEM/F12 medium (Invitrogen) was performed at passage 3.

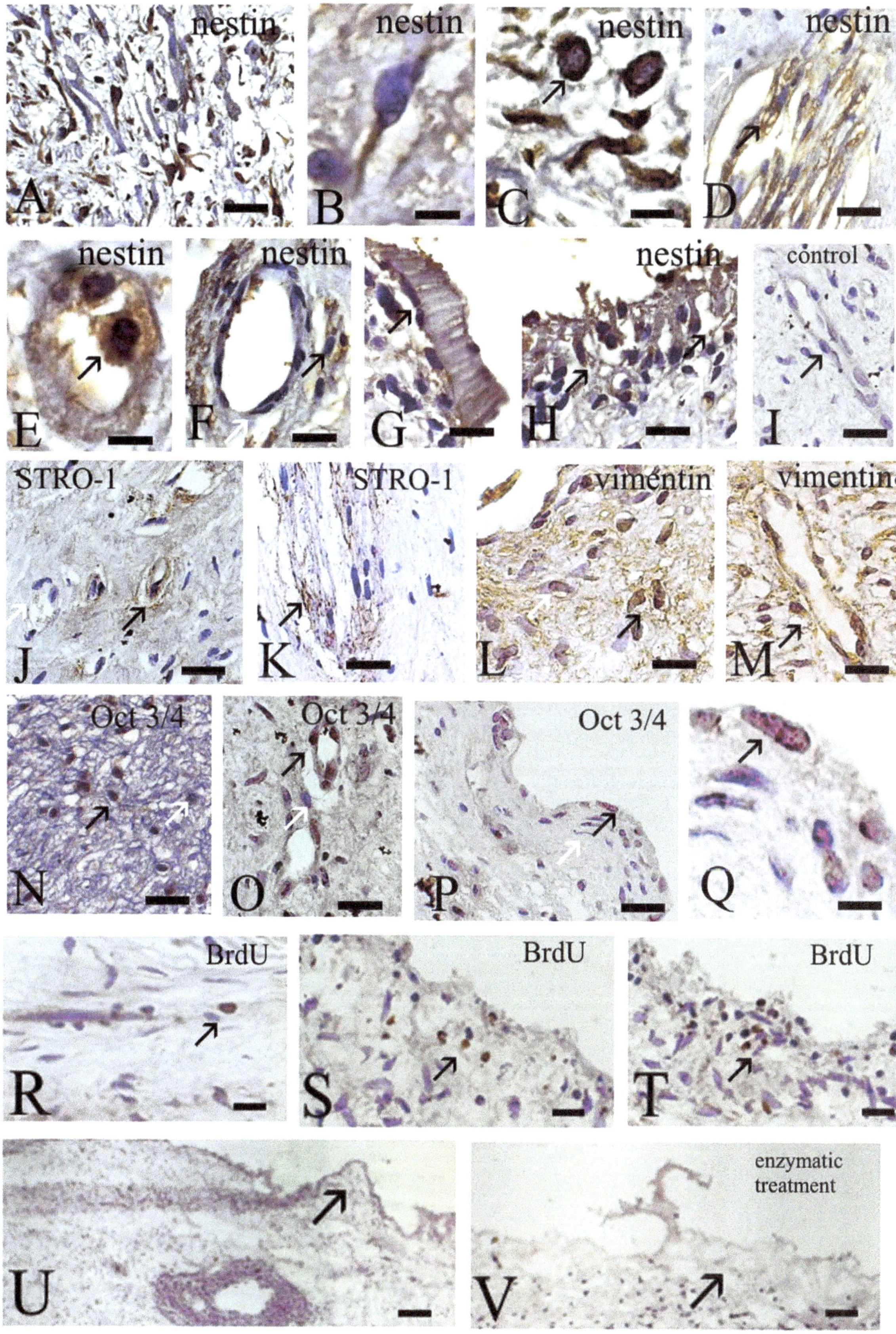

Figure 6. Expression of nestin, STRO-1, vimentin, Oct3/4 and BrdU in DP. A–H) Nestin expression. **A–C**) Cell rich zone. **A**) Multiple nestin positive cells can be observed. Here and below black arrows indicate immunopositive, while white arrows - immunonegative cells. **B**) Supposedly

undifferentiated MSC shows nestin cytoplasm localization. **C)** Nestin positive cells with two distinct morphologies round epithelial-like (ES-like) and fibroblast-like cells. **D–F)** Cell free zone. **D)** Nestin showing intermediate filament staining in nerve plexus. **E)** Small capillary with two intensively stained nestin positive cells. **F)** Same as in (E) with nestin positive cells in lateral of capillary (arrow). **G, H)** Odontoblastic layer. Nestin positive obontoblasts (G, H) can be observed. **I)** Negative control: only secondary antibody was used. **J, K)** Cell free zone. **J)** STRO-1 positive cells within capillaries (perivascular niche). **K)** Very poor STRO-1 immunostaining was observed within nerve plexus. **L, M)** Vimentin positive (black arrows) cells localization in cell rich (L) and cell free (M) zones. **N–Q)** Oct3/4 positive cells localization in cell rich (N) and cell free (O–Q) zones. **R–T)** BrdU immunostaining of DP. **R)** DP just after plating in culture medium. **S)** 48 hours after *in vitro* cultivation. **T)** 72 hours after *in vitro* cultivation. **U–V)** DP without (U) and with (V) enzymatic treatment. **V)** External cell layer of DP is destroyed by such treatment. A–V) Light Microscopy. Scale bars: A, D, F–P, R–T = 20 μm; B, C, Q = 5 μm; U, V = 50 μm.

Before harvesting, demecolcine (Sigma) at a final concentration of 0.1 μg/ml was added for 1 hour. Cells were harvested, washed in PBS and resuspended in 0.5 ml of medium and mixed with 0.075 M KCl (Sigma) to a volume of 10 ml. After incubation for 20 minutes at room temperature, cells were centrifuged at 400 g for five minutes and the pellet fixed in 5 ml three times (3:1) of cold methanol/acetic acid (Sigma). Three drops of cell suspension were fixed per slide. For chromosome counting, slides were stained in Giemsa for 15 minutes and; >200 cells were analyzed per cell line and reported on a Zeiss II microscope (Zeiss, Jena, Germany) according to the International System for Human Cytogenetic Nomenclature.

Transmission Electron Microscopy (TEM)

For TEM, EP and LP of IDPSCs were fixed in 2.5% glutaraldehyde (Sigma) for 48 h, post-fixed in 1% phosphate-buffered osmium tetroxide solution (pH 7.4) (Sigma) for 2 h at 4°C and embedded in Spurr's Resin (Sigma). Ultrathin sections were obtained using an automatic ultramicrotome (Ultracut R, Leica Microsystems, Germany). Sections were double-stained with uranyl acetate (Sigma) and lead citrate (Sigma) (2% and 0.5%, respectively) and analyzed using TEM (Morgagni 268D, FEI Company, The Netherlands; Mega).

Antibodies and Immunophenotyping

EP and LP of IDPSCs immunophenotyping was based on immunofluorescence and flow-cytometry analyses performed by using anti-human specific antibodies (vimentin, nestin, fibronectin, Oct3/4 (all from Santa Cruz Biotechnology, Santa Cruz, CA, USA), CD105/SH-2 and CD73/SH-3 (both from Case Western Reserve University, OH, USA). FITC-conjugated secondary antibodies (Chemicon, Temecula, CA, USA) were used and respective isotype matched controls. Immunofluorescence were analyzed using these aforementioned antibodies after cell fixation in 4% paraformaldehyde (Sigma) in PBS and permeabilization in 0.1% Triton X-100 (Sigma) in PBS. IDPSCs were incubated with 5% bovine serum albumin (BSA, Sigma) diluted in PBS for 30 minutes and further incubated for 1 h at room temperature with FITC-conjugated goat anti-mouse or anti-rabbit immunoglobulin (Chemicon) at a final dilution of 1:500 in PBS (Invitrogen). Microscope slides were mounted in Vectashield mounting medium with 4′,6-Diamidino-2-phenylindol (DAPI, Vector Laboratories, Burlingame, CA) and immunofluorescence was detected using a Carl Zeiss Axioplan fluoromicroscope (LSM 410, Zeiss, Jena, Germany) or Nikon Eclipse E1000 (Nikon, Kanagawa, Japan). Digital images were acquired with CCD camera (Applied Imaging model ER 339) and the documentation system used was Cytovision v. 2.8 (Applied Imaging Corp. - Santa Clara, CA, USA). Flow-cytometry was performed using EP and LP of IDPSC at passage 3. Cells were detached by using a 10 min treatment at 37°C with PBS 0.02% EDTA, pelleted (10 min at 400 g) and washed in 0.1% BSA in 0.1 M PBS at 4°C. Next, cells at a concentration of 10^5 cells/ml were stained with saturating concentration of aforementioned antibodies (10 μl). After 45 minute incubation in the dark at room temperature, cells were washed three times with PBS and resuspended in 0.25 ml of cold PBS. Flow-cytometry analysis was performed on a fluorescence-activated cell sorter (FACS; Becton, Dickinson, San Jose, CA) using the CELL Quest program (Becton, Dickinson). The flow cytometry and/or immunofluorescence analyses were repeated with all samples (n = 10), and one representative experiment is presented. All experiments have been done in triplicate and furthermore were repeated several times.

Cell Growth Rate

To evaluate the effect of different culture media on cell growth, freshly isolated and the same IDPSC frozen–thawed were equally divided in three groups (DMEM/F12, DMEM low-glucose (1000 mg/ml; DMEM-LG) and Minimum Essential Medium (MEM) Alpha Medium (MEM-alpha). All media (Invitrogen) were supplemented with 15% FBS (Hyclone), 100 units/ml penicillin, 100 μg/ml streptomycin, 2 mM L-glutamine, and 2 mM nonessential amino acids (all from Invitrogen). Cells were seeded at a density of 10^5/ cm^2 counted for at least fifteen consecutive days to evaluate the growth rate and the effect of cryopreservation. We also verified the capacity of DP tissue explant to produce IDPSC after consecutive rounds of cryopreservation and thawing. All experiments were performed in triplicate.

Data and Statistical Analysis

Growth curves were constructed using data from cell lines, passage number (P2 to P15), cryopreservation and growth medium. Cell number data were analyzed by using two-way analysis of variance ("cryopreservation" and "growth medium") complemented by Tukey post hoc multiple comparison tests. The significance level was set at 5% (SPSS 19.0, Chicago, IL, USA).

RNA Extraction and Reverse Transcription-polymerase Chain Reaction (RT-PCR)

EP and LP of IDPSCs were cultivated during seven passages in three distinct media (DMEM/F12, MEM-alpha, and DMEM-LG). To evaluate the effect of these different culture media on gene expression, total RNA was extracted using Trizol (Invitrogen): IDPSCs were washed in PBS and RNA extraction was performed according to manufactures instructions. cDNAs were synthesized from 1 μg of total RNA reverse transcribed with the RevertAid M-MuLV Reverse Transcriptase and oligo (dT) (Fermentas Life Science, Amherst, NY, EUA) according to the manufactures instructions. The final concentrations of reagents were: 20 μl of PCR reactions were prepared with 2 μl cDNA, 0,2 μM of each primer, 1 unit of Taq DNA Polymerase, 0,2 μM of dNTPs, 1,5 mM of magnesium chloride and buffer Taq DNA Polymerase (Fermentas Life Science). Primer sequences (forward and reverse), and the lengths of amplified products are summarized: Nestin FW 5′- CTCTGACCTGTCAGAAGAAT-3′, and RV 5′-GACGCTGACACTTACAGAAT-3′ (302 bp/54°C); Vimentin FW 5′-AAGCAGGAGTCCACTGAGTACC-3′, and RV

5′-GAAGGTGACGAGCCATTTCC-3′ (205 bp/55°C); Fibronectin FW 5′-GGATCACTTACGGAGAAACAG-3′, and RV 5′-GATTGCATGCATTGTGTCCT-3′ (386 bp/56°C); OCT3/4 FW 5′-ACCACAGTCCATGCCATCAC-3′, and RV 5′-TCCACCACCCTGTTGCTGTA-3′ (120 bp/61°C); SH2/CD105 FW 5′- TCTGGACCACTGGAGAATAC-3′, and RV 5′-GAGGCATGAAGTGAGACAAT-3′ (171 bp/56°C); SH3/CD73 FW 5′-ACACGGCATTAGCTGTTATT-3′, and RV 5′-AGTATTTGTTCTTTGGGCA-3′ (391 bp/56°C). For chondrogenic and myogenic differentiation, following primer sequences were used: COMP FW 5′-CCGACAGCAACGTGGTCTT-3′, and RV 5′-CAGGTTGGCCCAGATGATG-3′ (91 bp/53°C); ACTB FW 5′-TGGCACCACACCTTCTACAATGAGC-3′, and RV 5′ GCACAGCTTCTCCTTAATGTCACGC-3′ (395 bp/59°C); MYOD1 FW 5′-GCCGCCTGAGCAAAGTAAATGAGG-3′, and RV 5′-TAGTCCATCATGCCGTCGGAGC-3′ (280 bp/53°C). GADPH gene FW 5′- ACCACAGTCCATGCCATCAC-3′, and RV 5′-TCCACCACCCTGTTGCTGTA-3′ (463 bp/61°C) was used as control. Undifferentiated IDPSC were examined as negative control for differentiation specific primers. PCR reactions were performed under the following conditions: 1 cycle at 94°C for 5 minutes, followed by 35 cycles at 94°C for 1 minute, annealing temperature for 1 minute, and 72°C for 1 minute. Amplified products were resolved by electrophoresis on a 1.5% agarose gel (Sigma) and visualized using ethidium bromide (Sigma) staining.

Differentiation Assays

Chondrogenic differentiation. The differentiation was performed using pellet culture technique [19,20]. EP and LP of IDPSCs populations from sub confluent cultures (passage 3) were released by 0.5 g/L trypsin and 0.53 mmol/L EDTA, counted and used to generate micromass culture. Briefly, 4×10^6 cells were centrifuged at 500 g in 15 ml polypropylene conical tubes (Corning) and the resulting pellets were cultured for 4 weeks. Control cultures were grown in a serum-free chemically defined medium consisting of DMEM, high-glucose (4,500 mg/L; DMEM-HG) (Invitrogen) supplemented with 6.25 µg/ml insulin, 6.2 µg/ml transferrin, 6.25 µg/ml selenious acid, 5.33 µg/ml linoleic acid (ITS, Premix, BD, USA) and 1 mM sodium pyruvate (Invitrogen). To induce chondrogenic differentiation, control medium was supplemented with 0.1 µM dexamethasone and 50 µg/ml ascorbate-2-phosphate (both from Invitrogen), without transforming growth factor beta (TGF-β). Cultures were incubated for 4 weeks at 37°C in a humid atmosphere containing 5% CO_2; the medium was changed every day. Cell aggregates were harvested at 4 weeks for RT-PCR, with primer sequences aforementioned, and were also fixed for histology.

Muscle differentiation. EP and LP of IDPSCs at the passage 3 were seeded at a concentration of 5000 cells/cm^2. Control cultures were grown in a serum-free chemically defined medium consisting of DMEM-HG (Invitrogen) supplemented with 10% FBS (Hyclone). To induce myogenic differentiation, control medium was supplemented with 50 µM hydrocortisone (Sigma) and 5% horse serum (Invitrogen). Cultures were incubated for 21 days at 37°C in a humid atmosphere containing 5% CO_2; the medium was changed every three days. Then, cultured cells were fixed for histology and for immunohistochemistry with antibodies that recognize human muscle proteins. Aforementioned primers sequences were used to detect myogenic differentiation of IDPSCs.

Histological Analyses

Pellet culture analysis. Cell aggregates were fixed in 4% formaldehyde in PBS for 40 minutes and embedded in paraffin (Sigma). Paraffin sections were stained with Masson's trichrome (Sigma) and Toluidine Blue (Sigma) and analyzed using transmitted and polarized light microscopy Nikon Eclipse E1000 (Nikon), digital images were acquired by CCD camera (Applied Imaging).

Muscle analysis. Slices were fixed in 4% paraformaldehyde diluted in PBS for 40 min and analyzed using transmitted and polarized light microscopy Nikon Eclipse E1000 (Nikon), digital images were acquired by CCD camera (Applied Imaging).

Immunohistochemistry

Troponin I (Chemicon), titin (Chemicon), alpha-smooth muscle actinin (Sigma), MyoD1 (Chemicon) and sarcomeric myosin (Chemicon) proteins were analyzed using specific antibodies. First, cells were fixed in 4% paraformaldehyde diluted in PBS and permeabilized in 1% Triton X-100 diluted in PBS. EP and LP of IDPSC that were submitted to myogenic differentiation were incubated in 5% BSA diluted in PBS for 30 min and incubated for 1 h at room temperature with FITC-conjugated goat anti-mouse or anti-rabbit immunoglobulin (Chemicon) at a final dilution of 1:500 in PBS. Microscope slides were mounted in Vectashield mounting medium with DAPI and immunofluorescence was detected using a Carl Zeiss Axioplan fluoromicroscope (Zeiss) or Nikon Eclipse E1000 (Nikon), digital images were acquired by CCD camera (Applied Imaging) and the documentation system used was Cytovision v. 2.8 (Applied Imaging Corp).

Pulp Tissue Experiments

Tissue specimens. Ten deciduous teeth of ten healthy subjects (range 5–10 years), free of caries and restorations, were extracted and initially rinsed in PBS. Seven pulps were gently removed and fixed in 10% formalin solution for 48 h. The specimens were embedded in paraffin blocks and sections of 10 µm were obtained. The other three pulps were gently rinsed in PBS and sliced. Each slice was placed in different culture dish. Next, 5-bromo-2′-deoxyuridine (BrdU, Sigma) was added directly into basal culture medium. First pulp slice was fixed and processed after 6 h of treatment with BrdU. In the second culture dish with pulp slice, BrdU was added after 42 h and in the third - after 66 h. After 6 h of treatment with BrdU, all slices were fixed in 10% formalin solution for 48 h. The specimens were embedded in paraffin blocks and sections of 10 µm were obtained. All above specimens were treated by immunohistochemical methods.

Immunohistochemical

Paraffin sections of 10 µm were deparaffinized and then hydrated. Endogenous peroxidase activity was measured by incubating the sections for 30 min in a 0.1% solution of hydrogen peroxide (Sigma). For antigen retrieval, sections were incubated with trypsin for 10 min at 37°C. To inhibit nonspecific antigen binding, sections were incubated with blocking serum (5% fetal calf serum, Invitrogen) for 10 min. Sections were then incubated for 12–16 h with the primary antibody in a moist chamber at 4°C. Primary antibodies were the same used in immunophenotyping of IDPSCs and additionally anti-human STRO-1 (Santa Cruz) and mouse anti-BrdU IGg (Chemicon). The optimal dilution of the primary antibody was found to be 1:10. Slides were again rinsed with PBS and then incubated with biotinylated secondary antibody (DAKO, Glostrup, Denmark) in 1:200 dilution for 30 min. The samples were washed with PBST (PBS with 0.1% of Tween 20) and incubated with StrepABComplex/HRP (DAKO) at 1:100 dilution for 30 min. After one more wash with PBST, the colour was revealed by the chromogen 3 (3-diaminobenzidine DAB Kit, Zymed Laboratories, Inc.) for 5 min, followed by PBST washing, nuclear counterstaining with Harris haematoxylin for

45 s, dehydrated and mounted in Permount. Observation of the sections was conducted using a Carl Zeiss Axioplan fluoromicroscope (Zeiss). Negative control sections were treated identically, except the primary antibody which was substituted by PBS.

Acknowledgments

We are grateful to Dr Adriana da Costa Neves for her expertise in immunohistochemical assay and to Alexsander S. de Souza for his excellent assistance in confocal microscopy analysis. We wish to thank Dr. Graciela C. Pignatari for her assistance in molecular biology, Dr. Cristiane V. Wenceslau and Lisley I. Mambelli for their help during the development of this project.

Author Contributions

Conceived and designed the experiments: NFL IK AC. Performed the experiments: NFL CG. Analyzed the data: NFL IK AK JH CO. Contributed reagents/materials/analysis tools: NFL IK JH CO. Wrote the paper: NFL IK AC.

References

1. Gronthos S, Mankani M, Brahim J, Robey PG, Shi S (2000) Postnatal human DP stem cells (DPSCs) in vitro and in vivo. Proc Natl Acad Sci U S A 97: 13625–13630.
2. Miura M, Gronthos S, Zhao M, Lu B, Fisher LW, et al. (2003) SHED: Stem cells from human exfoliated deciduous teeth. Proc Natl Acad Sci U S A 100: 5807–5812.
3. Kerkis I, Caplan AI (2011) Stem cells in dental pulp of deciduous teeth. Tissue Eng Part B Rev Dec 28 [Epub ahead of print].
4. Kerkis I, Kerkis A, Dozortsev D, Stukart-Parsons GC, Gomes Massironi SM, et al. (2006) Isolation and characterization of a population of immature dental pulp stem cells expressing OCT-4 and other embryonic stem cell markers. Cells Tissues Organs 184: 105–116.
5. Fonseca SAS, Abdelmassih S, de Mello Cintra Lavagnolli T, Serafim RC, Clemente Santos EJ, et al. (2009) Human immature dental pulp stem cells' contribution to developing mouse embryos: production of human/mouse preterm chimaeras. Cell Prolif 42: 132–140.
6. de Mendonca Costa A, Bueno DF, Martins MT, Kerkis I, Kerkis A, et al. (2008) Reconstruction of large cranial defects in nonimmunosuppressed experimental design with human dental pulp stem cells. J Craniofac Surg 19: 204–210.
7. Gomes JA, Geraldes Monteiro B, Melo GB, Smith RL, Cavenaghi Pereira da Silva M, et al. (2010) Corneal reconstruction with tissue-engineered cell sheets composed of human immature dental pulp stem cells. Invest Ophthalmol Vis Sci 51: 1408–1414.
8. Kerkis I, Ambrosio CE, Kerkis A, Martins DS, Zucconi E, et al. (2008) Early transplantation of human immature DP stem cells from baby teeth to golden retriever muscular dystrophy (GRMD) dogs: Local or systemic? J Transl Med 6: 35.
9. Beltrão-Braga PCB, Pignatari GC, Maiorka PC, Oliveira NAJ, Lizier NF, et al. (2011) Feeder-free derivation of induced pluripotent stem cells from human immature dental pulp stem cells. Cell Transplant [Epub ahead of print]; DOI: 10.3727/096368911x566235.
10. Yan X, Qin H, Qu C, Tuan RS, Shi S, et al. (2010) iPS cells reprogrammed from human mesenchymal-like stem/progenitor cells of dental tissue origin. Stem Cells Dev 19: 469–480.
11. Kerkis I, Lobo SE, Kerkis A (2009) Dental pulp stem cells and perspectives of future application in cell therapy. In: Deb, K.D, and Totel, S.M., eds. Stem Cells. Basic and Applications. New Delhi: Tata McGraw Hill. 426–438.
12. Caplan AI (2009) Why are MSCs therapeutic? New data: new insight. J Pathol 217: 318–324.
13. Wagner J, Kean T, Young R, Dennis JE, Caplan AI (2009) Optimizing mesenchymal stem cell-based therapeutics. Curr Opin Biotechnol Oct;20(5): 531–536.
14. Stenderup K, Justesen J, Clausen C, Kassem M (2003) Aging is associated with decreased maximal life span and accelerated senescence of bone marrow stromal cells. Bone 33(6): 919–926.
15. Baxter MA, Wynn RF, Jowitt SN, Wraith JE, Fairbairn LJ, et al. (2004) Study of telomere length reveals rapid aging of human marrow stromal cells following in vitro expansion. Stem Cells 22(5): 675–682.
16. Rubio D, Garcia-Castro J, Martin MC, de la Fuente R, Cigudosa JC, et al. (2005) Spontaneous human adult stem cell transformation. Cancer Res 65(8): 3035–3039.
17. Wang Y, Huso DL, Harrington J, Kellner J, Jeong DK, et al. (2005) Outgrowth of a transformed cell population derived from normal human bone marrow mesenchymal stem cell culture. Cytotherapy 7(6): 509–519.
18. Spath L, Rotilio V, Alessandrini M, Gambara G, De Angelis L, et al. (2010) Explant-derived human dental pulp stem cells enhance differentiation and proliferation potentials. J Cell Mol Med. 2010 Jun;14(6B): 1635–44. Epub 2009 Jul 7.
19. Johnstone B, Hering TM, Caplan AI, Goldberg VM, Yoo JU (1998) In vitro chondrogenesis of bone marrow-derived mesenchymal progenitor cells. Exp Cell Res. Jan 10; 238(1): 265–272.
20. Yoo JU, Barthel TS, Nishimura K, Solchaga L, Caplan AI, et al. (1998) The chondrogenic potential of human bone-marrow-derived mesenchymal progenitor cells. J Bone Joint Surg Am. Dec; 80(12): 1745–1757.
21. Schofield R (1978) The relationship between the spleen colony forming cell and the hemopoietic stem cell. Blood Cells 4: 7–25.
22. Mitsiadis TA, Barrandon O, Rochat A, Barrandon Y, De Bari C (2007) Stem cell niches in mammals. Experimental Cell Research 313(16): 3377–3385.
23. De Bari C, Pringle S, Pitzalis C, Dell'Accio F (2006) The stem cell niche: a new target in medicine. Current Opinion in Orthopaedics 17(5): 398–404.
24. Abdallah BM, Boissy P, Tan Q, Dahlgaard J, Traustadottir GA, et al. (2007) dlk1/FA1 regulates the function of human bone marrow mesenchymal stem cells by modulating gene expression of pro-inflammatory cytokines and immune responserelated factors. J Biol Chem 282: 7339–7351.
25. Gronthos S, Brahim J, Li W, Fisher LW, Cherman N, et al. (2002) Stem cell properties of human dental pulp stem cells. J Dent Res 81: 531–535.
26. Pal R, Hanwate M, Jan M, Totey S (2008) Phenotypic and Functional comparison of optimum culture condition for upscaling of Bone marrow derived Mesenchymal stem cells. J Tissue Eng Regen Med 3: 163–174.
27. Papaccio G, Graziano A, d'Aquino R, Graziano MF, Pirozzi G, et al. (2006) Long-term cryopreservation of dental pulp stem cells (SBP-DPSCs) and their differentiated osteoblasts: a cell source for tissue repair. J Cell Physiol 208: 319–325.
28. Zhang W, Walboomers XF, Shi H, Fan M, Jansen JA, et al. (2006) Multilineage differentiation of stem cells derived from human dental pulp after cryopreservation. Tissue Eng 12: 2813–2823.
29. Arora V, Arora P, Munshi AK (2009) Banking stem cells from human exfoliated deciduous teeth (SHED): Saving for the Future. J Clin Pediatr Dent 33: 289–294.
30. Laino G, d'Aquino R, Graziano A, Lanza V, Carinci F, et al. (2005) A new source of human adult dental pulp stem cells: a useful source of living autologous fibrous bone tissue. J Bone Miner Res 20: 1394–1402.
31. Pierdomenico L, Bonsi L, Calvitti M, Rondelli D, Arpinati M, et al. (2005) Multipotent mesenchymal stem cells with immunosuppressive activity can be easily isolated from dental pulp. Transplantation 80: 836–842.
32. Jo YY, Lee HJ, Kook SY, Choung HW, Park JY, et al. (2007) Isolation and characterization of postnatal stem cells from human dental tissues. Tissue Eng 13: 767–773.
33. Huang AH, Chen YK, Lin LM, Shieh TY, Chan AW (2008) Isolation and characterization of dental pulp stem cells from supernumerary tooth. J Oral Pathol Med 37: 571–574.
34. Nam H, Lee G (2009) Identification of novel epithelial stem cell-like cells in human deciduous dental pulp. Biochem Biophys Res Commun 386: 135–139.
35. Waddington RJ, Youde SJ, Lee CP, Sloan AJ (2009) Isolation of distinct progenitor stem cell populations from dental pulp. Cells Tissues Organs 189: 268–274.
36. Shi S, Gronthos S (2003) Perivascular niche of post natal mesenchymal stem cells in human bone marrow and dental pulp. J Bone Miner Res 189: 696–704.
37. Crisan M, Yap S, Casteilla L, Chen CW, Corselli M, et al. (2008) A perivascular origin for mesenchymal stem cells in multiple human organs. Cell Stem Cell 3: 301–313.
38. About I, Laurent-Maquin D, Lendahl U, Mitsiadis TA (2000) Nestin expression in embryonic and adult human teeth under normal and pathological conditions. Am. J. Pathol. 157: 287–295.
39. Méndez-Ferrer S, Michurina TV, Ferraro F, Mazloom AR, Macarthur BD, et al. (2010) Mesenchymal and haematopoietic stem cells form a unique bone marrow niche. Nature 12; 466 (7308): 829–834.

17

Intracellular Freezing in the Infective Juveniles of *Steinernema feltiae*: An Entomopathogenic Nematode

Farman Ali*¤, David A. Wharton

Department of Zoology, University of Otago, Dunedin, New Zealand

Abstract

Taking advantage of their optical transparency, we clearly observed the third stage infective juveniles (IJs) of *Steinernema feltiae* freezing under a cryo-stage microscope. The IJs froze when the water surrounding them froze at −2°C and below. However, they avoid inoculative freezing at −1°C, suggesting cryoprotective dehydration. Freezing was evident as a sudden darkening and cessation of IJs' movement. Freeze substitution and transmission electron microscopy confirmed that the IJs of *S. feltiae* freeze intracellularly. Ice crystals were found in every compartment of the body. IJs frozen at high sub-zero temperatures (−1 and −3°C) survived and had small ice crystals. Those frozen at −10°C had large ice crystals and did not survive. However, the pattern of ice formation was not well-controlled and individual nematodes frozen at −3°C had both small and large ice crystals. IJs frozen by plunging directly into liquid nitrogen had small ice crystals, but did not survive. This study thus presents the evidence that *S. feltiae* is only the second freeze tolerant animal, after the Antarctic nematode *Panagrolaimus davidi*, shown to withstand extensive intracellular freezing.

Editor: Boris Rubinsky, University of California at Berkeley, United States of America

Funding: The Principal author acknowledges the support of a Postgraduate Otago University scholarship and a publishing bursary. The funders had no role in study design, data collection and analysis, decision to publish, or preparation of the manuscript.

Competing Interests: The authors have declared that no competing interests exist.

* E-mail: drfarman@gmail.com

¤ Current address: Department of Zoology, Abdul Wali Khan University Mardan, Mardan, Pakistan

Introduction

The entomopathogenic nematode *Steinernema feltiae* is a temperate species and is globally distributed [1]. In cold regions the infective free-living third-stage juveniles (IJs) of this nematode are exposed to varying and to sub-zero temperatures. Nematodes are aquatic [2] and are at risk of inoculative freezing by ice in their surroundings [3], penetrating through body openings such as the mouth or anus [4]. Some studies suggest that *S. feltiae* is a cold tolerant species [5]. However, the underlying mechanism of cold survival in entomopathogenic nematodes, including this species, is poorly understood.

In our previous experiments, we showed that *S. feltiae* IJs could survive sub-zero temperatures [6] but not whether the nematodes themselves or just the water surrounding them froze. Here we determine the mechanism of cold tolerance of *S. feltiae* IJs based on morphological changes at the cellular level after freezing them under various conditions. Two techniques have been used: cold stage microscopy and freeze substitution/transmission electron microscopy. Nematodes are transparent and freezing within their bodies can be observed under a cold microscope stage [5,7,8]. This allows us to observe whether or not the nematode freezes. Shrinkage of the nematode due to loss of water can also be observed. This occurs in situations where the water surrounding a nematode freezes but not that within its body; a process called 'cryoprotective dehydration' [9].

Freeze substitution is a process in which the ice within the frozen specimen is replaced by an organic solvent, such as methanol or acetone, at a very low temperature; below −70°C [10]. The technique is usually used for specimens that have been rapidly frozen and is followed by embedding in resin. Embedding permits cutting of ultrathin sections for electron microscopy. Rapid freezing of a biological specimen aims to prevent ice formation through water vitrification and thus provides better preservation of ultrastructure, and the retention of water-soluble components and antigenicity, than is achievable using more conventional chemical fixation [11]. However, the technique can also be used to visualize the position of ice formed in samples frozen more slowly [12,13,14]. We used methanol as a substitution solvent because this has a high capacity for water absorption, can dissolve sugar and is liquid at a substitution temperature (−90°C) which does not allow ice recrystallization [15]. Fixatives dissolved in the substitution fluid infiltrate into the sample and react with its proteins, lipids and other constituents when the temperature is raised sufficiently for rapid cross-linking to occur. Embedding in resin is then carried out, replacing the substitution fluid. The previous position of ice is visible as a white space. Freeze substitution provides accurate information about the distribution of ice in both extra and intracellular compartments.

We used freeze substitution of nematodes after different freezing regimes involving different temperatures and exposure times. The aim was not to prevent ice formation but to locate the position and pattern of ice crystals within the frozen nematodes, which may account for differences in their freezing survival.

Materials and Methods

Nematode culture

Steinernema feltiae was grown in *Galleria mellonella* larvae at 23°C using the method described by Kaya and Stock [16]. The fresh IJs were used for experiments on the same day they were collected.

Cold stage microscopy

The freezing process was observed on a microscope cold stage similar to that described by Wharton & Rowland [8]. This is based on a thermoelectric cooling (Peltier) module, the hot face of which is cooled by circulating fluid from a Haake F3-Q refrigerated circulator (Thermo Fisher Scientific, Waltham, MA) and which is mounted on a Zeiss Axiophot Photomicroscope (Carl Zeiss Inc., Thornwood, NY). Nematode samples in a 1 μl drop of artificial tap water (ATW) [17] were mounted between two small glass coverslips in the sample chamber of the cold stage. The temperature was monitored with a NiCr/NiAl thermocouple inserted in the metal finger of the cold stage close to the specimen via a Comark electronic thermometer. The electronic thermometer was calibrated against the melting point of deionized water. Samples were cooled from 2°C to various minimum temperatures (−1, −2, −3, −4, −5, −6°C) at 0.5°C min^{-1} and freezing was initiated by seeding with a small ice crystal at the edge of the coverslip. Nitrogen gas was supplied to the sample chamber of the cold stage to prevent condensation. The freezing process was visualized using a 10× objective lens and images were captured at regular intervals by a Canon Powershot A640 Digital Camera, controlled by Axio Vision v. 4.6 software (Zeiss) run on an Insite PC. Freezing of nematodes was observed as a sudden darkening. Unfrozen nematodes did not change their appearance and remained clear. After seeding, the sample was allowed to stay frozen for either 3 h at −1°C or 30 min at all other test temperatures, and then rewarmed to 2°C at 0.5°C min^{-1}. Nematodes were removed from the cold stage once it had reached room temperature and transferred to watch glasses in ATW. Survival was determined after 24 hours. The experiment was replicated three times.

Freezing regimes used for freeze substitution

To see the differences in morphology of IJs after freezing, different freezing regimes were used. Each sample consisted of 10 μl of nematode suspension in an Eppendorf tube (four replicates, two separate cultures) transferred to a cooling block, the temperature of which was controlled by fluid circulating from a Haake Phoenix II-C35P programmable refrigerated circulator (Thermo Fisher Scientific). No thermocouple was inserted into samples processed for freeze substitution. Fast-frozen samples were cooled from 2°C to −3°C or −10°C at 0.5°C min^{-1}, seeded with an ice crystal and held at −3°C or −10°C for 75 min. Slow-frozen samples were cooled from 2°C to −1°C at 0.5°C min^{-1}, seeded with an ice crystal and held at −1°C for either 75 min or overnight. At the end of the freezing time, it was checked that the samples were frozen and they were then transferred to liquid nitrogen for transportation before freeze substitution.

Freezing survival was determined from samples subjected to the same freezing regimes as above, but with a NiCr/NiAl thermocouple inserted in each tube. Thermocouples were interfaced to a Macintosh computer via a Powerlab A/D interface (ADInstruments Ltd., Dunedin, NZ). The temperature records were analysed using a computer programme (Chart v3.2.7, ADInstruments) to ensure that the exotherm, and hence the freezing of the sample was completed. The duration of the exotherm was used as an indication of its freezing rate. Samples were rewarmed at 0.5°C min^{-1} to 2°C and transferred to ATW in watchglasses at room temperature. Survival was determined after 24 hours by counting the proportion moving after a mechanical stimulus.

Nematode suspensions in Eppendorf tubes were also plunged directly into liquid nitrogen without prior freezing and processed via freeze substitution to serve as a frozen control. Nematodes were also prepared for electron microscopy using a conventional preparation technique to compare the normal nematode morphology with those that were frozen. This consisted of fixing them in 4% formaldehyde, 2.5% glutaraldehyde, 1% osmium tetroxide followed by dehydration in ethanol, and embedding in Spurr resin.

Freeze substitution and embedding

Samples in liquid nitrogen were transferred into a Reichert automatic freeze substitution (AFS) apparatus and substitution fluid (methanol containing fixatives) added. Freeze substitution and embedding were then carried out as reported previously [14]. Ultrathin (80 nm) sections were cut with a Reichert ultramicrotome, stained with uranyl acetate and lead citrate and observed using a Philips CM100 Transmission Electron Microscope, operated at 80 kV. The freeze substitution experiment was repeated three times.

Analysis of ice crystals

The following parameters were measured after transmission electron microscopy.

1. Areal Fractions (A_a) were calculated using a stereological method based on point counting [18]. An A4 size transparent sheet containing a grid of crosses (+) was superimposed on the entire transverse section image. Areal fraction was taken to be the ratio of the crosses hitting the ice crystals to the crosses hitting the section. This was determined by the formula $A_a = \sum P_c / \sum P_s$ (where $A_a =$ areal fraction, $\sum P_c =$ total points hitting the ice crystals and $\sum P_s =$ total points hitting the section).
2. Ice crystal diameter of the five largest ice crystals in each section of all the treatments was measured using ImageJ software [19].
3. Ice crystal pattern: frozen nematodes having all small, all large or a mixture of both were counted.
4. Shrinkage due to partial dehydration was compared between freezing regimes. This was visible in sections where the regular elliptical or circular profile was lost.

Nematodes were also examined for the presence of intracellular ice (ice surrounded by cytoplasm), extracellular ice (ice confined to the pseudocoel, intestinal lumen, and/or extracellular spaces), and those that were unfrozen. All images taken (13–46 per treatment) were used for quantifying ice crystal pattern and shrinkage of nematodes (parameters 3, 4). However, measurements of the diameter of the five largest ice crystals and areal fractions (parameters 1, 2) were confined to those sections which were well infiltrated with resin and were clear enough to measure these parameters accurately (13–24 per treatment).

Statistical analysis

The Statistical Package for Social Sciences (SPSS) ver. 15.0 was used for statistical analyses [20]. Areal fractions were arcsine transformed before analysis. Measurements of the five largest crystals from each section were averaged before analysis to give a single measurement. A one way analysis of variance (ANOVA) was conducted to see if the differences among the freezing regimes for various parameters were significant. Tukey's test was used to compare means. Where variances were not equal after running Leven's test of homogeneity of variances the Brown-Forsythe and Welch tests were used instead of ANOVA. Means were then separated using the Games-Howell post-hoc multiple comparison test.

Results

Cold stage observations

At −2°C and below, nematodes were clearly observed to freeze by exogenous ice nucleation by the surrounding ice, as indicated by a sudden darkening of the nematodes. Exogenous ice nucleation did not occur at −1°C and nematodes remained clear and unfrozen, although the medium surrounding them was frozen (Fig 1). Nematodes survived the freezing process down to −4°C. Nematode survival was better at high sub-zero temperatures (−1, −2, −3°C) than at −4°C (Fig. 2). There was no difference in survival between frozen and unfrozen nematodes at temperatures −1 to −3°C. Also there was no difference in the appearance of frozen nematodes between the test temperatures (−2 to −5°C).

Freeze substitution experiments

The morphology of an unfrozen *S. feltiae* IJ sectioned through the intestinal region after preparation for transmission electron microscopy by chemical fixation is shown in Fig. 3. The outermost layer is the cuticle, underneath which is a thin epidermis and then a layer of longitudinal muscles. Out of the four epidermal cords (two lateral, mid dorsal and mid ventral), the most prominent are the lateral epidermal cords. The intestinal lumen and the microvilli are clearly visible in the middle region. The dark circular structures are dense food storage granules. The cytoplasmic region is characterized by the presence of mitochondria. The pseudocoel is very thin and is not distinguishable in the image.

The morphology of specimens frozen in liquid nitrogen and prepared by freeze substitution for electron microscopy is similar to that of non-frozen controls except for its reticulated appearance due to the presence of small ice crystals throughout the tissues, including cell organelles (Fig. 3). The circular structures are dense food storage granules similar to those seen in unfrozen controls but they have a paler appearance after freezing and freeze substitution. A similar appearance observed in samples from other temperature

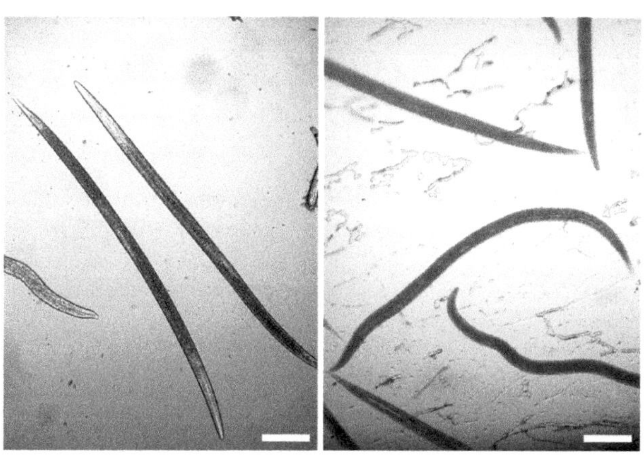

Figure 1. Infective juveniles of *Steinernema feltiae*: transparent nematodes are unfrozen 3 hours after seeding the ice crystal at −1°C (left); darkened nematodes are frozen 30 min after seeding at −2°C and below (right). Scale bars = 100 μm.

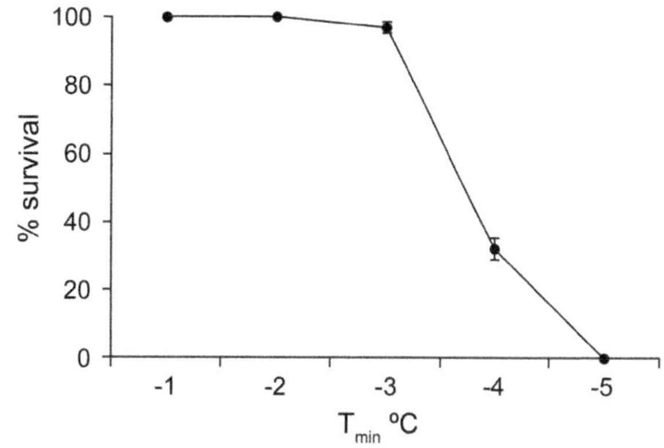

Figure 2. Survival of the infective juveniles of *S. feltiae* at various minimum temperatures (T_{min}) after freezing them on the microscope cold stage. Vertical bars represent the standard errors of the means of 3 measurements.

treatments to that of those frozen directly in liquid nitrogen was taken to indicate that nematodes were unfrozen before immersion in liquid nitrogen.

Nematode survival was high after freezing at −1°C or −3°C for 75 min and −1°C overnight but there was no survival after freezing at −10°C (Table 1). Three different ice crystal patterns were observed in nematodes after subjecting them to slow and fast freezing. Specimens with all small, all large or a mixture of small and large ice crystals were found (Fig. 4). Small ice crystals were found in nematodes frozen at −1°C. Ice crystals were larger in those frozen at −3°C and larger still in those frozen at −10°C (Fig. 4). The percentage of nematodes showing all small, all large or a mixture of ice crystals is shown in Fig. 5. Ice crystals were observed in both intracellular and extracellular compartments in all the freezing regimes and were prominent in the intestinal cells, pharynx, secretory-excretory cells and muscle cells. Sections with small ice crystals appeared slightly shrunken compared to those with large ice crystals.

Measurements on transverse sections

Overall there was a significant difference between the diameters of the five largest ice crystals measured between freezing regimes (Fig. 6: $F_{3, 58} = 16.45$; $P<0.05$). Post hoc comparison tests showed that ice crystal diameters in nematodes frozen at −10°C were significantly different from those frozen at −1°C and −3°C. However, ice crystal diameters did not differ significantly between nematodes frozen at −1°C overnight, −1°C and −3°C for 75 min (Fig. 6). Analysis of variance showed that areal fractions of the ice crystals did not vary significantly ($F_{3, 60} = 2.1$; $P>0.05$) among the freezing regimes (Fig. 7). No severe shrinkage of the tissues was noticed, since all the nematodes had ice crystals in their tissues. However, slight shrinkage of nematode tissues due to partial dehydration varied significantly among the treatments ($F_{3, 12} = 135.7$; $P<0.05$) and was more prominent in the slow freezing regimes (−1°C: 50–77.8%) than the fast freezing regimes (−3°C, −10°C: 33.3–38.5%) as shown in Fig. 8.

Discussion

Nematodes are transparent and thus the freezing process was clearly observed on the microscope cold stage. At −2°C and below the infective juveniles of *S. feltiae* froze by inoculative freezing from the external medium. None of the infective juveniles were frozen

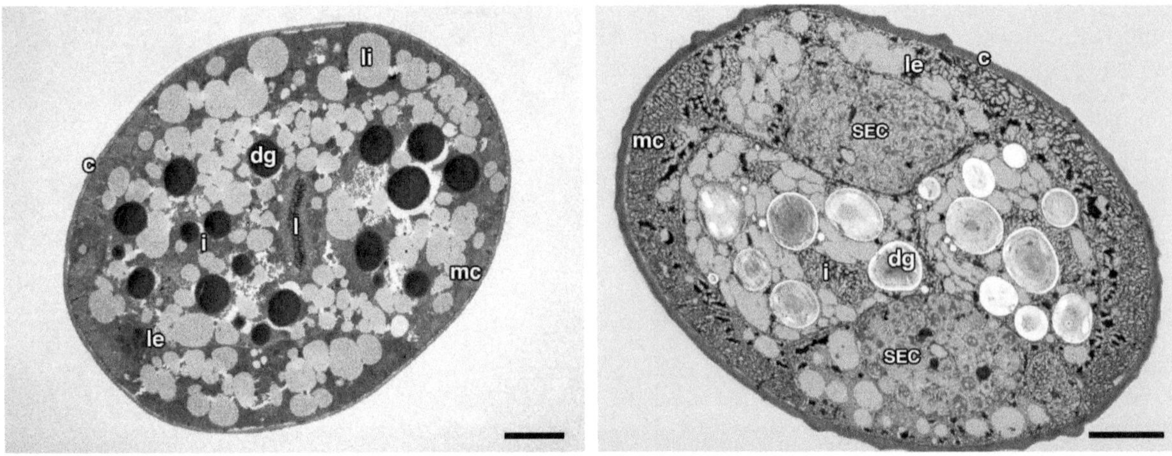

Figure 3. Transmission electron micrograph of transverse sections of the IJ of *Steinernema feltiae* prepared through conventional chemical fixation (left) and by freeze substitution after freezing in liquid nitrogen, note the small ice crystals throughout the tissues (right). c, cuticle; mc, muscle cell; dg, dense granules; le, lateral epidermal cord; i, intestinal cell; l, intestinal lumen; li, lipid droplet; SEC, secretary-excretory cell. Scale bars = 5 μm.

before the surrounding water froze. Inoculative freezing in nematodes is not uncommon, since they are always in contact with water and has been reported in several studies [7], [21], [5]. The sudden darkening coupled with immobilization of nematodes was an indication of freezing. The darkening may be due to the scattering of light by the ice. Darkening upon freezing in nematodes has been reported in several nematodes species such as *Aphelenchoides ritzemabosi* [22], *Meloidogyne* spp. [23], *Wetanema* sp. [21], *Steinernema* spp. and *Heterorhabditis* spp. [5], *Anisakis* sp. [24], *Panagrolaimus davidi* [9] and *Caenorhabditis elegans* [25].

Inoculative freezing did not occur at −1°C in *S. feltiae* and nematodes did not freeze at this temperature, although the water surrounding them was frozen. There was no change in appearance or any sign of darkening after seeding with an ice crystal, indicating that they did not freeze (Fig. 1). This indicates some low level of resistance by this species against inoculative freezing at high sub-zero temperature. The mechanism of this resistance at −1°C is not clear but it would facilitate cryoprotective dehydration, where the nematodes remain unfrozen and lose water to the surrounding ice [9]. Cryoprotectants such as glycerol and trehalose are accumulated by entomopathogenic nematodes before freezing [26] and could protect the nematodes from freezing at high sub-zero temperature (−1°C), since they raise the osmolality and thus lower the melting point of the body fluids. *S. feltiae* may have the ability to supercool its body fluids and be freeze avoiding at high sub-zero temperature (−1°C), similar to the cold tolerance strategy of *Heterorhabditis zealandica* [27].

Table 1. Exotherm durations and survival of *S. feltiae* samples ($N = 4$).

Treatment	Exotherm duration (min)	Survival (%)
−1°C overnight	65.1±1.2	100
−1°C for 75 min	64.6±0.1	100
−3°C for 75 min	24.6±0.4	97.5±0.9
−10°C for 75 min	4.5±0.8	0
Liquid nitrogen	—	0
Non-frozen control	—	100

means ± SE.

At −2°C, and lower temperatures, all the nematodes turned dark and stopped moving, suggesting that their survival in the cooling block experiments at these temperatures [6] was due to freezing tolerance. Cold stage observations on *S. feltiae* have been reported by Brown and Gaugler [5]. Our findings are in partial agreement with their results. They have reported that the infective juveniles of entomopathogenic nematodes, including *S. feltiae*, were freeze tolerant and froze as the temperature decreased after initiating freezing at −0.5°C. However, it is not clear from their study at which temperature the nematode froze, and they did not provide any pictures of the freezing event, so their nematodes may not have frozen at −1°C. The results of the present study are similar to those reported by Wharton et al. [9] from the Antarctic nematode *Panagrolaimus davidi*, which does not freeze if the medium is seeded at −1°C and undergoes cryoprotective dehydration, but freezes below −2°C. A small amount of shrinkage, that may indicate cryoprotective dehydration, was observed in *S. feltiae* frozen at −1°C. Some nematode species freeze even at −1°C and cannot avoid inoculative freezing [28].

Freeze substitution enabled the location of ice crystals to be determined, appearing as white spaces after the frozen water was replaced by substitution fluid and then by resin inside the frozen nematodes. Overall, fast-frozen samples had large and slow-frozen samples had small ice crystals. Ice crystals were observed both in intracellular and extracellular compartments. Nematodes frozen directly in liquid nitrogen had small ice crystals but did not survive (Fig. 4). They had a reticulated appearance indicating the formation of small ice crystals rather than an amorphous (vitrified or glassy) state [11]. None of the freezing regimes used in this study would be fast enough to produce vitrification. Water needs to be cooled at a rate greater than 10,000°C s^{-1} for vitrification to occur [29]. Nematodes frozen at −1°C for either 75 min or overnight had very small ice crystals similar to those frozen in liquid nitrogen. Nematodes do not freeze at −1°C (as observed earlier on the microscope cold stage) and their water would freeze rapidly when plunged into liquid nitrogen, producing the observed pattern of small ice crystals.

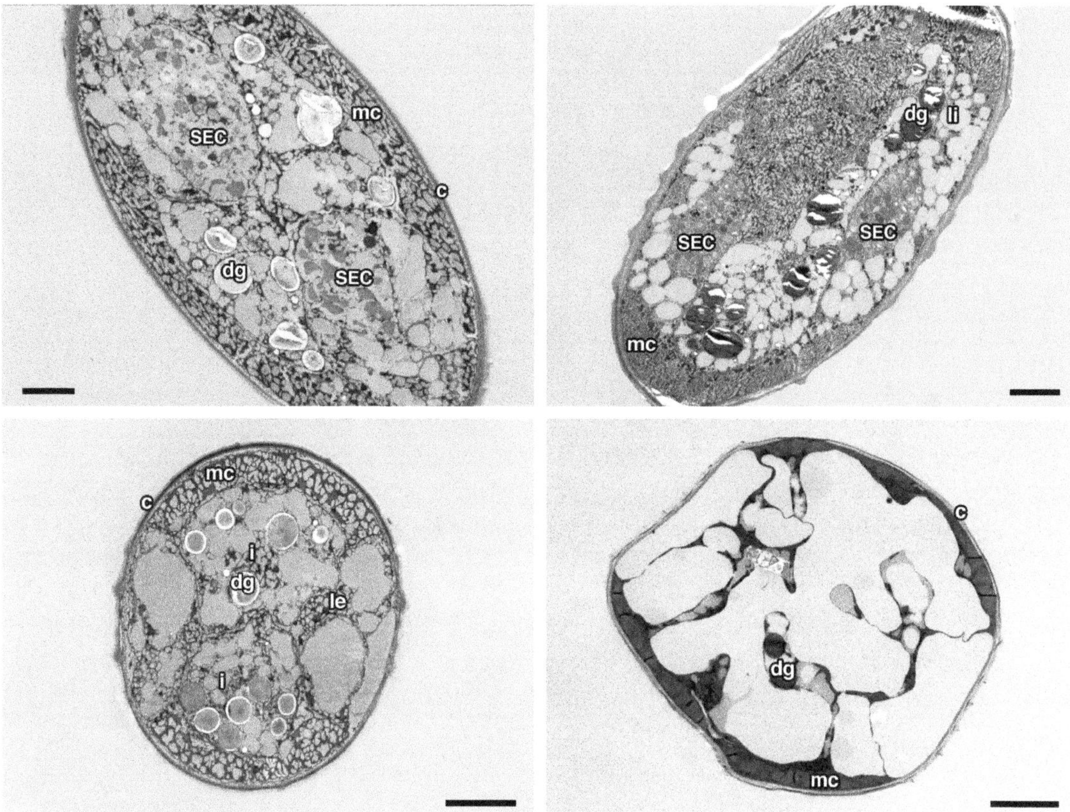

Figure 4. Transmission electron micrograph of a transverse section of the third stage IJ of *S. feltiae* frozen at −1°C for 75 min (top left), −1°C overnight (top right), −3°C for 75 min (bottom left) or −10°C for 75 min (bottom right); and prepared by freeze substitution. c, cuticle; mc, muscle cell; dg, dense granules; i, intestinal cell; le, lateral epidermal cord; li, lipid droplet; SEC, secretary-excretory cell. Scale bars = 5 μm.

Slight shrinkage is evident in some treatments. This indicates partial dehydration and was noticeably greater in the samples frozen overnight at −1°C than the rest of the treatments. This is an indication of cryoprotective dehydration as water is lost from liquid water inside the nematode to the surrounding ice, as a result of the difference in vapour pressure [30]. However, the evidence for cryoprotective dehydration in *S. feltiae* is not as strong as that described in an Antarctic nematode *P. davidi* [14] where no crystals were formed in slow frozen samples and the nematodes appeared severely shrunken. Unfrozen nematodes surrounded by ice also

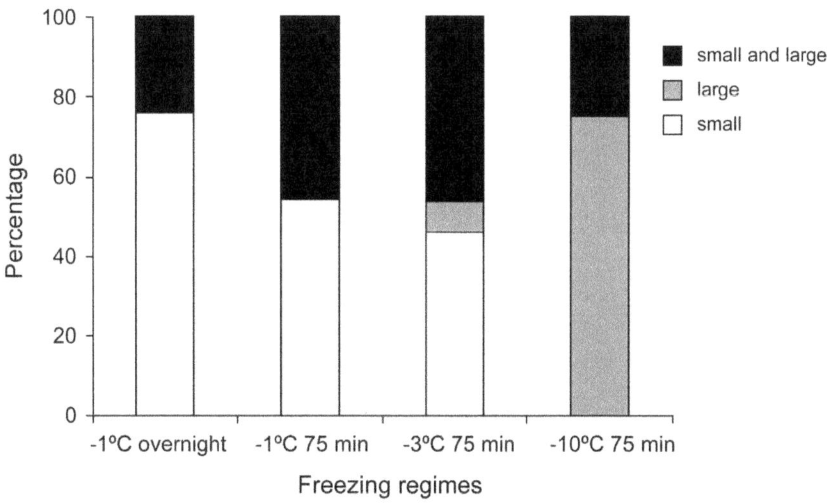

Figure 5. The percentage of IJs of *S. feltiae* having all small, all large and a mixture of small and large ice crystals in transverse sections.

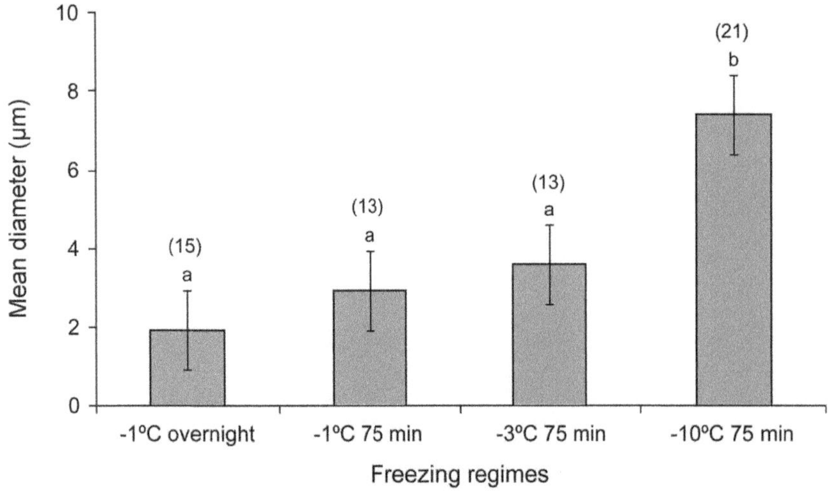

Figure 6. Diameters of the five largest ice crystals of transverse sections frozen overnight at −1°C, or for 75 min frozen at −1°C, −3°C, and −10°C. Vertical bars represent standard errors of means. The number of sections measured in each treatment is shown in parenthesis. Different small letters above each bar indicate that the treatments are significantly different ($P<0.05$).

have a shrunken appearance in a free living nematode, *Panagrellus redivivus* [28].

Some organisms, such as the freeze tolerant wood frog *Rana sylvatica*, produce cryoprotectants during the freezing process itself as soon as ice nucleation is initiated which helps the organism survive sub-zero temperatures [31]. This also could explain why slow freezing of the infective juveniles of *S. feltiae* for an extended time (overnight) at −1°C may improve survival by allowing them to synthesize cryoprotectants.

Nematodes frozen at −3°C and −10°C were observed to have ice crystals in both intracellular and extracellular compartments and a different appearance from samples frozen in liquid nitrogen. Nematodes survived freezing at −3°C but not at −10°C. Ice crystals were observed in most of the compartments of the infective juveniles, including the pharynx, muscle cells, intestinal cells, secretary-excretory cells and epidermis. The pseudocoel was not clearly visible.

After thawing the infective juveniles were motile, pathogenic to their insect host as before and produced the next generation [6]. This suggests that *S. feltiae* is one of the rare organisms that have been shown to survive intracellular freezing. Most freezing tolerant animals can survive only extracellular ice formation and for them intracellular freezing is lethal [32]. The Antarctic nematode, *P. davidi* is the only organism previously reported to survive extensive intracellular freezing [4]. During freezing 82% of its body water is converted into ice, a much higher ice content than can be frozen

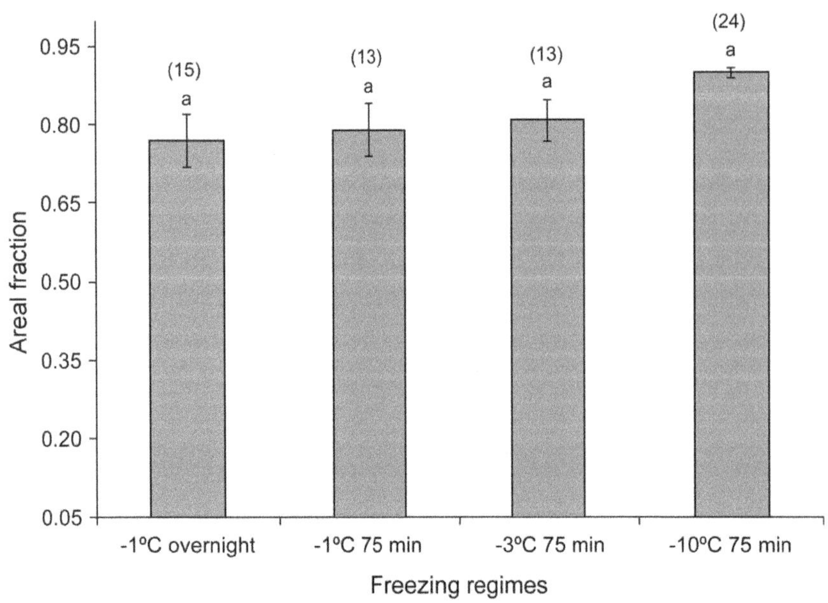

Figure 7. Changes in areal fraction of ice crystals in transverse sections frozen overnight at −1°C, or for 75 min frozen at −1°C, −3°C, and −10°C. Vertical bars represent standard errors of means. The number of sections measured in each treatment is shown in parenthesis. Different small letters above each bar indicate that the treatments are significantly different ($P<0.05$).

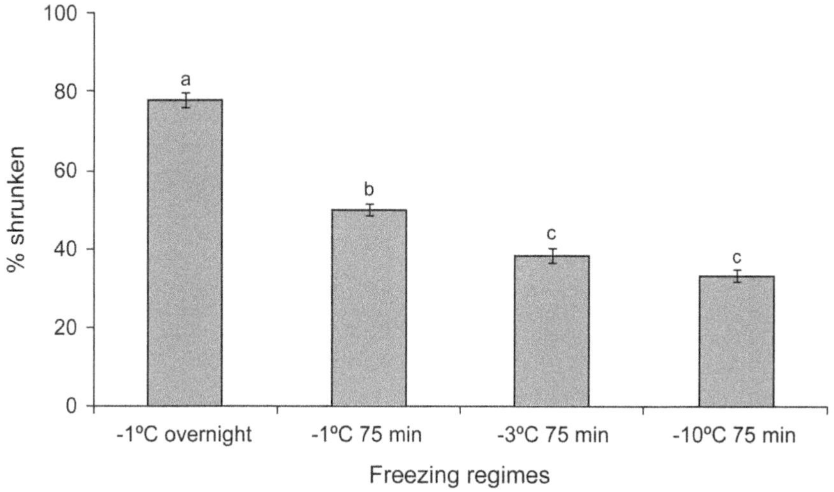

Figure 8. Percentage of the nematodes slightly shrunken after freezing at −1°C overnight or for 75 min frozen at −1°C, −3°C and −10°C. Vertical bars represent standard errors of means (N = 3). Different small letters above each bar indicate that the treatments are significantly different ($P<0.05$).

in most freeze tolerant animals [33]. Some fat body cells in insects have also been observed to survive intracellular freezing but survival does not occur throughout the tissues of the insect [34,35,36]. The cell membrane is thought to be a barrier to ice nucleation in

mammalian cells [37], and intracellular ice formation is considered lethal [38]. However, Toner et al. [39] report that intracellular ice formation is catalysed by the plasma membrane itself by acting as a nucleating agent in the presence of external ice. Nematodes being essentially aquatic [2] are

frozen predominantly by heterogeneous ice nucleation indicating this nematode may have a

specialized cell membrane. Survival of intracellular freezing may depend upon the pattern of ice formation. The size of ice crystals in samples frozen at −3°C was significantly smaller than those frozen at −10°C which may account for their survival. More than 97% of nematodes survived at −3°C, whereas all the nematodes died at −10°C where all or the majority of ice crystals formed were large.

The formation of small ice crystals at −3°C could be due to the presence of an ice active protein, such as a recrystallization-inhibiting protein that inhibits the growth of the crystals [40]. Recrystallization is the growth of larger ice crystals at the expense of smaller ones, which if it occurs in a frozen organism could be lethal. Recrystallization inhibition (RI) may assist the survival of intracellular freezing in *S. feltiae*. Recrystallization inhibition has been reported in another species of entomopathogenic nematode *S. carpocapsae*, which was second in RI activity only to *P. davidi* when compared to four other species [41].

Although ice crystal size varied across the treatments, areal fractions were not significantly different, indicating no effect on the proportion of the body frozen. Fewer but larger ice crystals appear to be lethal to the nematodes irrespective of the proportional area they occupy. The formation of small ice crystals allows *S. feltiae* infective juveniles to survive freezing.

In conclusion, infective juveniles of *S. feltiae* resist inoculative freezing at high sub-zero temperature (−1°C). At lower sub-zero temperature intracellular freezing occurs, which the nematodes can survive, but only down to about − 4°C. The species however, cannot completely control their ice crystal formation, since some large crystals, as well as small crystals, are formed even at −3°C. Survival of intracellular freezing is associated with the smaller size of ice crystals formed during slow freezing, which may be controlled by a substance that inhibits recrystallization.

Acknowledgments

We would like to thank Matthew Downes for technical assistance in electron microscopy and the Otago Centre for Electron Microscopy for access to their facilities.

Author Contributions

Conceived and designed the experiments: FA. Performed the experiments: FA. Analyzed the data: FA DAW. Contributed reagents/materials/analysis tools: FA DAW. Wrote the paper: FA.

References

1. Hominick WM, Reid AP, Bohan DA, Briscoe BR (1996) Entomopathogenic nematodes: Biodiversity, geographical distribution and the convention on biological diversity. Biocontrol Science and Technology 6: 317–331.
2. Wharton DA (1986) A functional biology of nematodes. London and Sydney: Croom Helm.
3. Wharton DA (1995) Cold tolerance strategies in nematodes. Biological Reviews of the Cambridge Philosophical Society 70: 161–185.
4. Wharton DA, Ferns DJ (1995) Survival of intracellular freezing by the Antarctic nematode *Panagrolaimus davidi*. Journal of Experimental Biology 198: 1381–1387.
5. Brown IM, Gaugler R (1996) Cold tolerance of Steinernematid and Heterorhabditid nematodes. Journal of Thermal Biology 21: 115–121.
6. Ali F, Wharton DA (2013) Cold tolerance abilities of two entomopathogenic nematodes, *Steinernema feltiae* and *Heterorhabditis bacteriophora*. Cryobiology 66: 24–29.
7. Wharton DA, Brown IM (1991) Cold tolerance mechanisms of the Antarctic nematode *Panagrolaimus davidi*. Journal of Experimental Biology 155: 629–641.
8. Wharton DA, Rowland JJ (1984) A thermoelectric microscope stage for the measurement of the supercooling points of microscopic organisms. Journal of Microscopy-Oxford 134: 299–305.
9. Wharton DA, Goodall G, Marshall CJ (2003) Freezing survival and cryoprotective dehydration as cold tolerance mechanisms in the Antarctic nematode *Panagrolaimus davidi*. Journal of Experimental Biology 206: 215–221.

10. Shiurba R (2001) Freeze-substitution: Origins and applications. In: Shiurba R, Kwang WJ, editors. International review of cytology - A survey of cell biology. San Diego: Academic Press Inc. pp. 45–96.
11. Robards AW, Sleytr UB (1985) Low temperature methods in biological electron microscopy. Amsterdam: Elsevier Science. 572 p.
12. Hunt CJ (1984) Studies on cellular structure and ice location in frozen organs and tissues: the use of freeze-substitution and related techniques. Cryobiology 21: 385–402.
13. Raymond Ml, Wharton D (2013) The ability of the Antarctic nematode *Panagrolaimus davidi* to survive intracellular freezing is dependent upon nutritional status. Journal of Comparative Physiology B 183: 181–188.
14. Wharton DA, Downes MF, Goodall G, Marshall C (2005) Freezing and cryoprotective dehydration in an Antarctic nematode (*Panagrolaimus davidi*) visualised using a freeze substitution technique. Cryobiology 50: 21–28.
15. Muhlfeld C (2009) High-pressure freezing, chemical fixation and freeze-substitution for immuno-electron microscopy. Methods in Molecular Biology 611: 87–101.
16. Kaya HK, Stock SP (1997) Techniques in insect nematology. In: Lacey LA, editor. Manual of techniques in insect pathology. San Diego, CA: Academic Press. pp. 281–324.
17. Greenaway P (1970) Sodium regulation in freshwater mollusc *Limnaea stagnalis* (L) (Gastropoda, Pulmonata). Journal of Experimental Biology 53: 147–163.
18. Gundersen HJG, Bendtsen TF, Korbo L, Marcussen N, Moller A, et al. (1988) Some new, simple and efficient stereological methods and their use in pathological research and diagnosis Acta Pathologica, Microbiologica et Immunologica Scandinavica 96: 379–394.
19. Rasband WS (1997-2009) ImageJ. US national institutes of health, Bethesda, Maryland, USA. Available: httpt://rsb.info.nih.gov/ij.
20. SPSS Inc (2006) SPSS for Windows, Release 15. SPSS Inc, Chicago, IL.
21. Tyrrell C, Wharton DA, Ramløv H, Moller H (1994) Cold tolerance of an endoparasitic nematode within a freezing-tolerant orthopteran host. Parasitology 109: 367–372.
22. Asahina E (1959) Frost-resistance in a nematode, *Aphelenchoides ritzemabosi*. Low Temperature Science, Series B 17: 51–62.
23. Sayre RM (1964) Cold-Hardiness of nematodes. I. Effects of rapid freezing on the eggs and larvae of *Meloidogyne incognita* and *M. hapla*. Nematologica 10: 168–179.
24. Wharton DA, Aalders O (2002) The response of *Anisakis* larvae to freezing. Journal of Helminthology 76: 363–368.
25. Hayashi M, Oguma T, Amino H, Kita K, Murase N (2009) Freezing survival of the nematode *Caenorhabditis elegans* in the presence of a cryoprotectant. Cryobiology and Cryotechnology 15: 115–119.
26. Jagdale GB, Grewal PS (2003) Acclimation of entomopathogenic nematodes to novel temperatures: trehalose accumulation and the acquisition of thermotolerance. International Journal for Parasitology 33: 145–152.
27. Wharton DA, Surrey MR (1994) Cold tolerance mechanisms of the infective larvae of the insect parasitic nematode *Heterorhabditis zealandica* Poinar. CryoLetters 15: 353–360.
28. Hayashi M, Wharton DA (2011) The oatmeal nematode *Panagrellus redivivus* survives moderately low temperatures by freezing tolerance and cryoprotective dehydration. Journal of Comparative Physiology B 181: 335–342.
29. Weimer RM (2006) Preservation of *C. elegans* tissue via high-pressure freezing and freeze-substitution for ultrastructural analysis and immunocytochemistry. In: Strange K, editor. Methods in molecular biology *C elegans*: methods and application. Totowa, New Jersy: Humana Press Inc. pp. 203–221.
30. Wharton DA (2011) Cold tolerance. In: Perry RN, Wharton DA, editors. Molecular and physiological basis of nematode survival. Wallingford: CABI Publishing. pp. 182–204.
31. Storey JM, Storey KB (1985) Triggering of cryoprotectant synthesis by the initiation of ice nucleation in the freeze tolerant frog, *Rana sylvatica*. Journal of Comparative Physiology B 156: 191–195.
32. Lee RE (1991) Principles of insect low temperature tolerance. In: Lee RE, Denlinger DL, editors. Insects at low temperature. New York/London: Chapman and Hall. pp. 17–46.
33. Storey KB, Storey JM (1988) Freeze tolerance in animals. Physiological Reviews 68: 27–84.
34. Lee RE, McGrath JJ, Todd Morason R, Taddeo RM (1993) Survival of intracellular freezing, lipid coalescence and osmotic fragility in fat body cells of the freeze-tolerant gall fly *Eurosta solidaginis*. Journal of Insect Physiology 39: 445–450.
35. Salt RW (1962) Intracellular freezing in insects. Nature 193: 1207–1208.
36. Salt RW (1959) Survival of frozen fat body cells in an insect. Nature 184: 1426–1426.
37. Grout BWW, Morris GJ (1987) Freezing and cellular organisation. In: Grout BWW, Morris GJ, editors. The Effects of Low Temperatures on Biological Systems. London: Edward Arnold. pp. 147–173.
38. Muldrew K, Acker JP, Elliott JAW, McGann LE (2004) The water to ice transition: implications for living cells. Life in the Frozen State: 67–108.
39. Toner M, Cravalho EG, Karel M (1990) Thermodynamics and kinetics of intracellular ice formation during freezing of biological cells. Journal of Applied Physics 67: 1582–1593.
40. Wharton DA, Barrett J, Goodall G, Marshall CJ, Ramløv H (2005) Ice-active proteins from the Antarctic nematode *Panagrolaimus davidi*. Cryobiology 51: 198–207.
41. Smith T, Wharton DA, Marshall CJ (2008) Cold tolerance of an Antarctic nematode that survives intracellular freezing: comparisons with other nematode species. Journal of Comparative Physiology B 178: 93–100.

18

Biophysical Characteristics of Successful Oilseed Embryo Cryoprotection and Cryopreservation Using Vacuum Infiltration Vitrification: An Innovation in Plant Cell Preservation

Jayanthi Nadarajan*, Hugh W. Pritchard

Seed Conservation Department, Royal Botanic Gardens Kew, Wellcome Trust Millennium Building, Ardingly, West Sussex, United Kingdom

Abstract

Heterogeneity in morphology, physiology and cellular chemistry of plant tissues can compromise successful cryoprotection and cryopreservation. Cryoprotection is a function of exposure time × temperature × permeability for the chosen protectant and diffusion pathway length, as determined by specimen geometry, to provide sufficient dehydration whilst avoiding excessive chemical toxicity. We have developed an innovative method of vacuum infiltration vitrification (VIV) at 381 mm (15 in) Hg (50 kPa) that ensures the rapid (5 min), uniform permeation of Plant Vitrification Solution 2 (PVS2) cryoprotectant into plant embryos and their successful cryopreservation, as judged by regrowth *in vitro*. This method was validated on zygotic embryos/embryonic axes of three species (*Carica papaya, Passiflora edulis* and *Laurus nobilis*) up to 1.6 mg dry mass and 5.6 mm in length, with varying physiology (desiccation tolerances) and 80°C variation in lipid thermal profiles, i.e., visco-elasticity properties, as determined by differential scanning calorimetry. Comparisons between the melting features of cryoprotected embryos and embryo regrowth indicated an optimal internal PVS2 concentration of about 60% of full strength. The physiological vigour of surviving embryos was directly related to the proportion of survivors. Compared with conventional vitrification, VIV-cryopreservation offered a ~ 10-fold reduction in PVS2 exposure times, higher embryo viability and regrowth and greater effectiveness at two pre-treatment temperatures (0°C and 25°C). VIV-cryopreservation may form the basis of a generic, high throughput technology for the *ex situ* conservation of plant genetic resources, aiding food security and protection of species from diverse habitats and at risk of extinction.

Editor: Joshua L. Heazlewood, Lawrence Berkeley National Laboratory, United States of America

Funding: This work was supported by the Millennium Commission, The Wellcome Trust and Orange plc. The Royal Botanic Gardens Kew receives grant-in-aid from DEFRA, UK. All funders acknowledged in our manuscript had no role in study design, data collection and analysis, decision to publish, or preparation of the manuscript.

* E-mail: j.nadarajan@kew.org

Introduction

The vitrification process plays a key role in cryopreservation for the long term conservation of plant genetic resources [1]. Vitrification is defined as a physical process by which a concentrated aqueous solution solidifies into a stable amorphous glass without the formation of ice crystals when the temperature is decreased [2]. Vitrification of plant specimens can be achieved in many ways, including air drying of embryos [3], and more recently through the use of highly concentrated plant vitrification solutions (PVS) that readily form glasses on cooling and inhibit crystallization [2,4]. However, exposure to PVS must be controlled to enable sufficient cellular dehydration whilst limiting injury from chemical toxicity and establishment of a simple and high-throughput cryopreservation method using cryoprotectant is highly desirable [5]. Plant vitrification solutions combine cryoprotectants that vary in permeability [e.g. dimethyl sulphoxide (DMSO) and glycerol], such that cellular water is replaced, cell viscosity is increased and the freezing behaviour of the remaining water is altered [6,7]. PVS2 [2] is probably the most commonly used cryoprotectant (CP) for plant cells, tissues and embryos; for example, the cryopreservation of embryonic axes of citrus [8] and *in vitro* shoot-tips of *Parkia speciosa*, a tropical species with recalcitrant (desiccation sensitive) seeds [9].

The conventional approach to vitrification generally involves a tissue pre-culture step on sucrose-enriched medium, followed by treatment with a loading solution and dehydration, before cooling, with highly concentrated vitrification solution for a period that varies with species, tissue and temperature [10]. For example, 60 min PVS2 treatment was optimal for cryopreserving embryonic axes of *Parkia speciosa* [9]. Moreover, protocorm-like bodies of *Dendrobium* orchid [11] and *Citrus madurensis* embryonic axes [8] required shorter PVS exposure times at 25°C than 0°C, i.e. 20 versus 60 min respectively. Whilst the time window for optimum PVS treatment is wider at lower temperatures, tropical species tend to respond better with warmer temperature treatment, e.g. *Colocasia esculenta* [12].

The physical dimensions (geometry) and permeability characteristics of the tissue under investigation profoundly affect the outcome of the cryoprotection and cryopreservation procedures.

Smaller apices (1.5 or 3 mm in diameter) of garlic displayed higher regeneration after cryopreservation than large ones (4.5 mm in diameter) [13]. Similarly, *Nephelium ramboutan-ake* shoot-tips of c. 2 mm tolerated cryopreservation well [14], as did 0.8 mm diameter axillary buds of *Colocasia esculenta* [12]. The cryopreservation of mature zygotic embryos of recalcitrant seeds generally requires a reduction in tissue mass to facilitate cryoprotectant uptake. Usually, this involves the excision of the embryonic axis. In axes of recalcitrant seeds of sweet chestnut (*Castanea sativa*) such surgical intervention results in a burst of superoxide ($O_2^{\cdot -}$), with further oxidative stress during subsequent desiccation [15]. In this context, the free radical scavenging capability of CPs is additionally crucial for survival of cryopreservation procedures, assuming that the protectants permeate sufficiently. Permeation of chemicals into the intercellular spaces and cells of plant tissues is compounded by many features (e.g. mass, morphology, cellular anatomy and chemical composition). To enable the rapid permeation of the viability stain, triphenyl tetrazolium chloride (TTC), into oily tissues of pine seed, we previously used vacuum infiltration [16]. Similarly, this system has been used to improve efficient gene transformation [17], the delivery of pathogenic bacteria into the intercellular spaces of plants to study pathogen-plant cell interactions [18,19] and the diffusion of an inhibitor of ethylene action so that pear fruits have prolonged storage [20]. In addition, preliminary studies have shown that vacuum-assisted glycerol cryoprotectant infiltration can preserve the normal histology of rat leg muscle with no ice crystal formation after 3 weeks storage at −80°C [21]. In this study, we developed and compared the efficacy of vacuum infiltration vitrification (VIV) using PVS2 for the cryopreservation of seed embryos of three species with varying morphology, stress physiology and chemistry: *Carica papaya* (Caricaceae); *Passiflora edulis* (Passifloraceae); and *Laurus nobilis* (Lauraceae).

These species have purported differences in seed storage characteristics. *C. papaya* has a high level of desiccation tolerance to about 5% moisture content [22], limited (months) storability at −20°C [23], but tolerance of cryopreservation [24]. *P. edulis* seeds may show reduced viability after drying to 5–6% moisture content, but the majority of dry seeds tolerate cryopreservation [24]. Both species have spatulate embryos in copious endosperm [25]. Finally, *L. nobilis* has seeds with a lowest safe moisture content of c. 24%, below which they are desiccation sensitive and successful moist storage at 0°C is limited to about 4 months [26]. *L. nobilis* embryos are linear and bigger (c. 5 mg dry mass) than both *C. papaya* and *P. edulis*. An unifying feature of these three species is that their seeds contain c. 25% oil. However, fatty acid composition varies greatly. Whilst *L. nobilis* seed has c. 50% saturated fat, mainly laurate (12:0), *C. papaya* and *P. edulis* seeds are about 80% unsaturated fat with either oleic (18:1) or linoleic (18:2) as the main (c. 80%) fatty acid, respectively [27,28,29,30,31,32]. Consequently, the thermal behavior of the lipids, and visco-elastic properties of the specimens, should vary between species and this has been correlated with poor dry seed storage performance at c. −20°C in some species, e.g. in an orchid [33] and a few *Cuphea* sp. [34]. Our primary objective in this study is to investigate whether vacuum infiltration vitrification (VIV)-cryopreservation is effective at preserving embryos of species from different provenances (i.e. Mediterranean to tropical), and with disparate morphology, chemistry and physiology. Given that seed desiccation sensitivity in plants correlates strongly with tropical moist forest habitats [35], in which the majority of the world's plant species grow and which are under continuing threat of deforestation, resolving the *ex situ* conservation options for such species is an urgent imperative.

Materials and Methods

Seed Material

Both *C. papaya* and *P. edulis* fruits were purchased from a local supermarket and *L. nobilis* seeds were acquired from a commercial seed supplier. Seeds were extracted from the fruits and surface sterilised with 20% v/v commercial bleach (Domestos) for 10 min before rinsing three times with sterile water.

Seed and Embryo Desiccation

Seeds of all three species were desiccated to and equilibrated at various moisture contents ranging from c. 40 to 5% (fresh weigh basis) above lithium chloride salt solutions, at various levels of saturation, for up to 2 weeks. Germination was subsequently assessed as radicle emergence (≥2 mm) on 1% (w/v) agar-water. For *L. nobilis*, the embryos (excised and sterilized using the method described below) were also subjected to desiccation to various moisture contents by air drying in a laminar air flow cabinet and germinated *in vitro* in Petri dishes containing MS medium [36]. Seed germination and embryo growth *in vitro* was carried out at 25±2 °C in a temperature-controlled growth chamber with a 16 h light/8 h dark photoperiod (50 μmol m^{-2} s^{-1}).

Embryo Excision and Sterilization

The embryos were excised aseptically and rinsed with 3% (w/v) boric acid for 10 min followed by 1 min in 50% (v/v) ethanol and rinsed again with sterile water. For *C. papaya* the whole embryo was used for cryopreservation, whereas embryonic axes were used for *P. edulis* and *L. nobilis* following removal of the cotyledons by cutting the cotyledonary petiole above the shoot tip. Details of explant size/dimension used are summarised in Table 1.

Preculture Treatment

Prior to vitrification treatment, surface disinfected explants were cultured on MS medium supplemented with 0.75 M of sucrose for 1 day at 25±2 °C in a temperature-controlled growth chamber with a 16 h light/8 h dark photoperiod (50 μmol m^{-2} s^{-1}). Embryos at this stage served as controls for the vitrification and cryopreservation treatments.

Vitrification Treatment

The vitrification procedure consisted of dehydrating explants in a loading solution containing 2 M glycerol and 0.4 M sucrose for 20 min at 25 °C, followed by exposure to PVS2 [30% (w/v) glycerol, 15% (w/v) ethylene glycol, 15% (w/v) dimethyl sulfoxide (DMSO) and 0.4 M sucrose] for different durations. For conventional vitrification (CV), explants were treated with PVS2 at 0 and 25°C for 0, 30, 60 and 90 min. For vacuum infiltration vitrification (VIV), the explants were incubated in PVS2 at 0 and 25°C for 0, 1, 2.5 and 5 min. The explants were immersed in 2 mL cryovials containing 1 mL of PVS2 solution in a vacuum chamber (10 L volume) and treated under vacuum (381 mm [15 in] of Hg) for the required period and the vacuum was then released.

Cryopreservation

Following vitrification treatment, the PVS2 solution was replaced with fresh PVS2 solution at the same temperature, the lid put on to the cryovial and the vial and explants plunged into liquid nitrogen. The explants were stored in liquid nitrogen for at least 24 h. Thereafter, the explants were rapidly warmed by placing the cryovials in a water bath at 38±2 °C for 2 min and

Table 1. Tissue physical and biophysical heterogeneity: dry mass of whole embryo, cotyledon and axis, explants dimension, oil content and tissue lipid melt mid-points for the species studied.

Species	*C. papaya*	*P. edulis*	*L. nobilis*
Whole embryo dry mass (mg)	1.4±0.1*	0.9±0.0	4.9±0.5
Cotyledon dry mass (mg)	1.2±0.1	0.7±0.0	3.3±0.5
Axis dry mass (mg)	0.2±0.0	0.1±0.0*	1.6±0.3*
Seed oil content (%)§	28Φ	23ϴ	28♀
Explants for cryopreservation	Embryo	Embryonic axis	Embryonic axis
Dimension (mm)	3.5×1.0	1.0×0.3	5.6×2.5
Total lipid melt enthalpy (J g^{-1})#	26.4±0.6	21.2±1.1	28.1±0.4
Lipid melt 1 mid-point (°C)	−45.2±1.6	−46.2±0.6	−32.4±1.4
Lipid melt 2 mid-point (°C)	−5.4±0.7	−23.8±0.2	−14.2±0.5
Lipid melt 3 mid-point (°C)	10.1±0.5	-	3.1±0.2
Lipid melt 4 mid-point (°C)	-	-	19.0±0.1
Lipid melt 5 mid-point (°C)	-	-	35.1±0.5

For explant dimension measurement, data represent mean ± SE for 4–10 individual embryos/axes; lipid melt peak calculation was based on DSC thermal analysis of embryos/embryonic axes with data shown as mean ± SE for 3 replicates. *Explants used for cryopreservation; Φ[50]; ϴ[31]; ♀[32]; §on fresh weight basis; #on dry weight basis.

then rehydrated for 20 min in liquid MS medium containing 1.2 M sucrose at room temperature.

Viability and Regrowth Assessments

The control, cryoprotected and cryopreserved explants were cultured on solid MS medium in a temperature-controlled growth chamber at 25±2 °C with a 16 h light/8 h dark photoperiod with light intensity of 50 μmol m^{-2} s^{-1}. Explants were transferred to fresh culture medium every day for the first 3 days to reduce any residual toxicity of PVS2. The embryos were considered viable when they turned green, became swollen and started to elongate. The embryos were also assessed for regrowth into seedlings with the first pair of leaves and an extending radicle. Embryo vigour *in vitro* was assessed as the mean time to germinate (MTG) as evidenced by elongation (i.e. viability) in days; $MTG = \frac{\Sigma T_i N_i}{\Sigma N_i}$, where N_i = number of embryos that have elongated at time T_i [37].

Thermal Analysis

Differential scanning calorimetry (DSC) was used to characterise the thermal signature of the seed and embryo lipid and to investigate the presence of free water in the cryoprotected explants in relation to the vitrification potential of the embryos. A DSC 7 (Perkin-Elmer, UK), controlled by a TAC-7 and Pyris 7 software, was calibrated using indium, zinc and pure water as standards for peak temperature and enthalpy. Between 5 to 10 embryos, weighing between 25–35 mg, were placed in pre-weighed aluminium pans, non-hermetically sealed (as recommended for water and lipid phase change studies) with the aid of a Perkin Elmer crimper and weighed to record fresh weight values. Samples were cooled from 25 to −100°C and then rewarmed to 25°C at a cooling/warming rate of ±10°C min^{-1}. Dry seeds and embryo/axes (as used in the cryopreservation experiments) with moisture contents ~4% and without any CP treatment were also subjected to DSC analysis to compare seed and embryo lipid thermal behaviours using the same temperature programme, except for *L. nobilis* which was re-warmed to 50°C. The enthalpies for lipid melting endotherms, and where relevant the enthalpies for ice and PVS2 melting endotherms, were calculated using the Pyris 7 software. When lipid melting events overlapped with the ice or PVS2 melting, the enthalpy for lipid melt (calculated from dry embryos/axes) was subtracted from the total enthalpy to calculate residual melting events. PVS2 solution was also analysed in the DSC during cooling from 25 to −150°C and rewarming to 0°C at a rate of ±10°C min^{-1}. All the thermal analyses were replicated three times.

Statistical Analysis

For each vitrification method, 120 embryos were used in three replicates of 40. The primary response was the percentage of viable (elongating) embryos/axes and the percentage of regrowing seedlings *in vitro*. These percentage values were transformed to logit values, and two-way ANOVA was used to test the level of significance. Means were compared using the Fisher-Pair Wise comparison test. Linear relations between embryo regrowth (%) and mean time for embryo viability (elongation) *in vitro* was assessed for all treatments in GENSTAT. A polynomial fit was used to define the optimal internal PVS2 concentration resulting in maximum post-cryopreservation regrowth using ORIGIN software. To ovoid skewing of the fit, zero values for survival due to ice crystallization and extreme toxicity were excluded.

Results

Seed and Axis Desiccation Tolerance

Seeds of *C. papaya* and *P. edulis* tolerated desiccation to around 4% moisture content, with germination falling slightly compared to undried material but remaining above 70% (Figure 1). On the contrary, desiccating *L. nobilis* seeds showed a steady decline in germination, with only 40% surviving after drying to 25% moisture content. Seeds lost viability completely by 15% moisture content. Aseptically excised and rapidly desiccated embryos of *L. nobilis* showed a similar trend in decreasing regrowth *in vitro* as the whole seed (Figure 1).

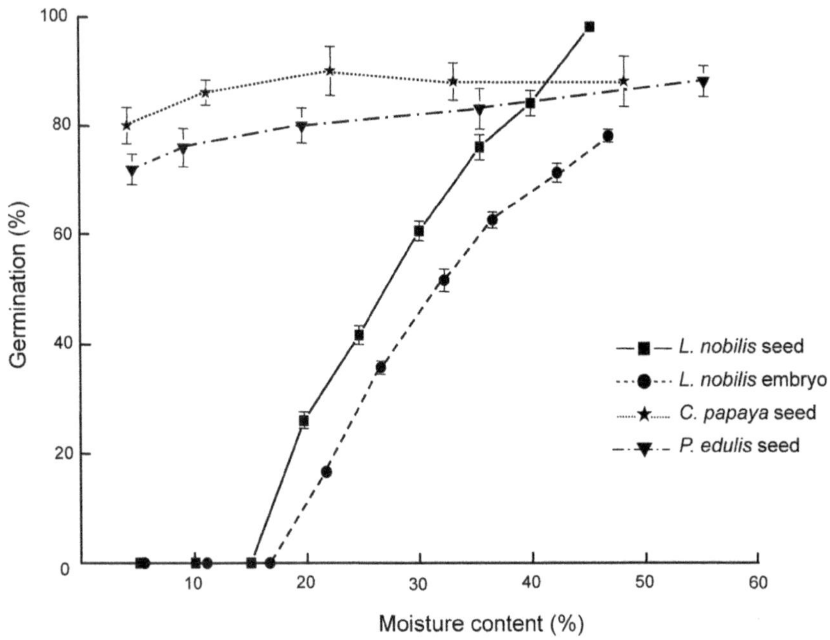

Figure 1. Desiccation sensitivity assessment for seeds and embryos. Legend: Germination percentage of *C. papaya*, *P. edulis* and *L. nobilis* seeds desiccated to various moisture contents and germinated on water agar. Also shown is the *in vitro* germination (regrowth) of *L. nobilis* embryos, following desiccation to various moisture contents, on MS medium at 25±2°C with a 16 h light/dark photoperiod.

Effects of Vitrification Method and Temperature

With the conventional vitrification (CV) method, embryos of *C. papaya* showed highest post-cryopreservation viability and regrowth (~25%) with 30 min PVS2 treatment at 0°C (Figure 2A). Increasing the treatment time to 60 and 90 min resulted in lower regrowth of around 18% and no regrowth, respectively. In *P. edulis* axes, highest post-cryopreservation viability and regrowth ~60% was recorded for 60 min PVS2 treatment (Figure 2B). For *L. nobilis* axes, following cryopreservation, highest viability (60%) and regrowth (52%) was achieved after 60 min PVS2 at 0°C (Figure 2C).

In comparison with CV, the vacuum infiltration vitrification (VIV) method resulted in successful cryopreservation after exposure times to PVS2 about 10-times shorter. Post-cryopreservation viability and regrowth of *C. papaya* embryos increased significantly ($p<0.05$) to about 55% following 5 min VIV treatment at 0°C (Figure 2G). For the smaller embryonic axes of *P. edulis*, 1 min VIV treatment at 0°C was sufficient for good viability and regrowth (above 55%) following cryopreservation (Figure 2H). Increasing the PVS2 exposure period to 2.5 min increased the regrowth significantly to >63% but then decreased to 50% when the treatment time with PVS2 was extended to 5 min (Figure 2H). In *L. nobilis*, 50% regrowth after cryopreservation was recorded with 1 min VIV at 0°C; regrowth increased significantly ($p<0.05$) to 60 and 70% following 2.5 and 5 min treatments respectively (Figure 2I).

Because of the subtropical/tropical origin of two of the species, the effect of PVS2 treatment at room temperature was assessed. For *C. papaya*, CV treatment of embryos at 25°C was optimal after 30 min treatment, similar to the findings at 0°C (Figure 2D). However, highest regrowth was still only 20%. VIV for 5 min at 25°C resulted in the highest viability and regrowth (>40%) but regrowth was significantly lower than when applied at 0°C (55%) (Figure 2J). In *P. edulis* axes, 60 min PVS2 CV treatment at 25°C gave the highest viability and regrowth following cryopreservation (Figure 2E); however only 43% regrowth resulted, which was significantly lower ($p<0.05$) than the 0°C treatment (i.e., 53%). VIV treatment at 25°C for 2.5 min resulted in ~50% viability and regrowth, but this was significantly ($p<0.05$) lower than the 60% achieved after 2.5 min exposure at 0°C and cryopreservation (Figure 2H, K). For *L. nobilis* axes, at 25°C, the highest regrowth following cryopreservation was ~30% after the 60 min CV treatment which was ~30% lower than that achieved at 0°C (Figure 2C, F). Whilst for this species VIV at 0°C and 25°C for 5 min were optimal, regrowth was significantly lower after the warm temperature treatment (Figure 2I, L).

Thermal Analysis of the Explants

Following PVS2 treatments, endotherms during warming were interpreted for different CV and VIV treatments, so that cryopreservation procedures could be optimised in relation to the presence/absence of freezable water and the permeation of PVS2. For all the species, there was a large ice melting peak observed in embryos not treated with PVS2 [e.g. for *C. papaya* (Figure 3)]. This peak was lost for *C. papaya* embryos, after 30, 60 and 90 min CV treatments at both temperatures, indicating freezable water had been removed from the embryos (Figure 3). However, whilst 60 and 90 min CV treatments avoided ice formation, the toxicity effect of the PVS2 was already evident in terms of reduced viability and regrowth (Figure 2A, D). For VIV at both temperatures, a small ice melting peak was observed for the 1 and 2.5 min treatments showing this exposure time was unlikely to achieve a vitrified state without any ice formation; consequently, regrowth was relatively low (Figure 2G, J). Extending the VIV treatment to 5 min resulted in no ice melting peak and the highest viability and regrowth following cryopreservation. For *P. edulis* axes, no ice melting was observed after all three exposure times for both CV and VIV at both temperatures. In the case of *L. nobilis* axes, 30 min CV treatment was insufficient to avoid the presence of an ice melt, but this was lost when 60 and 90 min exposure

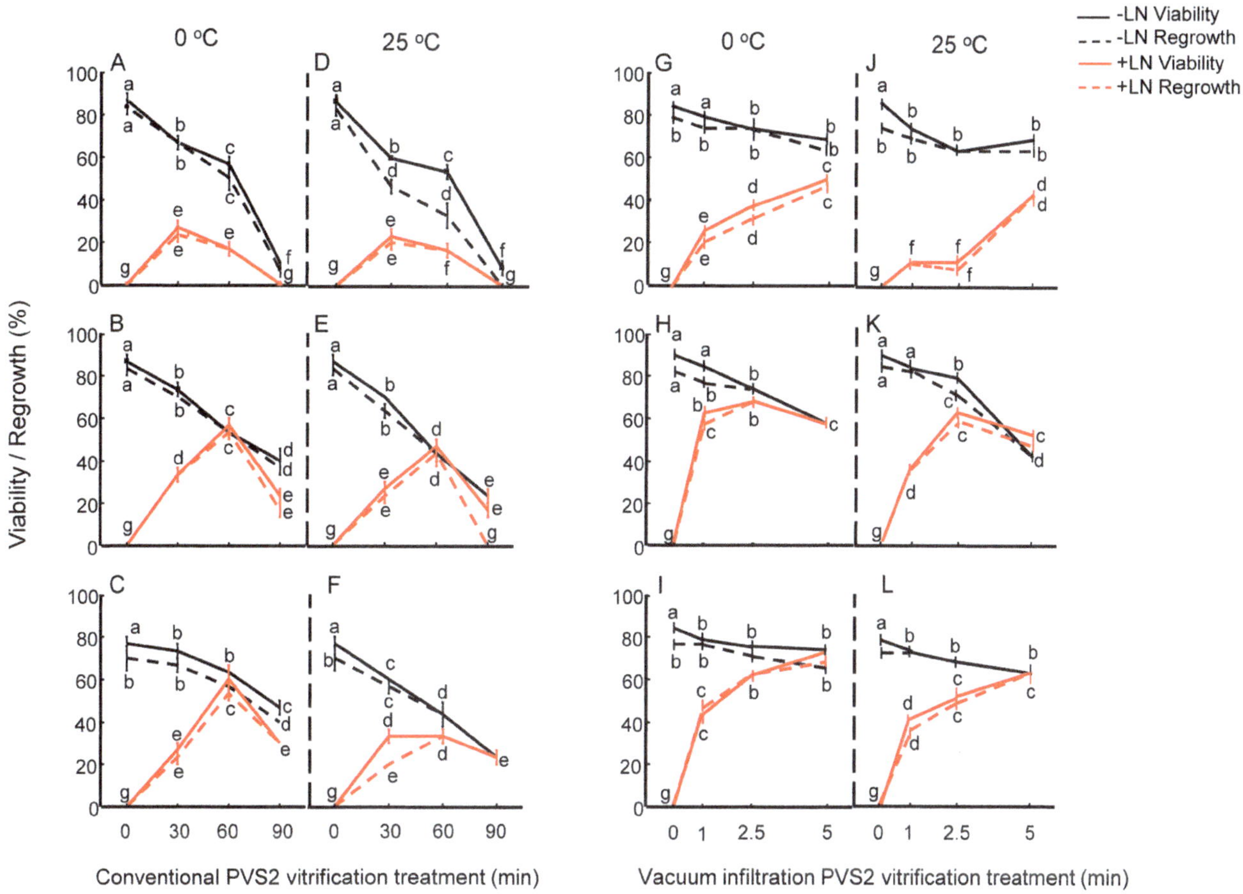

Figure 2. Viability and regrowth of embryos following cryoprotection and cryopreservation. Legend: Viability and regrowth of *C. papaya* (A, D, G, J), *P. edulis* (B, E, H, K) and *L. nobilis* (C, F, I, L) embryos/embryonic axes before (−LN) and after (+LN) cryopreservation using conventional (A–F) and vacuum infiltration (G–L) vitrification at 0°C. Data are means (± SE). Within species, means with the same letters are not significantly different ($p<0.05$) using Fisher's pair wise comparison.

times were used at both temperatures. For VIV treatment, all three exposure times showed thermal profiles with no ice melting peak.

The thermal profile of PVS2 solution during warming is presented as an inset in Figure 3. A glass transition (Tg) was observed at −115°C, followed by a devitrification (recrystallization) at around −75°C. The solution started to melt around −50°C with a midpoint of −37°C. The enthalpy of the PVS2 solution melt was 0.6 J g^{-1}. The enthalpy of PVS2 melt in CV-treated *C. papaya* embryos increased from 0.23 to 0.41 and to 0.49 J g^{-1} as PVS2 treatment time was extended from 30 to 60 and to 90 min (Figure 3 inset). A polynomial fit for PVS2 melt enthalpy in explants and their regrowth revealed that the estimated optimum internal PVS2 was approximately 55 to 67% (0.33 to 0.4 J g^{-1} melting) for all three species (Figure 4A–C).

As the visco-elastic properties of explants may affect cryoprotectant permeation and cryopreservation success, the thermal fingerprints for dry seeds and embryos were also characterised. Three prominent melting peaks were observed for *C. papaya* dry seeds and embryos at −45, −5 and 10°C indicating the embryos have relatively high lipid contents (Table 1; Figure 5A). Based on the published literature, the small melting peak at −45°C is consistent with linoleic acid (c. 3% of lipid in this species), and the combined peaks 2 and 3 at −10 to 10°C likely correspond to the presence of oleic acid (c. 65% of seed lipid). For the dry seed and embryonic axis of *P. edulis*, two main melts were observed at −45°C and −23°C, consistent with the presence of linoleic (74% of lipid) and oleic (17% of lipid) acids (Figure 5B). Several lipid melt peaks were evident for *L. nobilis* dry seed/axis (Figure 5C). Melting around 35°C was assigned to lauric acid (c. 38% of lipid). Other peaks, at −38°C and −15 to 15°C, correspond to the presence of linoleic (c. 40%) and oleic acids respectively. The results indicate potentially large differences in visco-elastic properties of the explants during the cryopreservation procedures at 0°C and 25°C.

Embryo Vigour

In addition to recording embryo viability and regrowth percentages following cryopreservation, embryo vigour was assessed by measuring embryos/axes elongation (viability) *in vitro* over time. Linear relations were found for all three species between the percentage of embryo regrowth and the mean times for elongation (MTG) in days (Figure 6A–C). Accumulation of stress following PVS2 and cooling/warming treatments were very evident for *C. papaya* embryos; for both vitrification treatments, MTG increased from 7 days for the control to 8 days following cryoprotection and 14 days following cryopreservation for both treatment temperatures (Figure 6A). However, for *P. edulis* and *L. nobilis* axes, the MTG increased from c. 6 days (control) to c. 15 days following CV. In contrast, MTG remained at 8 days after VIV cryoprotection and cryopreservation, irrespective of

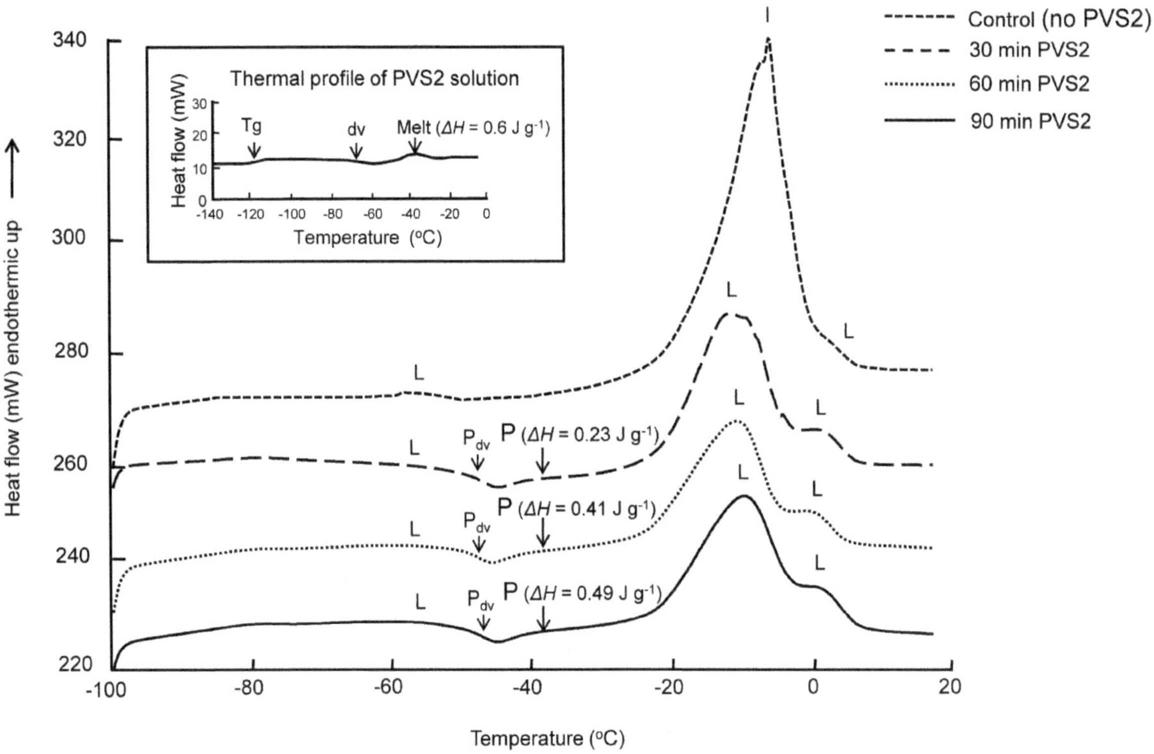

Figure 3. DSC warming thermograms for *C. papaya* embryos following cryoprotection. Legend: Representative DSC warming thermograms for *C. papaya* embryos untreated and treated with PVS2 under CV for 30, 60 and 90 min at 0°C showing ice melt (I), lipid melt (L), PVS2 melt (P) with melt enthalpies in brackets and PVS2 de-vitrification (P_{dv}). Samples were warmed from −100 to 25°C at 10°C min^{-1}. Inset: Warming thermal profile of PVS2 solution with melt enthalpy in bracket (warming program: from −150 to 0°C at 10°C min^{-1}).

treatment temperature, indicating the imposition of minimal stress by VIV (Figure 6B, C).

Discussion

The successful cryopreservation of seed embryos is determined by a complex interaction of factors that contribute to sample heterogeneity. These include structure (morphology and ultrastructure), cell physiology and stress (desiccation, cold) tolerance, and chemical composition and associated thermal properties, all of which impact on the optimisation of cryoprotection (permeation time and toxicity) and cryopreservation procedures (cooling/warming) [2,9,10,38,39].

Desiccation tolerance, as measured by seed germination and embryo recovery *in vitro*, varied for the three species studied. *C. papaya* tended to an orthodox response to drying, with 80% germination after desiccation to 5% moisture content and is in agreement with earlier findings [22,40]. However, the seeds store poorly under conventional seed bank conditions (−20°C) and therefore cryopreservation is recommended for long-term seed/embryo storage [23,24,41,42]. A similar level of drying tolerance was observed in *P. edulis* seeds, related species of which are also sensitive to dry, cold (−20°C) storage [43,44]. In comparison, *L. noblis* seeds are highly sensitive to drying, with reduction of germination to 50% (LD_{50}) at ~30% moisture content (Figure 1). Whilst desiccation sensitivity in *L. nobilis* seeds has been noted previously [45,46,47,48], and loss of viability within one year's moist storage [26], our study is the first detailed characterisation of seed and embryo desiccation sensitivity for this species.

Variation in biophysical properties of the embryos was also evident for the three species, specifically in their lipid thermal fingerprints, as determined by DSC, the peak temperatures for which spanned 83°C [−45°C (onset) to 38°C (endpoint)]. In accordance with Hor *et al.* [39] and Hamilton *et al.* [49] who assigned particular thermal properties in oilseeds (and their embryos) to individual lipid/fatty acids in *Citrus*, melting peaks at −45°C in all three species is assigned here to linoleic acid: 3% of seed oils in *C. papaya*, 74% in *P. edulis* and 40% in *L. nobilis* [29,50,51]. For the three species, melting peaks between about −20 and 15°C occur as single or complex peaks (i.e. with shoulders) and have been assigned previously to oleic acid, e.g. in *C. papaya* [50] and *L. nobilis* [51]. Oleic acid is generally present in plants/seeds as a mono-unsaturated omega-9 fatty acid (18:1 *cis* 9). However, this fatty acid is known to be susceptible to oxidation and to be less thermodynamically stable than the *trans* forms [52], perhaps explaining why we observed more than one peak in *C. papaya* and in *L. nobilis*. Finally, one of the main fatty acids in *L. nobilis* is lauric acid (38% of seed lipid; [53]) which we observed to melt at c. 35°C. This thermal response is similar to that seen in a number of *Cuphea* species that have high laurate and capric acid contents [54]. Consequently, seeds of these three species stored at any temperature from c. 38 to −45°C will be in different physical states, which may impact on their storage (in)stability [33,34,54] and also ensures variation in embryo visco-elastic properties during cryoprotection at the two pre-treatment temperatures used in this study.

The prepared explants (embryos and embryonic axes) of the three species differ in physical dimension, with *L. nobilis* axes being about 10 times heavier and larger than those of *P. edulis*.

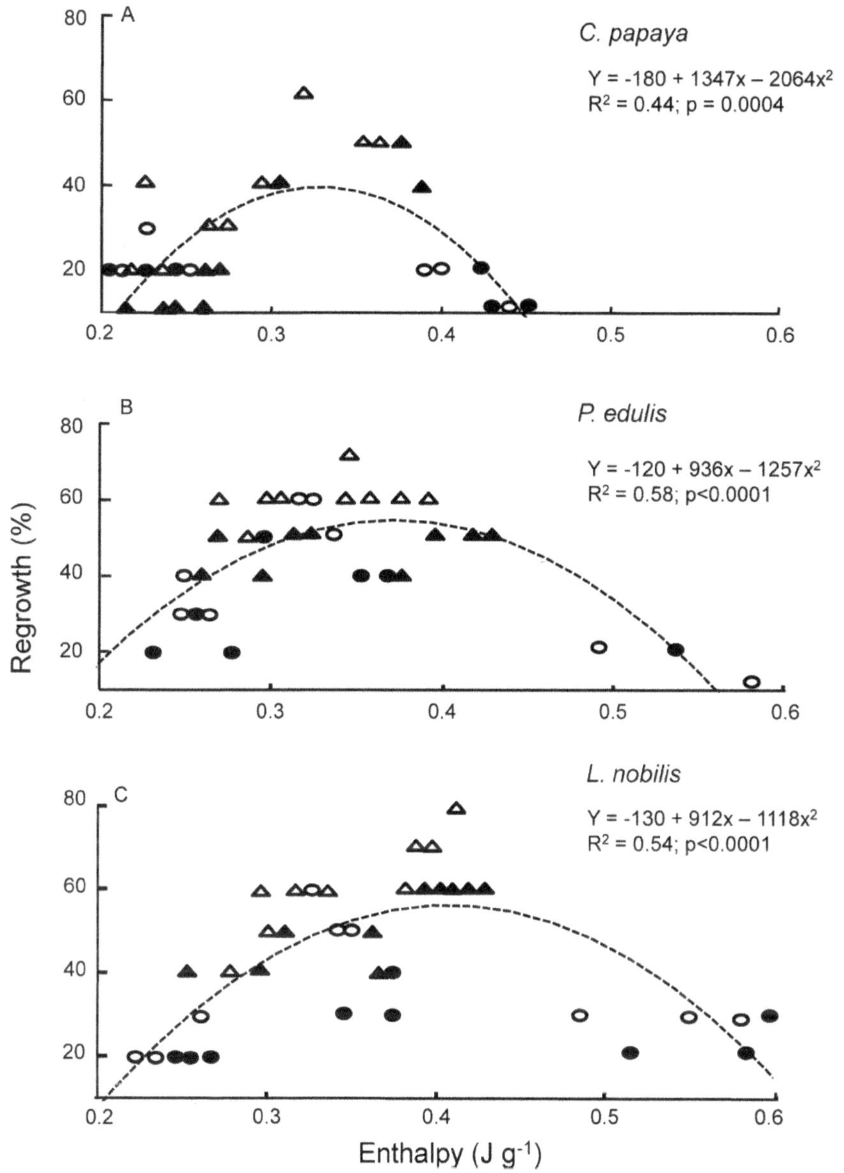

Figure 4. Embryos internal concentration of PVS2 and their corresponding regrowth. Legend: Regrowth following cryopreservation compared to internal concentration of PVS2 (defined by melt enthalpy using a DSC) of *C. papaya* (A), *P. edulis* (B) and *L. nobilis* (C) embryos/embryonic axes with a polynomial fit. (○ = CV at 0°C; ● = CV at 25°C; Δ = VIV at 0°C; ▲ = VIV at 25°C).

Consequently, a single treatment with cryoprotectant (e.g. time × temperature) is unlikely to result in an equal level of permeation and thus protection due to variable surface-area-to-volume ratios and diffusion path lengths. Often, PVS2 treatment of explants ≤ 0.5 mm diameter risks toxicity and explants ≥3 mm require much longer times for effective dehydration [55]. Our observation showed that for *L. nobilis* and *P. edulis* with large and small axes respectively, cryopreservation was maximal after 60 min CV PVS2 treatment at both 0 and 25°C. In comparison, whole *C. papaya* embryos, which are slightly larger than *P. edulis* axes, required a shorter PVS2 exposure time (i.e. 30 min). Overall the results show that tissue mass and volume are not the only determining factors for optimising PVS2 treatment duration. Possibly, not removing the spatulate cotyledons from *C. papaya* embryos may have enabled cryoprotectant permeation into the axis.

Maximal cryo-survival is presumed to coincide with an optimum internal concentration of cryoprotectant agent (CPA), which is generally not known for the sample under investigation, and the avoidance of ice formation in the tissues. According to our observation (Figure 3, inset) and Sakai *et al.* [56], PVS2 solution has a melting peak at ~−36°C. For *C. papaya*, with little linoleic acid (~3%) melting at −45°C, PVS2 melting during rewarming was evident as a small but distinct peak centred at −37°C (Figure 3). This peak was absent in control embryos (i.e. without PVS2 treatment). The enthalpy of the PVS2 melt in the embryos increased with treatment time (Figure 3). At the optimum PVS2 treatment time of 30 min for CV and 5 min for VIV, the enthalpies of the PVS2 melt in *C. papaya* embryos were approximated to 48% and 58% respectively of the maximal concentration, based on the total enthalpy of PVS2 solution (Figure 4A). Further increase in internal concentration of PVS2 to

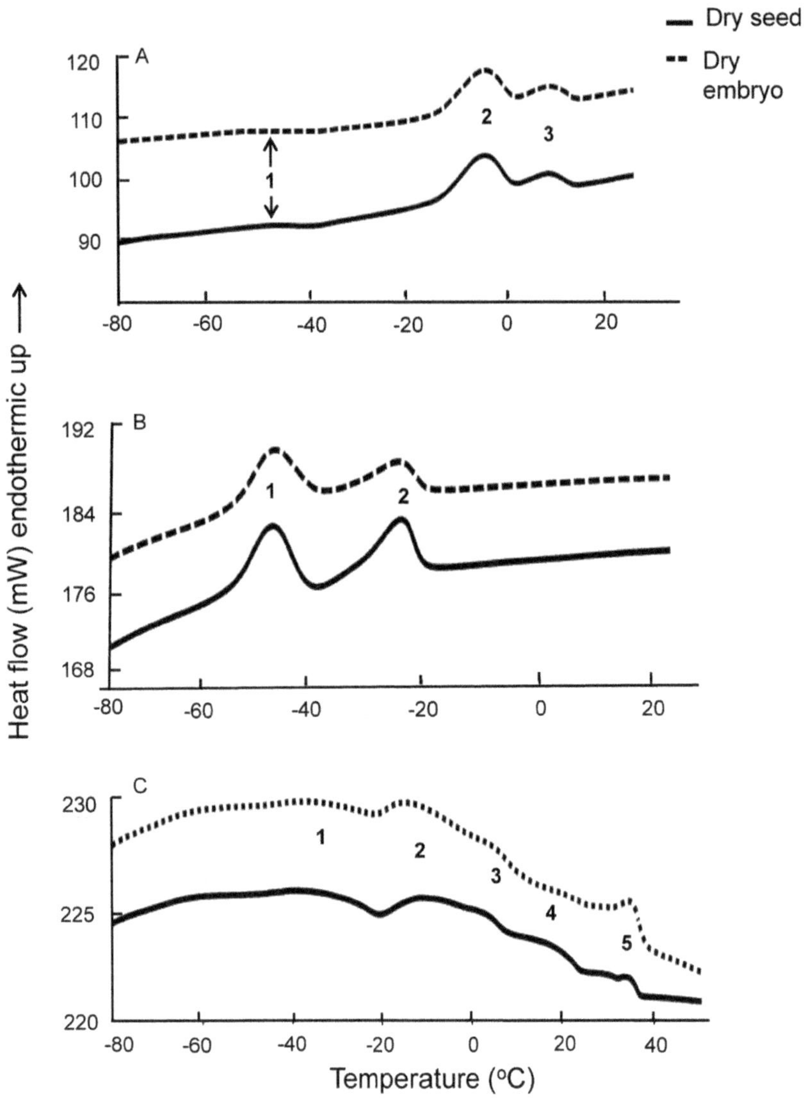

Figure 5. DSC thermograms for dry seeds and embryos. Legend: DSC warming thermograms for *C. papaya* (A), *P. edulis* (B) and *L. nobilis* (C) dry seeds and embryos/embryonic axes showing lipid melt endotherms. Samples were cooled from 25°C to −100°C and rewarmed to 25°C (50°C for *L. nobilis*) at ±10°C min^{-1}. Numbers on each peak indicating different lipid melt peaks corresponding to peaks in Table 1.

75% of the signal intensity of full strength PVS2, as a result of extending the treatment to 90 min, resulted in no post-cryopreservation viability, presumably as a consequence of cytotoxic effects. Interestingly, *C. papaya* embryos showed some post-cryopreservation survival following 1 and 2.5 min VIV treatments even though the DSC thermograms indicated small ice melting peaks for these cryoprotected embryos. This was also the case for *L. nobilis* axes treated with 30 min CV. In these cases, water crystallisation may have occurred in non-critical tissues, e.g. extracellularly, or ice formation limited to those embryos that failed to tolerate cryopreservation, i.e. a population effect. A similar phenomenon of ice formation was observed in surviving shoot-tips of *Parkia speciosa* following encapsulation, 60 min PVS2 exposure and cryopreservation [9] and in *Acer saccharinum* embryonic axes [57].

When the average total lipid melting enthalpies of *P. edulis* and *L. nobilis* axes were accounted for, it was possible to assign the residual signal to PVS2 melting. For the embryos/axes of the three species, the average internal optimum PVS2 concentration for cryopreservation success is estimated to correspond to 55–67% of maximal concentration (i.e. on a J g^{-1} basis c.f. 0.6 J g^{-1} for PVS2 solution). Personalisation of vitrification solution concentrations to balance osmotic and toxicity effects has been recommended by Kim and Lee [58], on the basis of improved cryopreservation of plant cells using diluted PVS. For the three species tested here, 60% concentration of PVS2 is equivalent to a water potential around −31 MPa. This approximates to the water potentials at which recalcitrant seeds have lost all free (and thus freezable) water, i.e. −30 MPa [59], and close to the optimum for the cryopreservation of desiccation sensitive seeds of four citrus species, i.e. −38 MPa [39]. Consequently, it appears that for the successful cryopreservation of diverse embryos of various plant taxa a precise internal concentration of cryoprotectant, of c. 60% PVS2, may be required.

The uptake of CPAs in animal and human tissues is enhanced with increasing temperature; for example in rat kidney cortex and rat liver [60], medaka fish embryos [61] and human oocytes [62]. For DMSO, enhanced uptake at higher temperatures may be due

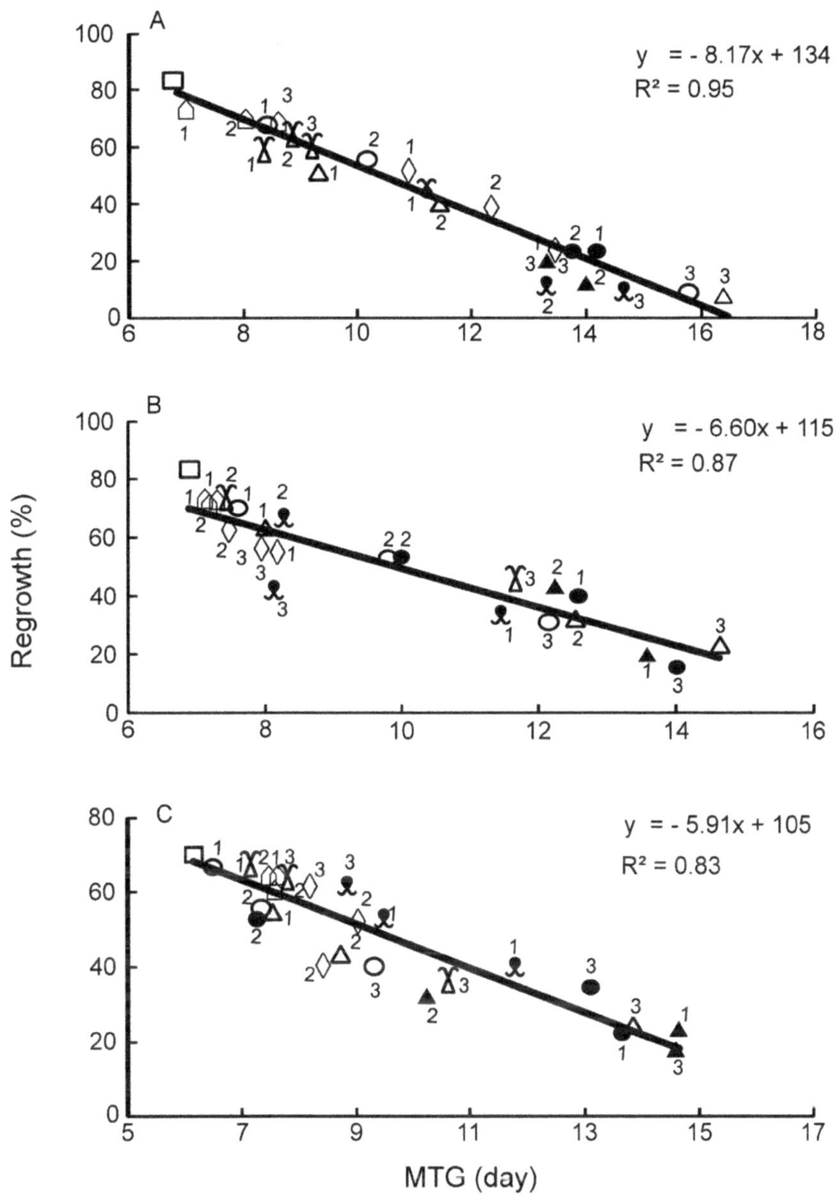

Figure 6. Seedling vigour and stress. Legend: Mean time to germinate (MTG) based on embryo viability (elongation) compared to regrowth for *C. papaya* (A), *P. edulis* (B) and *L. nobilis* (C) showing relationship between seedling vigour and stress. (□ = control; ○ = CV at 0°C -LN; ● = CV at 0°C + LN; Δ = CV at 25°C –LN; ▲ = CV at 25°C +LN; ⌂ = VIV at 0°C -LN; ◇ = VIV at 0°C +LN; ϒ = VIV at 25°C –LN; ϒ = VIV at 25°C +LN). Numbers by each symbol represent: 1 = 30 min; 2 = 60 min; 3 = 90 min for CV; and 1 = 1 min; 2 = 2.5 min; 3 = 5 min for VIV respectively.

to the progressive destabilization of the phospholipid bilayers [63], which increases the risk of toxicity [64], greater oxidation of sulfhydryl groups [65] and, in the presence of glycerol, the formation of formaldehyde by non-enzymatic reactions [65]. Nonetheless, in many tropical and sub-tropical plant species, regrowth is better after cryopreservation when PVS2 is applied at ambient temperature [55], as lower temperatures may cause chilling stress [8,66]. In contrast, most woody, herbaceous temperate and some sub-tropical plant tissues tolerate PVS2 at 0°C [2], but optimum exposure times are longer compared to room temperature application, e.g. 20 min at 25°C and 60 min at 0°C for *Citrus madurensis* embryonic axes [8]. Whilst the optimum PVS2 treatment of 60 min resulted in similar viability after cool or warm application in *C. papaya*, *P. edulis* and *L. nobilis* responded better to cryoprotection under cool conditions, both in terms of viability and vigour. Thus, the embryos of the three species investigated are relatively chilling tolerant over the relatively short, optimum time intervals used in the CV study (≤1 h).

One question we sought to answer was whether a more uniform and faster permeation of PVS2 was possible, such that the obvious inter-species and tissue-specific variables of time × temperature × concentration effects could be reduced and cryo-survival increased. Vacuum degassing at 96 kPa has been used on seeds to enable faster permeation of the viability stain triphenyl tetrazolium chloride into whole pine seeds [16]. A weaker, 'soft' vacuum (= 50 kPa) has also been used in gene transformation studies in one-week old tobacco seedlings [67]. We opted to use 'soft' vacuum for isolated embryos/axes, resulting in ~10-fold reduction in optimum PVS2 exposure times to 2.5 or 5 min for the VIV approach, compared with 30–60 min for the CV method (Figure 2). Moreover, the overall regrowth level was higher and recovery growth was faster (Figure 6) following VIV-cryopreservation. Generally, increasing pressure enhances permeability of cells and tissues, e.g. K+ permeability into human red blood cells [68] and DMSO uptake into 8-cell medaka fish embryos [61]. However, plant tissues contain interconnected networks of intra-cellular (symplast) and inter-cellular (apoplast) spaces, the latter being air-filled at most times in live tissue [69,70]. In detached roots of maize, the threshold pressure for infiltration of these spaces is about 40 kPa [71]. Over short exposure times we contend that the intercellular spaces could generate short-range thermodynamic disequilibration between the inside of the tissues and in the external solution by maintaining a (short-term) pressure difference (solute equilibrium pressure differential), as suggested on theoretical grounds by Elmoazzen *et al.* [72]. A vacuum environment would eliminate any pressure differential in the tissue, by removing air from the inter-cellular spaces, thus enabling tissues to expand rapidly with the inflow of CPA from the surrounding solution. For the embryos of the three species investigated, VIV permits successful cryopreservation after PVS2 treatments of a few minutes, and irrespective of the differing embryo morphologies (embryo vs axis, smaller vs larger). Indeed for the smallest embryo tissue used (axes of *P. edulis*), DSC showed that free water was removed from the tissues by 1 min VIV. Whilst a preliminary report has shown the preservation of normal histology of rat let muscle tissue after vacuum-assisted cryoprotectant infiltration [21], we believe our studies on three species provide the first report of fully functional tissue survival and regrowth after VIV-cryopreservation. There is an additional significance to this finding, as the VIV method may have overcome the further challenge of cryoprotectant permeation into oily, and thus relatively hydrophobic, embryos.

The restoration of tissue function is critical to successful cryo-conservation [73] with each step of the cryopreservation protocol having the potential to introduce stress: oxidative burst at embryo excision [15], shoot-tip damage by plasmolysing cryoprotection [74], cooling/warming and recovery procedures [9]. Ultimately, stress may be so great that only callus growth ensues after cryopreservation, e.g. embryos of some recalcitrant seeds [75]. Embryo regrowth *in vitro* after cryopreservation was observed in embryos/axes of all three species investigated (Figure 2) and speed of initiation of elongation (vigour) shown to be negatively related to regrowth level (Figure 6). The value of such an analysis of vigour is that the impact to viability of any single step in the cryopreservation protocol is observable as a continuum of stress effects across the population of samples.

In conclusion, VIV-cryopreservation has advantage over CV by shortening cryoprotectant treatment times to a few minutes, reducing the temperature dependency for cryoprotectant application and enabling higher post-cryopreservation regrowth (level and rate). Another advantage of this technology is the reduced risk of cross-contamination of specimens during liquid nitrogen storage compared to ultra-rapid cooling methods that bring the tissues into direct contact with the cryogen (e.g. the propelling of axes into cryogen slurry, droplet vitrification) as the specimens are enclosed in cryovials. As VIV-cryopreservation also overcomes the challenges of specimen heterogeneity (mass, volume, oil composition and thermal properties), the method shows considerable promise as the basis of a generic method for the preservation of various tissues of plant species.

Author Contributions

Conceived and designed the experiments: JN HWP. Performed the experiments: JN. Analyzed the data: JN HWP. Contributed reagents/materials/analysis tools: JN HWP. Wrote the paper: JN HWP.

References

1. Reed BM (2008) Plant Cryopreservation: A Practical Guide. New York: Springer.
2. Sakai A (2004) Plant cryopreservation. In: Fuller BJ, Lane N, Benson EE, editors. Life In The Frozen State. Florida: EE CRC Press LLC. 329–346.
3. Grout BWW, Shelton K, Pritchard HW (1983) Orthodox behaviour of oil palm seed and cryopreservation of the excised embryo for genetic conservation. Ann Bot 52: 381–384.
4. Taylor MJ, Song YC, Brockbank GM (2004) Vitrification in tissue preservation: new development. In: Fuller BJ, Lane N, Benson EE, editors. Life In The Frozen State. Florida: EE CRC Press LLC. 603–634.
5. Ogawa Y, Sakurai N, Oikawa A, Kai K, Morishita Y, et al. (2012) High-throughput cryopreservation of plant cell cultures for functional genomics. Plant Cell Physiol 53: 943–952.
6. Muldrew K, Acker JP, Elliott JAW, McGann LE (2004) The water to ice transition: implications for living cells. In: Fuller BJ, Lane N, Benson EE, editors. Life In The Frozen State. Florida: EE CRC Press LLC. 67–108.
7. Volk GM, Walters C (2006) Plant vitrification solution 2 lowers water content and alters freezing behavior in shoot tips during cryoprotection. Cryobiology 52: 48–61.
8. Cho EG, Hor YL, Kim HH, Rao VR, Engelmann F (2002) Cryopreservation of *Citrus madurensis* embryonic axes by vitrification: importance of loading and treatment with vitrification solution. CryoLetters 23: 317–324.
9. Nadarajan J, Mansor M, Krishnapillay B, Staines HJ, Benson EE, et al. (2008) Application of differential scanning calorimetry in developing cryopreservation strategies for *Parkia speciosa*, a tropical tree producing recalcitrant seeds. CryoLetters 29: 95–110.
10. Sakai A, Engelmann F (2007). Vitrification, encapsulation-vitrification and droplet-vitrification: a review. CryoLetters 28: 151–172.
11. Mohanty P, Das MC, Kumaria S, Tardon P (2012) High-efficiency cryopreservation of the medicinal orchid *Denrobium nobile* Lindl. Plant Cell Tiss Org Cult 109: 297–305.
12. Takagi H, Thinh NT, Islam OM, Senboku T, Sakai A (1997) Cryopreservation of *in vitro*-grown shoot tips of taro (*Colocasia esculenta* L. Schott) by vitrification. 1. Investigation of basic conditions of the vitrification procedure. Plant Cell Rep 16: 594–599.
13. Baek HJ, Kim HH, Cho EG, Chae YA, Engelmann F (2003) Importance of explant size and origin and of preconditioning treatments for cryopreservation of garlic shoot apices by vitrification. CryoLetters 24: 381–388.
14. Chua SP, Normah MN (2011) Effect of preculture, PVS2 and vitamin C on survival of recalcitrant *Nephelium ramboutan-ake* shoot tips after cryopreservation by vitrification. CryoLetters 32: 506–515.
15. Roach T, Ivanova M, Beckett R, Minibayeva FV, Green I, et al. (2008) An oxidative burst of superoxide in embryonic axes of recalcitrant sweet chestnut seeds as induced by excision and desiccation. Physiol Plant 133: 131–139.
16. Daws MI, Cousins C, Hall J, Wood CB (2006) Pressure - time dependency of vacuum degassing as a rapid method for viability assessment using tetrazolium chloride: a comparative study of 17 *Pinus* species. Seed Sci Technol 34: 475–489.
17. Ikram UH (2004) *Agrobacterium*-mediated transformation of cotton (*Gossypium hirsutum* L.) via vacuum infiltration. Plant Mol Biol Rep 22: 279–288.
18. Leelavathi S, Sunnichan VG, Mumria R, Vijaykanth GP, Bhatnagar RK, et al. (2004) A simple and rapid *Agrobacterium*-mediated transformation protocol for cotton (*Gossypium hirsutum* L.): embryogenic calli as a source to generate large number of transgenic plants. Plant Cell Rep 22: 465–470.
19. Song G, Yamaguchi K (2003) Efficient agroinfiltration-mediated transient expression system of assaying different parameters in rice. Plant Biotechnol J 20: 235–239.
20. Chen S, Zhang M, Wang S (2012) Physiological and quality responses of Chinese 'Suli' pear (*Pyrus bretschneideri* Rehd) to 1-MCP vacuum infiltration treatment. J Sci Food Agric 90: 1317–1322.
21. Kazutaka S, Katsumi S, Masaki T, Takashi Y, Taro K, et al. (2012) Preliminary report of novel technique for cryopreservation – vacuum-assisted cryoprotectant infiltration. Cryobiology 65: 21–26.
22. Wood CB, Pritchard HW, Amritphale D (2000) Desiccation-induced dormancy in papaya (*Carica papaya* L.) seeds is alleviated by heat shock. Seed Sci Res 10: 135–145.
23. Ellis RH, Hong TD, Roberts EH (1991) Effect of storage temperature and moisture on the germination of papaya seeds. Seed Sci Res 1: 69–72.
24. Becwar MR, Stanwood PC, Leonhardt KW (1983) Dehydration effects on freezing characteristics and survival in liquid nitrogen of desiccation-intolerant and desiccation-sensitive seeds. J Am Soc Hortic Sci 108: 613–618.

25. Mabberley DJ (1997) The Plant-Book: a portable dictionary of the vascular plants. Cambridge: Cambridge University Press. 874 pp.
26. Konstantinidou E, Takos I, Merou T (2008) Desiccation and storage behavior of bay laurel (*Laurus nobilis* L.) seeds. Eur J For Res 127: 125–131.
27. Beis SH, Dunford NT (2006) Supercritical fluid extraction of daphne (*Laurus nobilis* L.) seed oil. J Am Oil Chem Soc 83: 953–957.
28. Litchfield C, Miller E, Harlow RD (1967) The triglyceride composition of 17 seed fats rich in octanoic, decanoic, or lauric acid. Lipids 2: 345–350.
29. Nyanzi SA, Carstensen B, Schwack W (2005) A comparative study of fatty acid profiles of *Passiflora* seed oils from Uganda. J Am Oil Chem Soc 82: 41–44.
30. Nguyen H, Tarandjiiska R (2006) Lipid classes, fatty acids and triglycerides in papaya seed oil. Lipids 97: 20–23.
31. Liu S, Yang F, Li J, Zhang C, Ji H, et al. (2008) Physical and chemical analysis of *Passiflora* seed and seed oil from China. Int J Food Sci Nutr 59: 706–715.
32. Sedat HB, Nurhan TD (2006) Supercritical fluid extraction of daphne (*Laurus nobilis* L.) seed oil. J Am Oil Chem Soc 83: 953–957.
33. Pritchard HW, Seaton PT (1993) Orchid seed storage: historical perspective, current status, and future prospects for long-term conservation. Selbyana 14: 89–104.
34. Crane J, Miller AL, van Roekel JW, Walters C (2003) Triacylglycerols determine the unusual storage physiology of *Cuphea* seed. Planta 217: 699–708.
35. Tweddle JC, Dickie JB, Baskin CC, Baskin JM (2003) Ecological aspects of seed desiccation sensitivity. J Ecol 91: 294–304.
36. Murashige T, Skoog F (1962) A revised medium for rapid bioassay with tobacco tissue cultures. Physiol Plant 15: 473–497.
37. Ranal MA, Santana DG, Ferreira WR, Rodrigues CM (2009) Calculating germination measurements and organising spreadsheets. Brazilian J Bot 32: 849–855.
38. Panis B, Piette B, Swennen R (2005) Droplet vitrification of apical meristems: a cryopreservation protocol applicable to all *Musaceae*. Plant Sci 168: 45–55.
39. Hor YL, Kim YJ, Ugap A, Chabrillange N, Sinniah UR, et al. (2005) Optimal hydration status for cryopreservation of intermediate oily seeds: *Citrus* as a case study. Ann Bot 95: 1153–1161.
40. Azimi M, O'Brien C, Ashmore S, Drew R (2005) Cryopreservation of papaya germplasm. Acta Hort 692: 43–50.
41. Chin HF, Krishnapillay B (1989) Cryogenic storage of some horticultural species. Acta Hort 253: 107–112.
42. Magil W, Deighton N, Pritchard HW, Benson EE, Goodman BA (1994) Physiological and biochemical studies of seed storage parameters in *Carica papaya*. Proceedings of the Royal Society of Edinburgh 102B: 439–442.
43. Ospina JA, Guevara CL, Caicedo LE, Barney V (2000) Current research progress and application. In: Engelmann F, Takagi H, editors. Cryopreservation of Tropical Plant Germplasm. Rome: IPGRI. 378–381.
44. Gonzalez-Benito M, Aguilar N, Avila T (2009) Germination and embryo rescue from *Passiflora* species seeds post-cryopreservation. CryoLetters 30: 142–147.
45. Takos IA (2001) Seed dormancy in bay laurel (*Laurus nobilis* L.). New Forests 21: 105–114.
46. Takos IA, Konstantinidou E, Merou D (2002) The effect of desiccation on the seed germination of *Laurus nobilis* L. In: Thanos CA, Beardmore TL, Connor KF, Tolentino EL Jr., editors. Proceedings of the 2002 annual meeting of IUFRO. Chania, Crete, Greece, 11–15 Sept. 2002. 178–182.
47. Tilki F (2004) Influence of pretreatment and desiccation on the germination of *Laurus nobilis L.* seeds. J Environ Biol 25: 157–161.
48. Sari AO, Oguz B, Bilge A (2006) Breaking seed dormancy of laurel (*Laurus nobilis* L.). New Forests 31: 403–408.
49. Hamilton KN, Ashmore SE, Pritchard HW (2009) Thermal analysis and cryopreservation of seeds of Australian wild *Citrus* species (Rutaceae): *Citrus australasica, C. inodora and C. garrawayi*. CryoLetters 30: 268–279.
50. Puangsri T, Abdulkarim SM, Ghazali HM (2005) Properties of *Carica papaya* L. (papaya) seed oil following extractions using solvent and enzymatic methods. J Food Lipids 12: 62–76.
51. Sari A (2003) Thermal reliability test of some fatty acids as PCMs used for solar thermal latent heat storage applications. Energ Convers Manage 44: 2277–2287.
52. Rustan AC, Drevon CA (2005) Fatty Acids: Structures and Properties. London: Encyclopedia of Life Sciences, Nature Publishing, http://www.els.net.
53. Ozcan B, Esen M, Sangun MK, Coleri A, Caliskan M (2010) Effective antibacterial and antioxidant properties of methanolic extract of *Laurus nobilis* seed oil. J Environ Biol 31: 637–641.
54. Crane J, Kovach D, Gardner C, Walters C (2006) Triacylglycerol phase and 'intermediate' seed storage physiology: a study of *Cuphea carthagenensis*. Planta 223: 1081–1089.
55. Sakai A, Hirai D, Niino T (2008) Development of PVS-Based vitrification protocol. In: Reed BM, editor. Plant Cryopreservation: A Practical Guide. New York: Springer. 33–57.
56. Sakai A, Kobayashi S, Oiyama I (1990) Cryopreservation of nucellar cells of naval orange (*Citrus sinensis* Osb. var. *brasiliensis* Tanaka) by vitrification. Plant Cell Rep 9: 30–33.
57. Wesley-Smith J, Berjak P, Pammenter NW, Walters C (2014) Intracellular ice and cell survival in cryo-exposed embryonic axes of recalcitrant seeds of *Acer saccharinum*: an ultrastructural study of factors affecting cell and ice structures. Ann Bot 113: 695–709.
58. Kim HH, Lee SC (2012) Personalisation' of droplet-vitrification protocols for plant cells: a systematic approach to optimising chemical and osmotic effects. CryoLetters 33: 271–279.
59. Pritchard HW (1991) Water potential and embryonic axis viability in recalcitrant seeds of *Quercus rubra*. Ann Bot 67: 43–49.
60. Isbell SA, Fyfe CA, Ammons RLM, Pearson B (1997) Measurement of cryoprotective solvent penetration into intact organ tissues using high-filed NMR microimaging. Cryobiology 35: 165–172.
61. Routray P, Suzuki TM, Strussmann CA, Takai R (2002) Factors affecting the uptake of DMSO by the eggs and embryos of medaka, *Oryzias latipes*. Theriogenology 58: 1483–1496.
62. Paynter SJ, Cooper A, Gregory L, Fuller BJ, Shaw RW (1999) Permeability characteristics of human oocytes in the presence of the cryoprotectant dimethyl sulphoxide. Hum Reprod 14: 2338–2342.
63. Anchordoguy TJ, Carpenter JF, Crowe JH, Crowe LM (1992) Temperature dependent perturbation of phospholipid bilayers by dimethylsulphoxide. Biochim Biophys Acta Biomembrane 1104: 117–122.
64. Sum AK, de Pablo JJ (2003) Molecular simulation study on the influence of dimethylsulfoxide on the structure of phospholipid bilayers. Biophys J 85: 3636–3645.
65. Karran G, Legge M (1996) Fertilization and early embryology: non-enzymatic formation of formaldehyde in mouse oocyte freezing mixtures. Hum Reprod 11: 2681–2686.
66. Tiau KH, Rathinam X, Chan LK, Subramaniam S (2009) An assessment of early factors influencing the PVS2 vitrification method using protocorm-like bodies of *Dendrobium Sonia* 28. American-Eurasian J Sust Agric 3: 280–289.
67. Escudero J, Dulk-Ras A, Regensburg-Tuink TJG, Hooykaas PJJ (2003) VirD4-independent transformation by CloDF13 evidences an unknown factor required for the genetic colonization of plants via Agrobacterium. Mol Microbiol 47: 891–901.
68. Hall AC, Ellory JC, Klein RA (1982) Pressure and temperature effects on human red cell cation transport. J Membr Biol 68: 47–56.
69. Woolley JT (1983) Maintenance of air in intercellular spaces in plants. J Plant Physiol 72: 989–991.
70. Beck CB (2009) An Introduction to Plant Structure and Development: Plant Anatomy for the Twenty-First Century, Edition 2. Cambridge: Cambridge University Press. 431 pp.
71. Michael W, Cholodova VP, Ehwald R (1999) Gas and liquids in intercellular spaces of maize roots. Ann Bot 84: 665–673.
72. Elmoazzen HY, Elliott JAW, McGann LE (2005) Cryoprotectant equilibration in tissues. Cryobiology 51: 85–91.
73. Fang JY, Wetten A (2011) Importance of structural integrity of somatic embryos for long-term cryopreservation of cocoa (*Theobroma cacao* L.) germplasm. African J Agric Res 6: 3954–3961.
74. Volk GM, Caspersen AM (2007) Plasmolysis and recovery of different cell types in cryoprotected shoot tips of *Mentha x piperita*. Protoplasma 231: 215–226.
75. Pritchard HW, Prendergast FG (1986) Effects of desiccation and cryopreservation on the *in vitro* viability of embryos of the recalcitrant seed species *Araucaria hunsteinii* K. Schum. J Exp Bot 37: 1388–1397.

Cell Surface Profiling Using High-Throughput Flow Cytometry: A Platform for Biomarker Discovery and Analysis of Cellular Heterogeneity

Craig A. Gedye[1], Ali Hussain[2], Joshua Paterson[1], Alannah Smrke[1], Harleen Saini[1], Danylo Sirskyj[1], Keira Pereira[2], Nazleen Lobo[2], Jocelyn Stewart[2], Christopher Go[1], Jenny Ho[1], Mauricio Medrano[2], Elzbieta Hyatt[3], Julie Yuan[3], Stevan Lauriault[3], Maria Kondratyev[1], Twan van den Beucken[1], Michael Jewett[1], Peter Dirks[3], Cynthia J. Guidos[3], Jayne Danska[3], Jean Wang[1], Bradly Wouters[1,2], Benjamin Neel[1,2], Robert Rottapel[1], Laurie E. Ailles[1,2]*

1 Princess Margaret Cancer Centre, University Health Network, Toronto, ON, Canada, **2** Dept. of Medical Biophysics, University of Toronto, Toronto, ON, Canada, **3** Hospital for Sick Children Research Institute, Toronto, ON, Canada

Abstract

Cell surface proteins have a wide range of biological functions, and are often used as lineage-specific markers. Antibodies that recognize cell surface antigens are widely used as research tools, diagnostic markers, and even therapeutic agents. The ability to obtain broad cell surface protein profiles would thus be of great value in a wide range of fields. There are however currently few available methods for high-throughput analysis of large numbers of cell surface proteins. We describe here a high-throughput flow cytometry (HT-FC) platform for rapid analysis of 363 cell surface antigens. Here we demonstrate that HT-FC provides reproducible results, and use the platform to identify cell surface antigens that are influenced by common cell preparation methods. We show that multiple populations within complex samples such as primary tumors can be simultaneously analyzed by co-staining of cells with lineage-specific antibodies, allowing unprecedented depth of analysis of heterogeneous cell populations. Furthermore, standard informatics methods can be used to visualize, cluster and downsample HT-FC data to reveal novel signatures and biomarkers. We show that the cell surface profile provides sufficient molecular information to classify samples from different cancers and tissue types into biologically relevant clusters using unsupervised hierarchical clustering. Finally, we describe the identification of a candidate lineage marker and its subsequent validation. In summary, HT-FC combines the advantages of a high-throughput screen with a detection method that is sensitive, quantitative, highly reproducible, and allows in-depth analysis of heterogeneous samples. The use of commercially available antibodies means that high quality reagents are immediately available for follow-up studies. HT-FC has a wide range of applications, including biomarker discovery, molecular classification of cancers, or identification of novel lineage specific or stem cell markers.

Editor: Jörg D. Hoheisel, Deutsches Krebsforschungszentrum, Germany

Funding: Funding for this work was provided to LEA by the Ontario Institute for Cancer Research (http://oicr.on.ca/). CAG was supported by a Canadian Institute for Health Research/Kidney Cancer Canada Fellowship, and has been supported by a Royal Australasian College of Physicians CSL Fellowship and a National Health and Medical Research Council Overseas Postdoctoral Fellowship. This research was funded in part by the Ontario Ministry of Health and Long Term Care. The views expressed do not necessarily reflect those of the OMOHLTC. The funders had no role in study design, data collection and analysis, decision to publish, or preparation of the manuscript.

Competing Interests: The authors have declared that no competing interests exist.

* Email: lailles@uhnresearch.ca

Introduction

Cell surface proteins are of particular interest as biomarkers because they perform many important biological functions, including mediation of cell-cell communication and responses to external signals such as the presence of pathogens or chemical messengers. The cell "surfaceome" defines phenotypic and functional differences between cell types, and between normal and diseased cells, such as cancer cells. Cell surface proteins are useful as diagnostic markers or therapeutic targets in cancer, as evidenced by the large number of monoclonal antibodies (MAbs) currently approved for both diagnostic and therapeutic applications. Rapid characterization of the cancer cell surfaceome could not only lead to identification and development of new diagnostic markers and therapeutic targets, but also provide insight into the basic biology of disease, including environmental interactions and identification of important cellular subtypes and signaling pathways.

One approach to cell surfaceome characterization is to bioinformatically predict all membrane proteins in the human genome, and then identify subsets expressed in a given cell type using global gene expression data [1]. However, gene expression does not always correlate with protein expression [2,3] and not all expressed membrane proteins are present on the cell surface. Another approach has been to perform mass spectrometry-based proteomics, to sensitively and rapidly identify and quantify large numbers of peptides or proteins in a sample of interest. However, this is technically challenging due to the limited abundance of surface membrane proteins, and difficulty obtaining plasma membrane isolates and resolving and identifying hydrophobic proteins and peptides [4]. Recent technical advances have enabled "cell surface-capturing" for more accurate measurement of cell surface proteins by mass spectrometry [4–6]. Of note, both whole cell lysate and cell surface capture methods provide an average quantity of molecules measured over the entire sample, making analysis of tissue heterogeneity a challenge.

MAbs can provide reliable information about the expression of cell surface proteins, as well as the distribution of proteins within a heterogeneous tissue. Both immunohistochemistry (IHC) and flow cytometry (FC) utilize chemically- or fluorescently-tagged MAbs to detect proteins, including surface proteins. These assays are specific, sensitive and reproducible, and can provide information at the level of individual cells. However, IHC is limited by the small number of MAbs that can be simultaneously analyzed on a limited number of cells. FC is higher-throughput, allowing rapid analysis of proteins on large numbers cells in liquid suspension. FC is typically used to analyze up to 11 markers at a time, with complex analysis being possible through the use of overlapping panels [7] allowing identification and analysis of subpopulations of cells within complex mixtures. Indeed, such flow cytometry assays are now used clinically in several areas such as diagnosis and monitoring of hematological malignancies [8,9], illustrating the power of this approach.

Given the importance of cell surface proteins to a wide range of biological processes, their broad utility as research and clinical tools for the identification of specific cell types, their utility as biomarkers of disease, and their potential as therapeutic targets, we wished to develop a cell surface-targeted "array" as a discovery tool. As flow cytometry represents a robust and rapid platform for cell surface protein expression analysis, we assembled the broadest possible panel of fluorochrome conjugated cell-surface targeted antibodies. These antibodies were arrayed into 96-well plates, allowing high-througphut flow cytometry analysis of their expression using a commercially available high-speed sample loading device for flow cytometers (Becton Dickinson). This high throughput flow cytometry (HT-FC) assay provides highly reproducible results and can be used to answer a wide range of biological questions (see [10] for first published example). We have used the platform to identify cell surface antigens that are influenced by common cell preparation methods such as enzyme digestion, cryopreservation and fixation. By co-staining cells with up to 5 additional MAbs conjugated to complementary fluorochromes, we can segregate populations of cells to allow independent analysis of subpopulations of interest within complex cell mixtures such as primary tumors. Application of standard visualization methods can be used to generate heatmaps and perform supervised or unsupervised hierarchical clustering. We show that the 363-antigen cell surface profile provides sufficient molecular information to classify 119 samples from a variety of cancers and tissue types into biologically relevant clusters using unsupervised hierarchical clustering. Finally, we describe the identification of a candidate lineage-specific marker and its subsequent validation.

Methods

Cells

Commercially available cell lines used included MDA-MB-231, SCC4, SCC9, A549, CACO2, OVCAR3, OV-90, 786-O, 22RV1, A2058, H929, 8226, MCF7, and Jurkat, and were obtained from ATCC. Additional cell lines were established in-house in the Ailles Lab (clear cell renal cell carcinoma (ccRCC) cell lines, cancer-associated fibroblasts (CAFs)) or obtained from the labs of colleagues at the University Health Network (serous ovarian cancer (SOC) cell lines, Dr. Rob Rottapel; head and neck cancer (HNC) cell lines, Dr. Bradly Wouters; TEX acute myeloid leukemia cells [11], Dr. Aaron Schimmer). All samples used in this study were obtained from the University Health Network. Written informed consent was obtained for the collection of all human materials, which were used either to generate cell lines or to analyze primary samples directly, and the studies were approved by the Research Ethics Board of the University Health Network. Peripheral blood mononuclear cells (PBMCs) and normal dermal fibroblasts were purchased from Lonza. Cultured cells were quality controlled for cell line identity by short-tandem repeat analysis (AmpFℓSTR Identifiler, Life Technologies), performed by The Centre for Applied Genomics, The Hospital for Sick Children, Toronto, Canada, and tested negative for *Mycoplasma* infection by MycoAlert (Lonza) according to the manufacturer's instructions. Cell lines were cultured in Iscove's Modified Dulbecco's Medium (IMDM) supplemented with 10% fetal bovine serum (FBS; Sigma-Aldrich) and penicillin and streptomycin (Gibco) at final concentrations of 100 U/mL and 100 μg/mL, respectively. Cells were incubated at 37°C in 5% CO_2, 95% air for commercial cell lines, or 5% CO_2, 2% O_2 for primary ccRCC cell lines and CAFs. Adherent cells were passaged using 0.05% trypsin when confluent. Non-adherent cell lines were collected by aspiration and washing in PBS before re-suspension. Primary B-ALL and AML samples were obtained from consenting patients and processed by the Guidos Lab and the Wang Lab, respectively. HNC and SOC tissue samples were obtained from the University Health Network Tissue Bank and from the Cooperative Health Tissue Network. Tumors were procured within 2–24 hours of excision (Cooperative Health Tissue Network samples were shipped overnight from various sites in the United States). Tissues were minced with sterile scalpels and digested with 1X collagenase/hyaluronidase (Stem Cell Technologies) and DNase I (125 U/mL, Invitrogen) in Media199 at 37°C for a maximum of two hours with frequent gentle trituration. In the case of ascites samples from SOC patients, cells were harvested by centrifugation. Red blood cells were lysed by applying 1 mL of ammonium chloride lysis buffer (Gibco) to the cell pellet for 5 minutes on ice, followed by an immediate wash in IMDM/10% FBS. Following red cell lysis, cells were filtered through a 70-micron sterile nylon mesh and viable cells defined by trypan blue exclusion were enumerated at 10X magnification by hemocytometer. To generate CAFs, HNC and SOC single cell suspensions were cultured in IMDM/10% FBS in which CAFs selectively grew and were passaged by trypsinization when confluent. CAF cultures were validated by immunofluorescent staining for vimentin and pan-cytokeratin (CK), as well as by single nucleotide polymorphism array (for HNC CAFs) or p53 immunofluorescence in comparison to matched primary tumors (for SOC CAFs). Samples were either directly used for experiments or cryopreserved in 90% FBS/10% dimethyl sulfoxide.

The pool of cells for testing of enzymatic digestion, cryopreservation and fixation included the following: A549, CACO2, OVCAR3, A2058, SCC4, SCC9, H929, 8226, MCF7, Jurkat, LCL, RCC22, HNC CAFs, and PBMCs. Adherent cells were collected from culture by gentle cell scraping, followed by trituration and 70 µm mesh filtration to form a single cell suspension. Cells were mixed in equivalent doses (2 million each) to generate a mixed cell lineage. Aliquots of this cell mixture were then stained with no manipulation (control), stained and then fixed, subjected to enzymatic treatment then stained, or cryopreserved. Enzymatic digestion for the initial screen was performed by resuspending the cell aliquot in IMDM without serum and adding a mixture of collagenase type IV (Worthington Biochemical, 48J106-42, final concentration 200 U/mL), hyaluronidase type IV-S (Sigma Aldrich, H3884, final concentration 100 U/mL) and deoxyribonuclease I (Worthington Biochemical, 57S10106002, final concentration 100 U/mL). Cells were incubated at 37°C for 20 minutes whereupon dispase, (Becton Dickinson, 354235, final concentration 2.5 U/mL) was added followed 5 minutes later by 0.25% trypsin (Gibco, 25200-056, final concentration 0.05%) for a final 5 minutes. The cell suspension was then washed in IMDM/10% FBS to inactivate trypsin, followed by a further two washes in PBS. Digested cells were then resuspended in Hank's balanced salt solution+1% bovine serum albumin, 2 mM EDTA (FC buffer) and kept at 4°C during antibody staining. Cells were cryopreserved by resuspending in FBS containing 10% dimethyl sulfoxide and freezing in a CoolCell container (Biocision) to −80°C overnight before transferring to vapour phase liquid nitrogen storage at −140°C. Vials were stored for 4 weeks, then rapidly thawed and slowly diluted with PBS, washed twice with PBS and then resuspended in FC buffer and stained. Post-staining fixation with paraformaldehyde was performed by centrifuging the 96-well plates containing the stained cells (see below) and resuspending cell pellets in BD Cytofix (Becton Dickinson) diluted 1:3 with PBS. Cells were incubated on ice for 30 minutes, then washed with PBS and resuspended in FC buffer. Validation of antigens that appeared to be affected by a preparatory step was performed by generating a custom plate of the relevant antibodies, performing the preparatory steps, and staining as above. Antigens that appeared to be affected by digestion were tested against each enzyme individually.

HT-FC staining protocol

A schematic diagram of the staining protocol is shown in Figure S1. 363 commercially available directly fluorochrome-conjugated antibodies to cell surface antigens conjugated to PE, FITC or APC (see Table S1) were purchased and aliquoted into round bottom 96-well plates (2 µl per well into 48 µl of FC buffer). Control wells with FC buffer alone were included. Cell suspensions of 0.5 to 1 million cells/mL were aliquoted by multichannel pipette into pre-prepared HT-FC plates (50 µl per well), for a final volume of 100 µl per well and a final antibody dilution of 1:50. Plates were incubated for 20 minutes on ice in the dark, centrifuged for 5 minutes at 350×g, washed twice with 200 µl FC buffer, and resuspended in 50 to 80 µl FC buffer containing 0.1 µg/mL 4′,6-diamidino-2-phenylindole (DAPI; Sigma-Aldrich). In experiments where cells were fixed after antibody staining, viability was assessed with the LIVE/DEAD fixable near-IR cell stain according to the manufacturer's instructions (Invitrogen). In parallel, aliquots were stained in tubes for fluorescence-minus-one controls, which consisted of DAPI only staining if no co-staining was done. Antibodies used for co-staining of primary tumor samples included CD45-APC-Cy7 (1:200), CD31-PE-Cy7 (1:200), CD34-PerCP-Cy5.5 (1:50) (all BioLegend) and 1.2 µg/ml TE7-biotin (in-house production from hybridoma obtained from ATCC) followed by Streptavidin-eFluor450 (1:400; eBioscience). Fluorescence-minus-one controls were generated for each antibody used and compensations were set using BD Plus CompBeads and FACSDiva software. Data collection was performed on a Becton-Dickinson LSR II flow cytometer with ultraviolet (20 mW), violet (25 mW), blue (20 mW) and red (17 mW) lasers, with default filter configuration, utilizing the High Throughput Sampler attachment. At least 10,000 events were collected per well. The gating strategy based on fluorescence-minus-one controls is illustrated in Figures S2 and S4.

Data analysis

FCS 3.0 files were exported to FlowJo version 9.3. In all cases, dead cells and doublets were excluded prior to analyzing marker expression. Gates were set using fluorescence-minus-one controls, and percent-positive values were exported to Excel for use in further statistical analysis, generation of heat maps and hierarchical clustering. Spearman correlation coefficients and p values were calculated using GraphPad Prism software. Heat maps, hierarchical clustering and principle components analyses were performed using MultiExperiment Viewer software (www.tm4.org). Unsupervised hierarchical clustering was performed using a Pearson correlation distance metric with complete linkage clustering. Principle components analysis used a mean centering mode and 10 neighbors for *k*-nearest neighbors imputation.

Immunofluorescence

Frozen sections of SOC samples were co-stained for pan-CK and CD90, as follows: sections were fixed in ice-cold acetone, air-dried and blocked with 0.5% bovine serum albumin and 5% goat serum. Purified pan-CK antibody (1:200; Abcam clone C-11) was applied and incubated for 1 hour at room temperature. Goat-anti-mouse-Alexa488 (1:400; Life Technologies) was applied and slides were incubated in the dark for 30 minutes at room temperature. The slides were then washed in PBS/0.1% Tween-20 prior to incubation with CD90-biotinylated antibody (BD Pharmingen, 1:200) for one hour at room temperature. Finally, Streptavidin-Alexa594 (Life Technologies) was applied at 1:400 in the dark. The sections were then stained with Hoechst for one minute prior to imaging using a Leica DMI6000 fluorescent microscope. p53 staining on purified SOC populations was carried out by first staining cells in suspension with 1:200 each of CD45/CD31-PE-Cy7 as described above, plus 1:100 EpCAM-APC (BioLegend) and 1:200 CD90-PE (BD Pharmingen). The $CD45^{-}CD31^{-}$ population was fractionated into the $EpCAM^{+}$ population and the $EpCAM^{-}CD90^{++}$ population and cells were purified using fluorescence activated cell sorting on an Becton Dickinson FACS Aria II with blue (15 mW), red (15 mW) and violet (10 mW) lasers, with default filter configuration. Purified cells were cytospun onto glass slides, fixed in ice-cold methanol and stained for p53 as above, using clone DO-7 (1:50, Santa Cruz Biotech).

Results

Platform Reproducibility

To facilitate screening of large numbers of antibodies rapidly, we took advantage of the recent development of a high-speed sample loading device for flow cytometers that allows automated acquisition of samples from multi-well plates. We identified and purchased 363 antibodies (Table S1), which represented all that were commercially available at that time. These were aliquoted into 96-well plates to generate what is in essence an antibody array. We initially performed single-color staining on peripheral

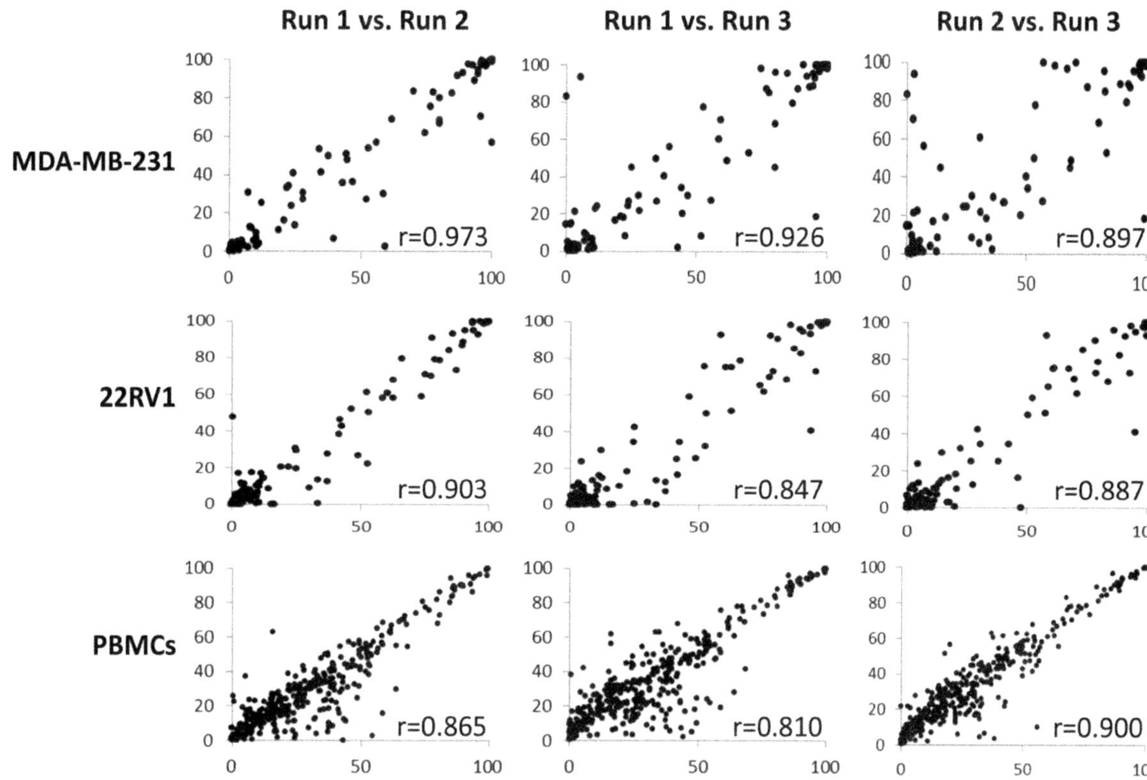

Figure 1. The HT-FC platform is highly reproducible. MDA-MB-231 breast cancer cells, 22RV1 prostate cancer cells, or PBMCs were each run 3 times on the HT-FC platform. When the same cell line or peripheral blood sample was analyzed 3 independent times, consistently reproducible results were obtained, with Spearman correlation coefficients between runs ranging from 0.847 to 0.973, and p<0.001 for all comparisons.

blood mononuclear cells and cell lines by simply adding cells into the wells and performing all antibody incubation and washing steps in the plates (see Supplementary Figure S1 for a schematic of the work flow). Control wells containing no antibody were included. Cell staining and data acquisition could be performed in a period of 4 to 5 hours total. Non-viable cells were detected by DAPI staining and cell doublets and clusters were eliminated by forward scatter and side scatter height *vs.* width plots. A gate was then set on the control "fluorescence-minus-one" sample (see Methods and Figures S1 and S2) and a table of percent positive values for all the antibodies was generated. To test the reproducibility of the HT-FC platform, we performed triplicate runs on a number of different sample types. Figure 1 shows the correlation of percent-positive values for multiple runs performed on MDA-MB-231 breast cancer cells, 22RV1 prostate cancer cells, and PBMCs. In all cases, the reproducibility of marker detection was very high, with Spearman correlation coefficients (r^2) ranging from 0.81 to 0.97 ($p<0.001$ for all comparisons).

Identification of antigens that are altered by common sample preparation methods

Flow cytometry is a technique that is commonly used to analyze cell surface antigens on cells that have been subjected to various experimental manipulations in the lab. These include enzyme digestion, such as trypsinization of cultured cells or collagenase digestion of solid tissues, and cryopreservation. In addition, it is common to fix cells in paraformaldehyde prior to analysis, in order to preserve cells until convenient to analyze and/or eliminate potential biohazards present in human samples. Cell surface proteins are potentially vulnerable to such treatments and may be negatively affected by non-specific enzymatic digestion and/or altered in response to stressful conditions such as freezing and thawing. We therefore sought to determine the influence of these common manipulations on cell surface antigen detection.

A variety of cell types were sourced and pooled in an attempt to generate the largest possible coverage of epitope detection. These included PBMCs, epithelial cancer cell lines, melanoma cell lines, leukemia cell lines, and cultured fibroblasts. All cell lines were confirmed to be *Mycoplasma* negative and identity was verified by short tandem repeat profiling [12]. This pool of cells was harvested without enzymes (adherent cells were detached using EDTA and gentle scraping and trituration), treated with a cocktail of trypsin, collagenase, hyaluronidase, dispase and DNAse with appropriate incubation times, or cryopreserved. Experiments were also done to compare unfixed cells to cells that were fixed in 2% paraformaldehyde in PBS after staining. At the time of these experiments, the panel contained 334 antibodies. Overall 258 of 334 antigens were expressed on ≥1% of the pooled cells (Figure 2A), therefore we could not determine the sensitivity of 76 antigens to the various preparation methods. Of the 258 expressed, a number of antigens showed altered expression after exposure to enzymatic digestion, fixation or cryopreservation, as seen in the heat map in Figure 2A (based on percent-positive values). Enzymatic digestion with a pooled enzyme cocktail influenced 79 antigens (30% of 258 antigens detectable) in our preliminary screen, and these were subsequently taken forward for validation against individual enzymes in a customized plate. Of these 79 antigens, 39 were validated as showing an absolute change in percent-positive cells of at least 5% with at least one enzyme, suggesting that many altered in the original screen were falsely influenced, likely due to

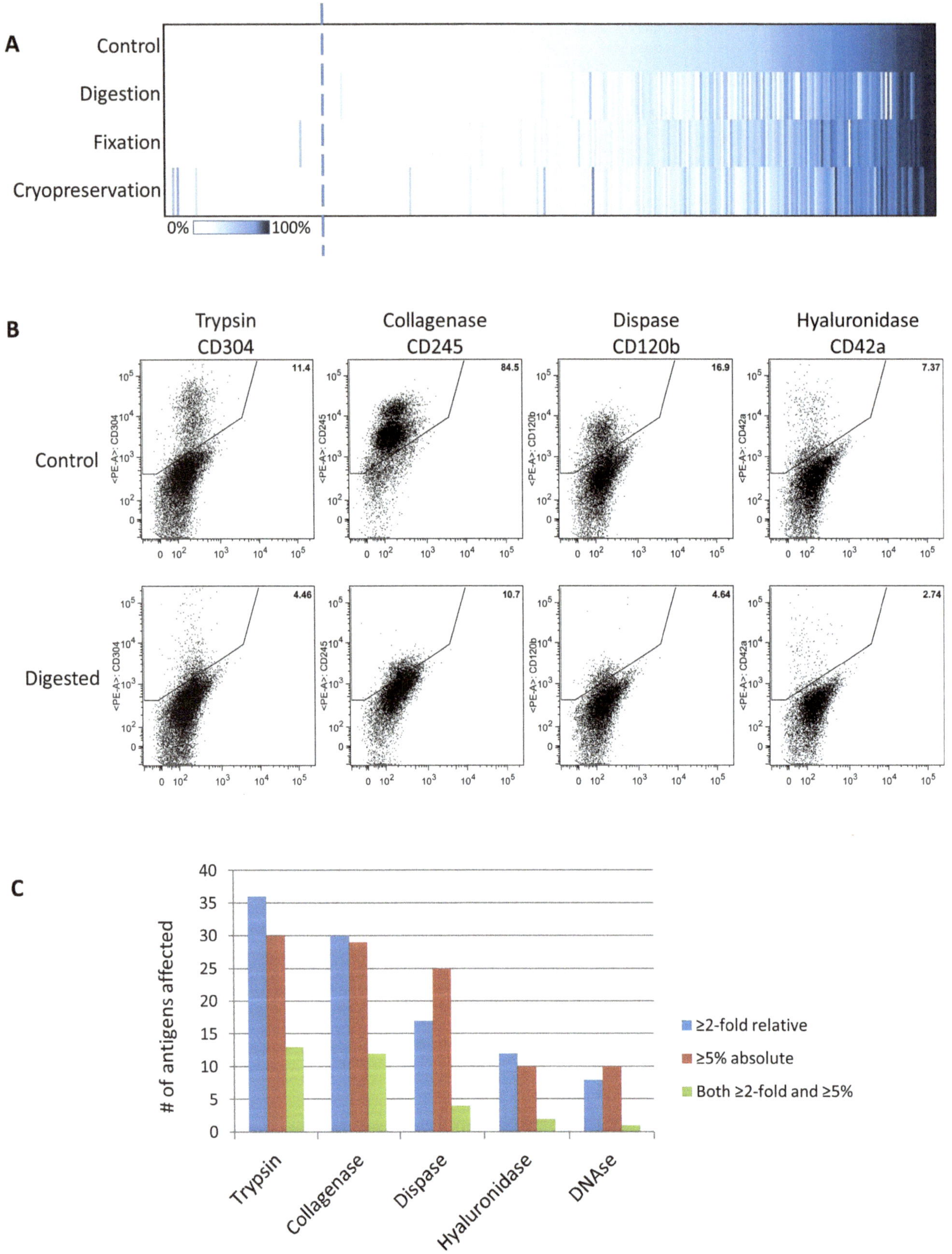

Figure 2. Effect of typical experimental procedures on HT-FC antigen detection. (**A**) 267 out of 344 antigens showed expression of greater than 1% on the "untreated" cell pool (nominal limit of detection of 1%; dashed line). Digestion with a cocktail of enzymes containing trypsin, collagenase, hyaluronidase, dispase and DNAse affected the largest number of antigens, while fixation and cryopreservation affected fewer antigens. (**B**) Examples of validated antigen loss after digestion with individual enzymes trypsin (CD304), collagenase (CD245), dispase (CD120b) and

hyaluronidase (CD42a). (**C**) Collagenase and trypsin were typically the most influential enzymes, but all had some influence. Graph shows the number of antigens altered by ≥2-fold relative to untreated cells (blue), or by ≥5 percent-positive cells (red), or both 2-fold and ≥5 percent-positive cells (green). Data for all enzymes is provided in Table S2.

combination treatment within the enzyme cocktail. In total, 22 antigens showed both an absolute detection change of at least 5% and a fold-change of at least 2-fold with at least one enzyme (Tables S2 and S3). Some specific examples of observed changes are shown in Figure 2B. Interestingly, many of these have known extracellular cleavage sites and are biologically active upon release from the cell surface (Table S3). Of the individual enzymes, trypsin and collagenase demonstrated the greatest influence, with dispase and hyaluronidase showing less influence (Figure 2C). However no enzyme was spared and even DNase alone appeared to show some modest alterations in detection of some antigens. Individual enzymes had different effects on different antigens (Table S2).

Fixation with paraformaldehyde after antibody staining, and cryopreservation showed the least overall perturbations of antigen detection in the preliminary screen; 31 and 40 antigens, respectively, were altered by either ≥5% positive cells or ≥2-fold. Upon validation, only 8 antigens showed significant variation of more than 5% absolute and 2-fold relative difference when fixed (Table S4). Similarly, only 9 antigens showed a significant change when cryopreserved (Table S5). Interestingly, only one antigen showed significant loss of expression upon cryopreservation, while the other 8 were increased.

Cell surface profiles stratify samples into biologically relevant clusters

We have run the HT-FC platform on a wide range of cell types, including many epithelial cancers (traditional and novel cell lines, primary samples), normal and malignant hematopoietic cells, fibroblasts from various sources, and a small selection of normal and malignant primary neural stem cell cultures. We identified only 10 antibodies that never stained any cell type analyzed; of these, 5 are predicted to stain very specific cell types that were not among those tested by us, and 5 should have stained cells that we tested and did not, suggesting that these 5 antibodies are suboptimal for this application (Table 1). Of the remaining antibodies, in addition to staining the predicted cell types, there were many that stained cell types not previously identified as expressing those markers. For example, some markers stained virtually every cell type we examined, while other antibodies were very cell-type or lineage specific.

Gene expression profiling has been widely used to generate "molecular signatures" that define a particular cell or tissue type or that can be used to define molecular subtypes within a given disease that are associated with clinical behaviors or outcomes [13–16]. To determine whether our 363-protein profile could similarly stratify samples based on their cell of origin, we compiled the large amount of HT-FC data obtained across many cell types using percent-positive values to perform unsupervised hierarchical clustering (Figure 3 and Figure S3). We found that the cell surface phenotype stratifies samples into clusters of related cell types or lineages. For example, we see the following major clusters: a hematopoietic cluster (red), which contains sub-clusters that include PBMCs, AML samples, B-ALL samples, and the CD45+ immune cell fraction of primary ccRCC samples; an epithelial cancer cell line cluster (green) composed of multiple sub-clusters; a fibroblast cluster (orange) which includes CAFs derived from SOC and HNC and normal dermal fibroblasts; and a ccRCC cluster (blue), which contains both early passage in-house cell lines and primary samples. Primary SOC samples form an independent cluster (yellow), whereas ccRCC cell lines and primary samples cluster together and are more closely associated with neuronal cells (pink) and fibroblasts (orange) than with epithelial cell lines. Principal components analysis revealed similar clustering patterns (Figure S3). These results indicate that the 363 cell surface protein profile provides sufficient molecular information to segregate human samples into biologically relevant groups, and could be used to rapidly identify cell surface proteins associated with specific cell types or disease states. Table S6 contains data obtained for

Table 1. Antibodies with no staining across all cell types analyzed.

Antigen	Expected Cell Type	Cell Type Analyzed?
CD103	intestinal intraepithelial lymphocytes (IEL)	No
CD41b	platelets	No
PAC-1	platelets	No
CD1a	cortical thymocytes, dendritic cells and Langerhans cells; DOES NOT react with PBMCs	No
CD209	dendritic cells, vascular endothelial cells	No
CD150	T cells, B cells, thymocytes, germinal center and dendritic cells	**Yes**
CD16b	neutrophils	**Yes**
FMC7	B lymphocytes	**Yes**
CD197	lymphocytes	**Yes**
CD129	low levels on eosinophils, mast cells, macrophages, B lymphocytes, T lymphocytes, and erythroid progenitors	**Yes**

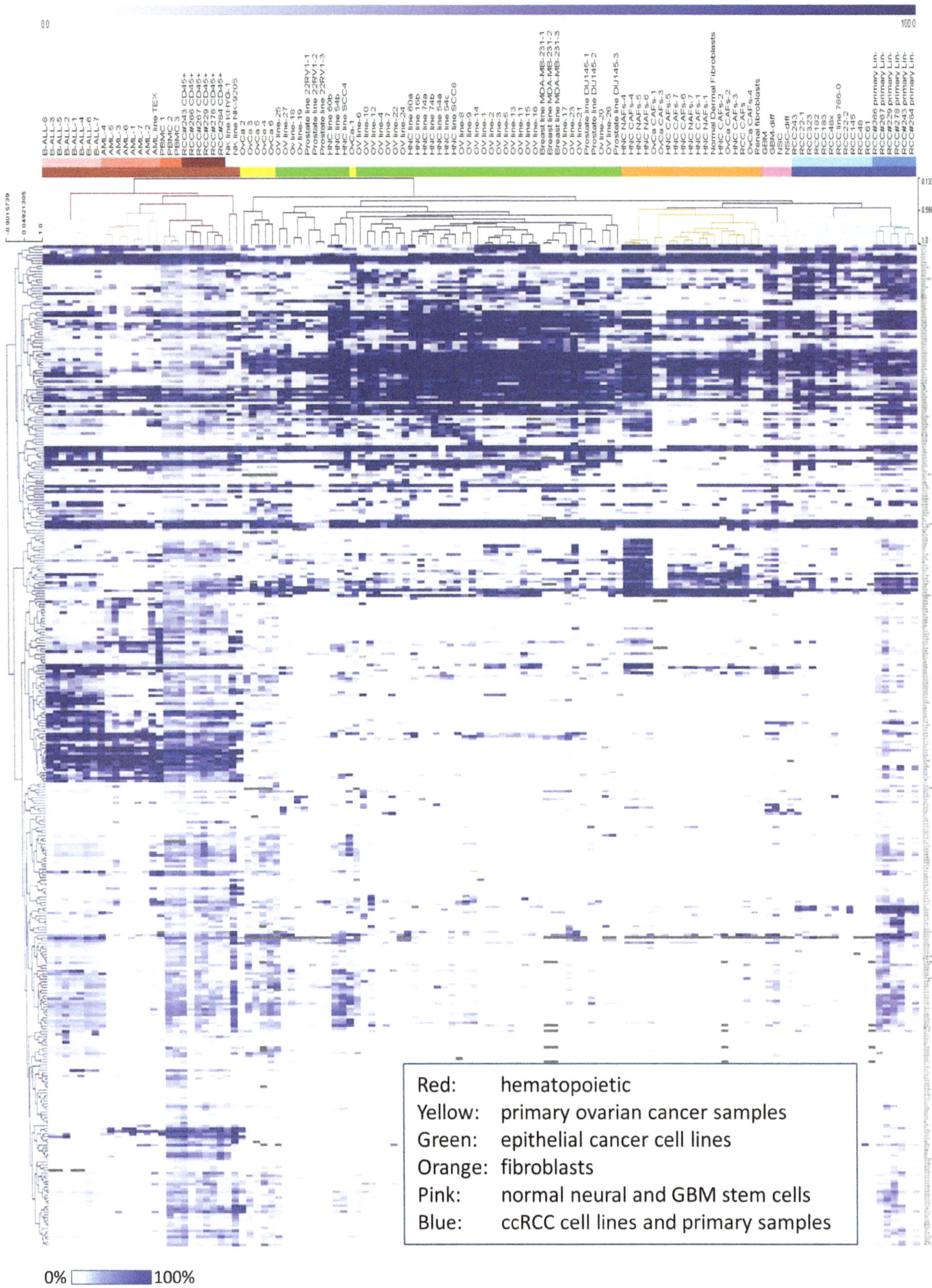

Figure 3. HT-FC allows phenotypic segregation and identification of cell samples from diverse lineages. Unsupervised hierarchical clustering of percent-positive marker expression values generated on 119 samples was performed. Colours indicate emergence of biologically related samples into clusters based on surface marker profiles. See Figure S2 for a magnified image of the dendrogram.

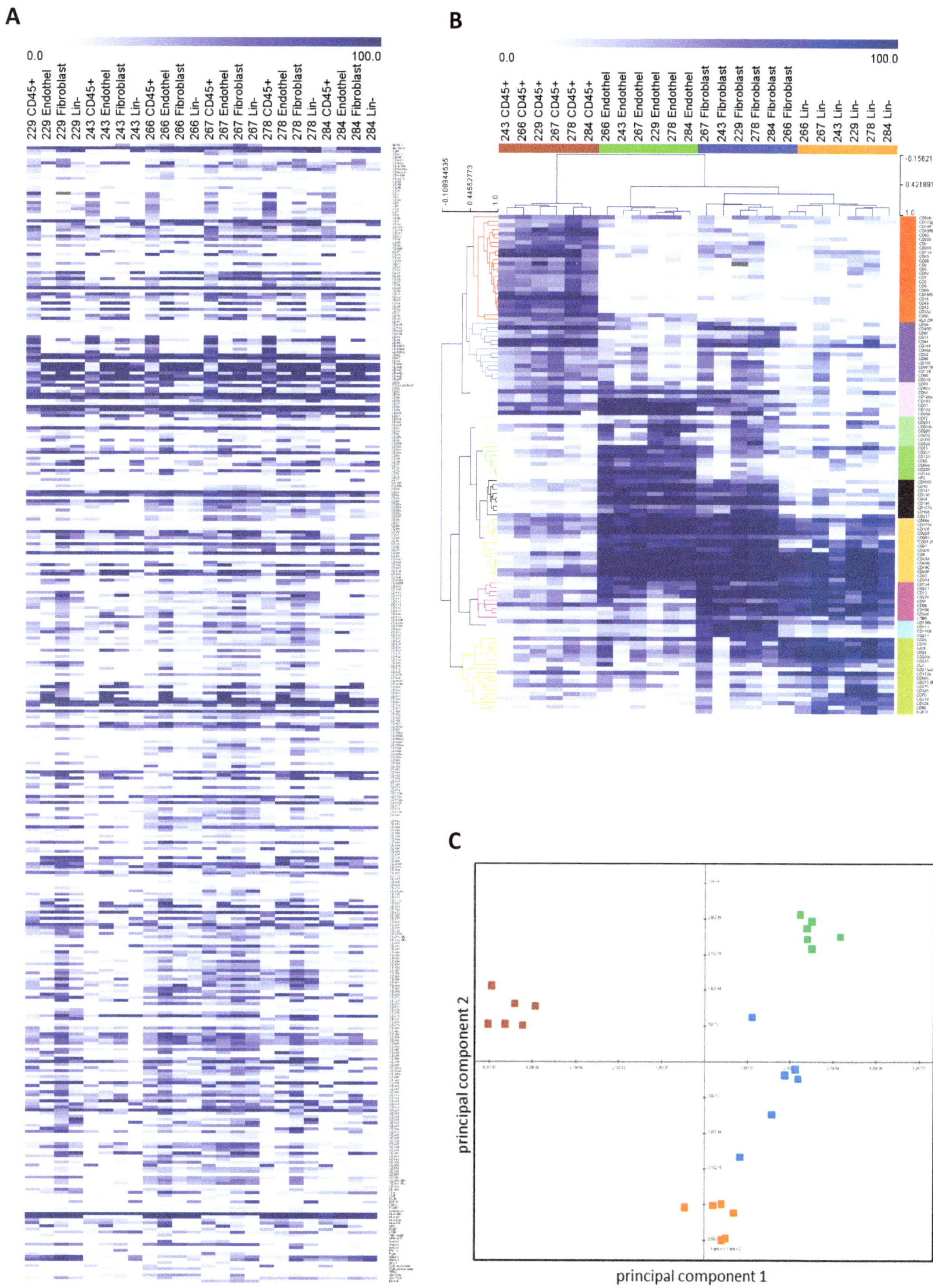
A
0.0
100.0
229 CD45+
229 Endothel
229 Fibroblast
229 Lin-
243 CD45+
243 Endothel
243 Fibroblast
243 Lin-
266 CD45+
266 Endothel
266 Fibroblast
266 Lin-
267 CD45+
267 Endothel
267 Fibroblast
267 Lin-
278 CD45+
278 Endothel
278 Fibroblast
278 Lin-
284 CD45+
284 Endothel
284 Fibroblast
284 Lin-
B
0.0
100.0
243 CD45+
266 CD45+
229 CD45+
267 CD45+
278 CD45+
284 CD45+
266 Endothel
243 Endothel
267 Endothel
229 Endothel
278 Endothel
284 Endothel
267 Fibroblast
243 Fibroblast
229 Fibroblast
278 Fibroblast
284 Fibroblast
266 Fibroblast
266 Lin-
267 Lin-
243 Lin-
229 Lin-
278 Lin-
284 Lin-
-0.1562165
0.42189175
1.0
-0.108944535
0.44552773
1.0
C
principal component 2
principal component 1

Figure 4. HT-FC allows intratumoral analysis of stromal and cancer cell subsets within primary ccRCC samples. (**A**) Heatmap showing expression of each of the 363 antibodies in four subpopulations of ccRCC samples: $CD45^+$ immune cells, $CD45^-CD31^+CD34^+$ vascular endothelial cells, $CD45^-TE7^+$ fibroblasts, and $CD45^-TE7^-CD31^-CD34^-$ cancer cells. Antibodies are simply arranged in alphabetical order on the vertical axis and the four populations for each sample are ordered across the top. A low resolution overview demonstrates a surprisingly reproducible "fingerprint" of tumor cell subpopulations from one sample to the next. (**B**) Supervised hierarchical clustering reveals clusters of antigens corresponding to specific cell subsets within tumors (see Table S6 for details). (**C**) Principal components analysis of the entire data set further illustrates how effectively the cell surface profile delineates the 4 distinct cell populations within primary ccRCC samples. Red: immune cells; Green: endothelial cells; Blue: fibroblasts; Orange: cancer cells.

PBMCs, MBA-MB231 cells, 22RV1 cells, renal cancer samples and cell lines, and CAFs derived from ovarian cancer and head and neck cancer primary samples.

Interrogation of cellular heterogeneity in primary tissue samples

Performing HT-FC on cell lines is straightforward, requiring simply aliquoting cells into 96-well plates, where each well contains a single antibody. However, primary tissues that are dissociated into single cell suspensions for FC analysis contain a complex mixture of cell types, and thus can't be accurately analyzed using a single marker at a time. For analysis of the primary SOC and ccRCC samples shown in Figure 3, samples were co-stained with additional markers to allow identification of abundant immune cells and other stromal cell types that are present in primary tumor samples, and selection of only the cancer cell population for analysis of the cell surface profile. However, co-staining of primary samples with multiple antibodies conjugated to additional fluorochromes not only allowed unwanted stromal cells to be excluded from the analysis of cancer cells within a tumor tissue, but also allowed us to obtain the cell surface profile of multiple populations within these complex samples. For example, the 6 primary patient-derived ccRCC samples were initially stained with four stromal-targeted antibodies prior to staining them with the full antibody panel (Figure 4). From this we were able to generate heat maps representing the cell surface profile of the hematopoietic ($CD45^+$), vascular endothelial ($CD31^+/CD34^+$), fibroblast ($TE7^+$) [17] and cancer ($CD45^-/CD31^-/CD34^-/TE7^-$) cell populations within each tumor. The gating strategy is provided in Figure S4. Figure 4A shows an expression heat map of the data (without hierarchical clustering). Antibodies are arranged in simple alphabetical order on the vertical axis for the four populations within each sample, grouped across the top. It is clear from the heat map that a remarkably similar pattern of staining is obtained from one sample to next, once again illustrating the reproducibility of the platform even when applied to individual patient-derived samples. Upon supervised hierarchical clustering, several antigen clusters were revealed that delineate the different cell lineages analyzed (Figure 4B). Some of the clusters contained predicted markers, such as CD3, CD4, CD9, and CD11b on the immune cell cluster (red on vertical axis); CD141, CD144 and CD146 on endothelial cells (green and black); and CAIX and CD10 on cancer cells (yellow). Interestingly, we also identified markers that have previously not been defined to be specific for the indicated lineages, or that stained multiple cell types (for example, a cluster specific for endothelial cells and fibroblasts, another for immune cells and endothelial cells, *etc*.). The markers for the identified clusters are listed in Table S7.

Identification of a cancer-associated fibroblast (CAF) cell surface marker in serous ovarian cancer

As with any high-throughput screen it is important to demonstrate its ability to identify specific candidates of interest, which are then validated upon follow-up investigations. Our laboratory has an interest in the role of CAFs in high-grade SOC. While it is possible to generate cultured CAFs from primary SOC samples that then can be passaged *in vitro*, we were interested in identifying cell surface markers for SOC stromal fibroblasts to facilitate their purification directly from primary tumors, allowing us to avoid artifacts associated with culturing cells for multiple passages. While many CAF markers are available (*e.g.* Vimentin, α-smooth muscle actin), the majority of them are intracellular, and thus cannot be used to purify viable cells using fluorescence-activated cell sorting (FACS). Fibroblast activation protein and platelet-derived growth factor receptor-β are reported to be cell surface fibroblast markers, but despite employing multiple different antibodies at high dilution we found the expression of both of these markers to be low and/or not to yield distinct cell populations that would allow for cell sorting of SOC samples. TE7, the fibroblast marker used to identify fibroblasts in ccRCC samples described above, was also found to be suboptimal for staining of fibroblasts in SOC. To identify new cell surface fibroblast markers, we performed the HT-FC screen on four cultured CAF lines we had generated, as well as on five primary SOC samples, co-stained with CD45 and CD31 to allow exclusion of contaminating immune and endothelial cell types. We then selected markers that were highly expressed on the pure fibroblasts but expressed at low levels on the CD45/CD31-negative fraction of primary SOC samples, which would be expected to contain predominantly cancer cells and contaminating fibroblasts. Based on this analysis, we identified 5 candidate CAF markers (Figure 5A). We chose to follow up on CD90, as it has been reported by others to be expressed on mesenchymal stem cells [18], on cells resembling mesenchymal stem cells that were cultured from SOC samples [19], and also to be a CAF marker in prostate cancer [20,21].

To validate CD90 as a candidate SOC CAF marker, we co-stained primary SOC samples with hematopoietic and vascular cell markers (CD45 and CD31), epithelial cell marker EpCAM (epithelial cell adhesion molecule), and CD90, and consistently identified a population of cells that was $CD45^-CD31^-EpCAM^-$ and very bright for CD90 ($CD90^{++}$; Figure 5B and Figure S5), which we hypothesized to represent the stromal fibroblasts. The proportion of $CD45^-CD31^-EpCAM^-CD90^{++}$ cells varied from sample to sample, ranging from <1% to 56% (mean 10.0%, n = 24). CD90 was also found to be variably expressed on the $EpCAM^+$ cancer cells, and thus must be used in combination with EpCAM to cleanly distinguish the stromal fibroblast population. Interestingly, the CD90 expression level was consistently higher on the EpCAM-negative fraction, as indicated by higher mean-fluorescence intensity of this population (Figure 5B). We confirmed the flow cytometry results using immunofluorescence co-staining with a pan-CK antibody (Figure 5C), where we saw brighter staining of stromal cells *vs.* epithelial cells. We also found that cell cultures initiated with FACS-purified $CD45^-CD31^-EpCAM^-CD90^{++}$ cells gave rise to fibroblasts that were positive for α-smooth muscle actin, a classical CAF marker (Figure 5D). Finally, we verified that the $EpCAM^-CD90^{++}$ cells were not

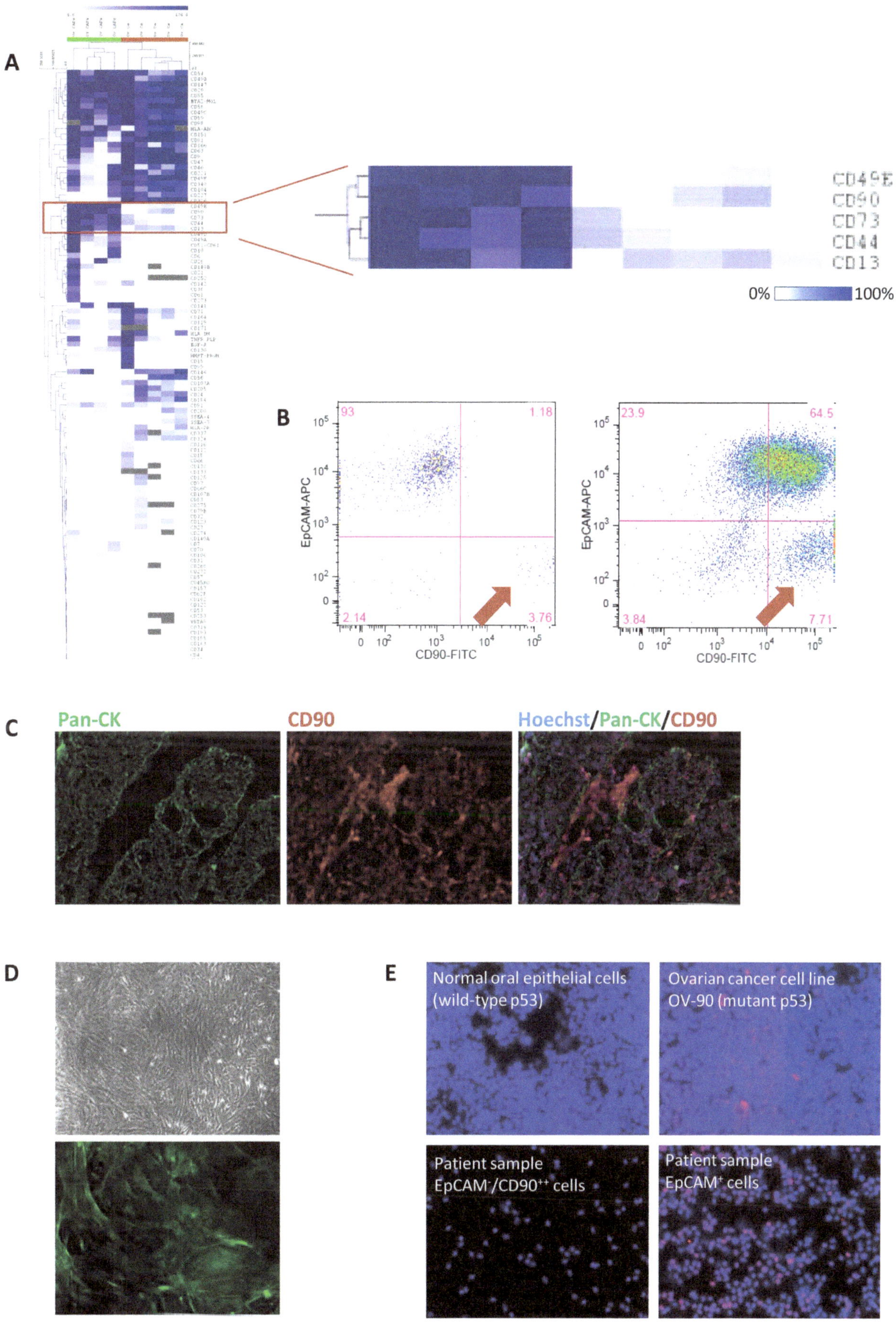
A
CD49E
CD90
CD73
CD44
CD13
0%
100%
B
EpCAM-APC
CD90-FITC
93
1.18
2.14
3.76
23.9
64.5
3.84
7.71
C
Pan-CK
CD90
Hoechst/Pan-CK/CD90
D
E
Normal oral epithelial cells (wild-type p53)
Ovarian cancer cell line OV-90 (mutant p53)
Patient sample EpCAM⁻/CD90⁺⁺ cells
Patient sample EpCAM⁺ cells

Figure 5. HT-FC identifies CD90 as a CAF marker in serous ovarian cancer. (**A**) HT- FC was performed on 4 SOC-derived CAF lines and 5 primary SOC samples. $CD45^{+}CD31^{+}$ cells were excluded from the analysis of the primary samples. A cluster of antigens was identified that were highly expressed in CAFs and expressed at low levels in primary cancer cells. (**B**) FACS plots of two primary SOC samples, gated on viable $CD45^{-}CD31^{-}$ cells, showing CD90 *vs.* EpCAM staining. A distinct population of $EpCAM^{-}CD90^{++}$ cells is consistently seen, whereas CD90 expression is variable on the $EpCAM^{+}$ cancer cells. Two representative samples are shown, out of a total of 19 samples analyzed. (**C**) Immunofluorescence staining of a frozen section of SOC (same sample shown in FACS plot on right in (B) above), stained with a pan-CK antibody (green) and CD90 (red). CD90 stains both the CK^{+} cells and the CK^{-} stromal cells, with more intense staining on the latter, thus recapitulating the staining seen by flow cytometry. Magnification = 100X. (**D**) FACS-purified $EpCAM^{-}CD90^{++}$ cells efficiently establish cultures that display a fibroblast phenotype (top; 40X) and stain positive for CAF marker α-smooth muscle actin (bottom; 100X). (**E**) Immunofluorescence staining for p53 is positive $EpCAM^{+}$ cancer cells, indicative of mutant p53 in this sample (bottom right), whereas the $EpCAM^{-}CD90^{++}$ putative fibroblast population from the same sample is p53 negative, and thus wild-type (bottom left). Controls: normal oral epithelial cells (p53 wildtype; top left), p53 mutant ovarian cancer cell line (OV-90; top right). Magnification = 100X.

cancer cells by preparing cytospins of the purified cell populations and staining them for p53. The majority of high-grade SOC samples are p53 mutant [22], which is indicated by high positive nuclear staining with a p53 antibody [23]. We therefore performed immunofluorescence staining for p53 on the $EpCAM^{+}$ and $EpCAM^{-}CD90^{++}$ fractions from 4 samples. Three of these samples were positive for p53 staining in the $EpCAM^{+}$ fraction, indicating their mutant status; the fourth sample was negative for p53 staining and thus was uninformative. In all three of the p53-mutant tumors, the matching $EpCAM^{-}CD90^{++}$ cells were negative and therefore p53 wild-type, verifying that these were not cancer-derived cells that had lost EpCAM expression and/or gained a mesenchymal phenotype *via* epithelial-to-mesenchymal transition. Thus the combination of EpCAM/CD90 staining together with CD45 and CD31 allows for isolation of stromal fibroblasts directly from primary patient samples for further characterization.

Discussion

Flow cytometry has long been a robust tool for reliable detection of cell surface proteins, and has many advantages over other immunological protein detection methods, such as Western blotting and IHC. These include ease of use, the ability to rapidly analyze very large cell numbers, analysis of rare populations of cells, and the ability to obtain multi-parameter information on individual cells, which is particularly important for heterogeneous cell samples. The robustness and utility of flow cytometry is illustrated by the large number of clinical applications for which it is now being used around the world [9]. Traditionally, flow cytometry assays are performed on individual samples with a panel composed of up to 11 antibodies at a time that are known to be useful for a particular diagnosis or identification of specific cell types. More recently mass cytometry or "cytometry time-of-flight" (CyTOF) assays, in which antibodies are labelled with lanthanide metals rather than fluorochromes and then detected by mass spectrometry, has been developed [24]. Due to the lack of spectral overlap between antibodies, this allows an even greater degree of multiplexing to be performed, currently allowing for simultaneous analysis of up to 35 markers at a time, but with the potential to increase this up to 100 in the future [24,25]. The HT-FC platform described here is complementary to these methods, as it takes a more unbiased, discovery-oriented screening approach; by screening cells of interest for a large panel of cell surface markers, it is possible to identify previously unknown proteins or protein combinations expressed on the surface of cells of interest, which once identified can then be developed into the more traditional multiplexed flow cytometry assays or translated into CyTOF assays. Thus HT-FC represents a valuable tool for new marker discovery, comparable to other discovery platforms such as gene expression microarrays, but with a focus on cell surface proteins. The cost per sample is similar to or less than gene expression microarrays (depending on the microarray platform), and is significantly less than similar commercially available cell surface protein screens. Each assay can be performed in an afternoon, and there is the potential to multiplex samples by pre-labelling with fluorescent dyes [26] to increase throughput and reduce costs even further. Data analysis is straightforward, as gates that are set on control samples are simply applied to the entire data set and a spreadsheet of expression values is generated that can then be used in standard bioinformatics software packages (we used Multi-Experiment Viewer). This is in contrast to CyTOF data, which requires antibodies to be conjugated to lanthanides and optimized by the individual investigator, and for which a simple and straightforward data analysis method is yet to be devised [27,28].

The use of well-established and validated commercially available antibodies means high quality protein-level data is obtained that will have a very high validation rate; indeed flow cytometry is a platform that is commonly used as the means of validation for other high-throughput methods such as gene expression profiling (where mRNA expression does not always correlate with protein expression) or mass spectrometry (where initial detection is typically validated by subsequent MAb-based assays). Thus this platform combines the advantages of a high-throughput screen with a detection method that is sensitive and highly reproducible, as demonstrated by the high correlation coefficients obtained between replicate runs. Furthermore, the use of commercially available antibodies means that high quality reagents are immediately available for follow-up studies and/or development of more focused panels for specific applications such as 11-plex flow cytometry and/or CyTOF. Notably, only 5 out of 363 antibodies were found to be potentially problematic, in that they did not stain cell types that are reported to express the corresponding antigens (Table 1), supporting the robustness of this approach.

A requirement for FC analysis is that cells must be in single cell suspension, thus cultured adherent cells are detached by trypsinization and solid tumors are enzymatically dissociated. In addition it is common to use cryopreserved cells or to fix cells prior to analysis. We were concerned that these experimental manipulations would lead to significant changes in cell surface antigen detection. It is reassuring that detection of the majority of antigens assayed was stable; however detection of a number of markers was significantly altered. For the purposes of this study we have chosen to define a significant change as at least 5% absolute change *and* at least a two-fold change in relative detection. With that definition, eight antigens were significantly altered by fixation after staining, though most often to reduce the mean fluorescence intensity rather than the overall pattern of antigen detection. Nine antigens were altered by cryopreservation and thawing; one antigen, CD138 (syndecan-1) was reduced, while the remainder were increased (Table S5). The latter are almost exclusively expressed on lymphoid cells, and we hypothesize that these changes in apparent antigen detection may be accounted for by differential ability of different cell types in the pool to survive the cryopreservation and thawing process. Finally, 22 antigens were significantly altered by enzymatic digestion. All enzymes caused some change; trypsin, dispase and collagenase caused the most changes and are known proteases, but hyaluronidase and DNase also caused some alterations in antigen detection, though we cannot exclude impurity in the cell culture grade products used in routine practice in this setting. Of the 22 antigens showing significant change, the majority are reported to have extracellular proteolytic cleavage sites, which in some cases may be integral to the mechanism of action of the molecule (Table S3). While the majority of influenced antigens showed decrease with enzymatic digestion, occasional antigens demonstrated increased detection after enzymatic digestion, *e.g.* TWEAK/CD255 where a soluble form is known to be generated from the membrane-associated protein by cleavage [29]. While we validated each individual enzyme in isolation, in everyday practice these are often used in combination, *e.g.* collagenase and hyaluronidase in tissue digestion [30]. In the preliminary screen, some epitopes showed almost complete abrogation of detection, including common antigens such as CD1d, CD4, CD25, CD27 and CD62P, a particular concern for the study of heterogeneity and immunology in solid tumors. Of all the antigens detected in this platform CD138/syndecan-1 is the most labile, with alteration in detection upon cryopreservation and thawing, fixation, and collagenase digestion. CD138 is strongly expressed in myeloma but is also shed from the cell surface [31], as well as being sensitive to proteolytic cleavage during cell processing [32], thus we hypothesize that various tissue processing steps can increase this antigenic loss. Overall, these data show that the majority of cell surface antigens are not significantly affected by the various manipulations tested, but we provide a list of those that are altered as a resource to the research community. This illustrates the need to evaluate possible antigenic changes caused by experimental manipulations when new targets are being analyzed using flow cytometry methods.

Through the course of offering the HT-FC platform as a core facility resource at our institute, we have processed a large number of cell lines and other sample types. As shown in Figure 3, upon unsupervised hierarchical clustering of the percent-positive data the samples stratify naturally into clusters of related cell types. For example, all hematopoietic lineages form a major cluster, with lymphoid and myeloid leukemias and normal hematopoietic cells forming individual sub-clusters. The clustering of the $CD45^+$ fraction of ccRCC samples with PBMCs nicely illustrates the consistent identification of a cell surface "signature" of hematopoietic cell lineages, even when derived from diverse sources (peripheral blood *vs.* surgically resected tumor samples). It also illustrates the consistent ability of the platform to identify a specific cell lineage, given that different samples were processed and analyzed by different individuals on different dates.

Epithelial cancer cell lines also form their own major cluster containing individual sub-clusters. Interestingly, the cell lines derived from a particular tumor type (*e.g.* ovarian cancer or head and neck cancer) do not perfectly segregate into type-specific clusters, but rather form sub-clusters that are intermingled with each other. The biological significance of this remains to be determined, but may relate to other properties such as the cell of origin of the individual cancer (*e.g.* stem cell or progenitor cell; ovarian surface epithelium or fallopian tube; primary tumor or metastasis; tongue *vs.* floor of mouth *vs.* other head and neck cancer sites; *etc.*), the mutational profile of the various cell lines, and/or the various culture conditions and selection pressures to which the cells have been exposed over the years. Notably, primary SOC samples form their own independent cluster, rather than clustering with ovarian cancer cell lines, suggesting an evolutionary divergence of the cell surface profile of passaged cell lines from that of the original primary tumors. It is also notable that cell lines in general express considerably fewer cell surface antigens overall than primary samples, suggesting that many cell surface proteins expressed *in vivo* are lost upon cell line establishment.

The other major cluster contained fibroblasts, and our in-house early passage ccRCC cell lines and primary ccRCC samples. In this case the cell lines did cluster with the primary samples, likely because they were in-house cell lines at early passage (passage 4 to 6), and thus perhaps had not yet diverged significantly in phenotype from the primary tumors from which they were derived. The clustering of ccRCC with fibroblasts is unexpected as ccRCC is considered to be an epithelial carcinoma. However, ccRCC has other evidence to suggest an intrinsically mesenchymal phenotype, such as its uniformly high expression of mesenchymal marker vimentin, lack of expression of epithelial structures such as cilia [33] and its propensity for differentiation along mesenchymal lineages as reported by Tun *et al* who demonstrated adipogenic and osteogenic transdifferentiation of ccRCC *in vitro* [34]. Thus our data provide further support to the concept that ccRCCs are phenotypically similar to other mesenchymal cell types, such as fibroblasts. Overall, our data suggests that the 363-marker cell surface profile is sufficient to molecularly define distinct cell types. This supports the utility of using the HT-FC platform for new lineage-specific marker discovery and molecular sub-classification of diseases, for biomarker discovery, and for research into biological mechanisms of disease. We are continuously expanding and refining the panel to include additional antibodies as they become available, and thus make the panel more comprehensive. The addition of more antibodies is done by simply adding new wells to the 96-well plates, and ultimately we may increase the plate number required from 4 to 5 as antibodies continue to become available. This does not significantly increase the cost or labor required to run the assay or analyse the data. We also have the flexibility to generate "custom plates" for those individuals who would prefer to assess a specific sub-panel of antibodies for a lower cost. Other innovations are also being developed, such as performing "bar coding" with intracellular fluorescent dyes that would allow multiple samples to be pooled and run simultaneously [26], thus increasing throughput and reducing costs even further.

One potential limitation of our assay is the approach of staining with one antibody at a time, rather than combining multiple antibodies together for multiplexing purposes. In some circumstances this is not the desired approach; however, the advantage of this format is that, as with other "array" type screening approaches, it can be applied to a wide range of cell types to rapidly provide a list of candidates, from which custom multiplexing panels can then be designed. This format also allows straightforward analysis of the expression profiles, and has the flexibility to allow co-staining with specific markers of interest to the individual investigator; thus multi-parameter information can still be obtained.

FC analysis of primary patient-derived tumor tissues can be used to identify cell subpopulations such as cancer-initiating cells, allowing their isolation for functional assays and molecular profiling. The process of identifying cell surface markers for such studies is challenging, and is often predicated on iterative examination of markers previously identified in other cancer or normal stem cell types. By co-staining with markers for various cell lineages present within primary tumors we were able to obtain marker profiles for the entire panel specifically on the cancer cell subset, providing a list of markers that are expressed on a small fraction of the cancer cells which might then be interrogated as markers of cell subpopulations, *e.g.* candidate cancer stem cell markers. Cell surface proteins are also excellent candidates for development of therapeutic antibodies, or biomarkers that can be prognostic, predictive of specific therapeutic responses, or used for disease detection and classification. Robust identification of markers in particular cellular subsets allows these hypotheses to be accurately examined. Furthermore, biomarkers may not be restricted only to the cancer cells; stromal cells such as fibroblasts, vascular endothelial cells and immune cells play important roles in tumor behavior and are even drug targets in some cases (*e.g.* vascular endothelial cells targeted by tyrosine-kinase inhibitors, the primary treatment modality for ccRCC patients with metastatic disease). The HT-FC platform allows for independent profiling of each individual population within the tumor for the entire 363 protein panel in a single assay, and thus represents a rapid and economical method for characterization of the cell populations present.

As with any high-throughput screening platform, it is essential to perform validation of candidates of interest that are identified with the screen. Follow-up studies should include MAb titration, and more standard in-tube staining and validation on larger cell numbers and/or additional samples. Here we present a case in which a marker discovered through the HT-FC screen was followed up and validated, and which will now provide a useful tool in the study of CAFs in SOC. CAFs are typically isolated by placing primary tumor samples into culture conditions in which the fibroblast population grows out, and can be passaged. However, this can introduce artifacts into analysis, as the CAFs may change their properties when no longer in the presence of cancer cells and/or may be influenced by the selection pressures of growing *in vitro*. A cell surface marker that would allow direct isolation of CAFs from tumor tissues by FACS would be extremely useful to facilitate their characterization. We found that surface markers reported for CAFs in other tumor types were not useful in our hands, so we performed HT-FC screening on cultured CAFs vs. primary SOC cells to find markers differentially expressed between the two. Several markers were found to be highly expressed on cultured CAFs and expressed at low levels on SOC cells, including CD90. When used in combination with epithelial cell marker EpCAM, a distinct population of $EpCAM^{-}/CD90^{++}$ cells was consistently observed. The use of CD90 is an improvement over using EpCAM alone and sorting the $EpCAM^{-}$ subset, as there are $EpCAM^{-}$ cancer cells present in some SOC samples, as previously determined by our group [35]. The addition of CD90 allowed the isolation of a population of cells that grows with the morphology of fibroblasts, expresses α-SMA, is negative for CK, and was shown to lack mutant p53 protein that was present in the matched cancer cells from the same patients. The latter verifies that the fibroblast population we are isolating does not arise as a result of epithelial-to-mesenchymal transition from the cancer cells, but represents a genetically normal stromal fibroblast population. CD90 was also found to be expressed to varying degrees on the $EpCAM^{+}$ population, with a lower intensity of expression. While this did not hinder CAF isolation due the distinct population of $EpCAM^{-}CD90^{++}$ cells present even when $EpCAM^{+}$ cells were CD90-positive, a more specific CAF marker would be useful. Several other markers were also identified in the screen (Figure 5), which can be tested for their ability to identify CAFs in SOC in the future.

Conclusions

We have developed a high-throughput flow cytometry platform that can be used to rapidly profile the cell surfaceome, providing robust and reproducible data that can be used for a wide range of applications, including biomarker discovery, molecular classification of cancers, or identification of novel lineage specific or stem cell markers.

Supporting Information

Figure S1 Schematic of HT-FC workflow.

Figure S2 Gating strategy for HT-FC analysis of PBMCs.

Figure S3 Enlargement of dendrogram from heat map in Figure 3 and principle components analysis of data from Figure 3.

Figure S4 Gating strategy for HT-FC analysis of immune cells, vascular endothelial cells, stromal fibroblasts, and cancer cells within primary ccRCC samples.

Figure S5 Gating strategy for FACS isolation of cancer-associated fibroblasts from primary serous ovarian cancer samples.

Table S1 Antibodies included in HT-FC panel.

Table S2 Validation of antigens influenced by enzymatic digestion.

Table S3 The 22 most-affected antigens appear predisposed to altered detection by enzymatic digestion.

Table S4 Fixation following antibody staining significantly alters detection of a small proportion of markers.

Table S5 Antigens with significant alteration in detection after cryopreservation and thawing.

Table S6 Cell surface profiling data of samples in Figure 3.

Table S7 Clusters of antigens expressed on primary ccRCC cancer and stromal cell populations.

Acknowledgments

We would like to acknowledge the patients who generously donated their cancer samples for research. We are most grateful for technical and logistic support from Laura Legere, Rehab Chanin, Ajla Meheljic, Hannah Chung, Colleen Simpson, the UHN BioBank, Pathology Research Program, UHN/SickKids Flow Cytometry Facility, the SickKids Centre for Applied Genomics, and the Cooperative Health Tissue Network.

Author Contributions

Conceived and designed the experiments: CAG AH BN RR LEA. Performed the experiments: CAG AH JP AS HS DS JS CG EH SL. Analyzed the data: CAG AH JP JY SL LEA. Contributed reagents/materials/analysis tools: KP NL MM JH MK TV MJ PD CG JD JW BW BN RR LEA CJG. Contributed to the writing of the manuscript: CAG AH LEA.

References

1. da Cunha JP, Galante PA, de Souza JE, de Souza RF, Carvalho PM, et al. (2009) Bioinformatics construction of the human cell surfaceome. Proc Natl Acad Sci U S A 106: 16752–16757.
2. Gygi SP, Rochon Y, Franza BR, Aebersold R (1999) Correlation between protein and mRNA abundance in yeast. Mol Cell Biol 19: 1720–1730.
3. Chen G, Gharib TG, Huang CC, Taylor JM, Misek DE, et al. (2002) Discordant protein and mRNA expression in lung adenocarcinomas. Mol Cell Proteomics 1: 304–313.
4. Wollscheid B, Bausch-Fluck D, Henderson C, O'Brien R, Bibel M, et al. (2009) Mass-spectrometric identification and relative quantification of N-linked cell surface glycoproteins. Nat Biotechnol 27: 378–386.
5. Kim Y, Elschenbroich S, Sharma P, Sepiashvili L, Gramolini AO, et al. (2011) Use of colloidal silica-beads for the isolation of cell-surface proteins for mass spectrometry-based proteomics. Methods Mol Biol 748: 227–241.
6. Wu CC, MacCoss MJ, Howell KE, Yates JR 3rd (2003) A method for the comprehensive proteomic analysis of membrane proteins. Nat Biotechnol 21: 532–538.
7. Streitz M, Miloud T, Kapinsky M, Reed MR, Magari R, et al. (2013) Standardization of whole blood immune phenotype monitoring for clinical trials: panels and methods from the ONE study. Transplant Res 2: 17.
8. Sewell WA, Smith SA (2011) Polychromatic flow cytometry in the clinical laboratory. Pathology 43: 580–591.
9. Virgo PF, Gibbs GJ (2012) Flow cytometry in clinical pathology. Ann Clin Biochem 49: 17–28.
10. Dubois NC, Craft AM, Sharma P, Elliott DA, Stanley EG, et al. (2011) SIRPA is a specific cell-surface marker for isolating cardiomyocytes derived from human pluripotent stem cells. Nat Biotechnol 29: 1011–1018.
11. Warner JK, Wang JC, Takenaka K, Doulatov S, McKenzie JL, et al. (2005) Direct evidence for cooperating genetic events in the leukemic transformation of normal human hematopoietic cells. Leukemia 19: 1794–1805.
12. Masters JR, Thomson JA, Daly-Burns B, Reid YA, Dirks WG, et al. (2001) Short tandem repeat profiling provides an international reference standard for human cell lines. Proc Natl Acad Sci U S A 98: 8012–8017.
13. Schadt EE (2009) Molecular networks as sensors and drivers of common human diseases. Nature 461: 218–223.
14. Golub TR, Slonim DK, Tamayo P, Huard C, Gaasenbeek M, et al. (1999) Molecular classification of cancer: class discovery and class prediction by gene expression monitoring. Science 286: 531–537.
15. van' t Veer LJ, Dai H, van de Vijver MJ, He YD, Hart AA, et al. (2002) Gene expression profiling predicts clinical outcome of breast cancer. Nature 415: 530–536.
16. Ramaswamy S, Tamayo P, Rifkin R, Mukherjee S, Yeang CH, et al. (2001) Multiclass cancer diagnosis using tumor gene expression signatures. Proc Natl Acad Sci U S A 98: 15149–15154.
17. Goodpaster T, Legesse-Miller A, Hameed MR, Aisner SC, Randolph-Habecker J, et al. (2008) An immunohistochemical method for identifying fibroblasts in formalin-fixed, paraffin-embedded tissue. J Histochem Cytochem 56: 347–358.
18. Nery AA, Nascimento IC, Glaser T, Bassaneze V, Krieger JE, et al. (2013) Human mesenchymal stem cells: from immunophenotyping by flow cytometry to clinical applications. Cytometry A 83: 48–61.
19. McLean K, Gong Y, Choi Y, Deng N, Yang K, et al. (2011) Human ovarian carcinoma-associated mesenchymal stem cells regulate cancer stem cells and tumorigenesis via altered BMP production. J Clin Invest 121: 3206–3219.
20. True LD, Zhang H, Ye M, Huang CY, Nelson PS, et al. (2010) CD90/THY1 is overexpressed in prostate cancer-associated fibroblasts and could serve as a cancer biomarker. Mod Pathol 23: 1346–1356.
21. Zhao H, Peehl DM (2009) Tumor-promoting phenotype of CD90hi prostate cancer-associated fibroblasts. Prostate 69: 991–1000.
22. Ahmed AA, Etemadmoghadam D, Temple J, Lynch AG, Riad M, et al. (2010) Driver mutations in TP53 are ubiquitous in high grade serous carcinoma of the ovary. J Pathol 221: 49–56.
23. Yemelyanova A, Vang R, Kshirsagar M, Lu D, Marks MA, et al. (2011) Immunohistochemical staining patterns of p53 can serve as a surrogate marker for TP53 mutations in ovarian carcinoma: an immunohistochemical and nucleotide sequencing analysis. Mod Pathol 24: 1248–1253.
24. Ornatsky O, Bandura D, Baranov V, Nitz M, Winnik MA, et al. (2010) Highly multiparametric analysis by mass cytometry. J Immunol Methods 361: 1–20.
25. Bendall SC, Simonds EF, Qiu P, Amir el AD, Krutzik PO, et al. (2011) Single-cell mass cytometry of differential immune and drug responses across a human hematopoietic continuum. Science 332: 687–696.
26. Krutzik PO, Clutter MR, Trejo A, Nolan GP (2011) Fluorescent cell barcoding for multiplex flow cytometry. Curr Protoc Cytom Chapter 6: Unit 6 31.
27. Qiu P, Simonds EF, Bendall SC, Gibbs KD Jr, Bruggner RV, et al. (2011) Extracting a cellular hierarchy from high-dimensional cytometry data with SPADE. Nat Biotechnol 29: 886–891.
28. Bendall SC, Davis KL, Amir el AD, Tadmor MD, Simonds EF, et al. (2014) Single-cell trajectory detection uncovers progression and regulatory coordination in human B cell development. Cell 157: 714–725.
29. Chicheportiche Y, Bourdon PR, Xu H, Hsu YM, Scott H, et al. (1997) TWEAK, a new secreted ligand in the tumor necrosis factor family that weakly induces apoptosis. J Biol Chem 272: 32401–32410.
30. Gedye C, Ailles L (2013) Isolation and characterization of cancer stem cells in vitro. Methods Mol Biol 946: 181–204.
31. Bayer-Garner IB, Sanderson RD, Dhodapkar MV, Owens RB, Wilson CS (2001) Syndecan-1 (CD138) immunoreactivity in bone marrow biopsies of multiple myeloma: shed syndecan-1 accumulates in fibrotic regions. Mod Pathol 14: 1052–1058.
32. Wijdenes J, Clement C, Klein B, Dore J-M (1997) New B-cell CD antigens. BC29: CD138 (syndecan-1) workshop panel report. In: Kishimoto T, Kutani H, von dem Borne A, Goyert S, Mason D et al., editors. Leucocyte Typing VI White Cell Differentiation Antigens. 6th ed. London: Garland Publishing Inc. 149–152.
33. Basten S, Willekers S, Vermaat J, Slaats G, Voest E, et al. (2013) Reduced cilia frequencies in human renal cell carcinomas versus neighboring parenchymal tissue. Cilia 2: 2.
34. Tun HW, Marlow LA, von Roemeling CA, Cooper SJ, Kreinest P, et al. (2010) Pathway signature and cellular differentiation in clear cell renal cell carcinoma. PLoS One 5: e10696.
35. Stewart JM, Shaw PA, Gedye C, Bernardini MQ, Neel BG, et al. (2011) Phenotypic heterogeneity and instability of human ovarian tumor-initiating cells. Proc Natl Acad Sci U S A 108: 6468–6473.

Characterization of the CD14^{++}CD16^{+} Monocyte Population in Human Bone Marrow

Manuela Mandl[1], Susanne Schmitz[1¤], Christian Weber[1,2], Michael Hristov[1*]

1 Institute for Cardiovascular Prevention, Ludwig-Maximilians-University (LMU), Munich, Germany, **2** Munich Heart Alliance, Munich, Germany

Abstract

Numerous studies have divided blood monocytes according to their expression of the surface markers CD14 and CD16 into following subsets: classical CD14^{++}CD16^{-}, intermediate CD14^{++}CD16^{+} and nonclassical CD14^{+}CD16^{++} monocytes. These subsets differ in phenotype and function and are further correlated to cardiovascular disease, inflammation and cancer. However, the CD14/CD16 nature of resident monocytes in human bone marrow remains largely unknown. In the present study, we identified a major population of CD14^{++}CD16^{+} monocytes by using cryopreserved bone marrow mononuclear cells from healthy donors. These cells express essential monocyte-related antigens and chemokine receptors such as CD11a, CD18, CD44, HLA-DR, Ccr2, Ccr5, Cx3cr1, Cxcr2 and Cxcr4. Notably, the expression of Ccr2 was inducible during culture. Furthermore, sorted CD14^{++}CD16^{+} bone marrow cells show typical macrophage morphology, phagocytic activity, angiogenic features and generation of intracellular oxygen species. Side-by-side comparison of the chemokine receptor profile with unpaired blood samples also demonstrated that these rather premature medullar monocytes mainly match the phenotype of intermediate and partially of (non)classical monocytes. Together, human monocytes obviously acquire their definitive CD14/CD16 signature in the bloodstream and the medullar monocytes probably transform into CD14^{++}CD16^{-} and CD14^{+}CD16^{++} subsets which appear enriched in the periphery.

Editor: Nathalie Signoret, University of York, United Kingdom

Funding: This work was supported by the Deutsche Forschungsgemeinschaft, (FOR809, HR18/1-1, SFB 1123 A1) August-Lenz-Stiftung. The funder had no role in study design, data collection and analysis, decision to publish, or preparation of the manuscript.

Competing Interests: The authors have declared that no competing interests exist.

* Email: michael.hristov@med.uni-muenchen.de

¤ Current address: Center of Allergy & Environment (ZAUM), Technical University of Munich, Munich, Germany

Introduction

Current knowledge defines three major monocyte subsets in human peripheral blood based on the expression patterns of CD16 (FcγRIII) and the LPS-receptor CD14: classical CD14^{++}CD16^{-}, intermediate CD14^{++}CD16^{+} and nonclassical CD14^{+}CD16^{++} monocytes [1]. These subsets differ essentially in their chemokine receptor expression, phagocytic activity and tissue distribution during inflammation or steady-state conditions [2]. Hence, the monocyte subsets are differentially involved in the pathophysiology of inflammation, atherosclerosis and regeneration after injury [3–5]. In particular, the unique features of the intermediate in contrast to the nonclassical subset (both previously referred as CD16^{+} monocytes) has become recently more evident and especially in cardiovascular disease this subset was shown to predict independently cardiovascular events at follow-up [6–8]. Thus, human monocyte subsets may represent a novel prognostic marker or therapeutic targets in clinical medicine and detailed investigation of their characteristics in bone marrow (BM) as compared to peripheral blood appears important for better understanding of the subset-specific development, maturation and functional specialization. However, the experimental research on resident human monocytes in BM together with the plasticity of subset development and trafficking to the periphery is highly restricted and only few published data exist. Therefore the aim of our study is to provide novel evidence on medullar CD14/CD16 monocytes by using commercially available cryopreserved BM samples from healthy donors as an alternative to fresh BM and to reveal further the effects of freezing/thawing on monocyte number, function and chemokine receptor expression.

Materials and Methods

Cells

Cryopreserved bone marrow mononuclear cells (BMCs; 25×10^{6} or 100×10^{6}) from ten healthy single donors were obtained by Lonza (Cologne, Germany). The BMCs were thawed and maintained overnight prior all further experiments in *Iscove's Modified Dulbecco's Medium* (IMDM; PAA, Pasching, Austria) supplemented with 10% FCS and antibiotics in 24-well plates (1×10^{6} cells per well). After written informed consent was obtained, 20 ml of citrate anticoagulated peripheral blood were collected by aseptic venipuncture from ten healthy volunteers. The procedure was approved by the Ethics Committee of the Medical Faculty of the LMU Munich. PBMCs were separated by *Biocoll* (Biochrom AG, Berlin, Germany) density gradient centrifugation. Some fresh PBMCs were immediately analyzed by flow cytometry as described below while the remaining cells were frozen in RPMI 1640 medium (PAA) with 10% FCS and 5% dimethyl sulfoxide

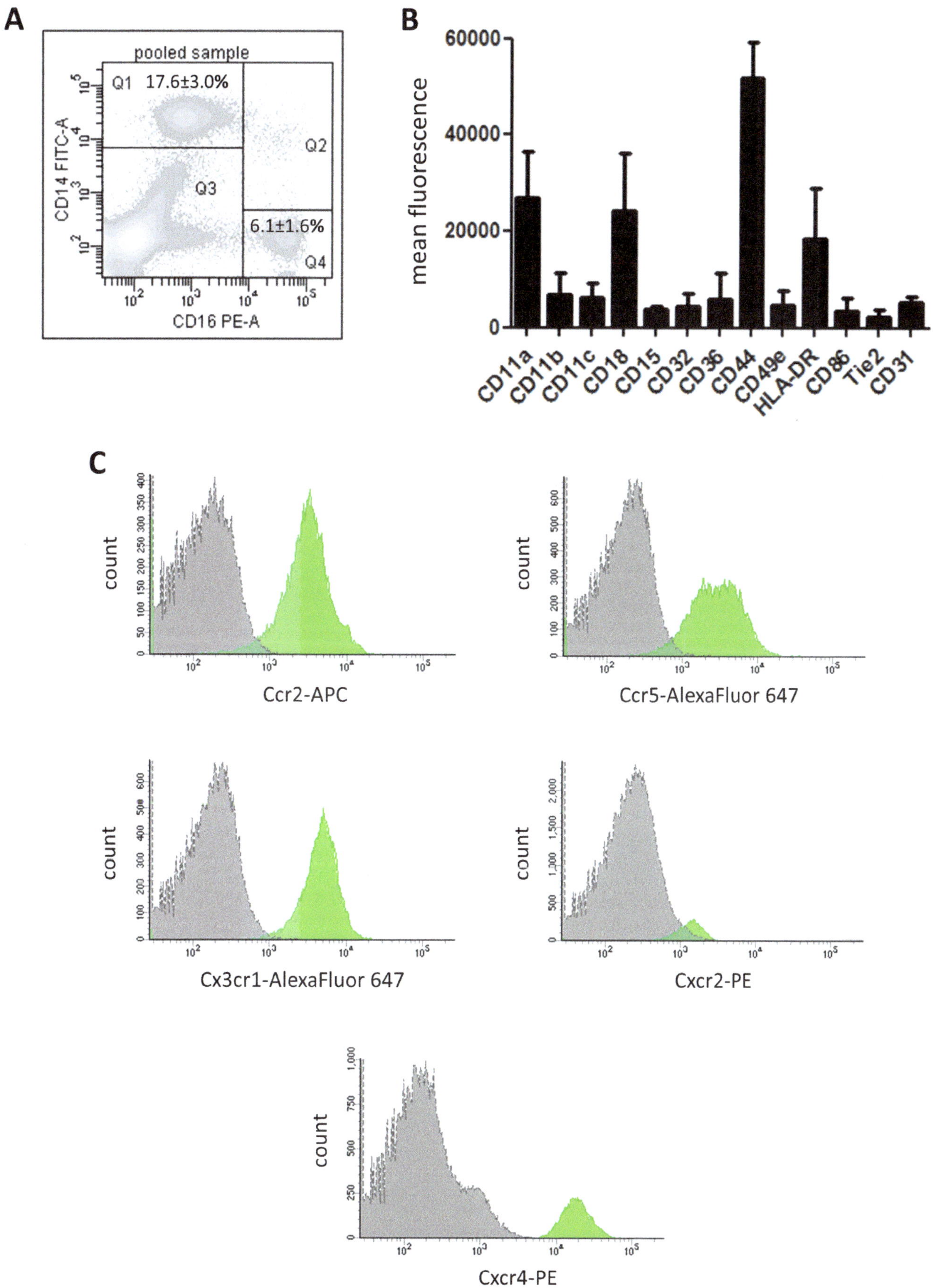
A
pooled sample
Q1 17.6±3.0%
Q2
Q3
6.1±1.6%
Q4
CD14 FITC-A
CD16 PE-A
B
mean fluorescence
60000
40000
20000
0
CD11a
CD11b
CD11c
CD18
CD15
CD32
CD36
CD44
CD49e
HLA-DR
CD86
Tie2
CD31
C
count
Ccr2-APC
Ccr5-AlexaFluor 647
Cx3cr1-AlexaFluor 647
Cxcr2-PE
Cxcr4-PE

Figure 1. Characterization of CD14^{++}CD16^{+} bone marrow monocytes by flow cytometry. (A) Representative contour plot of pooled BMC sample from three single donors showing a clear separation of two populations: CD14^{++}CD16^{+} (Q1) and CD14^{-}CD16^{++} (Q4). Gating was performed first in FSC/SSC dot plot following gating on CD45^{+} events in a SSC/CD45 dot plot. Finally, the expression of CD14 and CD16 was evaluated inside the CD45^{+} population. (B,C) Expression of monocyte-related antigens and representative histogram overlays with isotype control (gray filled) of crucial chemokine receptors on CD14^{++}CD16^{+} BMCs as analyzed by flow cytometry. Data are from 3 to 6 individual donors.

(DMSO; Sigma-Aldrich, Taufkirchen, Germany). The cells were cooled to −80°C in a *CoolCell Freezing Container* (BioCision, Mill Valley, CA) before being transferred to liquid nitrogen for storage. The cryopreserved PBMCs were thawed/cultured after one week of storage in the same fashion as the BMCs.

Flow cytometry analysis of surface antigens

Cells were collected by thorough resuspension and the flow cytometry analysis was performed as 4-color experiment by using the *Lyoplate Screening Panel* (BD, Heidelberg, Germany) following staining with anti-human CD45-PerCP (clone 2D1; BD), CD14-FITC (clone HCD14) and CD16-PE (clone 3G8) mAbs (both Biolegend, Fell, Germany). The control tube included anti-human CD45/CD14/CD16 mAbs and the AlexaFluor 647-conjugated secondary Ab of the screening panel. Further anti-human mAbs used for multicolor flow cytometry were as follows: CD117-APC (clone 104D2), CD16-Pacific Blue (clone 3G8), Ccr2-APC mouse IgG2a (clone K036C2, 100 µg/ml), Ccr5-AlexaFluor 647 rat IgG2a (clone HEK/1/85a, 100 µg/ml), Cx3cr1-AlexaFluor 647 rat IgG2b (clone 2A9-1, 100 µg/ml), Cxcr2-PE mouse IgG1 (clone 5E8, 50 µg/ml), Cxcr4-PE mouse IgG2a (clone 12G5, 50 µg/ml; all Biolegend); CD15-eFluor450 (clone HI98) and CD56-APC (clone MEM188; both eBioscience, Vienna, Austria); Tie2-APC (clone 83715; R&D Systems, Wiesbaden, Germany). Matching fluorescent isotype control Abs (clones MOPC-21/173 and RTK2758/4530; Biolegend) were used at same concentration as the respective mAb. Sample tubes were acquired on a FACSCanto II with Diva v8.0 software (BD) and appropriate CS&T bead calibration setting. The fluorescence compensation was automatically calculated using *OneComp eBeads* (eBioscience). At least 10.000 events were recorded within the target population. The specific mean fluorescence intensity (MFI) was assessed by subtracting the background of the respective isotype control. To determine chemokine receptor expression over time some BMCs were cultured in RPMI 1640 medium with 10% FCS and antibiotics on 6-well plates (4×10^6 cells per well). The flow cytometry analysis for expression of CD45, CD14, CD16, Ccr2, Ccr5 and Cx3cr1 was performed after one and seven days of culture as described above.

Functional assays

Single-donor BMCs (10^7) or pooled BMCs (75×10^6) of three individual donors were sorted (FACSAria III; BD) after staining for CD14/CD16/CD45 and plated on 24-well plates or 4-well chamber slides in RPMI 1640 medium with 10% FCS and antibiotics. Phagocytosis of FITC-conjugated latex beads (1:1000; Sigma-Aldrich) and hydroethidine (10 µg/ml; Invitrogen, Darmstadt, Germany) staining for spontaneous intracellular superoxide production in PBMC- or sorted BMC-derived macrophages were determined by flow cytometry after 24 hours in culture. Cytoskeletal filaments in cultured BMCs were visualized by fluorescence microscopy (Leica DM6000B, Wetzlar, Germany) after cell fixation/permeabilization (*Cytofix/Cytoperm*; BD) following incubation with anti-α-tubulin-FITC (Sigma-Aldrich) mAb and rhodamine-phalloidin (Biotium, Hayward, CA). *Vectashield* with DAPI (Vector Laboratories, Peterborough, UK) was used as mounting medium and to stain cell nuclei. In another experimental setup the sorted BMCs were cultured in angiogenesis µ-slide (IBIDI, Martinsried, Germany) on *Matrigel* (BD) supplemented with 50 ng/ml VEGF and 100 ng/ml Cxcl12 (both from Peprotech, Hamburg, Germany).

Statistics

Data were analyzed using *Prism 5* software (GraphPad Inc., La Jolla, CA) and presented as mean±SD. Data distribution was assessed by the *Shapiro-Wilk* normality test. Accordingly, we have used for comparison between two groups the unpaired *t test* or nonparametric *Mann-Whitney* test and for comparison between three groups one-way ANOVA with *Newman-Keuls* post-test. Differences with $p<0.05$ were considered statistically significant.

Results

Flow cytometry analysis of thawed human BMCs after staining for CD14, CD16 and CD45 showed clear separation of two scatter populations (Fig. 1A): a larger CD14^{++} population (Q1: 17.6±3.0% of CD45^{+} cells) with low-to-intermediate expression of CD16 (CD16^{+}) and a smaller CD14^{-}CD16^{++} population (Q4: 6.1±1.6% of CD45^{+} cells). The CD14^{-}CD16^{++} cells in Q4 were also CD15^{-}CD56^{++} thus probably referring to NK and NK-T cells (data not shown).

Further analysis of the CD14^{++}CD16^{+} population revealed expression of CD11a, CD18, CD44 and HLA-DR together with other essential monocyte-related antigens (Fig. 1B;S1). There was also expression of the chemokine receptors Ccr2, Ccr5, Cx3cr1, Cxcr2 and Cxcr4 as shown in Fig. 1C. Moreover, the expression of Ccr2 on medullar macrophages was significantly up-regulated after 7 days in culture (Table 1).

Table 1. Chemokine receptor expression on cultured medullar macrophages from six single donors as analyzed by flow cytometry.

Receptor	1 day	7 days	*p-value*
Ccr2	97±29%	351±192%	*0.04*
Ccr5	152±31%	340±136%	*0.06*
Cx3cr1	113±41%	328±159%	*0.06*

The MFI values of CD14^{++}CD16^{+} monocytes before plating served as control (=100%).

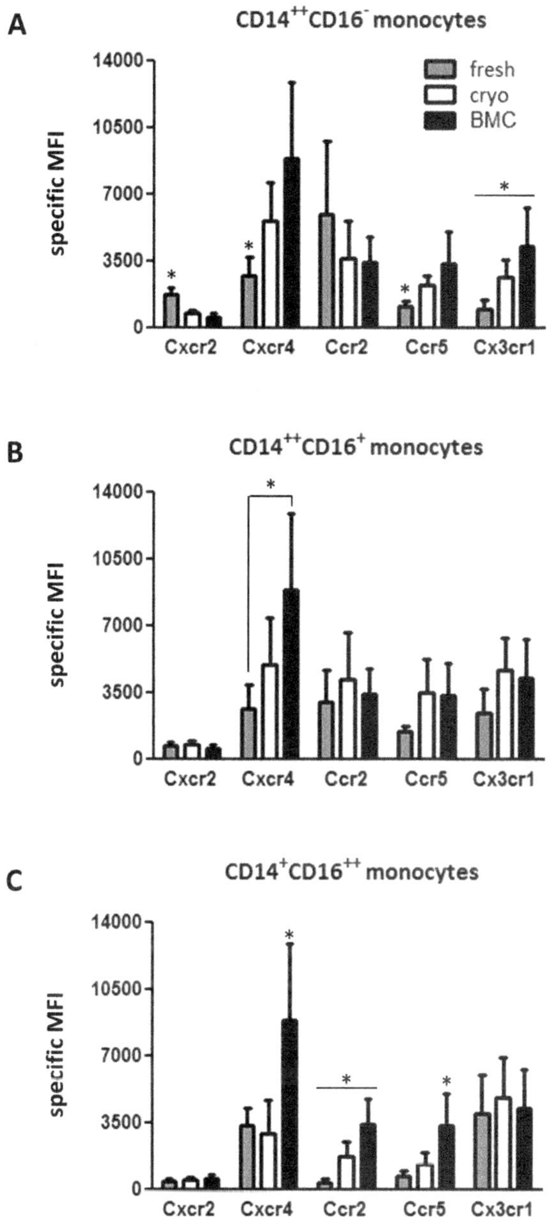

Figure 2. Side-by-side comparison of chemokine receptor expression on medullar with frozen or fresh blood monocytes. PBMCs from different donors were separated by density gradient centrifugation. Chemokine receptor expression was analyzed and compared on fresh monocyte subsets (gray bars) versus cryopreserved cells of the same donor (open bars) or unpaired medullar monocytes (black bars). $^{*}p<0.05$, $n=6$–10.

Side-by-side examination of chemokine receptor expression on $CD14^{++}CD16^{+}$ BM monocytes versus unpaired cryopreserved blood samples showed very similar levels for Ccr2, Ccr5 and Cx3cr1 as compared to intermediate monocytes together with almost identical levels of Ccr2 on classical and Cx3cr1 on nonclassical monocytes (Fig. 2A–C; Table S1). Further comparison with unpaired fresh blood monocytes revealed identical expression of Cx3cr1 as to the nonclassical subset (Fig. 2C). Overall, the BM monocytes showed highest levels for Cxcr4 (Fig. 2A–C) and more than 20× higher expression of the hematopoietic precursor marker CD117 (data not shown) whereas fresh classical monocytes had highest levels of Cxcr2 and Ccr2 (Fig. 2A–C). Another analysis of paired fresh versus cryopreserved blood monocytes demonstrated significant decrease of Cxcr2 next to increased expression of Cxcr4, Ccr5 and Cx3cr1 on classical monocytes while thawed nonclassical monocytes had higher expression only of Ccr2 as compared to freshly processed cells (Fig. 2A,C). Finally, cryopreservation failed to affect the percentages of all blood monocyte subsets (Table 2).

Functional analysis demonstrated phagocytic activity by 58.8±19.9% of sorted $CD14^{++}CD16^{+}$ BMCs as well as spontaneous generation of reactive oxygen species (ROS) at 61.5±8.9% after 24 hours in culture (Fig. 3A,D). By comparison, macrophages originating from frozen PBMCs showed phagocytic capacity of 51.7±9.5% and spontaneous superoxide production at 41.9±7.5% (Fig. 3B,E) while freshly isolated macrophages revealed 70.7±8.1% of bead uptake and 18.0±2.3% of ROS production after 24 hours of culture (Fig. 3C,F). Microscopic immunofluorescence analysis of cultured $CD14^{++}CD16^{+}$ BMCs further confirmed typical *fried egg-like* macrophage morphology with α-tubulin/F-actin-rich protrusions (Fig. 3G,H). The $CD14^{++}CD16^{+}$ BMCs displayed also angiogenic properties as they express CD31, Tie2 and Cxcr4, and fuse to clusters with shape change in Matrigel (Fig. 1B,C;3I).

Discussion

To our best knowledge, this is the first study characterizing CD14/CD16 monocyte subsets in cryopreserved human BM of healthy donors as a possible alternative to freshly prepared cells. We found a major pool of $CD14^{++}CD16^{+}$ BMCs that seems to correspond to intermediate blood monocytes. These cells clearly feature monocytes/macrophages as illustrated by their phenotype and function. Another important analysis mapped the effect of freezing/thawing on chemokine receptor expression, subset proportion and function of bloodstream monocytes. A limitation of our study is the lack of direct validation and comparison with fresh marrow samples, ideally obtained from the same donor. However, in contrast to venous blood it is not conceivable to collect routinely fresh BM of healthy volunteers for experimental research. Consequently, only one prior work has evaluated the immunophenotype of monocyte subsets in BM aspirates originat-

Table 2. Percentages of blood monocyte subsets before and after freezing/thawing ($n=7$–10).

Subset	fresh	thawed	*p-value*
$CD14^{++}CD16^{-}$	80.1±7.0%	82.8±3.0%	*0.54*
$CD14^{++}CD16^{+}$	3.7±2.0%	3.8±1.5%	*0.96*
$CD14^{+}CD16^{++}$	6.2±2.8%	6.3±1.7%	*0.93*

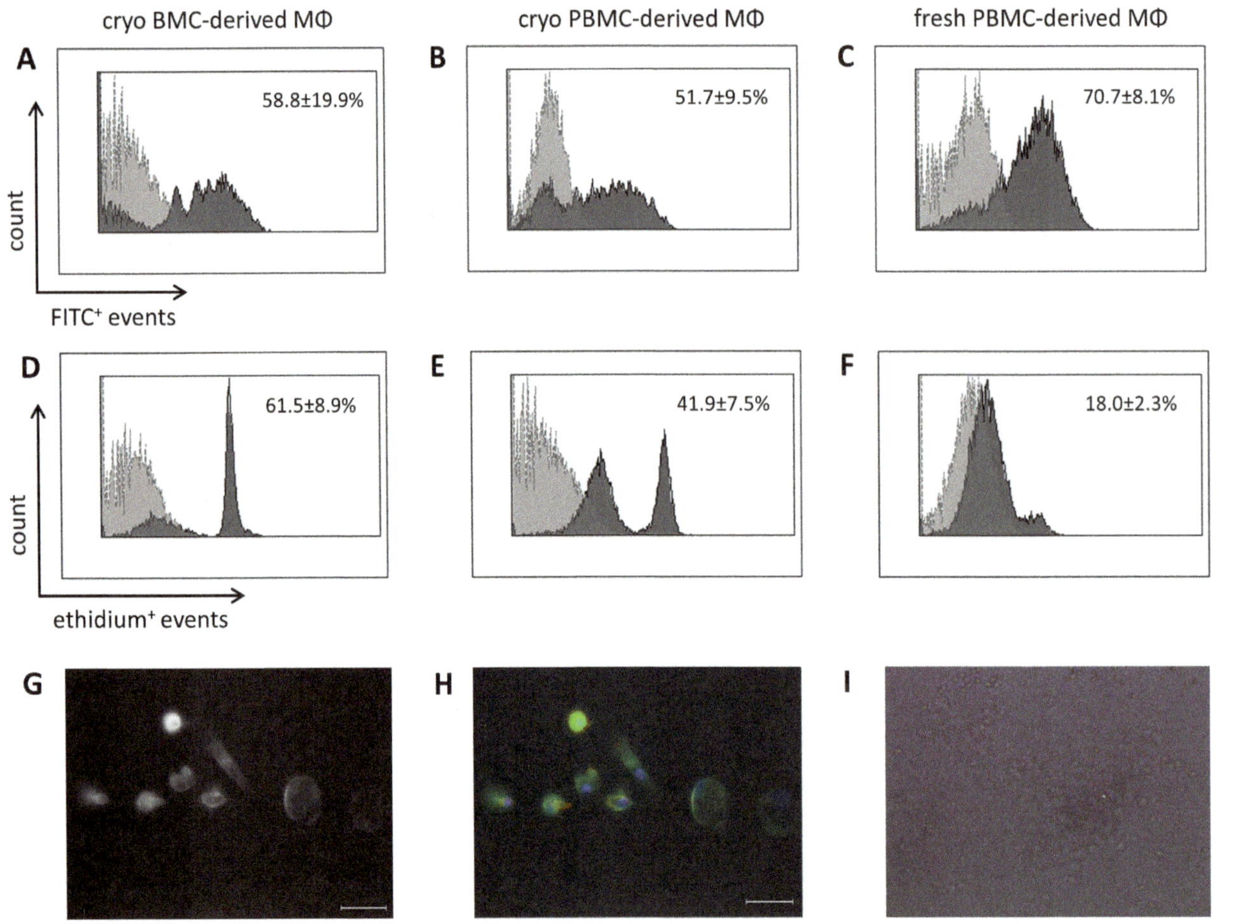

Figure 3. Functional analyses of CD14^{++}CD16^{+} BMC- and PBMC-derived macrophages. Representative flow cytometry histogram overlays for uptake of FITC-conjugated beads (A–C) and hydroethidine staining (D–F) in macrophages cultured from CD14^{++}CD16^{+} BMCs, fresh or cryopreserved PBMCs of three single donors. The gray filled histogram shows cells without beads or dye, respectively. (G,H) Phase contrast and immunofluorescent overlay of CD14^{++}CD16^{+} BMC-derived macrophages which were cultured in chamber slides and stained for α-tubulin-FITC/rhodamine-phalloidin after 7 days as described in methods. Cell nuclei were counterstained with DAPI. The scale bar indicates 40 μm. (I) Cluster formation and shape change of CD14^{++}CD16^{+} BM cells on Matrigel after 72 hours. Data in (G–I) are representative images of triplicates from pooled sample of three independent donors.

ing from healthy individuals with suspected peripheral lymphomas up to now [9].

Compared to humans, the recruitment of mouse monocyte subsets *in vivo* was investigated in detail by using several models of receptor knock-out, BM transplantation, cell depletion or adoptive transfer [10–12]. Up to now, there are two major murine monocyte subsets: the Gr1^{+}/Ly6C^{hi} cells are similar to human CD14^{++}CD16^{-} monocytes and Gr1^{-}/Ly6C^{low} cells are considered as counterparts of the CD14^{+}CD16^{++} monocytes [3,10,11]. During egress from BM, the CC-chemokine Ccl2 (MCP-1) plays a central role by either inducing Ccr2-dependent chemokinesis or chemotaxis of Ccr2^{+}Ly6C^{hi} mouse monocytes [11]. Our data demonstrated comparable expression levels of Ccr2 between medullar and thawed CD14^{++} blood monocytes. Remarkably, we found higher amount of Cxcr4 on BM monocytes in comparison with frozen or fresh bloodstream monocytes. As human CD14^{++}CD16^{+} BMCs express also the adhesion molecule CD44 they may be accessorily attracted by ligands for CD44 and Cxcr4 such as glycosaminoglycans and Cxcl12 to move towards the vascular niche of the BM and to enter the bloodstream. According to their origin BM monocytes still express the precursor marker CD117 which was absent on their terminally differentiated counterparts in the bloodstream. Further analysis of chemokine receptors revealed very similar expression profile between medullar and frozen intermediate monocytes. Fresh classical monocytes showed highest levels of Ccr2 while the expression of Cx3cr1 was similar between medullar and CD16^{++} blood subsets (fresh or frozen). Of note, the expression of Ccr2 on BMC-derived macrophages was inducible in culture. Thus, human monocytes probably acquire their differential chemokine receptor signature (e.g. high levels of Ccr2 on the classical subset) during or after recruitment to the periphery with simultaneous decline of Cxcr4 and CD117 in terms of differentiation. It is well conceivable that such maturation may occur more rapid *in vivo* than *in vitro*, especially during inflammatory conditions or infection with accompanying monocytosis and preferential shift of classical monocytes.

Our results next reveal higher expression of some chemokine receptors on blood monocytes shortly after thawing while the ratio of receptor expression together with subset frequencies remains unaffected as compared to fresh monocytes. Accordingly, cryopreservation may affect at least transiently the corresponding chemokine receptors on BMCs as well. Furthermore, macrophages that originated from frozen blood or BM monocytes showed similar phagocytic capacity but had significantly increased ROS production as compared to freshly obtained blood macrophages. One possible explanation for this difference could be transient cell activation following cryopreservation. In line with our findings previous studies have already reported that cryopreservation can rather influence activation status than frequency or function of mononuclear cells. In particular, human T-cells, monocytes or circulating angiogenic cells can be thawed without considerable alteration of their phenotype and function [13–15]. Cryopreservation is also routinely used for storage of autologous $CD34^{+}$ cells suggesting that cryopreserved blood cells usually match fine fresh cells [16]. Hereby, even if the cryopreserved cells in our study were not fully equal in phenotype and ROS generation to fresh cells (at least shortly after thawing) they may represent another alternative in terms of availability, storage, transportation and functionality.

Surprisingly, we also found that nonclassical monocytes do not certainly persist as population with clear CD14/CD16 boundaries among thawed BMCs. As mentioned above it is rather unlike that cryopreservation *per se* could impact the abundance of nonclassical monocytes since their proportion remains well-preserved after freezing/thawing as confirmed also by others [17]. As further shown in a previous study, the nonclassical monocytes in fresh marrow samples were not primary identified as separate $CD14^{+}CD16^{++}$ scatter population but first on the basis of their lower Ccr2 expression among gated $CD16^{+}$ events [9]. Hence, the proper detection of nonclassical monocytes in BM appears debatable and may indeed result from "contamination" with variable amounts of peripheral blood during the BM aspiration.

In contrast to the $Ccr2^{++}$ classical and intermediate subsets in the periphery, the nonclassical monocytes are $Cx3cr1^{++}$ but dimly express Ccr2 and function mainly as weak phagocytes that patrol the microvasculature in LFA-1 dependent manner to sense pathogens or cell remnants at steady-state [18]. Moreover, nonclassical monocytes express the angiopoietin receptor Tie2 and are further considered as essential paracrine players in tumor angiogenesis [5,19]. Similar pro-angiogenic characteristics were recently described for the intermediate subset as well [8]. Since the $CD14^{++}CD16^{+}$ BMCs express similar levels of Cx3cr1 and display pro-angiogenic properties, it seems that nonclassical monocytes could arise from this mixed population as well but specialize first in the periphery after coming across pathogens at homeostatic conditions, *e.g.* during patrolling the microvasculature of gut and lung.

In summary, a common $CD14^{++}CD16^{+}$ monocyte pool featuring mostly intermediate monocytes is primary enriched in human BM but depleted in peripheral blood where this subset exists at lowest percentage. The medullar monocytes seems to require Ccr2, Cxcr4 and possibly CD44 for mobilization and may give rise to more specialized peripheral $Ccr2^{++}$ classical monocytes with implication during inflammation next to $Cx3cr1^{++}$ nonclassical monocytes that are rather recruited at steady-state. The maturation of medullar monocytes obviously associates with decrease in the expression of Cxcr4 and c-kit. Thus, it is conceivable that human monocytes acquire their differential CD14/CD16 signature first in the bloodstream.

Supporting Information

Figure S1 Representative histogram overlays for surface markers of the BD Lyoplate screening panel.

Table S1 Specific MFI values of chemokine receptor expression on medullar (n = 6) and blood monocyte subsets (n = 10).

Author Contributions

Conceived and designed the experiments: MH. Performed the experiments: MM SS MH. Analyzed the data: MM SS MH. Contributed reagents/materials/analysis tools: MH CW. Wrote the paper: MH MM.

References

1. Ziegler-Heitbrock L, Ancuta P, Crowe S, Dalod M, Grau V, et al. (2010) Nomenclature of monocytes and dendritic cells in blood. Blood 116: e74–80.
2. Wong KL, Tai JJ, Wong WC, Han H, Sem X, et al. (2011) Gene expression profiling reveals the defining features of the classical, intermediate, and nonclassical human monocyte subsets. Blood 118: e16–31.
3. Hristov M, Weber C (2011) Differential role of monocyte subsets in atherosclerosis. Thromb Haemost 106: 757–62.
4. Rogacev KS, Ulrich C, Blömer L, Hornof F, Oster K, et al. (2010) Monocyte heterogeneity in obesity and subclinical atherosclerosis. Eur Heart J 31: 369–76.
5. De Palma M, Murdoch C, Venneri MA, Naldini L, Lewis CE (2007) Tie2-expressing monocytes: regulation of tumor angiogenesis and therapeutic implications. Trends Immunol 28: 519–24.
6. Rogacev KS, Cremers B, Zawada AM, Seiler S, Binder N, et al. (2012) CD14++ CD16+ monocytes independently predict cardiovascular events: a cohort study of 951 patients referred for elective coronary angiography. J Am Coll Cardiol 60: 1512–20.
7. Tapp LD, Shantsila E, Wrigley BJ, Pamukcu B, Lip GY (2012) The CD14++ CD16+ monocyte subset and monocyte-platelet interactions in patients with ST-elevation myocardial infarction. J Thromb Haemost 10: 1231–41.
8. Zawada AM, Rogacev KS, Rotter B, Winter P, Marell RR, et al. (2011) SuperSAGE evidence for CD14++CD16+ monocytes as a third monocyte subset. Blood 118: e50–61.
9. Shantsila E, Wrigley B, Tapp L, Apostolakis S, Montoro-Garcia S, et al. (2011) Immunophenotypic characterization of human monocyte subsets: possible implications for cardiovascular disease pathophysiology. J Thromb Haemost 9: 1056–66.
10. Weber C, Zernecke A, Libby P (2008) The multifaceted contributions of leukocyte subsets to atherosclerosis: lessons from mouse models. Nat Rev Immunol 8: 802–15.
11. Shi C, Pamer EG (2011) Monocyte recruitment during infection and inflammation. Nat Rev Immunol 11: 762–74.
12. Soehnlein O, Drechsler M, Döring Y, Lievens D, Hartwig H, et al. (2013) Distinct functions of chemokine receptor axes in the atherogenic mobilization and recruitment of classical monocytes. EMBO Mol Med 5: 471–81.
13. Van Hemelen D, Oude Elberink JN, Heimweg J, van Oosterhout AJ, Nawijn MC (2010) Cryopreservation does not alter the frequency of regulatory T cells in peripheral blood mononuclear cells. J Immunol Methods 353: 138–140.
14. Hiebl B, Fuhrmann R, Franke RP (2008) Characterization of cryopreserved CD14+-human monocytes after differentiation towards macrophages and stimulation with VEGF-A(165). Clin Hemorheol Microcirc 39: 221–8.
15. Sofrenovic T, McEwan K, Crowe S, Marier J, Davies R, et al. (2012) Circulating angiogenic cells can be derived from cryopreserved peripheral blood mononuclear cells. PLoS One 7: e48067.
16. Bakken AM (2006) Cryopreserving human peripheral blood progenitor cells. Curr Stem Cell Res Ther 1: 47–54.
17. Berg KE, Ljungcrantz I, Andersson L, Bryngelsson C, Hedblad B, et al. (2012) Elevated CD14++CD16− monocytes predict cardiovascular events. Circ Cardiovasc Genet 5: 122–31.
18. Cros J, Cagnard N, Woollard K, Patey N, Zhang SY, et al. (2010) Human CD14dim monocytes patrol and sense nucleic acids and viruses via TLR7 and TLR8 receptors. Immunity 33: 375–86.
19. Venneri MA, De Palma M, Ponzoni M, Pucci F, Scielzo C, et al. (2007) Identification of proangiogenic TIE2-expressing monocytes (TEMs) in human peripheral blood and cancer. Blood 109: 5276–85.

21

Residual Antibiotics in Decontaminated Human Cardiovascular Tissues Intended for Transplantation and Risk of Falsely Negative Microbiological Analyses

Marina Buzzi[1,9], Anna Guarino[2,9], Claudio Gatto[3], Sabrina Manara[1], Luca Dainese[2], Gianluca Polvani[2], Jana D'Amato Tóthová[3]*

1 Cardiovascular Tissue Bank of Emilia-Romagna, Azienda Ospedaliero-Universitaria Policlinico S. Orsola-Malpighi, Bologna, Italy, **2** Cardiovascular Tissue Bank of Lombardia, Centro Cardiologico Monzino, Milan, Italy, **3** Research and Development department, AL.CHI.MI.A. S.r.l., Ponte San Nicolò, Italy

Abstract

We investigated the presence of antibiotics in cryopreserved cardiovascular tissues and cryopreservation media, after tissue decontamination with antibiotic cocktails, and the impact of antibiotic residues on standard tissue bank microbiological analyses. Sixteen cardiovascular tissues were decontaminated with bank-prepared cocktails and cryopreserved by two different tissue banks according to their standard operating procedures. Before and after decontamination, samples underwent microbiological analysis by standard tissue bank methods. Cryopreserved samples were tested again with and without the removal of antibiotic residues using a RESEP tube, after thawing. Presence of antibiotics in tissue homogenates and processing liquids was determined by a modified agar diffusion test. All cryopreserved tissue homogenates and cryopreservation media induced important inhibition zones on both *Staphylococcus aureus*- and *Pseudomonas aeruginosa*-seeded plates, immediately after thawing and at the end of the sterility test. The RESEP tube treatment markedly reduced or totally eliminated the antimicrobial activity of tested tissues and media. Based on standard tissue bank analysis, 50% of tissues were found positive for bacteria and/or fungi, before decontamination and 2 out of 16 tested samples (13%) still contained microorganisms after decontamination. After thawing, none of the 16 cryopreserved samples resulted positive with direct inoculum method. When the same samples were tested after removal of antibiotic residues, 8 out of 16 (50%) were contaminated. Antibiotic residues present in tissue allografts and processing liquids after decontamination may mask microbial contamination during microbiological analysis performed with standard tissue bank methods, thus resulting in false negatives.

Editor: Ferruccio Bonino, University of Pisa, Italy

Funding: AL.CHI.MI.A. S.r.l., provided support in the form of salaries for authors C. Gatto and J. D'Amato Tothova, but did not have any additional role in the study design, data collection and analysis, decision to publish, or preparation of the manuscript. The specific roles of these authors are articulated in the 'author contributions' section.

Competing Interests: The authors C. Gatto and J. D'Amato Tothova are employed by AL.CHI.MI.A. S.r.l., whose company manufactures products discussed in the present paper. There are no further patents, products in development or marketed products to declare.

* Email: jtothova@alchilife.com

9 These authors contributed equally to this work.

Introduction

Tissue banks, including cardiovascular tissue banks, procure, process, store, and distribute human donor tissues intended for transplantation. The main clinical applications of heart valve allografts include valve replacement in patients with congenital or acquired heart disease and substitution of infected heart valve prosthesis [1,2]. Vascular allografts are used for vascular reconstruction, revascularization in absence of autogenic veins in patients with idiopathic or diabetic vasculopathies, and replacement of infected vascular prosthesis [1,3].

Decontamination is a critical phase of the processing of human tissues intended for transplantation. Tissue banks worldwide have established a wide variety of homemade antibiotic cocktails and procedures to eliminate microbial contaminants from tissues to maximize the safety of allografts [2,4–6]. Tissue decontamination may result in the presence of antibiotic residues in both decontaminated tissues and processing liquids [7].

Residual antibiotic concentrations may induce bacteriostasis of microorganisms eventually present on tissue samples and lead to false negative results during microbiological analysis [8]. According to the European and US Pharmacopoeias [9,10], the sterility test must be preceded by complete removal of any possible interfering antimicrobial agents.

No specific standards for microbiological analysis in tissue banking are available. Consequently, each tissue bank shall validate the method for tissue and liquid sample microbiological analysis, which is essential to determine whether the tissue can be released for transplantation [11–13]. The microbiological analysis is usually performed by means of automated blood culture systems or direct inoculum of growth media [13,14]. In general, these systems do not consider the presence of high antibiotic residues in tissues and processing liquids after decontamination, which is a possible cause of falsely negative microbiological results. This problem is further underestimated by the general assumption that antibiotics are rapidly degraded during the sustained tissue incubations necessary for sterility assessment.

To investigate the impact of antibiotic residues on microbiological analysis, we used human cardiovascular tissues, decontaminated with different antibiotic cocktails, two tissue bank microbiological methods and the RESEP tube method, which has been previously validated for the elimination of antibiotics from tissues and processing liquids, decontaminated with an industrial cocktail. All cryopreserved samples tested for sterility were also analyzed for the presence of antibiotic activity by a modified agar diffusion method.

Materials and Methods

Ethics statement

Donor tissues were collected and processed in compliance with the European Directives on setting quality standards and technical requirements for the processing of human tissues and cells. Only tissues unsuitable for transplantation were used for the study, and the permission was granted by the institutional review board of the Cardiovascular Tissue Bank of Emilia Romagna (CTBER) and the Cardiovascular Tissue Bank of Lombardia (CTBL). All clinical investigation has been conducted according to the principles expressed in the Declaration of Helsinki.

Tissues procurement and processing

Sixteen human cardiovascular tissues (5 aortic heart valves and 11 blood vessels graded as unsuitable for transplantation due to anatomical defects) were included in the study. Six and 10 tissues were collected and processed by CTBER and CTBL, respectively.

Tissues were transported to the tissue banks at 4°C in antibiotic-free transport media, Celsior (Genzyme Europe BV, Netherlands; CTBER) or Eurocollins (S.A.L.F. S.p.A., Italy; CTBL), and processed following tissue bank standard operating procedures. In particular, tissue were decontaminated with tissue bank-prepared antibiotic cocktails, containing cefoxitin, lincomycin, colimycin, vancomycin in RPMI1640 (cocktail 1, CTBER) or cefotaxime, lincomycin, vancomycin, polymyxin B sulfate in RPMI1640 (cocktail 2, CTBL) at 4°C for 24 h (valves) and 72 h (vessels). Tissues were then rinsed twice for 1 min with NaCl 0.9% (only for CTBER) and finally cryopreserved in RPMI 1640 (Sigma-Aldrich, St. Louis, MO, USA) containing 10% DMSO (Wak-Chemie, Germany), using programmable freezers with freezing rate of −1°C/min. The CTBER cryopreservation solution also contained 2% human albumin (Kedrion Biopharmaceuticals, Italy). Cryopreserved samples were then transported in dry ice to the R&D department of AL.CHI.MI.A. S.r.l. for further microbiological analysis and detection of antibiotic residues.

Microbiological analysis

Microbiological analysis of samples collected from processing liquids and tissues, before and after decontamination, was performed by the tissue banks, according to their standard operating procedures. At CTBER, 5 ml of the transport and cryopreservation solutions were introduced in the BacT/ALERT vials to detect aerobic and anaerobic bacteria and incubated at 37°C in the BacT/ALERT 3D automatic detection system (Biomérieux, France) for 5 days. Tissue samples were immersed in fluid thioglycollate medium (FTM) and tryptic soy broth (TSB) and incubated at 35°C for 7 days (direct inoculum method). At CTBL, 1 ml of transport and cryopreservation solutions and tissue samples were introduced in FTM and incubated at 37°C for 7 days (direct inoculum method). At R&D department of AL.CHI.MI.A S.r.l., 5 ml of thawed cryopreservation solutions and tissue samples were tested by incubation in TSB and FTM at 24°C and 33°C for 14 days, respectively (direct inoculum method, according to the European Pharmacopoeia [9]). In addition, separate samples from same solutions and tissues were introduced in the RESEP tube (AL.CHI.MI.A. S.r.l., Italy) containing a specific resin mixture, which was expressly formulated to eliminate the antibiotics contained in the BASE.128 decontamination medium (AL.CHI.MI.A. S.r.l., Italy). The RESEP tubes were filled with TSB and FTM and incubated at 24°C and 33°C for 14 days, respectively. Samples were considered contaminated (positive) when growth media appeared turbid at visual inspection as compared to sterile growth media (negative control), or when reported by the automatic system BacT/ALERT 3D. The presence of microorganisms in turbid samples was confirmed by subculture on agar plates or in fresh culture media.

Detection of antibiotic residues in tissue homogenates and cryopreservation media

Six cryopreserved tissue samples and cryopreservation media for each bank-prepared antibiotic cocktail were evaluated for antibiotic content, as previously described [7]. The agar diffusion test was adapted for the detection of antibiotics in tissue homogenates and cryopreservation media samples using *Staphylococcus aureus* ATCC6538 (SA)-, *Pseudomonas aeruginosa* ATCC 9027 (PA)-, and *Candida albicans* ATCC10231 (CA)-seeded agar plates. Each tissue homogenate and cryopreservation medium was assessed at least in triplicate for each microbial strain. Cryopreservation media and homogenates of six tissues decontaminated with BASE.128 and cryopreserved according to CTBL method were used as a control. For both the direct inoculum and RESEP tube methods, antibiotic activity of tissue samples was evaluated immediately after tissue thawing and at the end of the 14-day sterility testing; antibiotic activity of cryopreservation media was assessed immediately after thawing and after inoculation in culture broths. Antibiotic-free cryopreservation medium and 10% BASE.128 solution plated on each microorganism seeded agar plate served as negative and positive control, respectively.

Statistics

Tissue homogenates and cryopreservation media were grouped on the basis of antibiotic cocktail treatment (cocktail 1, cocktail 2, and BASE.128), and the mean inhibition zone diameters and standard errors were calculated for each microbial strain (SA, PA, CA).

Results

Presence of residual antibiotics in cryopreserved tissues and cryopreservation solutions

Immediately after thawing, all cryopreserved tissues homogenates, decontaminated with different antibiotic cocktails, induced similar inhibition zones in the range of 10.0–10.4 mm on SA-seeded plates. On the last day of sterility testing period, all tissues still induced 3.7–4.4 mm inhibition zones on SA plates.

Antibiotic elimination from tissue samples with the RESEP tube reduced inhibition zones by 81%–98% and 94%–100% on initial and final day of the 14-day sterility testing period, respectively (Figure 1A). Similarly, all cryopreservation media inhibited the

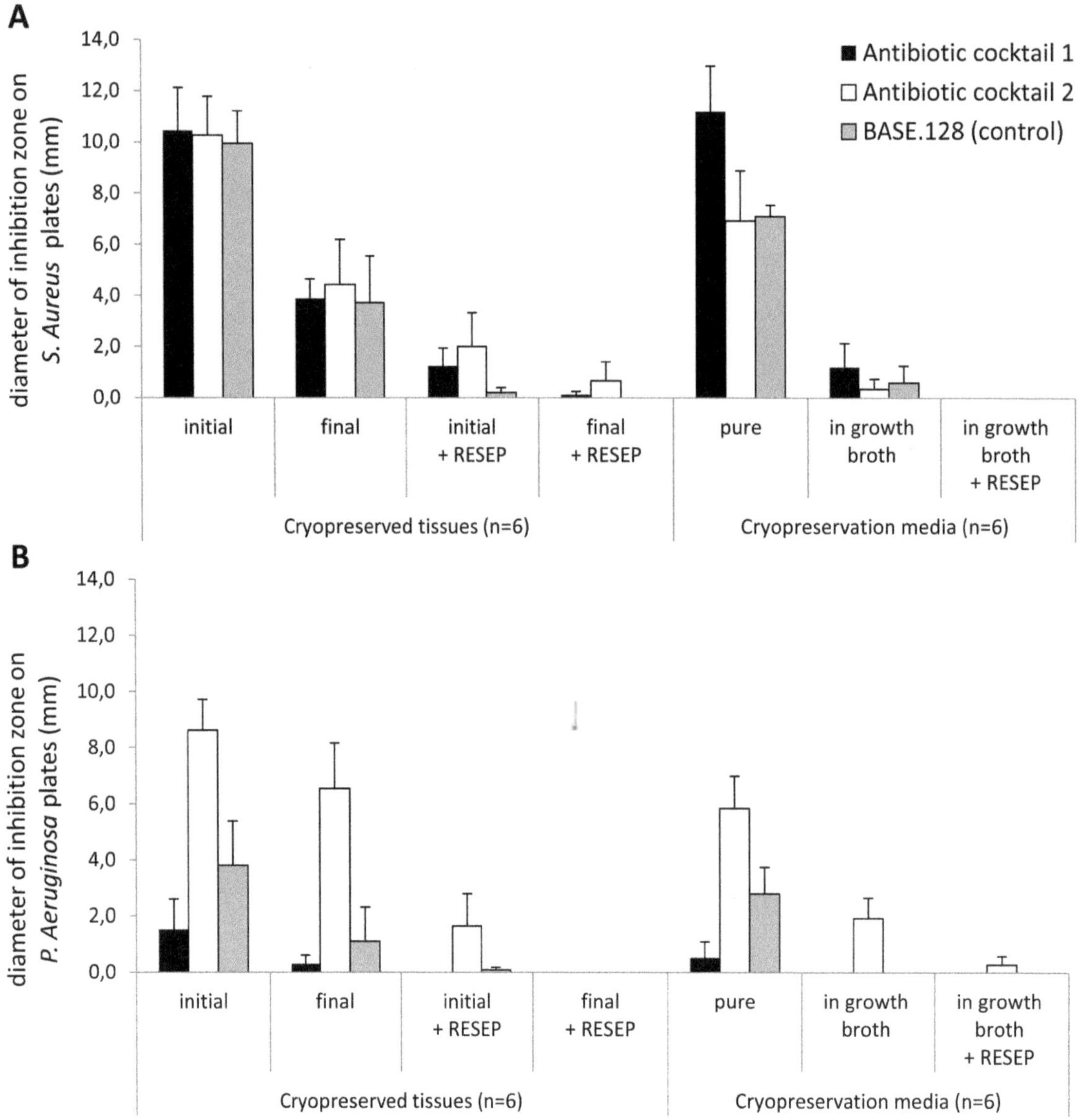

Figure 1. Presence of antibiotic residues in decontaminated and cryopreserved tissue homogenates and cryopreservation media. Agar diffusion assays performed using *Staphylococcus aureus*-seeded plates (A) and *Pseudomonas aeruginosa*-seeded plates (B), immediately after thawing (initial) and after 14 days of sterility testing (final) on tissues and media undergoing decontamination procedures with Cocktail 1 and Cocktail 2, prepared by tissue banks, and BASE.128 as a control. The plot legend on Figure 1A also applies to Figure 1B.

growth of SA on agar plates. Cocktail 1, cocktail 2, and BASE.128 residues induced 11.2-, 6.9-, and 7.1-mm inhibition zones on SA plates, respectively. All cryopreservation media inoculated in growth broths induced detectable SA inhibition activity that was totally abolished by the RESEP tube treatment (Figure 1A).

The width of inhibition zones on PA-seeded plates varied depending on the antibiotic cocktail used during tissue decontamination. Immediately after thawing, 1.5- and 8.6-mm zones were induced by cocktail 1- and cocktail 2-treated tissues, respectively. After the 14-day sterility testing period, cocktail 2-treated tissues still induced 6.6-mm inhibition zones, whereas cocktail 1- and BASE.128-treated tissues induced 0.3- and 1.1-mm inhibition zones on PA-seeded plates, respectively (Figure 1B).

The growth of PA on agar plates was also inhibited by all cryopreservation media. Cocktail 2 residues induced the largest inhibition zones (5.8 mm) on PA plates and 1.9-mm inhibition zones were still detected after inoculation of cryopreservation media in growth broths.

Inhibition zones of the RESEP tube-treated tissues and solutions were markedly reduced or absent on PA plates except for cocktail 2-treated tissues, which induced detectable inhibition zones despite the RESEP treatment (Figure 1B).

Inhibition zones were not detected on antimycotic-sensible CA plates for any investigated tissue homogenate or cryopreservation medium at any tested time point (Table S1).

Microbiological analysis

Tissue bank microbiological analysis of tissue samples and transport solution showed that 50% of the tissues were initially contaminated. These mostly corresponded to the blood vessels donated after cardiac death, collected by CTBER and were detected in both tissue and liquid samples. In CTBL, which procured tissues only from brain death donors, the initial contamination rate was 30% and contaminants were recovered exclusively from transport liquid samples (Table 1). Approximately half of the contaminated tissues were contaminated with *S. epidermidis*. The remaining contaminants were represented by other bacterial strains, including *Enterobacter cloacae, Enterococcus faecalis, Escherichia coli, Streptococcus mitis, Streptococcus viridans, Corynebacterium spp. and Bacillus spp.* One tissue was contaminated with *C. albicans* (Table S2).

After decontamination, 2 out of 6 samples tested by CTBER were still found positive for bacterial and fungal contaminants. One of these contaminations *(C. albicans)* was detected with the direct inoculum method on tissue sample and the other *(E. coli)* with the Bactalert method on liquid sample (Table 1). All samples tested by CTBL resulted negative after decontamination.

After thawing, the microbiological analysis of cryopreserved tissue samples and solutions performed with the direct inoculum method, i.e., in the presence of antibiotic residues, did not detect contamination in any of the tested tissue or liquid samples. Microbiological analysis of cryopreserved tissue samples and solutions performed after removal of antibiotic residues with the ResEP tube showed that 5 out of 10 (50%) CTBL samples and 3 out of 6 (50%) CTBER samples were positive. Altogether, RESEP method detected contaminations in 8 tissue samples and 5 related cryopreservation solutions (Table 1).

Discussion

We used a fast and simple modified agar diffusion method that allowed us to detect rapidly the presence of antibiotics in tissue homogenates and cryopreservation media at different time points of sterility testingwithout the need for additional extraction methods. We showed that the significant levels of antibiotic residues can be detected in decontaminated cardiovascular tissues and processing liquids after thawing, despite rinsing and cryopreservation. In fact, antimicrobial activity against both gram-positive and gram-negative bacteria was detected by SA- and PA-seeded plates, which varied depending on decontamination cocktail and protocol. We speculate that inhibition zones induced by cocktail 2, particularly marked on PA-seeded plates, were due to different composition and higher concentration of antibiotics active against PA. The absence of inhibition zones on CA-seeded plates confirmed that the tested samples did not contain antimycotics, as expected for tissues treated with bank-prepared cocktails lacking such components. Similarly, no inhibition zones on CA-seeded plates were observed for control tissues treated with BASE.128, indicating no carry-over of amphotericin B deoxycholate in tissue and liquid samples.

Consistently with a previous paper [7], the antibiotics were not only present in tissue homogenates, but also transferred from tissues to cryopreservation liquids and sterility test growth broths, where they remained active for up to 14 days of sample incubation. This strongly argues against the assumption that antibiotics are efficiently degraded under the standard conditions of microbiological analysis. Moreover, although some antibiotics such as beta-lactam antibiotics (cefoxime and cefotaxime) are relatively labile, others, including polymyxin B, lincomycin, vancomycin, and gentamicin, are stable enough to induce bacteriostatic effects under these conditions, which is in accordance with our results.

Our study highlighted a possible issue of validity of the microbiological method used for detection of microbial contaminations in tissue banking due to the presence of antibiotic residues in the tested samples.

Table 1. Results of microbiological analysis.

		Positive samples							
		Before decontamination		After decontamination		After thawing			
		Direct inoculum		Direct inoculum		Direct inoculum		RESEP tube	
		tissue	liquid	tissue	liquid	tissue	liquid	tissue	liquid
CTBER (Tissue Bank 1)	n = 6	5/6	5/6*	1/6	1/6*	0/6	0/6	3/6	2/6
CTBL (Tissue Bank 2)	n = 10	0/10	3/10	0/10	0/10	0/10	0/10	5/10	3/10
Total	n = 16	5/16	8/16	1/16	1/16	0/16	0/16	8/16	5/16
	%	32%	50%	6%	6%	0%	0%	50%	32%

Positive (contaminated) samples determined by microbiological analysis performed by standard tissue bank methods and the RESEP tube method on tissue and liquid samples of human cardiovascular allografts processed by two cardiovascular tissue banks and tested before and after decontamination and after thawing.

* *liquid samples were inoculated in BacT/ALERT blood culture vials.*

Tissue samples decontaminated with antibiotic cocktails are a specific type of microbiological sample that may present bacteriostatic and/or fungistatic properties [8,15]. The intrinsic properties of the tissues, such as enzymatic activity, immunologic events, and presence of antimicrobial peptides [16], which are well described in the skin [17], cornea [18], and amniotic membrane [19], may be exacerbated by the presence of residual antibiotics following tissue decontamination with antibiotic cocktails [7,8,20]. The European and US Pharmacopoeias indicate eliminating any factor that may interfere with microbial growth during sterility testing from samples [9,10].

In our study, the microbiological negativity of 14 out of 16 (88%) tissue and liquid samples was observed by standard tissue bank methods, after decontamination. The two bank-prepared cocktails used in the present study intentionally do not contain antimycotics and this may explain both the persistence of fungal contamination found in one tissue sample, despite the decontamination procedure, and its recovery by the direct inoculum method performed by the tissue banks. Fungal contaminations account for approximately 1% of total contaminations and correspond to the expected tissue discard rate of the two tissue banks.

All cryopreserved tissue and cryopreservation solution samples resulted negative with the direct inoculum method in the absence of antibiotics removal, whereas 50% of these samples resulted positive when tested with the RESEP tube method that comprises antibiotic residue removal. This clearly shows the false negative results obtained by the direct inoculum method.

Some automatic blood culture systems implement antibiotic neutralization features. However, they were specifically designed and validated for blood samples [13,21], where the expected antibiotic concentrations are relatively low. Therefore, the reliability of such system is questionable, when used with tissue processing liquids containing higher concentrations of decontaminating agents [13,15,21].

Although the RESEP tube is specifically designed to remove antibiotics of BASE.128, it eliminated significant amounts of antibiotic residues from tissue and liquid samples decontaminated with all investigated cocktails.

In addition, the RESEP tube method allowed ascertaining that as much as 50% of the tissue samples were contaminated by bacteria and/or fungi, after decontamination and cryopreservation. Such contamination was not detected with the direct inoculum method. Our results confirm that latent contaminations may be masked by bacteriostatic and fungistatic concentrations of antimicrobial agents [22]. Validated rinsing procedures to remove antibiotic residues from tissue samples might enhance bacterial recovery [23]. The dilution of antibiotics under minimal inhibitory concentration is an acceptable approach [9–11]; however, it requires high volumes of growth media in case of high antibiotic concentrations.

The tissue banks determine allograft microbiological contamination indirectly, through the analysis of tissues samples and processing liquids. The heart valve banks normally analyze small pieces (1 cm^3) of myocardium, aorta or pulmonary artery, and processing liquids with volumes ranging from 1 to 5 ml samples at different processing phases. These amounts usually correspond to less than 5% of the total tissue or processing liquid sample, and thus may not properly reflect the contamination profile of the allograft considering that at least 10% of the total volume for liquid samples should be tested according to European and US Pharmacopoeias [9,10]. In fact, we observed that the results obtained in tissue and related liquid samples did not always match. This could be partially explained by low volume sampling and/or low bacterial concentration, at the limit of detection threshold in tissue, which is mainly expected in brain-dead donors and may lead to increased variability in results. We also speculated that there is a heterogeneous distribution of some microbial strains in specific tissue sites, which were not included among tested samples.

Furthermore, we observed that one contamination was detected after tissue decontamination, but not after tissue thawing using the direct inoculum method. This could be due to the sampling differences described above and/or decrease in the number of viable microorganisms under the detection threshold of the direct inoculum method during the last processing phase.

It may be argued that the microbiological positivity observed only in cryopreserved and the RESEP tube-treated samples could be a consequence of additional manipulation. However, none of the control samples undergoing identical manipulations were contaminated.

Microbiological analysis is closely related to tissue safety as results are used to determine the suitability of the tissue for transplantation [11]. Moreover, in case of falsely negative microbiological results, the inefficacy of the decontamination procedure is not immediately apparent and can potentially lead to the release of contaminated tissue, thus compromising the safety of the allograft recipient [15,22].

The risks deriving from infected tissue transplants have been extensively documented [24,25] and early postsurgical infections may have severe clinical implications [15,26]. In addition to antibiotic prophylaxis used to prevent the risk of surgical site infection [27,28] and absence of immunosuppressant therapy, which is not routinely administered in tissue transplantation [3,29,30], thus maintaining immune defense against possible pathogens in allograft recipients, the elimination of false negative tissues could further improve the safety of transplantation procedures.

The policies concerning the management of microbiologically contaminated tissues differ considerably among Tissue Banks. Soo et al. discarded contaminated tissues before antibiotic treatment in order to not have false negatives, which probably results in higher than average discard rates [22].

Consistently with other reports [2,14,31], we observed higher contamination rate in donation after cardiac death donors as compared with brain-dead donors. This is likely due to agonal spread and postmortem translocation of microorganisms in cardiac death donors [31,32], in which the time from death to cardiectomy is usually longer than that in brain-dead donors [14]. This highlights the need for an efficient decontamination procedure of tissues from these donors and process validation in terms of antibiotic cocktail composition, time and temperature incubation conditions, and related microbiological analysis.

Our data showed that approximately 50% of the cardiovascular tissues decontaminated by routine methods were microbiologically positive. Even if no adverse reactions have ever been reported after transplantation of cardiovascular allograft distributed by our banks, the results of the present study prompted us to review and optimize our decontamination protocols and microbiological methods.

Conclusions

Antibiotic residues present in tissue samples and processing liquids after decontamination with antibiotic cocktails resulted in a high rate of falsely negative microbiological analysis. The RESEP tube method allowed a thorough elimination of residual antibiotics from tissue and processing liquid samples and a reliable microbiological assessment of tissue contamination. Although the results of the present study were observed in a relatively small group of tissues, they demonstrated the need for elimination of antibiotic residues from tested samples to improve the microbiological methods as required by current pharmacopeias.

Author Contributions

Conceived and designed the experiments: JDT AG MB CG GP. Performed the experiments: MB AG CG SM LD. Analyzed the data: MB AG CG JDT. Contributed reagents/materials/analysis tools: JDT MB GP. Wrote the paper: JDT AG MB.

References

1. Jashari R, Van Hoeck B, Ngakam R, Goffin Y, Fan Y (2013) Banking of cryopreserved arterial allografts in Europe: 20 years of operation in the European Homograft Bank (EHB) in Brussels. Cell Tissue Bank.14(4): 589–599.
2. de By TM, Parker R, Delmo Walter EM, Hetzer R (2012) Cardiovascular tissue banking in Europe. HSR Proc Intensive Care Cardiovasc Anesth 4(4): 251–260.
3. Mirelli M, Buzzi M, Pasquinelli G, Tazzari PL, Testi G, et al. (2005) Fresh and cryopreserved arterial homografts: immunological and clinical results. Transplant Proc 37(6): 2688–2691.
4. Pitt TL, Tidey K, Roy A, Ancliff S, Lomas R, et al. (2014) Activity of four antimicrobial cocktails for tissue allograft decontamination against bacteria and Candida spp. of known susceptibility at different temperatures. Cell Tissue Bank 15(1): 119–125.
5. Fan YD, Van Hoeck B, Holovska V, Jashari R (2012) Evaluation of decontamination process of heart valve and artery tissues in European Homograft Bank (EHB): a retrospective study of 1,055 cases. Cell Tissue Bank 13(2): 297–304.
6. Jashari R, Tabaku M, Van Hoeck B, Cochéz C, Callant M, et al. (2007) Decontamination of heart valve and arterial allografts in the European Homograft Bank (EHB): comparison of two different antibiotic cocktails in low temperature conditions. Cell Tissue Bank 8(4): 247–255.
7. Gatto C, Giurgola L, D'Amato Tothova J (2013) A suitable and efficient procedure for the removal of decontaminating antibiotics from tissue allografts. Cell Tissue Bank 14: 107–115.
8. Saegeman V, Ectors N, Lismont D, Verduyckt B, Verhaegen J (2008) Bacteriostasis testing on allograft tissue inoculated in Wilkins-Chalgren broth. J Hosp Infect 70(3): 278–283.
9. European Pharmacopoeia 8th Edition, chapter 2.6.1. Sterility, pp. 175–178.
10. U.S. Pharmacopeia, USP 36, chapter <71> Sterility tests.
11. Alexander K, Bryans T (2006) Evaluation of the sterility test for detection of microbial contaminants of allografts. Cell Tissue Bank 7(1): 23–28.
12. Varettas K (2013) Culture methods of allograft musculoskeletal tissue samples in Australian bacteriology laboratories. Cell Tissue Bank 14(4): 609–614.
13. Schroeter J, Wilkemeyer I, Schiller RA, Pruss A (2012) Validation of the Microbiological Testing of Tissue Preparations Using the BACTEC Blood Culture System. Transfus Med Hemother 39: 387–390.
14. Van Kats JP, Van Tricht C, Van Dijk A, Van der Schans M, Van den Bogaerdt A, et al. (2010) Microbiological examination of donated human cardiac tissue in heart valve banking. Eur J Cardiothorac Surg 37: 163–169.
15. Kainer MA, Linden JV, Whaley DN, Holmes HT, Jarvis WR, et al. (2004) Clostridium infections associated with musculoskeletal-tissue allografts. N Engl J Med 350(25): 2564–2571.
16. Choi KY, Chow LNY, Mookherjee N (2012) Cationic Host defense peptides: Multifaced Role in Immune Modulation and Inflamation. J Innate Immun 4: 361–370.
17. Brogden NK, Mehalick L, Fischer CL, Wertz PW, Brogden KA (2012) The emerging role of peptides and lipids as antimicrobial epidermal barriers and modulators of local inflammation. Skin Pharmacol Physiol 25: 167–181.
18. Huang LC, Petkova TD, Reins RY, Proske RJ, McDermott AM (2006) Multifunctional roles of human cathelicidin (LL-37) at the ocular surface. Invest Ophatalmol Vis Sci 47: 2369–2380.
19. Kim JY, Park SC, Lee JK, Choi SJ, Hahm KS, et al. (2012) Novel antibacterial activity of β(2)-microglobulin in human amniotic fluid. PLoS One 7(11): e47642.
20. Jashari R, Faucon F, Hoeck BV, Gelas SD, Fan Y, et al. (2011) Determination of Residual Antibiotics in Cryopreserved Heart Valve Allografts. Transfus Med Hemother 38: 379–386.
21. Mitteregger D, Barousch W, Nehr M, Kundi M, Zeitlinger M, et al. (2013) Neutralization of antimicrobial substances in new BacT/Alert FA and FN Plus blood culture bottles. J Clin Microbiol 51(5): 1534–1540.
22. Soo A, Healy DG, El-Bashier H, Shaw S, Wood AE (2011) Quality control in homograft valve processing: when to screen for microbiological contamination? Cell Tissue Bank 12(3): 185–190.
23. Pirnay JP, Verween G, Pascual B, Verbeken G, De Corte P, et al. (2012) Evaluation of a microbiological screening and acceptance procedure for cryopreserved skin allografts based on 14 day cultures. Cell Tissue Bank 13: 287–295.
24. Eastlund T (1995) Infectious disease transmission through cell, tissue, and organ transplantation: reducing the risk through donor selection. Cell Transplant 4(5): 455–477.
25. Eastlund T (2006) Bacterial infection transmitted by human tissue allograft transplantation. Cell Tissue Bank 7(3): 147–166
26. Kuehnert MJ, Clark E, Lockhart SR, Soll DR, Chia J, et al. (1998) Candida albicans endocarditis associated with a contaminated aortic valve allograft: implications for regulation of allograft processing. Clin Infect Dis 27(4): 688–691.
27. Ll Mangram AJ, Horan TC, Pearson ML, Silver LC, Jarvis WR (1999) Guidelines for prevention of surgical site infection. Hospital infection control Practise advisory committee. Infect Control Hosp Epidemiol 20(4): 250–278.
28. ASHP Therapeutic Guidelines on Antimicrobial Prophylaxis in surgery. American society of Health system Pharmacists. (1999) Am J Health Syst Pharm 56(18): 1839–1888.
29. Welters MJ, Oei FB, Vaessen LM, Stegmann AP, Bogers AJ, et al. (2001) Increased numbers of circulating donor-specific T helper lymphocytes after human heart valve transplantation. Clin Exp Immunol 124(3): 353–358.
30. Welters MJ, Oei FB, Witvliet MD, Vaessen LM, Cromme-Dijkhuis AH, et al. (2002) A broad and strong humoral immune response to donor HLA after implantation of cryopreserved human heart valve allografts. Hum Immunol 63(11): 1019–1025.
31. Tabaku M, Jashari R, Carton HF, Du Verger A, Van Hoeck B, et al. (2004) Processing of cardiovascular allografts: effectiveness of European Homograft Bank (EHB) antimicrobial treatment (cool decontamination protocol with low concentration of antibiotics). Cell Tissue Bank 5(4): 261–266.
32. Morris JA, Harrison LM, Partridge SM (2006) Postmortem bacteriology: a re-evaluation. J Clin Pathol 59(1): 1–9.

Sperm Flagellum Volume Determines Freezability in Red Deer Spermatozoa

José Luis Ros-Santaella[1,2]*, Álvaro Efrén Domínguez-Rebolledo[3], José Julián Garde[1]

1 SaBio, Instituto de Investigación en Recursos Cinegéticos (IREC), CSIC-UCLM-JCCM, Campus Universitario, Albacete, Spain, **2** Department of Animal Science and Food Processing, Faculty of Tropical AgriSciences, Czech University of Life Sciences, Prague, Czech Republic, **3** Instituto Nacional de Investigaciones Forestales, Agrícolas y Pecuarias (INIFAP), Mocochá, Yucatán, México

Abstract

The factors affecting the inter-individual differences in sperm freezability is a major line of research in spermatology. Poor sperm freezability is mainly characterised by a low sperm velocity, which in turn is associated with low fertility rates in most animal species. Studies concerning the implications of sperm morphometry on freezability are quite limited, and most of them are based on sperm head size regardless of the structural parts of the flagellum, which provides sperm motility. Here, for the first time, we determined the volumes of the flagellum structures in fresh epididymal red deer spermatozoa using a stereological method under phase contrast microscopy. Sperm samples from thirty-three stags were frozen and classified as good freezers (GF) or bad freezers (BF) at two hours post-thawing using three sperm kinetic parameters which are strongly correlated with fertility in this species. Fourteen stags were clearly identified as GF, whereas nineteen were BF. No significant difference in sperm head size between the two groups was found. On the contrary, the GF exhibited a lower principal piece volume than the BF (6.13 μm^3 vs 6.61 μm^3, respectively, $p = 0.006$). The volume of the flagellum structures showed a strong negative relationship with post-thawing sperm velocity. For instance, the volume of the sperm principal piece was negatively correlated with sperm velocity at two hours post-thawing ($r = -0.60$; $p < 0.001$). Our results clearly show that a higher volume of the sperm principal piece results in poor freezability, and highlights the key role of flagellum size in sperm cryopreservation success.

Editor: Chris D. Wood, Universidad Nacional Autónoma de México, Mexico

Funding: JLRS was supported by I3P program (Spanish National Research Council, CSIC, Spain), post-doctoral fellowship no. 99830/1181/1822 (Czech University of Life Sciences, Prague, Czech Republic), and by the project CIGA 20145001 (Czech University of Life Sciences, Prague, Czech Republic). The funders had no role in study design, data collection and analysis, decision to publish, or preparation of the manuscript.

Competing Interests: The authors have declared that no competing interests exist.

* Email: joseluis.ros@irec.csic.es

Introduction

It is well known that the cryopreservation process negatively affects the viability and fertility of reproductive cells. There are several factors caused by cryopreservation protocols that alter sperm integrity, such as a change in the temperature of the diluents [1]; osmotic and toxic stresses induced by cryoprotectants [1,2]; formation/reshaping of intracellular ice during freezing and thawing [3]; and dissolution of ice in the extracellular environment [2]. All of these factors induce sperm volumetric changes, plasma membrane alterations, flagellum morphological defects, as well as decrease mitochondrial membrane potential, sperm motility, viability, and fertility [2,4–10]. As a generalisation, some 40–50% of the sperm population does not survive cryopreservation even with optimised protocols [2]. Thus, many of the frozen-thawed spermatozoa show a shorter life span and have difficulties in reaching the oocytes and penetrating their vestments after conventional artificial insemination [10]. Variations between individuals in sperm freezability have been reported in numerous animal species. Within this context, semen donors have routinely been categorized as "good freezers" (GF) or "bad freezers" (BF). When experiments involve comparing bulk samples before freezing and after thawing, it is difficult to know which parts of the cryopreservation procedure may be causing problems and to detect differences between individual cells [11]. In boar sperm, it has been demonstrated that these consistent inter-individual variations in sperm freezability are genetically determined [12].

The understanding and prediction of the sperm functional response to cryopreservation is one of the major questions in sperm cryobiology [13]. For this reason, sperm morphometry classification has become an integral part of routine sperm analysis. In recent years, ASMA (Automated Sperm Morphometry Analysis) systems have been employed for determining morphometric parameters of the sperm head and midpiece to elucidate possible relations with sperm freezability [3,14]. These systems can only analyse the lengths and areas of the sperm head and midpiece, but they are unable, for example, to measure the sperm principal piece, which provides sperm motility. Thus, in order to study sperm midpiece morphometry with these systems, it is not easy to find a staining method which allows discerning this structure from the rest of the flagellum [15]. By contrast, with other software for analysis and image processing, it is possible to obtain the lengths of flagellum structures. Numerous studies argue the implications of sperm flagellum in several biological processes such as sperm velocity, male reproductive success, relationships with testicle size, and spermatogenesis [16–21]. On the other hand, studies concerning the role played by flagellum size on sperm freezability have not yet been reported, although there is evidence of the fragility of its internal and external structures when they are exposed to the cryopreservation process [5,7,22–24].

Curry et al. [25] reported that the sperm area and volume have great importance for determining optimal sperm cryopreservation protocols. For this purpose, stereology is able to accurately estimate surface areas and volumes for complex shapes from two-dimensional images [25]. Thereby, sperm volumes as well as the sperm head and flagellum structures can be examined. The sperm volumes of human and several mammalian species have been determined by stereological methods [19,25].

In the present study, we hypothesized that flagellum volume could predict freezability in red deer spermatozoa. For this reason, we focused on determining the volumes of the sperm flagellum, midpiece, and principal piece in fresh sperm using stereology under phase contrast microscopy. In this way, differences between males in sperm flagellum volume could be linked to sperm freezability, owing to the fragility of flagellum structures to the cryopreservation process [5,7,23], and the differences observed among red deer sperm to support cryopreservation [14]. In order to determine sperm freezability, sperm samples were classified as good or bad freezers using three kinetic parameters (VAP: average path velocity, VCL: curvilinear velocity, and VSL: straight linear velocity) assessed by a CASA (Computer Assisted Sperm Analysis) system at 2 hours post-thawing and with incubation at 37°C. Sperm viability, acrosomal status, and mitochondrial activity were also determined by flow cytometry.

Materials and Methods

Animals

The study was approved by the "Comité de Ética en Investigación de la Universidad de Castilla-La Mancha". All animal handling was done following the Spanish Animal Protection Regulation RD 53/2013, which conforms to the European Union Regulation 2003/65. Stags were legally culled and hunted in their natural habitat in accordance with the harvest plan of the game reserve. The harvest plans were made following the Spanish Harvest Regulation, Law 2/93 of Castilla-La Mancha, which conforms to European Union regulations. Thirty-three stags (age >4.5 years; body mass >130 kg) of red deer (*Cervus elaphus*) were used in this study. Landowners and managers of the red deer populations gave permission to the authors to use the samples.

Stags and testes collection

Testes were recovered from adult red deer culled during the 2008 hunting season in the south of Spain. Testes, together with the scrotum, were removed and transported at approximately 20–22°C to the laboratory. The time that elapsed between animal death and sperm recovery ranged from 3 to 6 hours, an adequate and reliable time interval for evaluating sperm parameters, because a decrease in the quality of sperm traits begins to take place 12 hours after the death of a male [26].

Chemicals and solutions

Unless otherwise stated, chemicals were obtained from Sigma-Aldrich (Madrid, Spain).

A Salamon's modified extender was prepared in two fractions, as previously described [27]. Fraction A contained: Tris (2.70%, w/v), fructose (1%, w/v), citric acid (1.4%, w/v), and clarified egg yolk (20%, v/v) (pH 6.8, osmolality 300 mOsm/Kg). Fraction B differed from the Fraction A in that water was replaced (12%, v/v) with the same volume of glycerol, with a final concentration =6% (v/v). Glutaraldehyde solution was composed of 2% glutaraldehyde (v/v), and 0.165 mol/L cacodylate/HCl buffer (pH 7.3), as previously described [27]. Bovine gamete medium (BGM-3) was composed of 87 mmol/L NaCl, 3.1 mmol/L KCl, 2 mmol/L $CaCl_2$, 0.4 mmol/L $MgCl_2$, 0.3 mmol/L NaH_2PO_4, 40 mmol/L HEPES, 21.6 mmol/L sodium lactate, 1 mmol/L sodium pyruvate, 50 µg/mL kanamicine, 10 µg/mL phenol red, and 6 mg/mL BSA (Bovine Serum Albumine) (pH 7.5), as previously described [28].

Cryopreservation of epididymal spermatozoa

Post-mortem seminal recovery is the most practical option to obtain sperm samples from wild populations of red deer, with hunting representing a constant source from harvested animals [29]. Spermatozoa were collected from the cauda of the epididymis by repeated longitudinal and transverse cuts with a surgical scalpel and placed in 0.5 mL of fraction A. The contents from both epididymides of each individual were pooled for processing. Afterward, a routine sperm evaluation was made. Sperm concentration was determined using a Bürker counter chamber. The percentage of motile sperm and the quality of motility (QM) were subjectively evaluated, the latter using a scale from 0 (lowest: immobile) to 5 (highest: progressive and vigorous movement). Then, the sperm motility index (SMI) was calculated according to the formula [30]:

$$SMI = \frac{\%\ motile\ sperm + (quality\ of\ motility \times 20)}{2}$$

Only samples with at least 60% motile sperm were used for this study (Table S1). The sperm cryopreservation protocol was performed as previously described [27]. Briefly, the sperm mass was diluted at room temperature to 400×10^6 spermatozoa/mL with fraction A, and then to 200×10^6 spermatozoa/mL with fraction B. The diluted samples were refrigerated for approximately 10 min to reach 5°C and then equilibrated at the same temperature for 2 h. After equilibration, the suspended sperm was loaded into 0.25 mL plastic straws and frozen for 10 min in nitrogen vapours, 4 cm above the level of liquid nitrogen (−120°C). The straws were then immediately immersed into liquid nitrogen (−196°C) for storage.

Morphometry assessment of fresh sperm

Sperm samples were directly recovered from the cauda of both epididymides. Spermatozoa were fixed in glutaraldehyde solution. A sub-sample of 2 µl was used to prepare the smears. Semen smears were air-dried for one day, then immersed in the glutaraldehyde fixative solution for 5 min and immediately mounted, sealing the edges with dibutyl phthalate xylene (DPX). This method avoids floating cells on the slide (Figure 1A), which greatly helps sperm morphometry analysis. Sperm samples were photographed using a high-resolution camera DXM1200 (Nikon, Tokyo, Japan) under a phase-contrast microscopy using an Eclipse E600 microscope (Nikon, Tokyo, Japan), and a 40X objective (Nikon, Tokyo, Japan). The resolution of the pictures was 3840×3072 pixels (TIFF format). A scale of 10 µm (181 pixels) was used for the measurements. The pixel size was 0.055 µm in the horizontal and vertical axes. Sperm lengths were assessed using ImageJ software (National Institutes of Health, USA). The main structures of red deer spermatozoon are shown in Figure 1B. The following sperm morphometry parameters were determined: head width, head length, proximal midpiece width, distal midpiece width, midpiece length, flagellum length, and terminal piece length (Table S1). From these measurements, we calculated other morphometric parameters such as total sperm and principal piece lengths. The head area was calculated using the formula for the area of an ellipse [31,32]:

$$Area = \pi \times L \times W$$

Head perimeter was calculated using the Ramanujan's formula for calculating the perimeter of an ellipse [33]:

$$Perimeter \approx \pi \left[3 \times (L+W) - \sqrt{(3L+W) \times (L+3W)} \right]$$

In both formulae, L and W are the semi-major and semi-minor axis of the sperm head, respectively. Twenty-five representative sperm were measured for each male as described by Malo et al. [16].

Stereology of the flagellum. Sperm flagellum volume and its structures were determined using a stereology method based on Anderson et al. [19] with some modifications. Anderson et al. [19] calculated the volumes of the sperm midpiece and flagellum using the formula for the volume of a cylinder. In this study, owing to the significant differences between the proximal and distal midpiece widths (0.94 μm and 0.74 μm, respectively; $p<0.0001$), we estimated sperm midpiece volume using the formula for the volume of a truncate cone:

$$Volume = \left(\frac{\pi \times L}{3} \right) \times \left(R^2 + r^2 + R \times r \right)$$

where L is the length of the midpiece, R is the half proximal

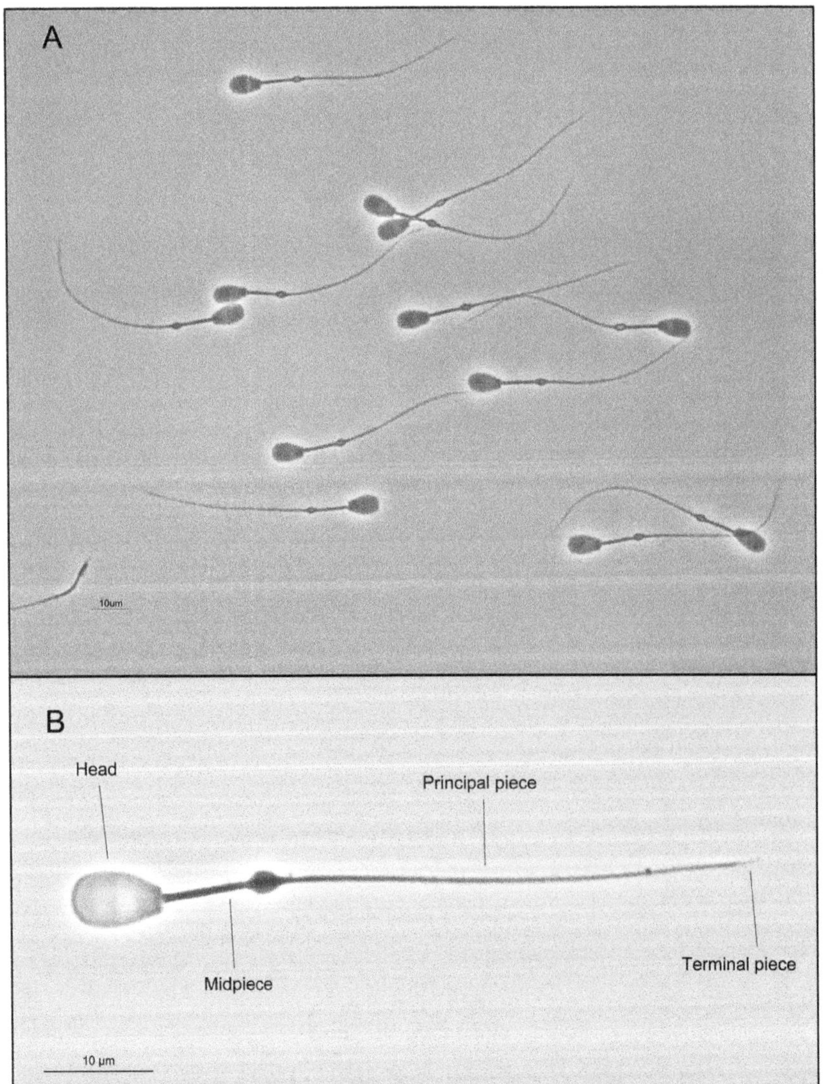

Figure 1. Red deer sperm. (A) It is remarkable how the method described in this study avoids floating cells. The picture was taken under phase contrast microscopy (40X objective). Scale bar, 10 μm. (B) Main structures of red deer spermatozoon. The picture was taken under phase contrast microscopy (100X objective) using the method described in the present study. Scale bar, 10 μm.

midpiece width, and r is the half distal midpiece width. On the other hand, the total flagellum and principal piece volumes were determined using the formula for the volume of a cone:

$$Volume = \frac{\pi \times R^2 \times L}{3}$$

where R is the half midpiece width (i.e., proximal or distal) and L is the length of the flagellum or the principal plus terminal piece.

Sperm thawing and sperm quality assessment

The sperm straws were thawed in a water bath with a saline solution at 37°C for 30 s, and each sample was poured in a tube. The samples were incubated at 37°C and analysed for motility (subjectively and with the CASA system), viability, acrosomal status, and mitochondrial status at 0 hours (after 10 min of incubation) and after 2 hours post-thawing (Table S2 and Table S3, respectively).

CASA analysis. Sperm were diluted down to 25–30×10^6 spermatozoa/mL with fraction A solution and loaded into a Makler counter chamber (Sefi-Medical instruments, Haifa, Israel) at 37°C. The CASA system consisted of a triocular optical phase contrast microscope Eclipse 80i (Nikon, Tokyo, Japan), equipped with a warming stage at 37°C and a Basler A302fs digital camera (Basler Vision Technologies, Ahrensburg, Germany). Images were captured and analysed using the Sperm Class Analyzer software (Microptic S.L., Barcelona, Spain). The analysis was carried out using a 10X negative phase-contrast objective (Nikon, Tokyo, Japan). A total of 4 descriptors of sperm motility were recorded analysing a minimum of 250 sperm per sample: average path velocity (VAP, μm/s), curvilinear velocity (VCL, μm/s), straight linear velocity (VSL, μm/s), and amplitude of lateral head displacement (ALH, μm). The standard parameters settings were as follows: 25 frames/s; 20 to 60 μm^2 for the head area.

Fluorescence probes for sperm viability, acrosomal status, and mitochondrial activity. Several physiological traits were assessed using fluorescent probes and flow cytometry, as previously described [28]. Briefly, the samples were diluted down to 10^6 spermatozoa/mL in BGM-3 solution and stained using four fluorophores. Sperm viability was assessed with 0.1 μmol/L YO-PRO-1 (Invitrogen, Barcelona, Spain) and 10 μmol/L PI (propidium iodide). Mitochondrial activity and acrosomal status were assessed with 0.1 μmol/L Mitotracker Deep Red (Invitrogen, Barcelona, Spain) and 4 μg/mL PNA-TRITC (peanut agglutinin), respectively. The spermatozoa stained in these two solutions were incubated 20 min in the dark before being run through a flow cytometer. The sperm populations shown in this work were: YO-PRO-1-/PI- (viable spermatozoa), MT+ (Mitotracker Deep Red, spermatozoa with active mitochondria), and PNA- (spermatozoa with intact acrosome).

Flow cytometry analysis. Samples were analysed as previously described [28]. Briefly, a Cytomics FC500 flow cytometer (Beckman Coulter, Brea, CA, USA) was utilized, with a 488-nm Ar-Ion laser (excitation for YOPRO-1, PNA-TRITC, and PI), and a 633-nm He-Ne laser (excitation for Mitotracker Deep Red). The FSC (forward-scattered light) and SSC (side-scattered light) signals were used to gate out debris (non-sperm events). Fluorescence from YO-PRO-1 was read using a 525/25BP filter, PNA-TRITC was read using a 575/20BP filter, PI was read using a 615DSP filter, and Mitotracker Deep Red was read using a 675/40BP filter. Fluorescence captures were controlled using the RXP software provided with the cytometer. All of the parameters were read using logarithmic amplification. For each sample, 5000 spermatozoa were recorded, saving the data in flow cytometry standard (FCS) v. 2 files. The analysis of the flow cytometry data was carried out using WEASEL v. 2.6 (WEHI, Melbourne, Victoria, Australia).

Statistical analysis

All statistical analyses were performed using the SPSS 20.0 statistical software package (SPSS Inc, Chicago, IL, USA). Sperm samples from thirty-three (N = 33) red deer were used for statistical analysis. The Kolmogorov-Smirnov test was used to check the normal distribution of the data. A repeated measures one-way ANOVA test was used to compare subjective motility between fresh and thawed sperm (0 and 2 hours) using the Mauchly's sphericity test (Greenhouse-Geisser correction) to verify homogeneity of variance. The quality of motility (QM) was not normally distributed in some groups, therefore, non-parametric Friedman test (repeated measures) was used. On the other hand, to check differences in sperm parameters (kinetics-CASA and flow cytometry) between 0 and 2 hours post-thaw we used a paired-samples t-student test (repeated measures). In order to determine sperm freezability we performed a hierarchical clustering analysis using the Euclidean distance measure after determining automatically the number of conglomerates by cluster analysis in two phases. For this purpose, we used three sperm kinetic parameters (VAP, VCL, and VSL) because they are highly related with fertility in red deer [34] and are good indicators of post-thawing sperm quality [35]. Sperm freezability was determined at 2 hours post-thawing to test spermatozoa thermo-resistance. The independent-samples student-t test (Levene's test to verify homogeneity of variance) and the Mann-Whitney U test were used to check for differences in sperm parameters between the GF and the BF.

On the other hand, since the variables of sperm velocity (VAP, VCL, VSL, ALH, and SMI) were highly correlated among themselves, we reduced the number of predictor variables using principal component analysis (PCA) to obtain an overall sperm velocity at 0 and 2 hours post-thawing, respectively. Bartlett sphericity and Keiser-Meyer-Olkin tests were assessed as measures of sampling adequacy [36]. Pearson's correlation test and the linear regression model were used to assess the relationship between sperm velocity and morphometric parameters. We used the RMA software to reduced major axis regression [37]. The within male coefficient of variation (CV) in sperm morphometry was calculated to show the intra-male variability in sperm design.

Results

Effects of sperm cryopreservation

After the freezing-thawing process, a remarkable decrease in subjective motility parameters was observed between fresh and thawed sperm, but also between 0 and 2 hours post-thawing: motile sperm, $F(1.49, 47.55) = 94.27$, $p<0.001$; QM, $\chi^2(2, N = 33) = 37.83$, $p<0.001$; and SMI, $F(1.51, 48.22) = 116.02$, $p<0.001$ (Table 1). In the same way, there were significant differences for all of the CASA kinetic parameters between 0 and 2 hours post-thawing: VAP, $t(32) = 9.77$, $p<0.001$; VCL, $t(32) = 10.46$, $p<0.001$; VSL, $t(32) = 10.29$, $p<0.001$; and ALH, $t(32) = 10.18$, $p<0.001$ (Table 2). Furthermore, sperm viability and organelle functionality showed a decrease after thawing, displaying highly significant differences between 0 and 2 hours post-thawing: YOPRO-1-/PI-, $t(32) = 11.47$, $p<0.001$; MT+, $t(32) = 6.26$, $p<0.001$; and PNA-, $t(32) = 11.18$, $p<0.001$ (Table 2). For instance, the mean percentage of sperm with active mitochondria ranged from 51.13% to 28.63% during incubation (at 0 and 2 hours, respectively).

Table 1. Subjective motility of fresh and thawed red deer spermatozoa (N = 33).

	Sperm motility		
	Motile sperm (%)	**QM (0–5)**	**SMI (%)**
Fresh sperm	81.82 ± 11.44^{a}	2.24 ± 0.48^{a}	63.33 ± 9.11^{a}
0 hours post-thaw	55.91 ± 16.32^{b}	1.86 ± 0.28^{a}	46.51 ± 9.96^{b}
2 hours post-thaw	42.57 ± 13.52^{c}	1.43 ± 0.21^{b}	35.61 ± 7.71^{c}

Different superscript letters within the same column differ significantly ($p<0.001$). Data are shown as mean ± SD (standard deviation). QM, quality of motility; SMI, sperm motility index.

Deer sperm freezability (GF and BF)

The cluster dendrogram analysis of sperm freezability is shown in Figure 2. Fourteen stags were clearly identified as GF, whereas nineteen were BF. Not only did the GF and the BF show clear differences in the three parameters (VAP, VCL, and VSL) used for the cluster analysis, but they also showed differences in the other sperm functionality parameters. Indeed, at 0 hours post-thawing we found significant differences in the following sperm parameters: motile sperm, $t(31) = -2.23$, $p = 0.033$; VAP, $t(31) = -3.58$, $p = 0.001$; VCL, $t(31) = -4.18$, $p<0.001$; VSL, $t(31) = -2.66$, $p = 0.012$; ALH, $t(31) = -4.30$, $p<0.001$; YOPRO-1-/PI-, $t(27.97) = -2.82$, $p = 0.009$; and MT+, $t(26.75) = -2.30$, $p = 0.029$ (Table 3). The differences between the two groups of males in sperm quality were more evident across sperm incubation (Table 3). Thus, sperm kinetics, viability, and organelle functionality were significantly different between the GF and the BF at 2 hours post-thawing as follows: motile sperm, $t(31) = -3.87$, $p< 0.001$; QM, $U = 66.00$, $p = 0.003$; SMI, $t(31) = -4.62$, $p<0.001$; VAP, $t(31) = -9.66$, $p<0.001$; VCL, $t(19.24) = -8.28$, $p<0.001$; VSL, $t(31) = -8.08$, $p<0.001$; ALH, $t(31) = -7.60$, $p<0.001$; YOPRO-1-/PI-, $t(28) = -2.67$, $p = 0.013$; MT+, $t(31) = -3.43$, $p = 0.002$; and PNA-, $t(31) = -2.69$, $p = 0.011$ (Table 3). On the other hand, the quality of motility, SMI, and acrosomal status were not significantly different between the GF and the BF at 0 hours post-thawing (Table 3).

Sperm quality before freezing

There were no significant differences between the GF and the BF before freezing in any of the subjective kinetics parameters evaluated. Indeed, the GF and the BF showed similar subjective motility and, therefore, were expected to have the same freezability. Motility parameters in fresh spermatozoa were: motile sperm (83.21 ± 13.24 vs 80.79 ± 10.17, $t(31) = -0.60$, $p = 0.556$), QM (2.25 ± 0.51 vs 2.24 ± 0.48, $U = 127.50$, $p = 0.825$), and SMI (64.11 ± 10.63 vs 62.76 ± 8.07, $t(31) = -0.41$, $p = 0.682$). Data are shown as the mean ± SD for the GF and the BF, respectively.

Sperm freezability and morphometry of fresh sperm

The descriptive statistics for the morphometric parameters are shown in Table 4. There were no differences between the GF and the BF in any of the sperm head measurements (Table 4). On the other hand, highly significant differences were observed between groups in regard to sperm flagellum morphometry (Table 4). Thus, the GF exhibited a lower mean principal piece volume than the BF ($6.13 \pm 0.42\ \mu m^3$ vs. $6.61 \pm 0.49\ \mu m^3$, $t(31) = 2.93$, $p = 0.006$). Also, the GF exhibited a smaller mean distal midpiece width than the BF ($0.73 \pm 0.02\ \mu m$ vs. $0.75 \pm 0.03\ \mu m$, $t(31) = 2.37$, $p = 0.024$). The data are shown as the mean ± SD for the GF and the BF, respectively. Moreover, the flagellum and midpiece volumes together with some sperm lengths (e.g., total sperm and flagellum) showed lower values in the GF than the BF, although the differences were not statistically significant (Table 4). Sperm freezability according to the sperm principal piece volume is shown in Figure 3.

Table 2. Kinetics, viability, and organelle status of red deer spermatozoa (N = 33) at 0 and 2 hours post-thaw.

Assessed parameters	**0 hours post-thaw**		**2 hours post-thaw**	
	Mean ± SD	**Range**	**Mean ± SD**	**Range**
Sperm kinetics				
VAP (μm/s)	60.60 ± 13.48^{a}	26.08-87.21	40.09 ± 10.41^{b}	23.18–57.16
VCL (μm/s)	97.58 ± 21.52^{a}	55.49–144.36	66.60 ± 14.14^{b}	43.92–95.45
VSL (μm/s)	34.17 ± 7.17^{a}	15.56–49.03	22.17 ± 5.00^{b}	14.11–29.44
ALH (μm)	3.78 ± 0.80^{a}	2.49–5.47	2.69 ± 0.43^{b}	2.05–3.55
Sperm viability and organelle status				
Viability (YOPRO-1-/PI-, %)	35.81 ± 8.91^{a}	18.46–51.16	29.70 ± 8.28^{b}	13.20–41.32
Active mitochondria (MT+, %)	51.13 ± 13.93^{a}	26.42–78.51	28.63 ± 19.00^{b}	0.88–56.08
Intact acrosome (PNA-, %)	83.96 ± 6.82^{a}	67.40–93.96	69.67 ± 11.16^{b}	50.74–90.72

Different superscript letters within the same row differ significantly ($p<0.001$) between 0 and 2 hours post-thaw. SD, standard deviation; VAP, average path velocity; VCL, curvilinear velocity; VSL, straight linear velocity; ALH, amplitude of lateral head displacement.

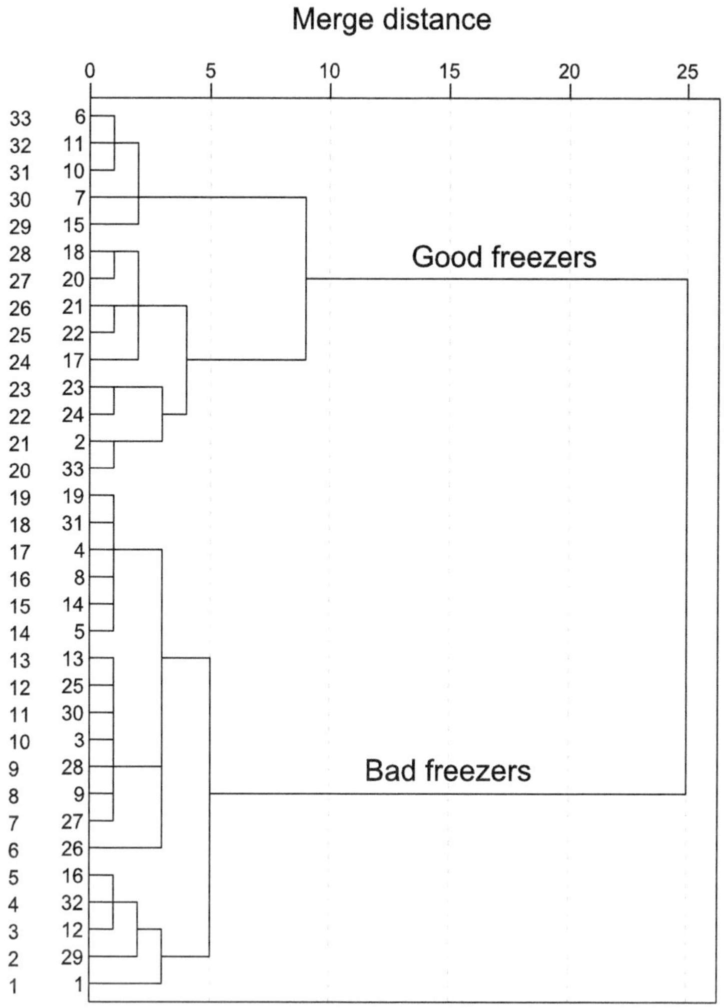

Figure 2. Cluster dendrogram analysis showing red deer sperm freezability. Fourteen males were identified as good freezers, whereas 19 as bad freezers.

Sperm velocity post-thaw and its relationship with morphometry of fresh sperm

Principal component analysis rendered only one component both at 0 and 2 hours post-thawing. The components explained 91.68% and 83.98% of the variance in sperm velocity at 0 and 2 hours post-thawing, respectively. The PCA are shown in Table 5.

None of the morphometric parameters of the sperm head were related with sperm velocity. In contrast, sperm velocity showed strong and negative relationships with sperm flagellum volumes, particularly with the sperm principal piece volume. Indeed, at 0 hours post-thawing, the principal piece volume was the only parameter showing a significant relationship with sperm velocity ($r = -0.36$; $p = 0.038$) (Figure 4A). In addition, such a relationship reached the highest values at 2 hours post-thawing ($r = -0.60$; $p < 0.001$) (Figure 4B). On the other hand, sperm velocity was also negatively correlated with distal midpiece width ($r = -0.55$; $p = 0.001$) (Figure 4C) and with midpiece and flagellum volumes ($r = -0.44$; $p = 0.011$ and $r = -0.36$; $p = 0.038$) at 2 hours post-thawing (Figures 4D–E, respectively).

Discussion

In this study, for the first time, the volumes of the flagellum structures in fresh sperm have been determined to predict sperm freezability in red deer. Our results clearly show that sperm with a higher principal piece volume freeze worse, that is, the GF have a lower principal piece volume than the BF. We also found that sperm velocity is strongly and negatively related with the volumes of the flagellum structures. Sperm velocity is crucial in the process of fertilization in a large number of taxa (including fish [38], birds [39], and mammals [34,40]). Moreover, VAP, VCL, and VSL have been proven to be good indicators of sperm freezability in red deer as previously described in canine sperm [35], and in turn, these parameters are closely related with fertility in red deer using thawed sperm [34].

The evaluation of sperm function throughout post-thawing and sperm incubation provides additional information about the quality of the spermatozoa [41,42] and is more closely related to sperm fertility than those sperm assessed immediately after thawing [43,44]. According to this assumption, we found more and stronger relationships between sperm velocity and the volumes of the flagellum structures at 2 hours than at 0 hours post-thawing. For example, we did not find any differences between the GF and

Table 3. Sperm kinetics, viability and organelle status in good freezers (GF, n = 14) and bad freezers (BF, n = 19) at 0 and 2 hours post-thaw.

Assessed parameters	0 hours post-thaw		2 hours post-thaw	
	GF	BF	GF	BF
Sperm kinetics				
VAP (μm/s)	$68.96 \pm 12.40^{a**}$	$54.44 \pm 10.83^{b**}$	$50.43 \pm 5.38^{a***}$	$32.47 \pm 5.21^{b***}$
VCL (μm/s)	$112.40 \pm 20.13^{a***}$	$86.65 \pm 15.30^{b***}$	$80.37 \pm 9.70^{a***}$	$56.45 \pm 5.56^{b***}$
VSL (μm/s)	$37.71 \pm 7.02^{a*}$	$31.56 \pm 6.24^{b*}$	$26.90 \pm 2.27^{a***}$	$18.70 \pm 3.25^{b***}$
ALH (μm)	$4.34 \pm 0.78^{a***}$	$3.36 \pm 0.53^{b***}$	$3.08 \pm 0.32^{a***}$	$2.40 \pm 0.19^{b***}$
Motile sperm (%)	$62.86 \pm 15.53^{a*}$	$50.79 \pm 15.30^{b*}$	$51.43 \pm 9.89^{a***}$	$36.05 \pm 12.20^{b***}$
QM (0–5)	1.89 ± 0.19^{a}	1.83 ± 0.33^{a}	$1.55 \pm 0.11^{a**}$	$1.34 \pm 0.22^{b**}$
SMI (%)	50.36 ± 9.14^{a}	43.68 ± 9.80^{a}	$41.25 \pm 5.07^{a***}$	$31.45 \pm 6.63^{b***}$
Sperm viability and organelle status				
Viability (YOPRO-1-/PI-, %)	$40.10 \pm 5.00^{a**}$	$32.65 \pm 9.91^{b**}$	$33.50 \pm 4.70^{a*}$	$26.91 \pm 9.29^{b*}$
Active mitochondria (MT+, %)	$56.72 \pm 7.44^{a*}$	$47.01 \pm 16.20^{b*}$	$40.06 \pm 18.20^{a**}$	$20.22 \pm 15.05^{b**}$
Intact acrosome (PNA-, %)	86.26 ± 6.30^{a}	82.26 ± 6.84^{a}	$75.24 \pm 9.87^{a*}$	$65.57 \pm 10.45^{b*}$

Different superscript letters within the same row differ significantly between GF and BF at 0 and 2 hours post-thaw, respectively (*$p<0.05$; **$p<0.01$; ***$p<0.001$). Data are shown as mean ± SD (standard deviation). GF, good freezers; BF, bad freezers; VAP, average path velocity; VCL, curvilinear velocity; VSL, straight linear velocity; ALH, amplitude of lateral head displacement; QM, quality of motility; SMI, sperm motility index.

the BF in acrosomal status at 0 hours post-thawing, but we did at 2 hours post-thawing. This is probably because if the sperm membrane or other structures (e.g. axoneme) were disrupted, this damage was not manifested immediately upon thawing, but occurred during post-thaw re-warming within specific temperatures [23]. Thus, the damage in this structure would be higher due to thermal stress during sperm incubation, inasmuch as sperm thawing is more deleterious than sperm freezing [23,45] and can result in more morphological damage [46].

In the present work, we did not find any significant differences between the GF and the BF in sperm head size, and also none of the morphometric parameters of this structure showed any relationship with sperm velocity. By contrast, Esteso et al. [14] found that sperm head size is related with sperm freezability in red deer (i.e., increased head size entails a poor sperm freezability). However, Esteso et al. [14] classified sperm donors as a GF or BF using the sperm motility index, instead of a CASA system, together with acrosomal status and membrane stability. Within this context, we found that the intact acrosome is negatively related with sperm head size at 0 hours post-thawing (Figure S1). On the other hand, the sperm principal piece volume and distal midpiece width showed significant differences between the GF and the BF, and

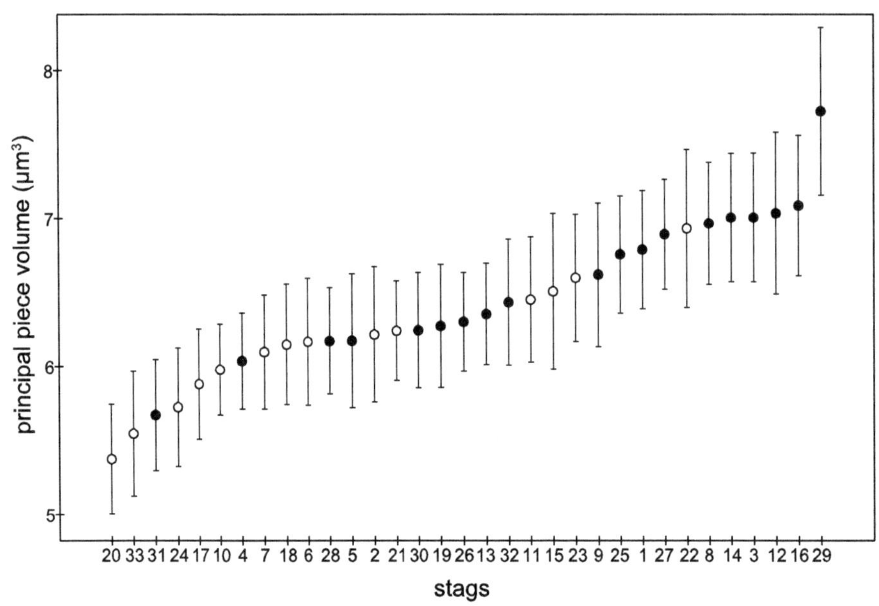

Figure 3. Classification of sperm principal piece volume according to red deer sperm freezability. Good freezers are shown in open circles and bad freezers in closed circles.

Table 4. Overall morphometry of fresh sperm (data derived from 825 spermatozoa from 33 red deer) and differences between good (GF) and bad freezers (BF) in sperm morphometry parameters.

Assessed parameters	Overall (N = 33)			GF (n = 14)	BF (n = 19)	*P*
	Mean ± SD	Range	CV	Mean ± SD	Mean ± SD	
Sperm head						
Width (μm)	5.17±0.12	4.82–5.35	2.35	5.17±0.11	5.16±0.13	0.766
Length (μm)	8.75±0.24	8.35–9.30	2.79	8.77±0.25	8.74±0.25	0.730
Area (μm^2)	35.51±1.17	33.09–37.94	3.30	35.64±1.30	35.41±1.10	0.597
Perimeter (μm)	22.23±0.43	21.36–23.23	1.92	22.27±0.46	22.20±0.41	0.653
Flagellum and sperm length						
Midpiece width, proximal (μm)	0.94±0.03	0.85–1.03	3.56	0.93±0.05	0.94±0.02	0.382
Midpiece width, distal (μm)	0.74±0.03	0.68–0.80	3.63	0.73±0.02	0.75±0.03	**0.024**
Midpiece length (μm)	12.06±0.26	11.50–12.60	2.12	12.02±0.18	12.08±0.30	0.506
Principal piece length (μm)	41.47±1.26	38.92–44.39	3.04	41.07±1.42	41.76±1.07	0.122
Terminal piece length (μm)	2.68±0.26	2.05–3.21	9.89	2.61±0.28	2.72±0.25	0.248
Flagellum length (μm)	56.20±1.29	53.56–59.05	2.29	55.71±1.41	56.57±1.09	0.057
Midpiece volume (μm^3)	6.73±0.48	5.68–7.70	7.09	6.57±0.56	6.85±0.38	0.097
Principal piece volume (μm^3)	6.41±0.51	5.37–7.72	7.98	6.13±0.42	6.61±0.49	**0.006**
Flagellum volume (μm^3)	12.95±1	10.36–15.22	7.70	12.67±1.24	13.16±0.73	0.161
Sperm length (μm)	64.96±1.29	62.35–67.52	1.99	64.48±1.34	65.31±1.17	0.068

Bold letters show significant differences between GF and BF. SD, standard deviation; CV, coefficient of variation.

they are also negatively correlated with sperm velocity. According to this fact, Peña et al. [3] found that midpiece width is a predictor of post-thawing boar sperm motility. In our study, the principal piece volume is mainly determined by the distal midpiece width (Figure S2), which can explain their similar relationships with sperm velocity, and also the differences found between the GF and the BF in these sperm measures. The damage caused by sperm freezing protocols to the sperm head and tail membranes may occur independently: an intact tail membrane does not necessarily indicate an intact sperm head membrane and vice versa [47]. For example, after the freezing-thawing process, the disruption of the sperm head membrane occurs more easily than the tail in human and ram sperm [23,47], whereas the flagellum membrane is more vulnerable to the cryopreservation process in equine sperm [7]. Sperm with an intact head membrane but a damaged flagellum are most likely immotile, explaining the low fertilization rates with frozen/thawed sperm [7,48]; therefore, these cells should be included in the dead rather than the alive category [48]. In our study, the sperm flagellum is likely more sensitive to the freezing/thawing process than the sperm head or, at least, a damaged flagellum negatively affects sperm kinetics more than a damaged sperm head.

The fragility of the sperm flagellum and its ability to withstand the freezing/thawing process has been reported in many studies. The addition of glycerol as a cryoprotectant agent alters sperm functionality, mainly during the post-thawing sperm incubation stage [19], showing an increase in the proportion of epididymal spermatozoa with axonemal vacuoles, damaged plasma mem-

Table 5. Results of principal component analysis (PCA) to determine overall sperm velocity at 0 and 2 hours post-thaw.

PCA	0 hours	2 hours
Variables	PC1	PC1
VAP	0.986	0.951
VCL	0.975	0.967
VSL	0.960	0.920
ALH	0.957	0.927
SMI	0.907	0.809
Variance explained (%)	91.676	83.978
Eigenvalue	4.584	4.199
Bartlett's test of sphericity	0.000	0.000
Kaiser-Meyer-Olkin test	0.795	0.744

VAP, average path velocity; VCL, curvilinear velocity; VSL, straight linear velocity; ALH, amplitude of lateral head displacement; SMI, sperm motility index.

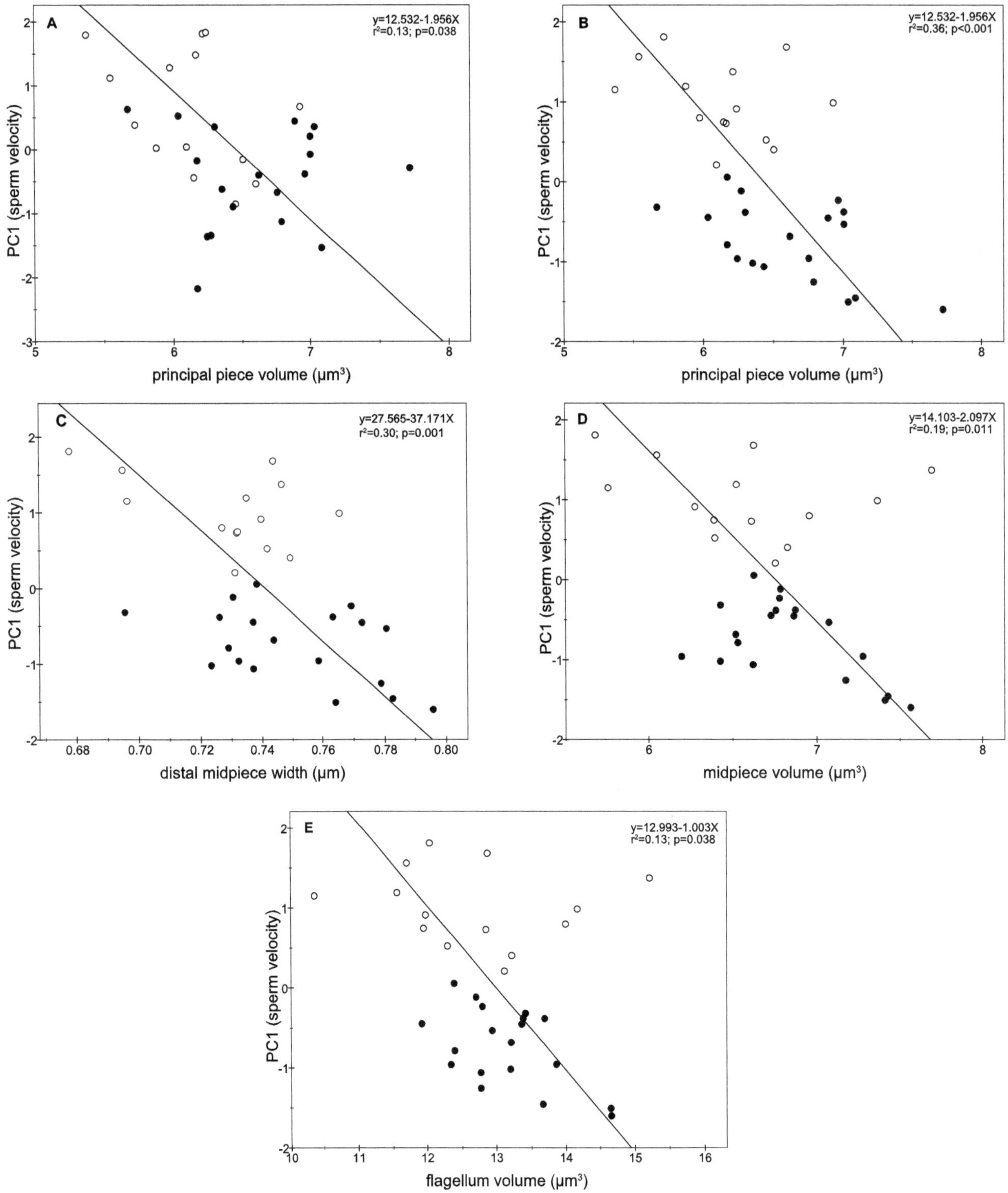

Figure 4. Relationships between sperm velocity and flagellum morphometry. Good freezers are shown in open circles and bad freezers in closed circles: A (0 hours post-thaw); B, C, D, and E (2 hours post-thaw).

branes, and abnormal mitochondria when compared to media without glycerol [50]. Furthermore, the cytoskeleton is responsible for the appropriate cell volume regulation and its stability is highly altered by the cryopreservation process, causing its proteins (e.g., F-actin) to become more fragile [13,24,51,52]. Thus, in red deer, a higher volume of the sperm flagellum might result in an increased amount of ice crystals and increased formation of axonemal vacuoles during the freezing/thawing process, adversely affecting flagellum integrity and consequently, cell volume regulation and sperm velocity. Supporting this hypothesis, Correa et al. [53]

suggested a direct connection between cell volume regulation, flagellum morphology, motility, and the actin cytoskeleton in the sublethal damage that occurs during osmotic stress and, potentially, during cryopreservation. On the other hand, we did not find any significant differences between the GF and the BF in the midpiece or flagellum size which contain sperm mitochondria and the whole sperm axoneme, respectively. On the contrary, the sperm principal piece volume differs between the GF and the BF. Such a result might be related to the fibrous sheath, which is located along the principal piece and linked with sperm kinetics [54]. Thus, differences among males in the principal piece volume could differently affect their sperm freezability. Furthermore, it is thought that the fibrous sheath plays a mechanical role in sperm motility, providing a rigid support to the flagellum and determining its planar beat [55,56]. Additionally, glycolysis is carried out along the length of the principal piece and this, instead of oxidative phosphorylation in the midpiece, is the most important source of ATP for the tail [57]. At least one fibrous sheath protein may act to protect sperm from oxidative stress, which could interfere with sperm motility or cause DNA damage [58]. Therefore, the proteome and the general structure of the fibrous sheath could be potentially damaged by the sperm freezing and thawing process, particularly in those sperm with a higher sperm principal piece volume, causing a decrease in sperm function mainly during sperm incubation.

In conclusion, our results provide evidence that the volumes of the flagellum structures are a determinant to predict post-thaw sperm velocity in red deer, and the BF have a higher sperm principal piece volume than the GF. In contrast, sperm head size is not a good predictor of post-thaw sperm velocity in red deer spermatozoa. However, further studies, including additional analyses such as freeze-fracture electron microscopy observations of spermatozoa [50,59] and the evaluation of sperm tail membrane integrity by light microscopy [48] during sperm cooling and especially during sperm warming, are necessary in order to confirm our findings. On the other hand, our subsequent studies will be directed towards the use of electron microscopy, to assess sperm flagellum morphometry in more depth, and also to measure their internal structures (axoneme, fibrous sheath, etc.). Our results clearly show that a higher principal piece volume results in poor sperm freezability, and highlights the key role of the volume of flagellum structures in sperm cryopreservation success.

Supporting Information

Figure S1 Relationships between sperm head perimeter and intact acrosome at 0 hours of sperm thawing ($r = -0.365$; $p = 0.037$).

Figure S2 Relationships between sperm principal piece volume and distal midpiece width ($r = 0.93$; $p < 0.0001$).

Table S1 Individual mean morphometry and subjective motility of fresh red deer spermatozoa ($N = 33$).

Table S2 Individual mean kinetics, viability, and organelle status of red deer spermatozoa at 0 hours post-thaw ($N = 33$).

Table S3 Individual mean kinetics, viability, and organelle status of red deer spermatozoa at 2 hours post-thaw ($N = 33$).

Acknowledgments

The assistance provided by Alfonso Bisbal, Enrique del Olmo, Mari Cruz Sotos, and Zandra Maulen is gratefully acknowledged. Eliana Pintus is acknowledged for her insightful comments. Landowners, managers, and rangers are acknowledged for facilitating access to samples.

Author Contributions

Conceived and designed the experiments: JLRS JJG. Performed the experiments: JLRS AEDR. Analyzed the data: JLRS. Contributed reagents/materials/analysis tools: JJG. Wrote the paper: JLRS. Sperm morphometry assessment: JLRS. Revised the manuscript: JLRS AEDR JJG.

References

1. Watson PF (1995) Recent developments and concepts in the cryopreservation of spermatozoa and the assessment of their post-thawing function. Reprod Fertil Dev 7: 871–891.
2. Watson PF (2000) The causes of reduced fertility with cryopreserved semen. Anim Reprod Sci 60–61: 481–492.
3. Peña FJ, Saravia F, Garcia-Herreros M, Nuñez-Martinez I, Tapia JA (2005) Identification of sperm morphometric subpopulations in two different portions of the boar ejaculate and its relation to postthaw quality. J Androl 26: 716–723.
4. Parks JE, Graham JK (1992) Effects of cryopreservation procedures on sperm membranes. Theriogenology 38: 209–222.
5. Billard R, Cosson J, Linhart O (2000) Changes in the flagellum morphology of intact and frozen/thawed Siberian sturgeon *Acipenser baerii* (Brandt) sperm during motility. Aquac Res 31: 283–287.
6. Holt WV (2000) Basic aspects of frozen storage of semen. Anim Reprod Sci 62: 3–22.
7. Domes U, Stolla R (2001) Staining of stallion spermatozoa with damaged tail membrane after freezing/thawing. Anim Reprod Sci 68: 329–330.
8. Devireddy RV, Swanlund DJ, Alghamdi AS, Duoos LA, Troedsson MH, et al. (2002) Measured effect of collection and cooling conditions on the motility and the water transport parameters at subzero temperatures of equine spermatozoa. Reproduction 124: 643–648.
9. Brouwers JF, Silva PF, Gadella BM (2005) New assays for detection and localization of endogenous lipid peroxidation products in living boar sperm after BTS dilution or after freeze-thawing. Theriogenology 63: 458–469.
10. Cremades T, Roca J, Rodriguez-Martinez H, Abaigar T, Vazquez JM, et al. (2005) Kinematic changes during the cryopreservation of boar spermatozoa. J Androl 26: 610–618.
11. Holt WV, Medrano A, Thurston LM, Watson PF (2005) The significance of cooling rates and animal variability for boar sperm cryopreservation: insights from the cryomicroscope. Theriogenology 63: 370–382.
12. Thurston LM, Siggins K, Mileham AJ, Watson PF, Holt WV (2002) Identification of amplified restriction fragment length polymorphism markers linked to genes controlling boar sperm viability following cryopreservation. Biol Reprod 66: 545–554.
13. Petrunkina AM, Gröpper B, Günzel-Apel AR, Töpfer-Petersen E (2004) Functional significance of the cell volume for detecting sperm membrane changes and predicting freezability in dog semen. Reproduction 128: 829–842.
14. Esteso MC, Soler AJ, Fernández-Santos MR, Quintero-Moreno AA, Garde JJ (2006) Functional significance of the sperm head morphometric size and shape for determining freezability in iberian red deer (*Cervus elaphus hispanicus*) epididymal sperm samples. J Androl 27: 662–670.
15. Tuset VM, Dietrich GJ, Wojtczak M, Słowińska M, de Monserrat J (2008) Comparison of three staining techniques for the morphometric study of rainbow trout (*Oncorhynchus mykiss*) spermatozoa. Theriogenology 69: 1033–1038.
16. Malo AF, Gomendio M, Garde J, Lang-Lenton B, Soler AJ (2006) Sperm design and sperm function. Biol Lett 2: 246–249.
17. Voordouw MJ, Koella JC, Hurd H (2008) Intra-specific variation of sperm length in the malaria vector *Anopheles gambiae*: males with shorter sperm have higher reproductive success. Malar J 7: 214.
18. Anderson MJ, Dixson AF (2002) Sperm competition: motility and the midpiece in primates. Nature 416: 496.
19. Anderson MJ, Nyholt J, Dixson AF (2005) Sperm competition and the evolution of sperm midpiece volume in mammals. J Zool 267: 135–142.
20. Parapanov RN, Nusslé S, Crausaz M, Senn A, Hausser J (2009) Testis size, sperm characteristics and testosterone concentrations in four species of shrews (Mammalia, Soricidae). Anim Reprod Sci 114: 269–278.

21. Ramm SA, Stockley P (2010) Sperm competition and sperm length influence the rate of mammalian spermatogenesis. Biol Lett 6: 219–221.
22. Keates RA (1980) Effects of glycerol on microtubule polymerization kinetics. Biochem Biophys Res Commun 97: 1163–1169.
23. Holt WV, North RD (1994) Effects of temperature and restoration of osmotic equilibrium during thawing on the induction of plasma membrane damage in cryopreserved ram spermatozoa. Biol Reprod 51: 414–424.
24. Felipe-Pérez YE, Valencia J, Juárez-Mosqueda Mde L, Pescador N, Roa-Espitia AL (2012) Cytoskeletal proteins F-actin and β-dystrobrevin are altered by the cryopreservation process in bull sperm. Cryobiology 64: 103–109.
25. Curry MR, Millar JD, Tamuli SM, Watson PF (1996) Surface area and volume measurements for ram and human spermatozoa. Biol Reprod 55: 1325–1332.
26. Garde JJ, Ortiz N, García AJ, Gallego L, Landete-Castillejos T (1998) Postmortem assessment of sperm characteristics of the red deer during the breeding season. Arch Androl 41: 195–202.
27. Fernández-Santos MR, Esteso MC, Montoro V, Soler AJ, Garde JJ (2006) Cryopreservation of Iberian red deer (*Cervus elaphus hispanicus*) epididymal spermatozoa: effects of egg yolk, glycerol and cooling rate. Theriogenology 66: 1931–1942.
28. Domínguez-Rebolledo AE, Fernández-Santos MR, Bisbal A, Ros-Santaella JL, Ramón M, et al. (2010) Improving the effect of incubation and oxidative stress on thawed spermatozoa from red deer by using different antioxidant treatments. Reprod Fertil Dev 22: 856–870.
29. Garde JJ, Martínez-Pastor F, Gomendio M, Malo AF, Soler AJ, et al. (2006) The application of reproductive technologies to natural populations of red deer. Reprod Domest Anim 41: 93–102.
30. Comizzoli P, Mauget R, Mermillod P (2001) Assessment of *in vitro* fertility of deer spermatozoa by heterologous IVF with zona-free bovine oocytes. Theriogenology 56: 261–274.
31. Hossain AM, Barik S, Kulkarni PM (2001) Lack of significant morphological differences between human X and Y spermatozoa and their precursor cells spermatids exposed to different prehybridization treatments. J Androl 22: 119–123.
32. Sanchez L, Petkov N, Alegre E (2005) Statistical approach to boar semen head classification based on intracellular intensity distribution. Computer Analysis of Images and Patterns, Proceedings 3691: 88–95.
33. Ramanujan S (1914) Modular equations and approximations to π. Quart J Pure Appl Math 45: 350–372.
34. Malo AF, Garde JJ, Soler AJ, Garcia AJ, Gomendio M (2005) Male fertility in natural populations of red deer is determined by sperm velocity and the proportion of normal spermatozoa. Biol Reprod 72: 822–829.
35. Nuñez-Martínez I, Morán JM, Peña FJ (2006) Two-step cluster procedure after principal component analysis identifies sperm subpopulations in canine ejaculates and its relation to cryoresistance. J Androl 27: 596–603.
36. Budaev SV (2010) Using principal components and factor analysis in animal behaviour research: caveats and guidelines. Ethology 116: 472–480.
37. Bohonak AJ, Linde K (2004) RMA: Software for reduced major axis regression, Java version. Website: http://www.kimvdlinde.com/professional/rma.html.
38. Gage MJ, Macfarlane CP, Yeates S, Ward RG, Searle JB (2004) Spermatozoal traits and sperm competition in Atlantic salmon: relative sperm velocity is the primary determinant of fertilization success. Curr Biol 14: 44–47.
39. Birkhead TR, Martinez JG, Burke T, Froman DP (1999) Sperm mobility determines the outcome of sperm competition in the domestic fowl. Proc R Soc B-Biol Sci 266: 1759–1764.
40. Holt WV, Shenfield F, Leonard T, Hartman TD, North RD (1989) The value of sperm swimming speed measurements in assessing the fertility of human frozen semen. Hum Reprod 4: 292–297.
41. Bollwein H, Fuchs I, Koess C (2008) Interrelationship between plasma membrane integrity mitochondrial membrane potential and DNA fragmentation in cryopreserved bovine spermatozoa. Reprod Domest Anim 43: 189–195.
42. Anel-López L, Alvarez-Rodríguez M, García-Álvarez O, Alvarez M, Maroto-Morales A, et al. (2012) Reduced glutathione and Trolox (vitamin E) as extender supplements in cryopreservation of red deer epididymal spermatozoa. Anim Reprod Sci 135: 37–46.
43. Saacke RG, White JM (1972) Semen quality tests and their relationship to fertility. Proceedings of the Fourth Technical Conference on Artificial Insemination and Reproduction National Association of Animal Breeders 2–7.
44. Del Olmo E, Bisbal A, Maroto-Morales A, García-Alvarez O, Ramon M, et al. (2013) Fertility of cryopreserved ovine semen is determined by sperm velocity. Anim Reprod Sci 138: 102–109.
45. Medrano A, Watson PF, Holt WV (2002) Importance of cooling rate and animal variability for boar sperm cryopreservation: insights from the cryomicroscope. Reproduction 123: 315–322.
46. Woolley DM, Richardson DW (1978) Ultrastructural injury to human spermatozoa after freezing and thawing. J Reprod Fertil 53: 389–394.
47. Zhu WJ, Liu XG (2000) Cryodamage to plasma membrane integrity in head and tail regions of human sperm. Asian J Androl 2: 135–138.
48. Nagy S, Házas G, Papp AB, Iváncsics J, Szász F (1999) Evaluation of sperm tail membrane integrity by light microscopy. Theriogenology 52: 1153–1159.
49. Macías García B, Ortega Ferrusola C, Aparicio IM, Miró-Morán A, Morillo Rodriguez A, et al. (2012) Toxicity of glycerol for the stallion spermatozoa: effects on membrane integrity and cytoskeleton, lipid peroxidation and mitochondrial membrane potential. Theriogenology 77: 1280–1289.
50. McClean RV, Holt WV, Johnston SD (2007) Ultrastructural observations of cryoinjury in kangaroo spermatozoa. Cryobiology 54: 271–280.
51. Petrunkina AM, Hebel M, Waberski D, Weitze KF, Töpfer-Petersen E (2004) Requirement for an intact cytoskeleton for volume regulation in boar spermatozoa. Reproduction 127: 105–115.
52. Petrunkina AM, Radcke S, Günzel-Apel AR, Harrison RAP, Töpfer-Petersen E (2004) Role of potassium channels, the sodium-potassium pump and the cytoskeleton in the control of dog sperm volume. Theriogenology 61: 35–54.
53. Correa LM, Thomas A, Meyers SA (2007) The macaque sperm actin cytoskeleton reorganizes in response to osmotic stress and contributes to morphological defects and decreased motility. Biol Reprod 77: 942–953.
54. Schlingmann K, Michaut MA, McElwee JL, Wolff CA, Travis AJ (2007) Calmodulin and CaMKII in the sperm principal piece: evidence for a motility-related calcium/calmodulin pathway. J Androl 28: 706–716.
55. Fawcett DW (1975) The mammalian spermatozoon. Dev Biol 44: 394–436.
56. Lindemann CB, Orlando A, Kanous KS (1992) The flagellar beat of rat sperm is organized by the interaction of two functionally distinct populations of dynein bridges with a stable central axonemal partition. J Cell Sci 102: 249–260.
57. Turner RM (2006) Moving to the beat: a review of mammalian sperm motility regulation. Reprod Fertil Dev 18: 25–38.
58. Fulcher KD, Welch JE, Klapper DG, O'Brien DA, Eddy EM (1995) Identification of a unique mu-class glutathione S-transferase in mouse spermatogenic cells. Mol Reprod Dev 42: 415–424.
59. Holt WV, North RD (1984) Partially irreversible cold-induced lipid phase transitions in mammalian sperm plasma membrane domains; freeze fracture study. J Exp Zool 230: 473–483.

23

Encapsulating Non-Human Primate Multipotent Stromal Cells in Alginate via High Voltage for Cell-Based Therapies and Cryopreservation

Oleksandr Gryshkov[1]*, Denys Pogozhykh[1], Nicola Hofmann[1], Olena Pogozhykh[2], Thomas Mueller[2], Birgit Glasmacher[1]

1 Institute for Multiphase Processes, Leibniz Universitaet Hannover, Hannover, Germany, **2** Institute for Transfusion Medicine, Hannover Medical School, Hannover, Germany

Abstract

Alginate cell-based therapy requires further development focused on clinical application. To assess engraftment, risk of mutations and therapeutic benefit studies should be performed in an appropriate non-human primate model, such as the common marmoset (*Callithrix jacchus*). In this work we encapsulated amnion derived multipotent stromal cells (MSCs) from *Callithrix jacchus* in defined size alginate beads using a high voltage technique. Our results indicate that i) alginate-cell mixing procedure and cell concentration do not affect the diameter of alginate beads, ii) encapsulation of high cell numbers (up to 10×10^6 cells/ml) can be performed in alginate beads utilizing high voltage and iii) high voltage (15–30 kV) does not alter the viability, proliferation and differentiation capacity of MSCs post-encapsulation compared with alginate encapsulated cells produced by the traditional air-flow method. The consistent results were obtained over the period of 7 days of encapsulated MSCs culture and after cryopreservation utilizing a slow cooling procedure (1 K/min). The results of this work show that high voltage encapsulation can further be maximized to develop cell-based therapies with alginate beads in a non-human primate model towards human application.

Editor: Giovanni Camussi, University of Torino, Italy

Funding: This work was supported by Deutsche Forschungsgemeinschaft (DFG - German Research Foundation, http://www.dfg.de/) by the Cluster of Excellence REBIRTH (DFG EXC 62/1, http://www.rebirth-hannover.de) and Open Access Publishing Fund of Leibniz Universitaet Hannover. The funders had no role in study design, data collection and analysis, decision to publish, or preparation of the manuscript.

Competing Interests: The authors have declared that no competing interests exist.

* Email: gryshkov@imp.uni-hannover.de

Introduction

Cell-based therapies are under development to treat a wide range of acute and chronic diseases. To date, they have been successfully applied in treatments of the central and peripheral nervous system [1], bone and cartilage regeneration, hepatic fibrosis and cardiac insufficiencies [2,3]. The main challenge in such allogenic therapies is the suppression of the host immune system prior to and during the treatment. In addition, drug-based immune system suppression has many side effects for the patient [4]. One strategy to avoid harmful immunosupression of the host is the suppression of the major histocompatibility complex I (MHC I), a major obstacle in transplantation, in the transplanted cells by small hairpin RNA (shRNA) method [5]. Alternatively, cells can be encapsulated into polymer matrices with semi-permeable properties; these shield transplanted cells from immune responses, while allowing controlled release of drugs and cellular products [6]. Interestingly, most matrices mimic the extra-cellular matrix and therefore provide the cells with a niche-like environment during post-transplantation (Figure 1A).

Alginate is known to be a linear block co-polymer containing sequences of (1–4)-linked β-D-mannuronate (M-residue), its C-5 epimer α-L-guluronate (G-residue) and alternating M and G residues (MG-residues). It can be produced from brown algae and bacteria. However, alginate extracted from different sources has variable properties and alginate beads produced by a range of cross-linking methods display a wide range of final biological and physical properties, affecting the mechanical properties of a bead and cell response *in-vitro* and *in-vivo* [7]. The features of cross-linking include: ionic cross-linking with divalent cations, such as barium and calcium, that provides living cells with a niche-like environment and ease of gelation, forming so called egg-box structure [8]; whereas covalent cross-linking with polyethylene glycol induced by laser irradiation enhances the degradation rate facilitating a more homogeneous cross-linking [9]. Cell cross-linking utilizing arginine-glycine-aspartate (RGD) coupled alginate improves attachment of adherent mammalian cell types [10], whereas thermal gelation with thermo-responsible materials allows cross-linking at body temperature [11]. In summary, by varying the type and density of cross-linking, the physical and biological properties of the bead can be controlled and matched with the required application.

To date, three major methods are suitable for encapsulation of living cells in alginate beads: 1) electro-spraying (ES), utilizing a

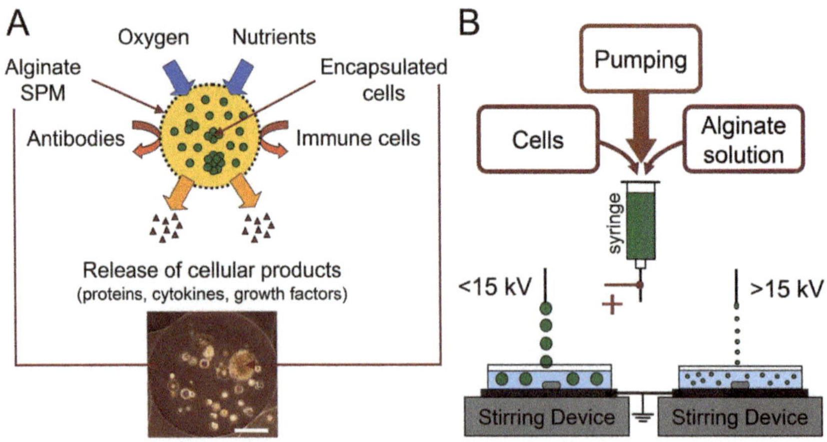

Figure 1. Schematic presentation of alginate high voltage encapsulation. (**A**) Application of encapsulation of cells in alginate using high voltage (**B**) in cell-based therapy for immunoisolation, controllable drug release through semi-permeable membrane (SPM) and long-term storage of cells. Scale bar is 100 μm.

voltage in order to detach alginate beads from the nozzle, 2) air-flow (AF), allowing bead production using air motion under pressure, and 3) ink-jet fabrication, a method based on piezoelectric effects [12,13]. However, encapsulation methods require further optimization in order to reduce bead size while avoiding ruptured structures. Smaller alginate beads are preferable in cell culture and transplantation due to increased surface-to-volume ratio and enhanced heat and mass transfer to encapsulated cells. Here, the application of poly-cationic coatings, such as poly-L-lysine, has been shown to stabilize the structure of alginate beads and control molecular weight cut-off [14], but bead biocompatibility was significantly lower when compared to uncoated and unmodified calcium and barium beads [15,16].

The terminology "mesenchymal stem cells" is misleading because the mesenchyme is embryonic connective tissue from mesodermal origin, differentiating in hematopoietic and connective tissue, which this cell type cannot. Therefore, "multipotent stromal cells" (MSC) is a far more accurate term for this interesting cell population which will be tested prior to clinical application in the common marmoset monkey (*Callithrix jacchus*) as a relevant preclinical non-human primate model. For future application in regenerative medicine, the introduction of such a model is more important than widely used rodent models due to high phylogenetic similarity of a marmoset to a human and derivation of embryonic (ESC), induced pluripotent (iPS) and adult stem cells [17–20]. In our experiments, MSCs were derived from the placental amnion membrane of the animals, offering a non-invasive strategy for retrieval and theoretical availability for each (future) patient. This is due to the fact that the amnion membrane is generated from the embryonal epiblast, whereas the chorion is originated from the trophoblast and the decidua from maternal origin [21]. Immediate availability of these cells can be assured by their long-term storage at low temperatures with appropriate cryopreservation procedures. This is currently the only possible technique for the long term storage of rare cell types. The preservation of stem cells with high viability, proliferation and yet preserving their differentiation potential called "stemness" still poses challenges. One strategy to improve viability and proliferation after cryopreservation deals with the encapsulation of cells in small-sized alginate beads before freezing. The gel-like structure, mild environment inside alginate beads and improved heat and mass transfer due to increased surface-to-volume ratio may protect encapsulated cells from cryo-injury and resist the reorganization of ice crystals during thawing.

Therefore, in this work, we applied high voltage ES to encapsulate MSCs in small alginate beads with defined diameter. We investigated the efficiency of the ES method to encapsulate high cell numbers and the effects of cell concentration and cell-mixing procedure on bead diameter. Furthermore, we evaluated the effect of high voltage encapsulation procedure and cryopreservation on viability, proliferation and differentiation of MSCs after encapsulation, using the AF method as a comparative approach.

Materials and Methods

All chemicals used in the study were purchased from Sigma, unless indicated otherwise.

Alginate: source and sterilization

Medium viscosity alginate sodium salt from brown algae (molecular weight 80000–120000 g/mol, viscosity 2000 cP (2%, 25°C)) was dissolved in 10 mM 4-(2-hydroxyethyl)-1-piperazineethanesulfonic acid (HEPES, Carl Roth GmbH, Karlsruhe, Germany) buffered saline (pH 7.4) at a concentration of 1.6% (w/v) with gentle shaking overnight. Aliquots of alginate solution were kept at 4°C until use. Raw alginate solution was sterilized using 0.8-0.45-0.2 μm filters. Membrane filtration of alginate solution was used to sterilize the product, as this the only type of sterilization that does not cause significant changes in molecular weight and viscosity of alginate [22]. Sterile alginate solution was transferred to 10 ml sterile syringes and kept at 4°C. Aliquots of alginate solution were left to warm up at room temperature (RT) one day prior to the experiments.

Encapsulation methods

We have previously designed the high voltage ES encapsulation method and optimized the process parameters. It applies a high voltage to overcome the surface tension of the alginate solution (Figure 1B). Briefly, the alginate solution is held in a syringe and passed through an adaptor under controlled flow rate to the nozzle (blunt needle, outer/inner diameters 0.4/0.2 mm, length 25 mm, BBraun, Melsungen, Germany). Alginate gel is detached from the tip of the nozzle by applying a high voltage between the needle and the grounded electrode with following cross-linking in the bath containing stirred solution with Ca^{2+} ions. Calcium ions replace sodium ions in alginate due to their higher affinity to the alginate, forming a gel-like structure. We compared our approach to the commonly used AF method that does not apply a high voltage to generate alginate beads. The detailed description of the AF encapsulation method has been published previously [12]. In brief, a high velocity air flow overcomes the surface tension of the alginate solution resulting in bead detachment. The AF method has previously been optimized for the production of alginate beads with a narrow size distribution compared to the beads obtained using ES approach (bead diameter 300–400 μm) [23].

The size of generated alginate beads was determined to investigate the encapsulation efficiency. Briefly, the alginate beads (with or without encapsulated cells) were cross-linked after encapsulation for 30 min (unless stated otherwise). After cross-linking, the diameter of alginate beads generated by the ES or AF method was measured using an Axiovert 200M microscope (Carl Zeiss, Germany). Pictures were taken in bright-field at 5X magnification and analyzed using AxioVision software (Rel. 4.7) (8–10 images, 10–15 beads per image). The average bead diameter of one hundred beads was determined.

Cell derivation, characterization, culture and encapsulation

Ethics statement. No animal experiments were performed in this study. Donation of the placenta material after natural birth's was carried out from the Center for Reproductive Medicine and Andrology Muenster (CeRA) by institutional animal care takers and approved by the Institutional Animal Care. The ethical husbandry of *Callithrix jacchus* in this institute was in accordance with the guidelines of the Ministry for Environment and Nature Protection, Agriculture and Consumer Protection (LANUV), North-Rhine Westphalia, Germany.

Cell derivation and isolation. Placental samples were collected after natural birth from healthy marmosets in the CeRA. Placentas were washed with phosphate buffered saline (PBS, Biochrom AG, Germany) with 1% Ciprobay (Fresenius Kabi, Bad Homburg, Germany). Then the amnion membrane of each part of placenta was removed and dissected into smaller pieces. The amnion membrane pieces were incubated with 0.25% trypsin for 1 h at 37°C, filtered through 100 μm cell strainers and centrifuged at 1,200 rpm for 5 min. The cell pellet was re-suspended in MSC growth medium and plated into 10 cm cell dishes. All cell samples were tested for mycoplasma contamination.

Cell culture and encapsulation. MSCs derived from the common marmoset monkey *Callithrix jacchus* were cultured under sterile conditions in Dulbecco's modified Eagle's medium (DMEM, Biochrom AG, Germany) containing 15% (v/v) fetal bovine serum (FBS, Biochrom AG, Germany), 1% penicillin/streptomycin, 1% ascorbic acid in 10 cm tissue culture dishes (TPP, Biochrom AG, Germany) in a humidified incubator at 37°C and 5% CO_2. The cells were harvested using 0.05%/0.02% (w/v) trypsin/EDTA solution followed by centrifugation at 1,200 rpm for 5 min. Membrane integrity of MSCs was assessed by trypan blue staining using Vi-CELL XR cell viability analyzer (Beckman Coulter GmbH, Krefeld, Germany). After counting, MSCs were centrifuged and re-suspended in 10 mM HEPES buffered saline (pH 7.4) to adjust the desired cell concentration before encapsulation in alginate. The encapsulation was performed by the AF and ES method; here, the cells were gently mixed with 1.6% alginate in 10 ml syringe and pumped at 10 ml/h flow rate into 100 mM $CaCl_2$ cross-linking solution (in 10 mM HEPES buffered saline). The obtained alginate beads containing MSCs were washed twice with washing solution (WS), containing 10 mM HEPES, 1.5 mM $CaCl_2$ buffered saline (pH 7.4), and either immediately used for experimental purposes after encapsulation ("Immediate") or after a day of culture in a humidified incubator at 37°C and 5% CO_2 ("Incubated"). The presence of calcium ions in WS prevented alginate beads from swelling [24]. The following process parameters were kept constant during encapsulation using both the AF and ES method: spraying distance 10 cm, alginate flow rate 10 ml/h, concentration of $CaCl_2$ solution 100 mM, applied voltage 20 kV, air flow 150 l/h, unless stated otherwise.

Flow cytometry marker analysis. After trypsinization and fixation in 4% paraformaldehyde (PFA, Merck, Germany), cells were aliquoted in equal numbers for flow cytometry (FACS Calibur, Becton Dickinson) and stained with primary and secondary antibodies at RT for 1 h, respectively (Table S1). Cells incubated with only secondary antibody were used as negative controls.

Immunohistochemical staining. For immunohistochemical staining Dako LSAB+System-HRP kit was used (Dako North America, California, USA). For the examination for mesenchymal markers, 2×10^4 cells/well of each sample were seeded on a glass slide (20 mm) in a 12-well plate (Greiner Bio-One GmbH, Frickenhausen, Germany) and then fixed with 4% PFA. After washing steps with PBS, six drops of 3% hydrogen peroxide were added to each well for 5 min to block endogenous peroxidase activities. The cells were incubated with appropriate antibody over night at 4°C (Table S1). Then the link solution was applied to the glass slides for 30 min followed by streptavidin peroxidase for 30 min. The substrate solution was added to each well and incubated for 7–15 min until the antibody horseradish peroxidase conjugate (AB-HRP) catalyzed the 3,3′-diaminobenzidine (DAB) substrate into the desired brown to black reporter color for a positive signal. Afterwards, cells were washed twice with 1 ml distilled water (dH_2O) and treated with 300 μl haematoxylin for 3–5 min providing counterstaining of the nuclei. The glass slides were additionally rinsed with dH_2O, transferred upside down onto an object plate in one drop of Mowiol 4–88 (Carl Roth GmbH, Karlsruhe, Germany) and dried over night in the dark. Images were taken using a Keyence Biozero microscope (Keyence Germany GmbH, Neu-Isenburg, Germany).

Total RNA Isolation and Reverse Transcription Polymerase Chain Reaction (RT-PCR). RNA was extracted using the peqGOLD Total RNA Kit (Peqlab GmbH, Erlangen, Germany) according to the manufacturer's instructions. Briefly, the cell pellet was lysed in 400 μl RNA Lysis Buffer, transferred to a provided DNA removing column and centrifuged at 12,000 g for 1 min. Next, 400 μl 70% ethanol were added to the flow through, the lysate was loaded onto a provided Perfect Bind RNA Column and centrifuged at 10,000 g for 1 min. After three washing steps (1x 500 μl RNA Wash Buffer I and 2×600 μl RNA Wash Buffer II), the column was dried by centrifugation at 10,000 g for 2 min and the RNA eluted from the column by applying 50 μl sterile RNAse-free dH_2O. Concentration of the obtained RNA was measured by a Nanodrop photometer (Spectrophotometer ND1000, Peqlab GmbH, Erlangen, Germany). Extracted RNA was transcribed into complementary DNA (cDNA) using the High Capacity cDNA Reverse Transcription Kit (Life technologies, Carlsbad, USA). By adding Oligo-17-dT primers (TIB Molbiol, Berlin, Germany) only the messenger RNA (mRNA) with intact poly-A tail was transcribed. For the analysis of the mesenchymal markers and immunorelevant molecules, a RT-PCR reaction of 30 μl per sample was set up with standard conditions (24 μl dH_2O, 3 μl 1X PCR buffer (NEB, Frankfurt, Germany), 0.5 Units Taq Polymerase (NEB, Frankfurt, Germany), 100 mM nucleoside triphosphates containing deoxyribose (dNTPs, Fermentas, St. Leon-Rot, Germany), 20 pmol/µl of each primer and 1 µg cDNA). The RT-PCR reaction was performed in a Thermal cycler C1000 (Bio-Rad Laboratories GmbH, Munich, Germany). The cycling conditions were as follows: 1x initial denaturation step at 95°C for 180 s, followed by 35x denaturation at 95°C for 45 s, annealing at 58°C for 45 s, extension at 72°C for 90 s and a final extension step at 72°C for 600 s. In all amplifications, ribosomal protein S29 (RPS29) was utilized as a housekeeper gene; exact oligonucleotide sequences, corresponding fragment sizes and annealing temperatures are enlisted in Table S1.

Optimization of cell number

Cell concentration (1×10^6, 5×10^6, 10×10^6 cells/ml), alginate-cell mixing procedure and applied voltage (10, 15, 20, 25, 30 kV) was also optimized, since the incorporation of living cells may also change the viscosity and surface tension of final alginate-cell solution resulting in increased bead size, thus affecting the heat and mass transfer to the encapsulated cells. Concentration of alginate (1.6% in 10 mM HEPES buffered saline, w/v), cross-linking solution (100 mM in 10 mM HEPES buffered saline), spraying distance (10 cm), and alginate flow rate (10 ml/h) were kept constant. The diameter of alginate beads was measured as described above.

Shrinking rate and encapsulation efficiency

Shrinking rate (SR) of alginate in presence of gelling agents (Ca^{++}) was analyzed by comparing the initial volume of alginate droplets detached from the nozzle via gravity (manual dropping, MD), AF and high voltage. The volume of alginate beads was calculated by the diameter after cross-linking for 40 min. The volume at t = 0 revealed the initial volume of alginate droplet. The MD, AF and ES method was utilized to generate beads with different surface-to-volume ratios (S/V) as a result of initial bead diameter. The initial volume (also respective S/V ratio) was calculated by dropping the alginate solution on the collector and calculating the amount of generated droplets *vs.* dropping time. This preliminary experiment was performed ten times providing the resulted S/V ratios (assuming the alginate drop having a form of a sphere) of 2.47, 9.15, 8.28 and 11.47 mm^{-1} for 1.5%-MD, 1.2%-AF, 1.5%-AF and 1.5%-ES of experimental settings, respectively. The volume of alginate beads during gelling was calculated from measuring their diameter at defined time points during gelling: from 0 to 5 min with a step of 1 min, from 5 to 40 min with a step of 5 min. In general, SR can be presented as follows:

$$SR_t = \frac{V_0}{V_t} \tag{1}$$

whereas V_0 – initial volume of alginate drop, V_t – volume of alginate bead at time t of cross-linking. The value of SR calculated for 1.5%-ES and 10 min of gelling was used to analyze the efficiency of encapsulation of MSCs in alginate at a concentration of 1×10^6, 5×10^6 and 10×10^6 cells/ml. In this particular case (alginate concentration 1.5%, cross-linking time 10 min), assuming no release of cells from the alginate bead during spraying and gelling, the final number of cells encapsulated in alginate bead was evaluated as follows:

$$Cell\ Number = C_0 \times SR_t \times V_t \tag{2}$$

whereas C_0 – initial cell concentration (cells/ml), SR_t – shrinking rate of alginate after time t of gelling, V_t – volume of alginate bead after time t of gelling (ml).

Cell viability

MSCs were encapsulated into 1.6% alginate using AF and ES under high voltage (15, 20, 25, 30 kV) followed by washing steps. After washing, alginate beads were transferred to 6-well culture dishes (TPP, Biochrom AG, Germany) and cultured for 10 days in a humidified incubator at 37°C, 5% CO_2 with daily cell culture medium change. At defined time points (day 0, 1, 3, 5, 7 and 10), approximately 20 beads were taken, washed twice with WS and stained for live-dead viability based on previously published protocol with some modifications [25]. CalceinAM indicated viable cells upon their enzymatic activity by green fluorescence (excitation/emission: 494/517 nm), whereas Ethidium Homodymer (CalceinAM/EthD, Sigma, USA) indicated dead cells with ruptured cell membranes in red (excitation/emission: 528/617 nm) by intercalating to the nuclear DNA of the cell nuclei. Equal volumes of beads containing cells in HEPES and working solution of the live-dead staining were mixed to get the final concentrations of 2 µM and 4 µM of CalceinAM and Ethidium Homodymer, respectively, and incubated for 10 min in the dark at RT. After staining beads were washed twice with WS and immediately imaged using an Axiovert 200M fluorescence microscope at 5× magnification to detect viable and dead MSCs. Separate images of live and dead cells were analyzed using AxioVision software (Rel. 4.7).

Cell recovery

To evaluate the proliferation and differentiation capacity of differently encapsulated MSCs, the cells were allowed to release from the alginate beads by 55 mM sodium citrate in 10 mM HEPES buffered saline (pH 7.4) with gentle shaking for 5 min, thus replacing calcium ions by sodium ions. This resulted in formation of diluted sodium alginate and the release of MSCs from the beads. The cells were further collected by centrifugation at 1,200 rpm for 5 min and the proliferation and differentiation potential after encapsulation, incubation and cryopreservation evaluated. Because the recovery of cells might not be finished in only 24 h, we cultured the MSCs in 6-well culture flasks at a concentration of 1×10^5 cells/well in a humidified incubator at 37°C and 5% CO_2 for 5 days. After this period of time, we assessed the cell number, membrane integrity and further seeded cells for proliferation assay.

Cell proliferation

The MSCs were seeded in 96-well culture flasks (TPP, Biochrom AG, Germany) at a concentration of 1×10^4 cells/well for the CellTiter 96 Non-radioactive Cell Proliferation Assay (G4000, Promega, USA) following manufacturer's protocol. Briefly, after 24 h in culture flasks in a humidified incubator (37°C/5% CO_2), 15 µl dye solution of the proliferation assay was added to each well and incubated for 4 h to let living cells convert the tetrazolium component of the dye solution into formazan crystals. After the addition of 100 µl stop solution, the well was incubated for 1 h in the dark to solubilize the formed formazan crystals and the absorbance was measured at 570 nm wavelength using 96-well micro-plate reader (Multiscan Ex, Thermo Scientific, China). Absorbance values were normalized with controls of cultured cells without encapsulation and cryopreservation.

Characterization of differentiation potential

MSCs were encapsulated using the high voltage (20 kV) and AF method, released from the beads, tested for viability as described above and seeded at a density of 5×10^4 cells/well in 6-well culture dishes. The differentiation of MSCs towards adipogenic and osteogenic lineages was performed following the protocol published by our group [26].

Adipogenic. Adipogenic differentiation of MSCs was induced by culture of the cells for 5 days in medium containing 1 μM dexamethasone, 0.2 mM indomethacin, 0.5 mM 3-Isobutyl-1-methylxanthin and 10 μg/ml insulin in DMEM supplemented with 10% FBS. Detection of formed lipid vacuoles was assessed using Oil Red O stain as described earlier [26]. Briefly, the cells were washed with PBS, fixed in 4% PFA for 20 min and rinsed twice with tap water followed by one final wash with 50% ethanol (ROTH, Germany). Formed lipid vacuoles were detected by incubation for 10 min in Oil Red O in acetone/50% ethanol (Merck, Germany) and a final rinse with tap water. All differentiation procedures were repeated four times, compared with undifferentiated control cells (n = 4) and visualized with an Axiovert 200M microscope in bright field.

Osteogenic. The cells were cultured in osteogenic differentiation medium supplemented with 0.1 μM dexamethasone, 10 mM glycerophosphate, 0.05 mM L-ascorbic acid-2-phosphate and 1% ITS (BD Biosciences, Germany) in DMEM with 15% FBS. To detect mineralization of differentiated osteoblasts, Von Kossa staining was performed as described earlier [26]. Briefly, MSCs with the mineral deposit were washed twice in PBS and fixed with 10% PFA, washed once with PBS and twice with dH_2O followed by addition of 1% silver nitrate (Ridel de Haen GmbH, Germany). In the following, the dish was exposed to sunlight for 30 min, washed with bi-distilled water (bi-dH_2O), stained with 5% sodium thiosulfate for 5 min and again rinsed with bi-dH_2O.

All differentiation procedures were repeated four times, compared with undifferentiated control cells (n = 4) and visualized with an Axiovert 200M microscope in bright field.

Cryopreservation

The MSCs were encapsulated and cryopreserved at a concentration of 0.5×10^6 cells/sample either immediately after encapsulation ("Immediate") or after cultivation in culture medium in a humidified incubator at 37°C, 5% CO_2 for 24 h ("Incubated"). Cells encapsulated in alginate under high voltage (15, 20, 25 kV) and using the AF method were washed twice with WS and exposed to 2X freezing medium at 4°C drop-wise (0.5 ml/min), containing 20% (v/v) dimethyl sulfoxide (DMSO, Carl Roth GmbH, Germany) and 20% FBS to obtain the final concentration of 10% DMSO and 17.5% FBS. Samples were equilibrated at 4°C for 15 min in 1.5 ml cryo-vials (TPP, Biochrom AG, Germany) and cooled down at 1 K/min (slow cooling protocol) down to −80°C with Mr. Frosty Freezing Container (Thermo Fisher Scientific, Schwerte, Germany), without active control over the temperature of extracellular ice formation (spontaneous ice formation). Frozen samples were transferred to −150°C and kept for 7 days before thawing at 37°C in a water bath for 2 min. The thawing rate was as follows: from −150°C to −25°C with 243 K/min, from −25°C to 4°C with 18 K/min (Figure S1). In order to reduce the negative effect of osmotic stress which takes place during fast dilution procedure, the thawed samples were further diluted by drop-wise addition of fresh culture medium at 4°C. The cells were released from the beads as described above, checked for membrane integrity by trypan blue staining and seeded for proliferation assay either immediately after thawing or after being recovered for 5 days in culture. As control, non-encapsulated MSCs in suspension were also cryopreserved under the same conditions.

Data analysis and statistics

The data are shown as a mean±SD. Here, "n" represents the number of independent experiments with "N" samples each. Significance level was analyzed using one-way ANOVA with a mean comparison using Tukey method with Microsoft Excel 2007 software. The level of $p<0.05$ was assumed as significant.

Results

Characterization of the MSCs

Primary culture obtained from amnion membrane displayed high proliferation capacities until passage 26 (data not shown). Flow cytometry analysis demonstrated 87.56±4.92% of the cells were positive for MSC maker CD105 (ENG) and 85.00±4.28% – for CD73 (NT5E, Figure 2A). Immunohistochemistry showed positive brown to black reporter color for CD90 (THY-1), CD105 (ENG), Brachyury (Bra) and Snail-1 (Figure 2B–E). MSC characteristic genes like THY-1, ITGA6, GFRa1, CD73, CD105, ALCAM and CD44 were detected also in RT-PCR, whereas hematopoietic marker CD34 was absent (Figure 2F).

Optimization of cell encapsulation

Effect of cell encapsulation on bead diameter. For transplantation and pre-clinical studies, the size of alginate beads is one of the main factors, limiting diffusion of nutrients to encapsulated cells and vice versa the release of cell produced substances. The initial concentration of cells and alginate-cell mixing procedure may affect the resulting bead diameter and the performance of alginate beads after transplantation. In order to investigate these parameters, we encapsulated MSCs into 1.6% alginate with three different cell concentrations and under the range of voltage from 10 to 30 kV (Figure 3A). At the region below 10 kV, the strength of electric field was much lower than the surface tension, resulting in dropping mode and a bead size inapplicable for cell culture (>1 mm). At non stable region (10–15 kV), the slight increase in applied voltage had a pronounced effect on bead diameter, allowing to generate beads sized from 1,000 μm to less then 300 μm, respectively (dripping mode). Further increase in voltage did not significantly alter bead diameter (jetting mode). Such behavior has previously been reported, however, without encapsulated cells [27,28]. In addition, concentration of 1.5% alginate was simulated by diluting appropriate volumes of 1.6% alginate with HEPES solution. Interestingly, an increase of cell concentration from 1×10^6 to 10×10^6 cells/ml of initial alginate did not have noticeable influence on bead diameter (Figure 3A), indicating that large number of cells can be encapsulated without altering the size of alginate beads.

Encapsulation efficiency. It is especially important in therapies with a cell-based drug delivery system to precisely calculate the amount of encapsulated cells in alginate beads. We investigated the efficiency of encapsulation by analyzing the shrinking rate of alginate during cross-linking. The shrinking rate of alginate (1.5% and 1.2%, w/v) in cross-linking solution without encapsulated cells depending on initial surface-to-volume ratio at defined time intervals and for a maximum of 40 min is shown in Figure 3B. Interestingly, the alginate beads possessing higher S/V ratio showed very fast and uncontrollable shrinking over the first 5 min of gelling, whereas those obtained by MD were shrinking stable and slow (Figure 3B, AF-ES *vs.* MD). In opposite, alginate drops generated via ES method and possessing the highest value of

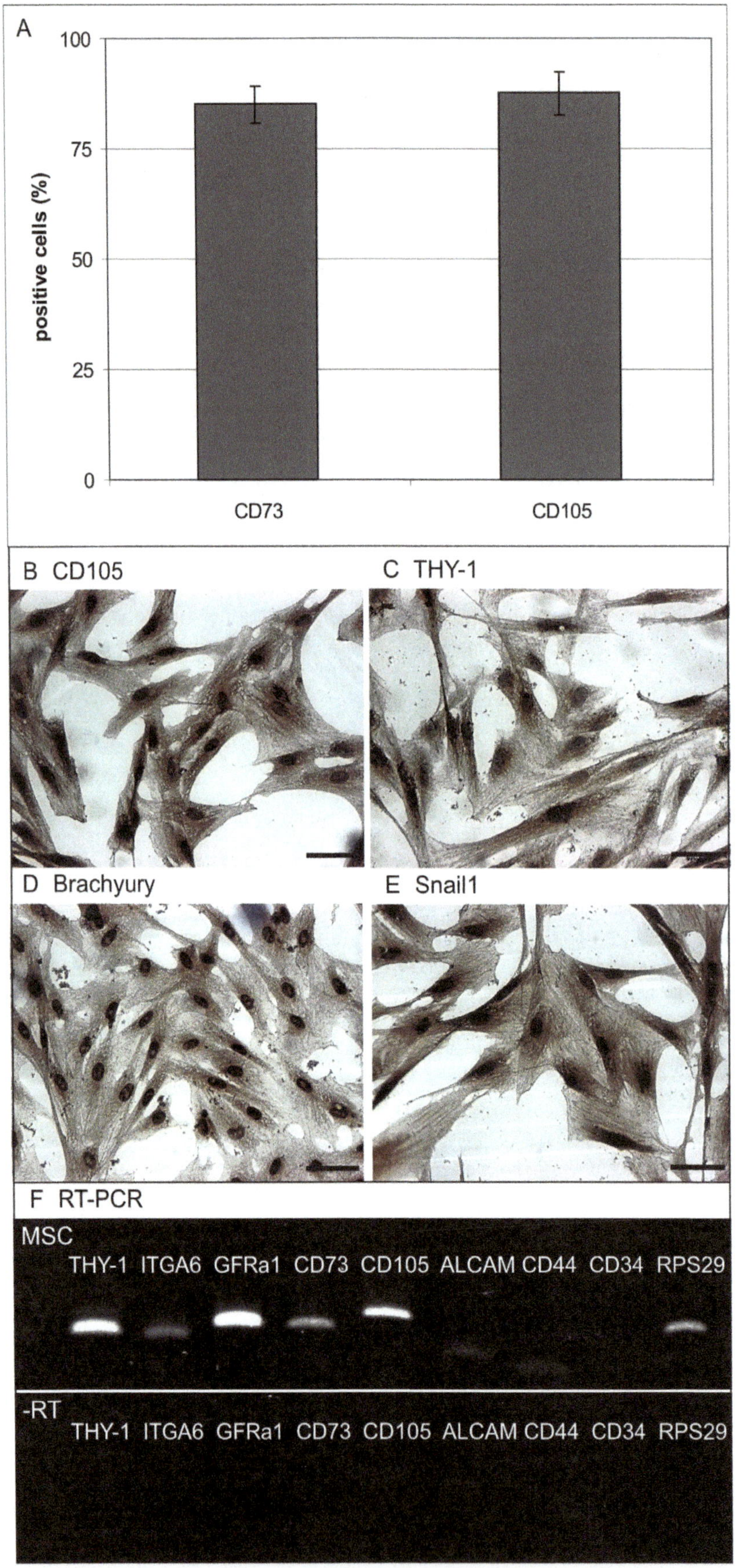
A
100
75
50
25
0
positive cells (%)
CD73
CD105
B CD105
C THY-1
D Brachyury
E Snail1
F RT-PCR
MSC
THY-1 ITGA6 GFRa1 CD73 CD105 ALCAM CD44 CD34 RPS29
-RT
THY-1 ITGA6 GFRa1 CD73 CD105 ALCAM CD44 CD34 RPS29

Figure 2. Characterization of amnion derived multipotent stromal cells. (**A**) FACS analysis with characteristic markers CD105 and 73 (mean±SD, n = 4). (**B–E**) Immunohistochemistry stainings with MSC markers CD105, CD90, Bra and Snail1. (**F**) RT-PCR analysis displays presence of characteristic MSC markers CD90, ITGA6, GFRa1, CD73, CD105, ALCAMm CD44, whereas CD34 as marker for hematopoietic progenitors is absent (housekeeper RPS29). Scale bars are 50 μm.

S/V ratio (11.47 mm^{-1}) showed extremely fast shrinking during the first 5 min of cross-linking.

Calculating the number of cells to be entrapped in one bead (Figure 3C) with equation (2), the alginate beads contained 1×10^6, 5×10^6 and 10×10^6 cells/ml and showed virtually homogeneous distribution of cells throughout the bead (Figure 3C, C1–C3). The cell concentration barely affected the size of alginate bead for a wide range of applied high voltage (Figure 3A). The average diameter of alginate beads was equal to 300 μm for each cell concentration. In exact numbers, we could show that it is possible to entrap 48 to 483 cells in an alginate bead for 1×10^6 and 10×10^6 cells/ml, respectively.

Viability after encapsulation

In order to evaluate if the high voltage of the ES method affects the viability of cells especially by membrane damages, we utilized the trypan blue assay as a solid marker for membrane integrity of cells; however, the release of cells from the beads by immersing in sodium citrate and additional centrifugation may also lead to cell death by apoptosis. To reveal this effect, we stained encapsulated MSCs additionally at day 0, 1, 5, 10 with CalceinAM/EthD. We observed no noticeable effect of high voltage on viability of cells both post-encapsulation and incubation of MSCs inside alginate beads for 10 days as compared to AF (Figure 4, AF-30 kV). However, we noticed the negative effect of incubation time on the viability of encapsulated cells as compared to shorter incubation periods (Figure 4, day 10 *vs.* day 1). Despite this, most of cells were

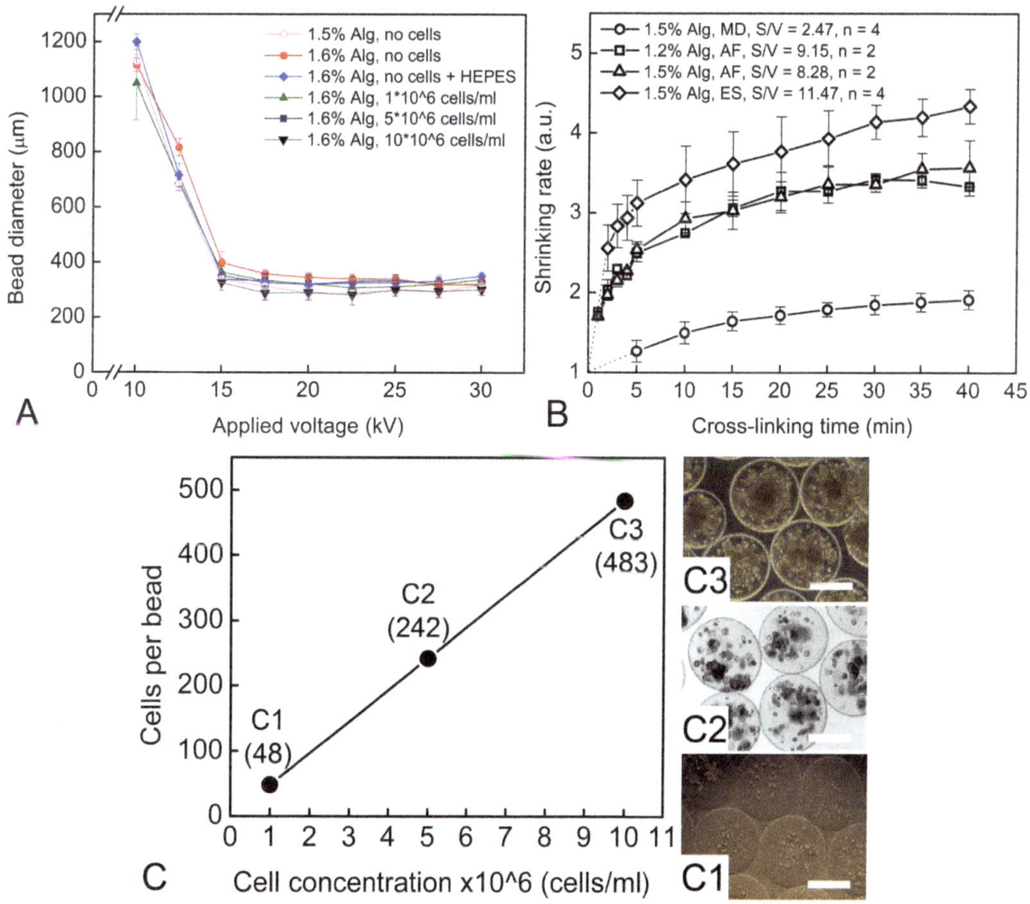

Figure 3. Optimization of cell concentration and cell number. (**A**) Effect of applied voltage, initial cell concentration and alginate-cell mixing procedure on diameter of alginate beads (mean±SD). (**B**) Shrinking of alginate in presence of 100 mM $CaCl_2$ depending on gelling time for different initial surface-to-volume ratio (S/V) and (**C**) calculation of encapsulation efficiency of MSCs in alginate beads. Shrinking rate was analyzed based on decrease in volume during gelling caused by water release for: MD – manual dropping of alginate solution, AF – using air-flow method and ES – using electro-spraying; dotted lines show the tendency of shrinking outside the studied cross-linking time. (**C**) Effect of initial cell concentration on number of encapsulated cells and respective photographs of alginate encapsulated MSCs at initial concentration of 1×10^6 (**C1**), 5×10^6 (**C2**) and 10×10^6 (**C3**) cells/ml. The next process parameters were kept constant: both for AF and ES – spraying distance 10 cm, concentration of gelling solution 100 mM $CaCl_2$, alginate flow rate 10 ml/h, and alginate concentration from 1.2% to 1.6% (w/v); for AF – air flow 150 l/h; for ES – applied voltage 20 kV. The data are presented as a mean±SD (n = 3, N = 10). Scale bars are 200 μm.

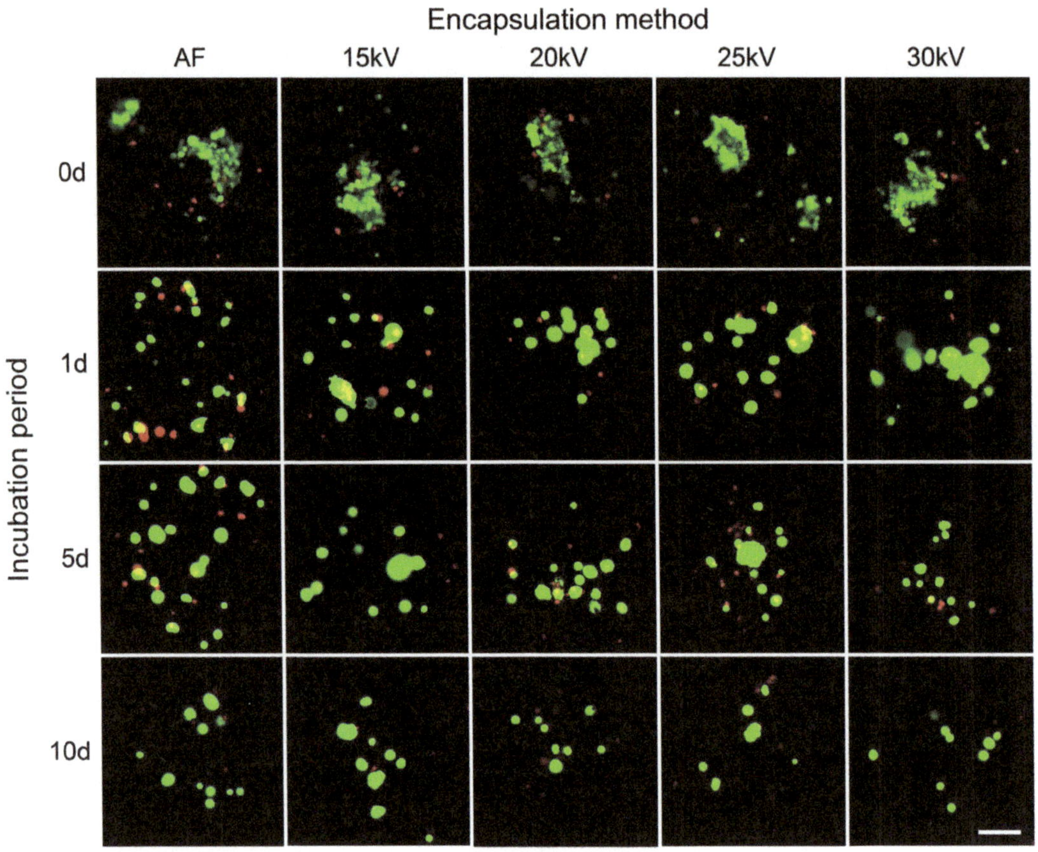

Figure 4. Viability of encapsulated MSCs *vs.* applied voltage and incubation time. Live-dead staining with CalceinAM (2 μM) and Ethidium Homodymer (4 μM). Note: beads obtained under 15 kV and using air flow (AF) had slightly bigger diameter as compared to the others. Viable cells stained in green, cells with massive membrane damages – in red. No visible difference in viability of encapsulated cells depending on applied voltage can be observed. The next process parameters were kept constant: both for AF and electro-spraying – spraying distance 10 cm, concentration of gelling solution 100 mM $CaCl_2$, alginate flow rate 10 ml/h, and alginate concentration 1.6% (w/v); for AF – air flow 150 l/h. Scale bar is 100 μm.

still viable at day 10. It should be noted that the effect of high-voltage on viability of encapsulated cells cannot be properly evaluated, since the beads obtained under 20, 25 and 30 kV had a smaller diameter as compared to AF and 15 kV. Thus, fewer viable cells were detected.

Cell proliferation

Proliferation of encapsulated MSCs. We evaluated the effect of high voltage and incubation time of bead-enclosed cells on proliferation after encapsulation (Figure 5A1) and after recovery (Figure 5A2). Here, the encapsulation procedure and applied high voltage does not significantly affect the viability of encapsulated MSCs, as indicated by similar proliferation efficiency and metabolic activity as controls. Moreover, MSCs derived from alginate beads immediately after encapsulation recovered with the same rate as controls, but proliferation decreased significantly after 24 h of incubation. The recovery of MSCs post-encapsulation was dependent on pre-incubation time; such non-incubated cells (seeded for proliferation assay immediately after encapsulation) recovered at the same rate as a native control (Figure 5A2, "Immediate"). However, MSCs incubated in beads for 24 h showed a significantly lower recovery rate (Figure 5A2, "Incubated").

Proliferation of encapsulated MSCs after cryopreservation. For long-term storage it is necessary to provide alginate encapsulated cells with high viability and proliferation after cryopreservation. The proliferation of frozen cells inside alginate structures incubated for 24 h (Figure 5B1, "Incubated") was significantly lower in comparison with native control (Figure 5B1, "CN"). Interestingly, MSCs frozen inside alginate beads immediately after encapsulation revealed significantly higher proliferation rate as compared to the cells incubated for 24 h inside beads; whereas the proliferation rate was slightly lower for encapsulated and immediately frozen cells as compared to frozen control cells (Figure 5B1, "CF"). The recovery of MSCs post-cryopreservation was also affected by incubation time (Figure 5B2). In addition, MSCs frozen inside alginate beads possessed higher recovery rate as compared to frozen control cells and the same rate as fresh control cells (Figure 5B2, "CF"–"CN"). Application of high voltage to generate cell-containing alginate beads did not alter the metabolic activity of cells after thawing, as for "Immediate" and "Incubated" groups we did not observe the negative effect of high voltage on MSCs as compared to AF encapsulated ones (Figure 5B1–B2, 15–25 kV *vs.* AF).

In summary, the cryopreservation procedures reduced the number of cells with intact membrane. We showed that 62±5% of the cells were viable after thawing of cryopreserved MSCs directly after encapsulation, as compared to initially encapsulated ones (n = 3, N = 5). On the other hand, only 30±7% of cells were alive after applying pre-freeze incubation of MSCs inside alginate beads in culture for 24 h. We further detected the number of not encapsulated frozen MSCs of 76±3% after thawing, as compared

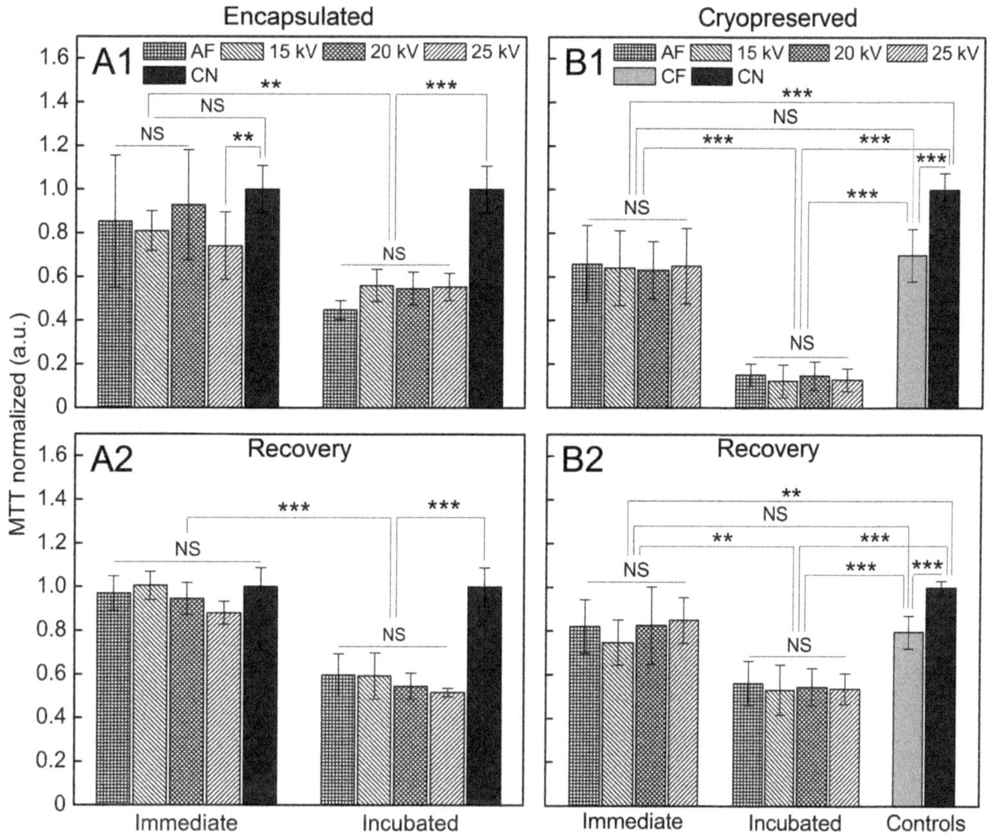

Figure 5. Effect of applied encapsulation methods and incubation on proliferation of MSCs post-encapsulation and cryopreservation. Proliferation of MSCs: (**A1**) after encapsulation in alginate; (**A2**) after recovery post-encapsulation; (**B1**) cryopreserved in alginate; (**B2**) after recovery post-thawing. "Recovery" – incubation of cells in culture (37°C, 5% CO_2 in a humidified incubator) for 5 days after encapsulation and/or cryopreservation; "Immediate" – immediately after encapsulation; "Incubated" – after 24 h of incubation inside alginate beads in culture; AF – air flow; CF – frozen not encapsulated cells; CN – not encapsulated and not frozen MSCs. The next process parameters were kept constant during encapsulation: both for AF and electro-spraying – spraying distance 10 cm, concentration of gelling solution 100 mM $CaCl_2$, alginate flow rate 10 ml/h, and alginate concentration 1.6% (w/v); for AF – air flow 150 l/h. In B1–B2 freezing parameters included cooling rate 1 K/min down to −80°C and freezing medium with 10% DMSO and 10% FBS. The data are shown as a mean±SD (n = 3, N = 5). One-way ANOVA: NS – not significant ($p = 0.05$); *, **, *** - significantly different with $p<0.05$, $p<0.01$, $p<0.001$, respectively.

to initially frozen cells. The number of cells recovered after thawing increased in 4.9 and 2.9 times after 5 days for immediately frozen and incubated for 24 h before cryopreservation (as compared to the number of cells seeded for recovery assay), respectively. According to our data, the number of not encapsulated frozen MSCs and fresh control increased in 4.6 and 4.8 times, respectively.

Proliferation of MSCs cultured inside beads for 7 days. During evaluation of the effect of high voltage on proliferation of cells as compared to the AF method for a period of 7 days, we found that the viability and proliferation of encapsulated cells was decreasing during the incubation which resulted in significant decrease at day 7 independent from encapsulation method (Figure 6, day 7 *vs.* day 0). The cells seeded for proliferation immediately after encapsulation at 20 and 25 kV of high voltage showed slightly lower proliferation as compared to AF encapsulated MSCs. However, further incubation did not reveal significantly different effect of high voltage on the viability and proliferation of cells post-incubation as compared to AF method. In summary, the encapsulated cells proliferated slower as compared to non-encapsulated control cells.

Differentiation capacity

We studied the effect of high voltage on the capacity of encapsulated MSCs to differentiate into adipogenic and osteogenic lineages. MSCs were able to form adipocytes (Oil Red O staining) and osteocytes (Von Kossa staining) post-encapsulation (Figure 7). We detected no visible difference in the plasticity for MSCs encapsulated in alginate using high voltage (20 kV) and AF method to form lipid vacuoles (Figure 7, Oil Red O/Positive) and mineralize calcium phosphate (Figure 7, Von Kossa/Positive). Non-encapsulated MSCs also showed the capacity to differentiate into adipocytes and osteocytes (Figure 7, Fresh), whereas untreated control cells without differentiation media did not display vacuoles or mineralization.

Discussion

High voltage has been widely used in bioengineering and regenerative medicine. Such applications include a) electrophoresis to extract DNA and proteins and b) electroporation to permeabilize the cell membrane to deliver genes and liposome into the cells and tissues to treat rare diseases and store rare cell types for long term [29–31]. In these approaches, the strength of electric field ranges from 1–10 V/cm of constant voltage for gel

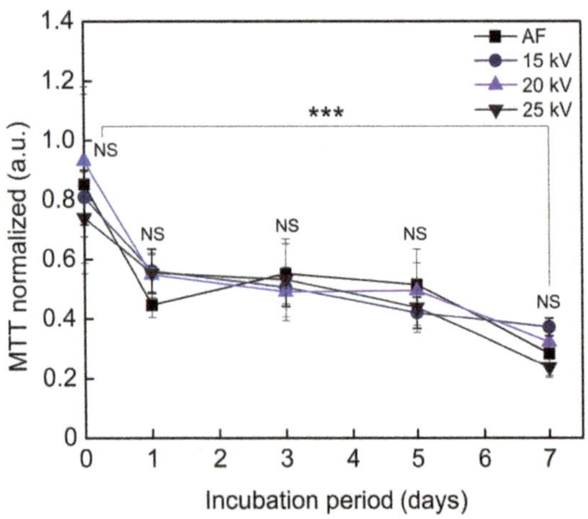

Figure 6. Effect of encapsulation method and period of incubation of MSCs inside beads on proliferation of cells. MSCs were encapsulated in alginate beads using air flow (AF) and high voltage (15, 20, 25 kV) and incubated for 0, 1, 3, 5 and 7 days inside beads in culture (37°C, 5% CO_2 in a humidified incubator). Culture medium was replaced each second day to a fresh one. The next process parameters were kept constant: both for AF and electro-spraying – spraying distance 10 cm, concentration of gelling solution 100 mM $CaCl_2$, alginate flow rate 10 ml/h, and alginate concentration 1.6% (w/v); for AF – air flow 150 l/h. The data are shown as a mean±SD (n = 2, N = 5). One-way ANOVA: NS – not significant (p = 0.05); *** – significantly different (p<0.001).

electrophoresis up to 100 kV/cm of pulse high voltage (the pulse duration can vary depending on application) for conventional electroporation (the pulse duration is in microsecond range) [32] and supra-electroporation (nanosecond pulsed electric field) causing high density nanopore formation in all cell membranes [33]. In our experiments we generated small-sized alginate beads with encapsulated MSCs applying the strength of electric field ranging from 1 kV/cm to 3 kV/cm. Comparing the high voltage method with the commonly used AF, we could show that the range of applied high voltage from 15 to 25 kV did not alter the viability, proliferation and differentiation capacity of MSCs after encapsulation in alginate. Although high voltage showed a significant effect on the proliferation rate of MSCs immediately after encapsulation depending on the method utilized (Figure 5A1, 25 kV *vs.* "CN"), there was no significant difference observed for 7 days in culture (Figure 6, "NS").

The findings mentioned above correlate with our recent data on cell counts in studies after encapsulation and cryopreservation [34]. MSCs encapsulated in alginate beads may require longer periods of time to recover post-encapsulation; maybe due to limited diffusion of serum molecules through the negatively charged alginate porous structure - a problem that might get solved by applying higher FBS concentrations. It has previously been shown that an application of 3 kV/cm did not alter the viability of cells [35,36]. In addition, the high voltage encapsulation of living cells supports cells with a mild environment during encapsulation that has also been suggested, but for impulse voltage [37].

Encapsulation of cells in alginate has been known since late 1960s and has been widely applied in mice, rats and rabbits as preclinical models [8]. However, the genotype and phenotype of these animals is very different compared to the human, especially in the field of stem cells research with an application of very experimental cell types like ESCs and iPS cells with very early features of cellular maturity. In the human, multipotent stromal cells are to date the only cell type already utilized in clinical trials [38]. This made them ideal assets in our experiments for extraction, cultivation, encapsulation and differentiation. True multipotent stromal cells with high "stemness" from the placental amnion of the common marmoset monkey can be natively obtained for personalized therapies and possible option of transplantation in our non-human primate model without ethical concerns or clinical safety problems.

The proliferation of encapsulated MSCs reduced after 7 days in culture as compared to not encapsulated control cells. It may be related to the restriction of space as well as the effects of hampered diffusion of nutrients and FBS to encapsulated cells, especially to centrally positioned ones, as it has been reported previously [39]. In addition, the negative effect of washing steps and liquefying procedures (release of cells from alginate beads) can not be excluded. One of the possible solutions could be an increase of concentration of FBS, along with a decrease in bead diameter, which could improve heat and mass transfer through the bead. A similar solution has also been suggested, showing that alginate has to be modified to allow attachment and proliferation of adherent mammalian cell types [40–42]. This could support our observation on a decrease of the viability and metabolic activity of encapsulated cells over the incubation period. On the other hand, the metabolic activity of encapsulated adherent mammalian cell types may be improved by modifying the initial alginate with RGD peptides [10]. These modifications have been shown to improve survival of cells after cryopreservation as compared to cells encapsulated in unmodified alginate. In this work, we showed that non-human primate MSCs encapsulated using high voltage can be readily stored at low temperatures for long term without altering the metabolic activity of cells after thawing, as compared to not encapsulated MSCs. In addition, by studying the shrinking rate of alginate we were also able to quantify the number of cells to be entrapped in one bead for the respective S/V ratio of 11.47 mm^{-1}.

To our knowledge, there is to date no single study with a complete and comparative characterization of amnion derived encapsulated and cryopreserved MSCs including optimization of cell number, evaluation of metabolic activity of cells after encapsulation and cryopreservation and evaluation of the effect of encapsulation method on the differentiation potential of alginate encapsulated MSCs.

Conclusions

The MSCs derived from a marmoset amnion proved to be true multipotent stromal cells with "stemness" and plasticity. They can be encapsulated in unmodified alginate beads at high concentration using high-voltage encapsulation with narrow size distribution and preserving the viability and proliferation of cells after encapsulation. In particular, optimization of process parameters of the ES method revealed only a slight influence of cell number on bead size. This indicates the possibility for large-scale generation of encapsulated cells that has previously been shown to be the critical point in translation of encapsulation technology to clinical application requiring high cell numbers [43]. The cells encapsulated using the high voltage method proliferated at the same rate as non-encapsulated cells and the cells encapsulated utilizing the traditional AF method. Incubation of MSCs in alginate in culture showed a significant decrease in proliferation 5 days after encapsulation, but MSCs were still viable at day 10. In addition, we showed that cryopreservation of MSCs inside alginate beads should be performed immediately after encapsulation to

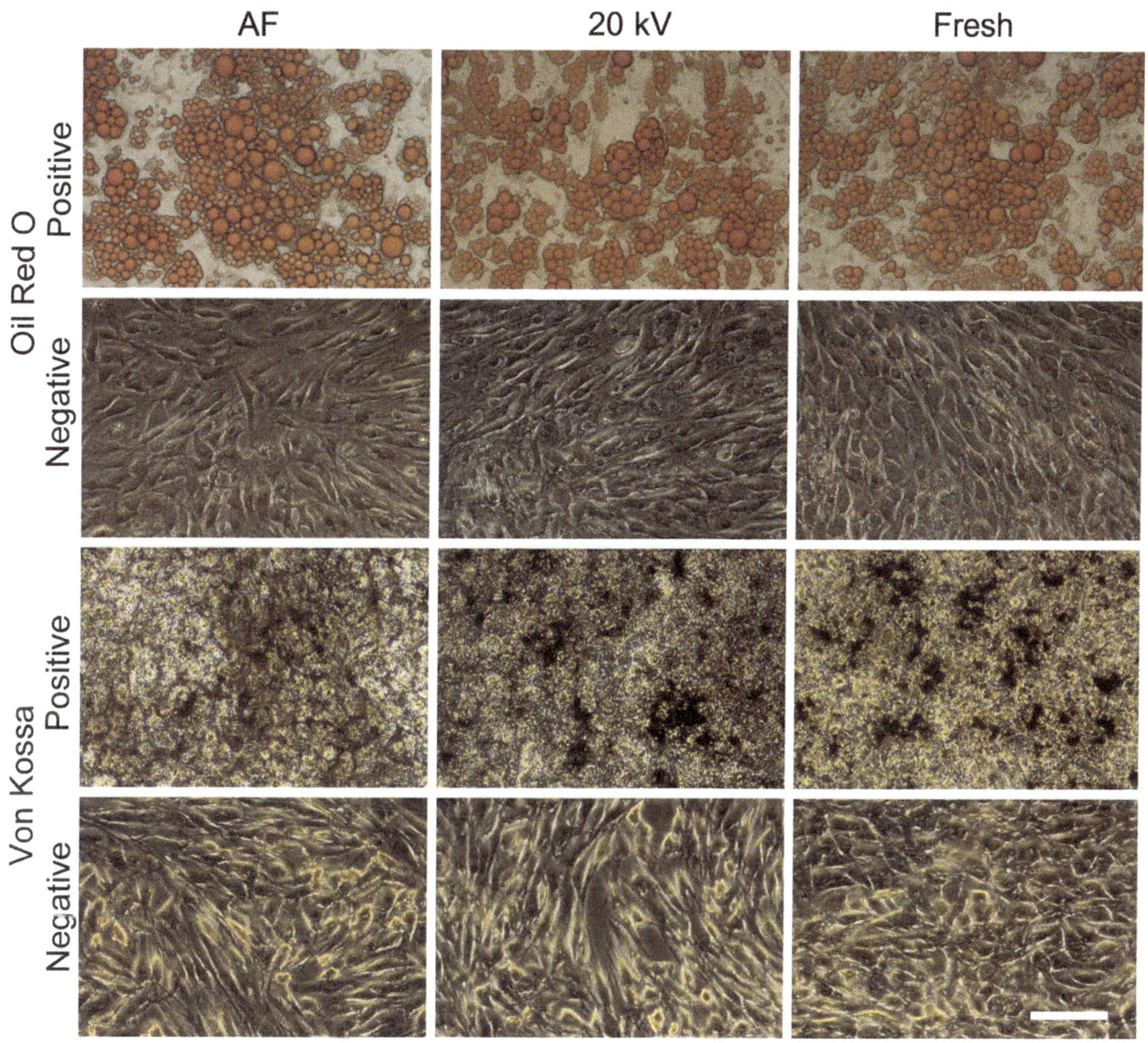

Figure 7. Differentiation capacity of encapsulated MSCs. MSCs were differentiated into adipogenic (Oil Red O) and osteogenic direction (Von Kossa) after encapsulation using air flow (AF) and high voltage (20 kV) followed by dilution of alginate with 55 mM sodium citrate for 5 min, or as not encapsulated MSCs. Cells expanded in regular culture medium represent the negative control (37°C, 5% CO_2 in a humidified incubator) and stained the same as positive cells of respective lineage. The next process parameters were kept constant: both for AF and electro-spraying (ES) – spraying distance 10 cm, concentration of gelling solution 100 mM $CaCl_2$, alginate flow rate 10 ml/h, and alginate concentration 1.6% (w/v); for AF – air flow 150 l/h; for ES – applied voltage 20 kV. Scale bar is 100 μm.

ensure high proliferation rates of cells after thawing. Further work should be aimed at establishing of alternative cryopreservation protocols for freezing of encapsulated MSCs with a control over the temperature of ice formation, which could improve the cryopreservation outcome. This work may give a background to further develop a preclinical therapy with high voltage alginate encapsulated MSCs in our non-human primate model towards human application.

Supporting Information

Figure S1 Thawing rate of alginate encapsulated MSCs. DMSO was added to the sample at 4°C at a rate of 0.5 ml/min (drop-wise). The thermocouple was frozen at the center of a cryovial so that it was positioned at the point which separates equal volumes of freezing medium containing alginate encapsulated cells. The frozen sample was removed from −150°C and thawed immediately in a water bath at 37°C with gentle shaking until a small ice crystal was visible. The temperature increase was monitored each 3 seconds. Each point represents the results of measuring the thawing rate of 3 frozen samples.

Table S1 Oligonucleotides and antibodies utilized for MSCs characterization.

Acknowledgments

The authors acknowledge Julia Struss for technical assistance, Holger Zernetsch for the help with figure design and Prof. S. Schlatt for the kind donation of placental material.

Author Contributions

Conceived and designed the experiments: OG. Performed the experiments: OG DP OP. Analyzed the data: OG. Contributed reagents/materials/analysis tools: NH TM BG. Contributed to the writing of the manuscript: OG TM. Critically revised the manuscript: DP OP NH TM BG. Gave final approval of the version to be published: TM BG.

References

1. Portero A, Orive G, Hernandez RM, Pedraz JL (2010) Cell encapsulation for the treatment of central nervous system disorders. Rev Neurol 50: 409–419.
2. Lee MK, Bae YH (2000) Cell transplantation for endocrine disorders. Adv Drug Deliv Rev 42: 103–120.
3. Wollert KC, Drexler H (2006) Cell-based therapy for heart failure. Curr Opin Cardiol 21: 234–239.
4. Orive G, Hernandez RM, Gascon AR, Calafiore R, Chang TM, et al. (2003) Cell encapsulation: promise and progress. Nat Med 9: 104–107.
5. Figueiredo CS, Wedekind D, Müller T, Vahlsing S, Horn PA, et al. (2013) MHC universal cells survive in an allogeneic environment after incompatible transplantation. Biomed Res Int, doi: 10.1155/2013/796046.
6. Li RH (1998) Materials for immunoisolated cell transplantation. Adv Drug Deliv Rev 33: 87–109.
7. Pawar SN, Edgar KJ (2012) Alginate derivatization: A review of chemistry, properties and applications. Biomaterials 33: 3279–3305.
8. Chang TMS (1964) Semipermeable microcapsules. Science 146: 524–526.
9. Lee KY, Rowley JA, Eiselt P, Moy EM, Bouhadir KH, et al. (2000) Controlling mechanical and swelling properties of alginate hydrogels independently by cross-linker type and crosslinking density. Macromolecules 33: 4291–4294.
10. Rowley JA, Madlambayan G, Mooney DJ (1999) Alginate hydrogels as synthetic extracellular matrix materials. Biomaterials 20: 45–53.
11. Rzaev ZMO, Dincer S, Piskin E (2007) Functional copolymers of Nisopropylacrylamide for bioengineering applications. Progr Polym Sci 32: 534–595.
12. Chan ES, Lim TK, Ravindra P, Mansa RF, Islam A (2012) The effect of low air-to-liquid mass flow rate ratios on the size, size distribution and shape of calcium alginate particles produced using the atomization method. J Food Eng 108: 297–303.
13. Huang SB, Wu MH, Lee GB (2010) Microfluidic device utilizing pneumatic micro-vibrators to generate alginate microbeads for microencapsulation of cells. Sens Actuators B Chem 147: 755–764.
14. Ma X, Vacek I, Sun A (1994) Generation of alginate-poly-l-lysine-alginate (APA) biomicrocapsules: the relationship between the membrane strength and the reaction conditions. Artif Cells Blood Substit Immobil Biotechnol 22: 43–69.
15. Rokstad AM, Brekke OL, Steinkjer B, Ryan L, Kolláriková G, et al. (2011) Alginate microbeads are complement compatible, in contrast to polycation containing microcapsules, as revealed in a human whole blood model. Acta Biomater 7: 2566–2578.
16. Zimmermann H, Zimmermann D, Reuss R, Feilen PJ, Manz B, et al. (2005) Towards a medically approved technology for alginate-based microcapsules allowing long-term immunoisolated transplantation. J Mater Sci Mater Med 16: 491–501.
17. Nature Publishing Group (2009) The common marmoset: biomedical supermodel Nature 459: 7246.
18. Mansfield K (2003) Marmoset models commonly used in biomedical research. Comp Med 53: 383–392.
19. Müller T, Fleischmann G, Eildermann K, Mätz-Rensing K, Horn PA, et al. (2009) A novel embryonic stem cell line derived from the common marmoset monkey (Callithrix jacchus) exhibiting germ cell-like characteristics. Hum Reprod 24: 1359–1372.
20. Wiedemann A, Hemmer K, Bernemann I, Göhring G, Pogozhykh O, et al. (2012) Induced pluripotent stem cells generated from adult bone marrow-derived cells of the nonhuman primate (Callithrix jacchus) using a novel quad-cistronic and excisable lentiviral vector. Cell Reprogram 14: 485–496.
21. Miki T, Lehmann T, Cai H, Stolz DB, Strom SC (2005) Stem cell characteristics of amniotic epithelial cells. Stem Cells 23: 1549–1559.
22. Vandenbossche GMR, Remon GP (1993) Influence of the sterilization process on alginate dispersion. J Pharm Pharmacol 45: 484–486.
23. Gryshkov O, Hofmann N, Glasmacher B (2013) Electropsraying as a promising technique for cell encapsulation in semi-permeable micro-capsules for immunoisolation. Cryo Letters 34: 186.
24. Mørch YA, Donati I, Strand BL, Skjåk-Braek G (2006) Effect of Ca2+, Ba2+, and Sr2+ on alginate microbeads. Biomacromolecules 7: 1471–1480.
25. Xie J, Wang CH (2007) Electrospray in the dripping mode for cell microencapsulation. J Colloid Interface Sci 312: 247–255.
26. Bernemann I, Mueller T, Blasczyk R, Glasmacher B, Hofmann N (2011) Colonization of collagen scaffolds by adipocytes derived from mesenchymal stem cells of the common marmoset monkey. Biochem Biophys Res Commun 411: 317–322.
27. Klokk TI, Melvik JE (2002) Controlling the size of alginate gel beads by use of a highelectrostatic potential. J Microencapsul 19: 415–424.
28. Shi P, He P, Teh TKH, Morsi YS, Goh JCH (2011) Parametric analysis of shape changes of alginate beads. Powder Technology 210: 60–66.
29. Neumann E, Schaefer-Ridder M, Wang Y, Hofschneider PH (1982) Gene transfer into mouse lyoma cells by electroporation in high electric fields. EMBO J 1: 841–845.
30. Hanahan D, Weinberg RA (2000) The hallmarks of cancer. Cell 100: 57–70.
31. Kroemer G, Pouyssegur J (2008) Tumor cell metabolism: cancer's Achilles' heel. Cancer Cell 13: 472–482.
32. Neumann E, Rosenheck K (1972) Permeability Changes Induced by Electrical Impulses in Vesicular Membranes. J Memb Biol 10: 279–290.
33. Stewart DA, Gowrishankar TR, Weaver JC (2004) Transport lattice approach to describing electroporation: Use of local asymptotic model. IEEE Transact Plasma Sci 32: 1696–1708.
34. Gryshkov O, Pogozhykh D, Zernetsch H, Hofmann N, Mueller T, et al. (2014) Process engineering of high voltage alginate encapsulation of mesenchymal stem cells. Mater Sci Eng C Mater Biol Appl 36: 77–83.
35. Zhang W, He X (2009) Encapsulation of living cells in small (~100 μm) alginate microcapsules by electrostatic spraying: a parametric study. J Biomech Eng, doi:10.1115/1.3153326.
36. Im do J, Noh J, Yi NW, Park J, Kang IS (2011) Influences of electric field on living cells in a charged water-in-oil droplet under electrophoretic actuation. Biomicrofluidics, doi: 10.1063/1.3665222.
37. Lewinska D, Rosinski S, Werynski A (2004) Influence of process conditions during impulsed electrostatic droplet formation on size distribution of hydrogel beads. Artif Cells Blood Substit Immobil Biotechnol 32: 41–53.
38. Ng TK, Fortino VR, Pelaez D, Cheung HS (2014) Progress of mesenchymal stem cell therapy for neural and retinal diseases. World J Stem Cells 26: 111–119.
39. Huang X, Zhang X, Wang X, Wang C, Tang B (2012) Microenvironment of alginate-based microcapsules for cell culture and tissue engineering. J Biosci Bioeng 114: 1–8.
40. Berrier AL, Yamada KM (2007) Cell–matrix adhesion. J Cell Physiol 213: 565–73.
41. Koo LY, Irvine DJ, Mayes AM, Lauffenburger DA, Griffith LG (2002) Co-regulation of cell adhesion by nanoscale RGD organization and mechanical stimulus. J Cell Sci 115: 1423–1433.
42. de Vos P, Bučko M, Gemeiner P, Navrátil M, Svitel J, et al. (2009) Multiscale requirements for bioencapsulation in medicine and biotechnology. Biomaterials 30: 2559–2570.
43. Serra M, Brito C, Correia C, Alves PM (2012) Process engineering of human pluripotent stem cells for clinical application. Trends Biotechnol 30: 351–359.

Digital Microfluidic Processing of Mammalian Embryos for Vitrification

Derek G. Pyne[1]☯, **Jun Liu**[1]☯, **Mohamed Abdelgawad**[2]*, **Yu Sun**[1]*

1 Department of Mechanical and Industrial Engineering, University of Toronto, Toronto, Ontario, Canada, **2** Department of Mechanical Engineering, Assiut University, Assiut, Egypt

Abstract

Cryopreservation is a key technology in biology and clinical practice. This paper presents a digital microfluidic device that automates sample preparation for mammalian embryo vitrification. Individual micro droplets manipulated on the microfluidic device were used as micro-vessels to transport a single mouse embryo through a complete vitrification procedure. Advantages of this approach, compared to manual operation and channel-based microfluidic vitrification, include automated operation, cryoprotectant concentration gradient generation, and feasibility of loading and retrieval of embryos.

Editor: David T. Eddington, University of Illinois at Chicago, United States of America

Funding: This work was supported by the Grants from NSERC (Natural Sciences and Engineering Research Council of Canada) via a Discovery Grant (http://www.nserc-crsng.gc.ca/index_eng.asp); and the Canada Research Chairs Program (http://www.chairs-chaires.gc.ca/home-accueil-eng.aspx). The funders had no role in study design, data collection and analysis, decision to publish, or preparation of the manuscript.

Competing Interests: The authors have declared that no competing interests exist.

* Email: mohamed.abdelgawad1@eng.au.edu.eg (MA); sun@mie.utoronto.ca (YS)

☯ These authors contributed equally to this work.

Introduction

In 1983, the first successful pregnancy following cryopreservation was reported [1]. Three decades later, stem cells [2], sperms [3], and embryos [4] are now routinely frozen and preserved for use at a later time. Patients who undergo therapeutic procedures that can place their fertility at risk, such as chemotherapy, have the option of preserving their reproductive cells (sperms or oocytes/embryos) for future use through in vitro fertilization techniques (IVF) [5–8]. Furthermore, extra fertilized embryos after an IVF procedure can also be frozen for use at a later time. The length of time an embryo is frozen has been shown not to have a significant impact on clinical pregnancy, miscarriage, implantation, or live birth [9].

The two commonly used cryopreservation techniques for freezing embryos are the slow freezing method and the vitrification method. Both techniques aim to minimize cell damage caused by freezing that is largely due to the formation of intracellular ice crystals [10]. Conventionally, cells are frozen through the slow freezing method where cells are placed in a large freezer that can accurately control the freezing rate down to liquid nitrogen temperatures, with low concentrations of cryoprotectants [11]. During slow freezing extracellular water freezes away from the embryo, using a seeding technique, which creates an osmotic gradient that draws water out of the cell until it finally freezes without the formation of intracellular ice crystals [12]. This procedure requires sophisticated equipment to control the freezing rate, which ranges between 0.3 and 1.0°C/min, and produces a relatively poor survivability rate [13].

On the other hand, vitrification offers an alternative approach in which cells are frozen at extremely high rates, usually by directly plunging the sample into liquid nitrogen, after bathing them in a sequence of high concentration cryoprotectants [14]. Vitrification reduces intracellular ice formation, which is the primary cause of cell death, by freezing the sample in a glass-like state before the molecules have a chance to form crystal structures. This results in a higher cell survival rate after thawing compared to conventional slow freezing without the need for a seeding procedure or a programmable freezer [13,15]. However, vitrification requires precise washing sequences and timings in each cryoprotectant medium since higher concentrations are used and there are risks of toxicity if overexposed. The process is expensive in terms of technical skills required. In IVF clinics, processing an embryo/oocyte in cryoprotectant medium typically costs a highly skilled embryologist 10 to 15 minutes.

Digital microfluidics, which enables the manipulation of liquid droplets over an array of electrodes [16], is a useful tool for sequential sample processing and has been used in many biological applications such as PCR, cell culture, and immunoassays [17–19]. It was also used for testing cryoprotectant concentrations in slow-freezing cryopreservation [20]. Given its capabilities, digital microfluidics is well poised to automate embryo processing for embryo vitrification. The key to automating the vitrification process is to replicate the washing and timing steps of a given protocol while also keeping complete control of the embryo. Droplets on the digital microfluidic platform can act as micro-vessels to move an embryo and subject it to a series of cyroprotectants of different concentrations, as required by IVF vitrification protocols (Figure 1).

Here we developed digital microfluidic devices, for the first time, to automate embryo preparation for the vitrification procedure, lowering the high labour cost and ultimately helping

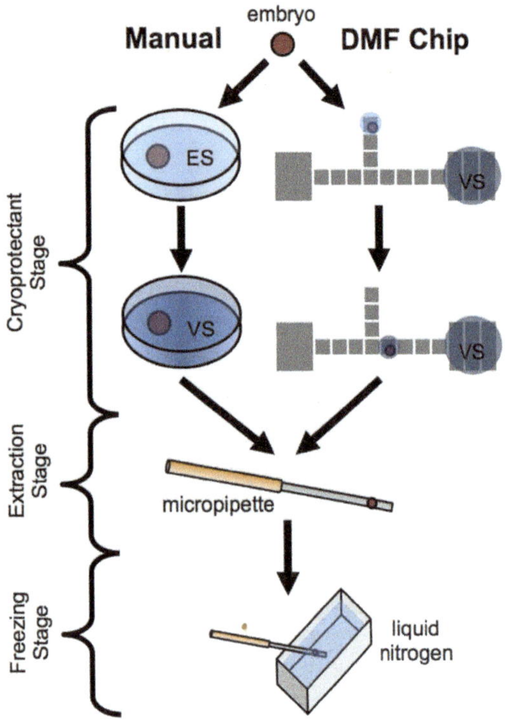

Figure 1. Manual and digital microfluidic vitrification workflow. Schematic showing differences between manual vitrification approach, which requires manual pipetting between cryoprotectant mediums, and the digital microfluidic (DMF) approach, which moves the embryo between mediums on chip. The chip automates the high skill portion of the procedure providing labor cost savings and opportunities for parallel processing.

further spread the use of vitrification in IVF clinics. Although vitrification was demonstrated inside microchannels [21–23], the device reported in this paper does not undesirably 'park' the embryo to a confined area, which could expose the embryo less uniformly to cryoprotectant medium and has less intricacy of embryo introduction and retrieval onto and from the device. By keeping the embryo in a single droplet, as opposed to microchannels or wells, we are also able to constantly track and control the embryo's locations within the field of view of our imaging system throughout the procedure to avoid cell loss.

Materials and Methods

Materials

All mouse experiments were performed in accordance with Canadian Council on Animal Care (CCAC) guidelines for Use of Animals in Research and Laboratory Animal Care under protocols (permit or protocol #:AUP0015) approved by the animal care committees of Toronto Centre for Phenogenomics (TCP). Mouse embryos were gathered from the Canadian Mouse Mutant Repository (CMMR; Toronto, ON). Embryos were produced by superovulating a female and were gathered 2.5 days past conception, which corresponds to most embryos being in the 8-cell stage. Vitrification solution (VS) typically contains antifreezing agents or cryoprotectants, such as dimethyl sulfoxide (DMSO), some small molecular sized glycols (e.g., ethylene glycol), or sucrose [24]. A combination of DMSO and sucrose was used in this work according to the protocol provided by CMMR. The vitrification solution (VS) was made by diluting DMSO in serum-free KSOM medium (EMD Millipore, Billerica, US) at 33% concentration, with 1.0 M sucrose. The equilibrium solution (ES) was at half concentration of VS (i.e., 16.5% DMSO+0.5 M sucrose). VS was preloaded on the DMF chip before each experiment. The first mixing step, which mixes VS with embryo culture medium (i.e., serum-free KSOM), generates the equilibrium solution ES.

Device construction

Devices were fabricated in the cleanroom facilities of the Toronto Nanofabrication Centre (TNFC). Pre-coated chromium glass slides (Deposition Research Labs Inc., MO) were first primed with Hexamethyldisilazane (HMDS) P-20 (Shin-Etsu MicroSi, Phoenix, AZ) by spin coating (3,000 rpm, 30 sec) before spin coating Shipley S1811 photoresist (3000 rpm, 30 sec). Substrates were then baked to remove solvents (115°C, 2 min) and UV exposed through a transparency mask for 10 sec. Substrates were next developed in MF-321 (2 min), hard baked (115°C, 1 min), etched with CR-4 (2 min) and then photoresist was removed with AZ-300T stripper (15 min in ultrasonic bath). Electrodes (1 mm × 1 mm) were separated by a 20 μm gap.

A 2 μm thick dielectric layer of Parylene C was then deposited. Lastly, a hydrophobic coating of Teflon AF 1600 (Dupont, Mississauga, ON) was spin coated on the device (1% w/w in Fluorinert FC-40, 2000 rpm, 1 min) and baked (160°C, 10 min). A second glass slide coated with un-patterned ITO (Deposition Research Labs Inc., MO) and Teflon AF was used as the ground electrode. Finally, the devices (Figure 2(a)) were assembled by placing the top ITO slide over the patterned device using two pieces of double-sided tape as a spacer. This produces a gap of approximately 100 μm between the top and bottom structures.

Device design

Voltages applied to actuate droplets were 55–75 Vrms at 15 kHz. Cyroprotectant droplets were actuated inside silicone oil (2.0 cSt, Gelest Inc., Morrisville, PA) to reduce friction and evaporation. Different regions of the device were designed to achieve the general digital microfluidic fucntions of transporting, mixing, dispensing, and splitting droplets. A large reservoir was used to hold and dispense the high concentration cryoprotectant medium, as shown in Figure 2(b). This reservoir was split up into many sections to handle variations in liquid volume in the reservoirs as droplets are dispensed during the vitrification protocol. The second reservoir was used as a waste reservoir and, thus, was split into two large electrodes only. A central inverted T-shaped array of electrodes was used for droplet transport, mixing, and splitting. Electrodes in this array were interdigitated to allow droplet overlap with adjacent electrode and increase electrodynamic forces applied on droplets. Top electrode in the leg of the T-shape was an input/output region where half of the electrode was exposed out of the ITO slide to enable embryo loading.

Embryo loading and retrieval

As shown in Figure 2(c), to input an embryo, a small droplet (200 nL) containing the embryo was pipetted onto the loading electrode and then actuated into the device, as described in [25]. This technique minimized exposure of the embryo to outside air. Extraction of the embryo was completed in the opposite manner by transporting the embryo-containing droplet to the edge of the device and retrieving it by a micropipette. Once the embryo was extracted from the device, it was directly frozen in the micropipette inside liquid nitrogen.

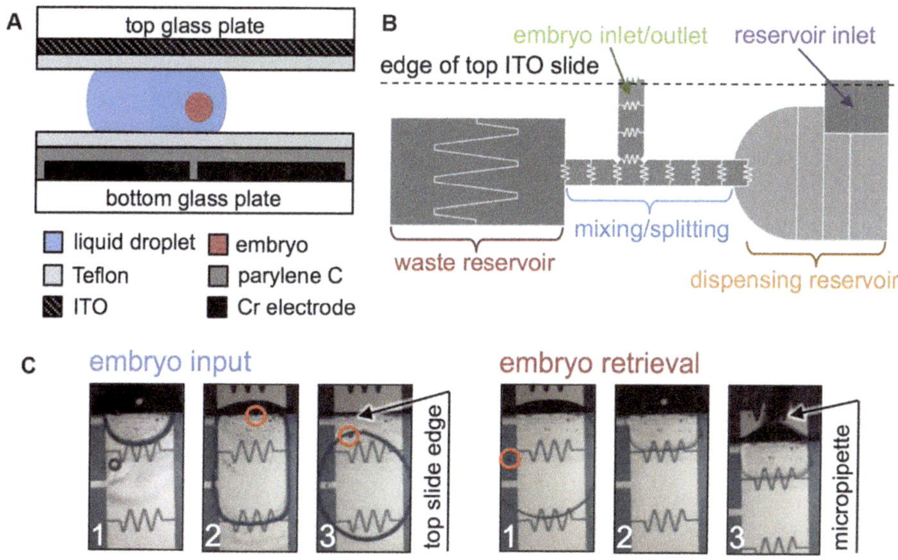

Figure 2. Key device design elements and fabrication layers. (A) Device structure and fabrication. Electrodes (1 mm×1 mm) were separated by a 20 μm gap. (B) Regions for vitrification medium dispensing and for embryo loading/extraction. The top ITO slide is placed on the device in a manner that exposes portions of the top electrodes in the dispensing reservoir and exposes portions of the leg of the T-shaped electrodes, for medium and embryo loading, respectively. (C) Embryo is input and extracted by actuating electrodes at edge of top glass slide.

Cryoprotectant mixing

An embryo is input into the device in a small droplet of embryo culture medium, and 100% cryoprotectant is input into the device in larger volumes ('reservoir inlet' in Figure 2(b)). The cryoprotectant bathing procedure is then performed through a serial mixing/splitting process (see Figure 3). This is accomplished by mixing the embryo-containing droplet with a vitrification solution droplet (VS), thus increasing the concentration of cryoprotectant around the embryo. The resulting droplet is then split into two daughter droplets with the one containing the embryo identified and kept, while the other droplet is moved to the waste reservoir (see Video S1).

After the first mixing step the droplet reaches 50% cryoprotectant concentration (i.e., equilibrium solution or ES), the embryo is kept in the ES droplet as long as the specific protocol requires. Then the cryoprotectant concentration is increased again by droplet mixing and splitting. Contrary to its state in ES medium, embryo volume sharply decreases in VS medium and does not recover. The overall mixing profile generated with a single dispensing reservoir is $C(n) = 100(1 - \frac{1}{2^n})\%$,where $C(n)$ is the concentration of the droplet, and n is the number of mixing steps.

This protocol mimics a typical two-step human embryo/oocyte protocol. However, mouse embryos are typically frozen with only a single step protocol where the embryo is directly transferred to VS medium and then frozen. On our device, this corresponds to simply removing the waiting time in the ES medium step. Both protocols were performed; however, the results presented were done with the mouse embryo timings to follow the protocol provided by CMMR.

After complete transfer of the embryo into the VS medium, the droplet containing the embryo is moved toward the edge to be collected by a micropipette (Figure 2(c)), and then plunged into liquid nitrogen. Contrary to conventional vitrification protocols and manual operation, which subject embryos to sudden changes

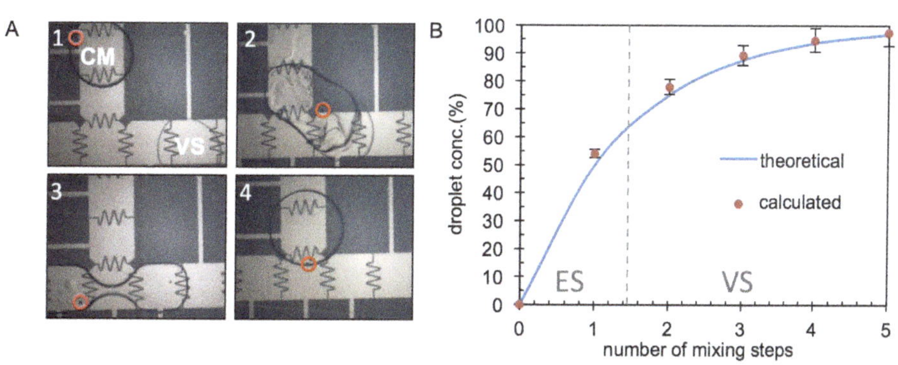

Figure 3. Droplet mixing protocol and resulting concentration profile. (A) 1: Embryo (red circle) contained in culture medium (CM) droplet. 2: Embryo droplet mixed with VS droplet. 3: Droplet split into two droplets (left contains embryo). 4: Droplet containing embryo is kept and other droplet is sent to waste. Process is repeated to increase VS concentration. (B) Mixing profile showing the generation of ES medium and VS medium. Droplet volumes were calculated by imaging droplet boundaries and modeling as cylinders. Concentrations were then calculated using these volumes before and after each mixing step.

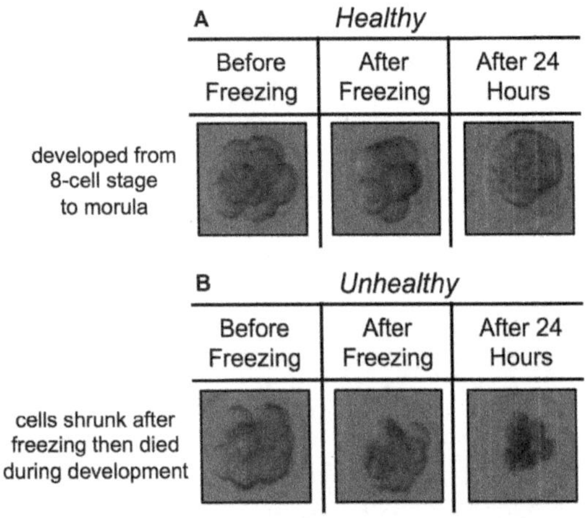

Figure 4. Healthy and unhealthy embryo morphology before and after freezing, and after 24 hours culturing. (A) Healthy and (B) unhealthy examples showing morphology changes before and after freezing. Survival rate was determined on the basis of embryo morphology after thawing vitrified embryos. Cell leakage, abnormal shapes, and membrane damage, as commonly used in vitrification studies, were counted as failure cases. Development rate was quantified by the embryo stage reached after culturing. For instance, the 8-cell stage embryo shown in (A) after vitrification, thawing, and culturing successfully developed to the blastocyst stage.

in medium concentration, the digital microfluidic approach gradually increases the VS medium concentration, (Figure 3), which is generally accepted by IVF practitioners to be more benign to embryos due to lower osmotic stress [26].

Thawing

To verify success of the vitrification process using digital microfluidics, embryos vitrified on device were thawed back and confirmed to have recovered in volume and have healthy morphology. Embryos were thawed by plunging in a bath of culture medium with 1.0 M of sucrose, which helps draw the cryoprotectants out of the embryo to minimize toxicity. The embryo is left in this bath for approximately 10 minutes over which its volume slowly increases back to its original size. After this point the embryo is transferred to culture medium without sucrose and is returned to the incubator. The same thawing procedure was used for manual and digital microfluidic operation.

Results and Discussion

Survival rate and development rate were used to evaluate the performance of digital microfluidic vitrification. Survivability was measured by examining the morphology of the embryo before and after freezing, as commonly performed in the literature (e.g., [27]). Embryos were considered unhealthy if they had an abnormal shape, membrane damage, leakage of cellular content or degeneration of their cytoplasm [28]. The development rate was determined by culturing survived embryos for an additional 24 hours after thawing (Figure 4). If the cell number within the embryo increased or it developed to the blastocyst stage, it was counted as developed. Control samples of non-vitrified embryos were also cultured to identify the base development rate of the mouse embryo population. Only embryos morphologically judged to be healthy were used for either manual or digital microfluidic testing. Only embryos that had healthy morphology after freezing were cultured following similar procedures to other vitrification studies [29].

Table 1 summarizes the experimental results, showing comparable survival and development rates between manual and automated processing. Additionally, since the embryos are constantly imaged on chip, its volume can be measured throughout the procedure and used to measure the quality of both the embryo and the protocol (Figure 5). Embryo relative volume response to the VS of different concentrations is an interesting and complex issue for developing new vitrification protocols [30,31]. The measurement of cell volume is difficult because embryos may collapse or flatten in VS instead of shrinking symmetrically [32].

Previous studies of embryo vitrification (4–16 cell stages) reported survival rates in the range of 80–100% [4]. Based on the limited sample size, vitrification using digital microfluidics in our work produced a survival rate of 77%, and our own manual vitrification trials resulted in a survival rate of 73%. These lower survival rates, compared to the results in the literature, can be mainly attributed to our use of a micropipette (vs. vitrification straw) inside liquid nitrogen. The micropipette tip is a standard plastic pipette for embryo manipulation and had an inner diameter of 125 μm (The STRIPPER micropipetter, Origio). Much research has gone into developing different mechanical structures (e.g., straw-type carriers [33], cryotube [34], Cryotop [35], and mesh-type carriers [36]) to increase the heat transfer rate [12]. Using these devices in liquid nitrogen requires the transfer of a processed embryo onto the device (e.g., vitrification straw) with minimal liquid volume remained on the vitrification straw. Since this embryo transfer process is not within the capability of our present microfluidic device and would introduce an extra manipulation step by hand, this work focused on proving the feasibility of using digital microfluidics for embryo processing. Therefore, the microfluidically processed embryos were directly frozen inside the micropipette tip in our experiments.

We are aware that using micropipette tips in liquid nitrogen was not ideal, which should have negatively affected the survival rate; however, since it was held constant between manual and DMF trials, its affect should be systematic. Additionally, the development rate, which was measured using embryos that survived freezing as commonly practiced in the literature, was high and comparable with other vitrification studies. This development rate measurement desirably removes the effect of our limited manual

Table 1. Summary of vitrification results.

	Survival Rate	Development Rate
control (non-vitrified)	100% (14/14)	93% (13/14)
manual	73% (11/15)	91% (10/11)
DMF Chip	77% (10/13)	90% (9/10)

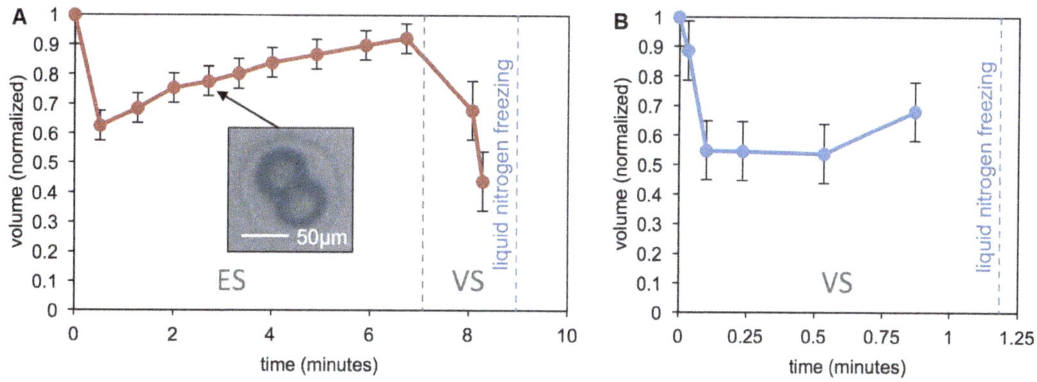

Figure 5. Embryo cell volume monitoring. Mouse embryo cell volume change measured on chip. (A) Vitrified using a two-step human embryo vitrification protocol (Irvine in Figure 6 with the addition of sucrose in the ES stage). (B) Vitrified using a one-step mouse embryo vitrification protocol (CMMR in Figure 6). Snapshots from recorded videos, at instances when the droplet was not moving, were used to measure volume. Volumes were calculated by modeling cells as spheres and were normalized to initial volume (this could result in errors as cells may collapse or flatten instead of shrinking symmetrically [32]. The initial volume dip in the human protocol matches the volume dip over the mouse protocol. For this experiment of volume measurement, 2-cell embryos were used to simplify image analysis. Error bars in (B) are relatively large because for this vitrification protocol, droplets are required to move quickly, which did not leave sufficient time to switch to higher microscope magnification for imaging on our digital microfluidic platform.

embryo handling skills and shows the potential of our microfluidic device for automated processing of embryos for vitrification.

Due to its programmability, one key benefit to the digital microfluidic approach is the ability to implement/test a number of vitrification protocols for efficacy comparisons. Figure 6 lists example human and mouse vitrification protocols, all using different cryoprotectants, number of mediums, and timings. However, at their core, all of these protocols involve the controlled increase in cryoprotectant concentration over a given period of time with the initial equilibrium solution typically containing half the cryoprotectant concentration as the full strength vitrification solution. This is convenient for DMF device design as it always produces a 50% concentration after the first mixing (neglecting the time required to homogenize concentration of the resulting droplet which is in the order of 2~3 seconds when the droplet is rolled over a few electrodes to enhance mixing). This means that a typically two-step protocol involving an initial 50% concentration ES step, following by a short 100% VS step, can be easily realized by following the mixing curve shown in Figure 3 with a pause after the first mixing step to allow the embryo to reach equilibrium.

Figure 7 shows how most two-step protocols follow the same mixing curve, only differing by their timings. This shows that although the protocols in Figure 6 involve significantly different amounts of manual manipulation, when performed on the digital microfluidic chip, they follow the same mixing procedure with the only difference lying in their timings. For our current trials, a mouse protocol provided by CMMR was used which involved a single mixing step. This allowed us to better conduct our manual trials as it required less manipulation; however, more complicated protocols can be readily performed by adding more dispensing reservoirs on chip and filling them with lower concentrations of cryoprotectant. This would allow for a high number of producible concentrations, especially in the low concentration range. Multiple reservoirs could also be used to implement protocols using different cryoprotectant compositions throughout their procedure [37].

One limitation of the digital microfluidic platform was handling culture mediums containing high serum concentrations. Serum contains a mixture of proteins that can adsorb on the Teflon coated surfaces of the device, eventually accumulating to the point that the surface becomes hydrophilic, making droplets immovable [38]. Some strategies have been developed to help overcome this problem, such as the use of Pluronic additives [39], silicone oil baths [40], and superhydrophobic surfaces [41]. Pluronic additives were avoided in our work as embryos are highly sensitive to additives. Superhydrophobic surfaces were also avoided as they require significant additional fabrication efforts. A silicone oil bath was used which did increase droplet movability; however, this

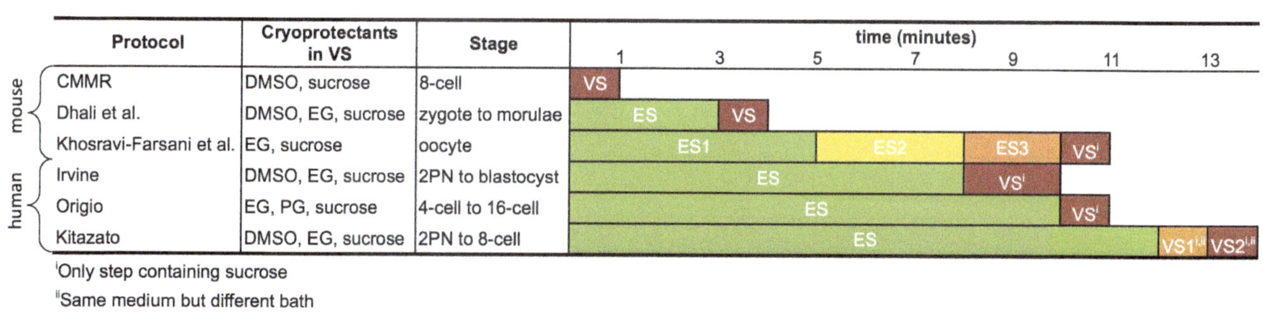

Figure 6. Comparisons of mouse and human vitrification protocols [28,29,42–47]. For those protocols that specify a timing range, the average value is used.

A

Protocol	c_1	t_1	t_2	t_3	t_4
CMMR	100%	0.5	1	-	-
Dhali et al.	50%	1	3	3.5	4
Irvine	50%	1	8	9	10
Origio	50%	1	10	10.5	11
Kitazato	50%	1	12	13	14

B

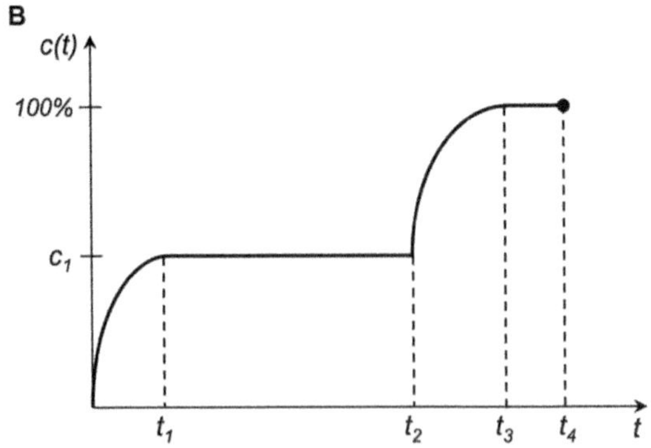

Figure 7. Vitrification protocol implementation on digital microfluidic chip. Implementation of common vitrification protocols on a digital microfluidic chip with a single dispensing reservoir. Timings and concentrations are shown in (A), and the generalized mixing curve is shown in (B).

approach did not work with sufficient effectiveness for conventional embryo culturing mediums that contain high serum concentrations. Therefore, in this work, a serum free culture medium was chosen for this proof-of-principle study. In further work, we will explore droplet movability improvement for serum-containing culture medium and the automation of transferring embryos onto vitrification devices (e.g., straw or Cryotop).

Conclusion

This paper described a digital microfluidic device that shows feasibility to perform automated embryo processing for vitrification applications. The results demonstrated that the embryo survival and development rate achieved by using the automated approach are comparable to manual operation, based on the limited sample size tested. Advantages of this approach, compared to manual operation and channel-based microfluidic vitrification, include automated operation, cryoprotectant concentration gradient generation, and feasibility of loading and retrieval of embryos. The device permits one to readily modify/test vitrification protocols with reduction in labor costs. Further development could possibly facilitate new vitrification protocol development and clinical IVF practice.

Author Contributions

Conceived and designed the experiments: YS MA. Performed the experiments: DP JL. Analyzed the data: DP JL. Contributed reagents/materials/analysis tools: DP JL. Wrote the paper: DP JL MA YS.

References

1. Trounson A, Mohr L (1983) Human pregnancy following cryopreservation, thawing and transfer of an eight-cell embryo. Nature 305: 707–709.
2. Reubinoff BE, Pera MF, Vajta G, Trounson a O (2001) Effective cryopreservation of human embryonic stem cells by the open pulled straw vitrification method. Hum Reprod 16: 2187–2194.
3. Medeiros C, Forell F, Oliveira A, Rodrigues J (2002) Current status of sperm cryopreservation: why isn't it better? Theriogenology: 327–344.
4. Tsang WH, Chow KL (2010) Cryopreservation of mammalian embryos: Advancement of putting life on hold. Birth Defects Res C Embryo Today 90: 163–175.
5. Oktay K, Newton H, Aubard Y, Salha O, Gosden R (1998) Cryopreservation of immature human oocytes and ovarian tissues: an emerging technology? Fertil Steril 69: 1–7.
6. Sonmezer M, Oktay K (2004) Fertility preservation in female patients. Hum Reprod Update 10: 251–266.
7. Lee SJ, Schover LR, Partridge AH, Patrizio P, Wallace WH, et al. (2006) American Society of Clinical Oncology recommendations on fertility preservation in cancer patients. J Clin Oncol 24: 2917–2931.
8. Wallace WHB, Anderson RA, Irvine DS (2005) Fertility preservation for young patients with cancer: who is at risk and what can be offered? Lancet Oncol 6: 209–218.
9. Riggs R, Mayer J, Dowling-Lacey D, Chi T-F, Jones E, et al. (2010) Does storage time influence postthaw survival and pregnancy outcome? An analysis of 11,768 cryopreserved human embryos. Fertil Steril 93: 109–115.
10. Pegg DE (2010) The relevance of ice crystal formation for the cryopreservation of tissues and organs. Cryobiology 60: S36–44.
11. Whittingham D (1971) Survival of mouse embryos after freezing and thawing. Nature 233: 125–126.
12. Saragusty J, Arav A (2011) Current progress in oocyte and embryo cryopreservation by slow freezing and vitrification. Reproduction 141: 1–19.
13. Vajta G, Nagy ZP (2006) Are programmable freezers still needed in the embryo laboratory? Review on vitrification. Reprod Biomed Online 12: 779–796.
14. Rall WF, Fahy GM (1985) Ice-free cryopreservation of mouse embryos at -196 degrees C by vitrification. Nature 313: 573–575.
15. AbdelHafez FF, Desai N, Abou-Setta AM, Falcone T, Goldfarb J (2010) Slow freezing, vitrification and ultra-rapid freezing of human embryos: a systematic review and meta-analysis. Reprod Biomed Online 20: 209–222.
16. Pollack MG, Fair RB (2000) Electrowetting-based actuation of liquid droplets for microfluidic applications. Appl Phys Lett 77: 1725–1726.
17. Barbulovic-Nad I, Au SH, Wheeler AR (2010) A microfluidic platform for complete mammalian cell culture. Lab Chip 10: 1536–1542.
18. Chang Y-H, Lee G-B, Huang F-C, Chen Y-Y, Lin J-L (2006) Integrated polymerase chain reaction chips utilizing digital microfluidics. Biomed Microdevices 8: 215–225.
19. Sista RS, Eckhardt AE, Srinivasan V, Pollack MG, Palanki S, et al. (2008) Heterogeneous immunoassays using magnetic beads on a digital microfluidic platform. Lab Chip 8: 2188–2196.
20. Park S, Wijethunga PAL, Moon H, Han B (2011) On-chip characterization of cryoprotective agent mixtures using an EWOD-based digital microfluidic device. Lab Chip 11: 2212–2221.
21. Heo YS, Lee H-J, Hassell BA, Irimia D, Toth TL, et al. (2011) Controlled loading of cryoprotectants (CPAs) to oocyte with linear and complex CPA profiles on a microfluidic platform. Lab Chip 11: 3530–3537.
22. Lai D, Ding J, Smith GD, Takayama S (2013) Automated microfluidic gradient cryoprotectant exchange platform for murine oocyte and zygote vitrification reduces osmotic stress and improves embryo developmental competence. Fertil Steril 100: S107.
23. Song YS, Moon S, Hulli L, Hasan SK, Kayaalp E, et al. (2009) Microfluidics for cryopreservation. Lab Chip 9: 1874–1881.
24. Ali J, Shelton JN (1993) Design of vitrification solutions for the cryopreservation of embryos. J Reprod Fertil 99: 471–477.
25. Berthier J, Clementz P, Roux J, Fouillet Y, Peponnet C (2006) Modeling microdrop motion between covered and open regions of EWOD microsystems. NSTI Nanotechnology Conference and Trade Show. Boston, USA, Vol. 1. pp. 685–688.
26. Swain JE, Lai D, Takayama S, Smith GD (2013) Thinking big by thinking small: application of microfluidic technology to improve ART. Lab Chip 13: 1213–1224.

27. Hogan B, Costantini F, Lacy E (1986) Manipulating the mouse embryo: a laboratory manual. 3rd ed. Cold Spring Harbor, NY: Cold spring harbor laboratory.
28. Dhali A, Anchamparuthy VM, Butler SP, Pearson RE, Mullarky IK, et al. (2009) Effect of droplet vitrification on development competence, actin cytoskeletal integrity and gene expression in in vitro cultured mouse embryos. Theriogenology 71: 1408–1416.
29. Zhang J, Cui J, Ling X, Li X, Peng Y, et al. (2009) Vitrification of mouse embryos at 2-cell, 4-cell and 8-cell stages by cryotop method. J Assist Reprod Genet 26: 621–628.
30. Newton H, Pegg DE, Barrass R, Gosden RG (1999) Osmotically inactive volume, hydraulic conductivity, and permeability to dimethyl sulphoxide of human mature oocytes. J Reprod Fertil 117: 27–33.
31. Mullen S, Li M, Li Y, Chen Z, Critser J (2008) Human Oocyte Vitrification: The permeability of metaphase II oocytes to water and ethylene glycol and the appliance toward vitrification. Fertil Steril 89: 1812–1825.
32. Shaw JM, Jones GM (2003) Terminology associated with vitrification and other cryopreservation procedures for oocytes and embryos. Hum Reprod Update 9: 583–605.
33. Otoi T, Yamamoto K, Koyama N, Tachikawa S, Suzuki T (1998) Cryopreservation of mature bovine oocytes by vitrification in straws. Cryobiology 37: 77–85.
34. Nakao K, Nakagata N, Katsuki M (1997) Simple and efficient vitrification procedure for cryopreservation of mouse embryos. Exp Anim.
35. Kuwayama M, Vajta G, Ieda S, Kato O (2005) Comparison of open and closed methods for vitrification of human embryos and the elimination of potential contamination. Reprod Biomed Online 11: 608–614.
36. Martino A, Songsasen N, Leibo SP (1996) Development into blastocysts of bovine oocytes cryopreserved by ultra-rapid cooling. Biol Reprod 54: 1059–1069.
37. Vitrolife (2011) Rapid-i ™ Vitrification of Cleavage stage embryos. Available: http://www.vitrolife.com/en/Fertility/Procedures/Vitrification/. Accessed 5 December 2013.
38. Au SH, Kumar P, Wheeler AR (2011) A new angle on pluronic additives: advancing droplets and understanding in digital microfluidics. Langmuir 27: 8586–8594.
39. Luk VN, Mo GC, Wheeler AR (2008) Pluronic additives: a solution to sticky problems in digital microfluidics. Langmuir 24: 6382–6389.
40. Srinivasan V, Pamula VK, Fair RB (2004) An integrated digital microfluidic lab-on-a-chip for clinical diagnostics on human physiological fluids. Lab Chip 4: 310–315.
41. Jönsson-Niedziółka M, Lapierre F, Coffinier Y, Parry SJ, Zoueshtiagh F, et al. (2011) EWOD driven cleaning of bioparticles on hydrophobic and super-hydrophobic surfaces. Lab Chip 11: 490–496.
42. Canadian Mouse Mutant Repository (2004) Modified Rapid Embryo Freezing Protocol D for ENU and Non-Enu Mice Embryos at Eight-Cell Stage. Available: http://www.cmmr.ca/protocols/embryo_cryo_D.pdf. Accessed 5 December 2013.
43. Khosravi-Farsani S, Sobhani A, Amidi F, Mahmoudi R (2010) Mouse oocyte vitrification: the effects of two methods on maturing germinal vesicle breakdown oocytes. J Assist Reprod Genet 27: 233–238.
44. Irvine Scientific (2011) Simplified Embryo Vitrification Protocol. Available: http://www.irvinesci.com/products/90133-dso-vitrification-freeze-solutions-for-embryo-pn-to-blastocyst-stage?dpt=Assisted+Reproductive+Technology. Accessed 3 December 2013.
45. Origio (2011) MediCult Vitrification Cooling. Available: http://www.origio.com/products/medicult media/medicult vitrification cooling.aspx. Accessed 5 December 2013.
46. Kitazato Corp. (2012) Cryotop Safety Kit. Available: http://www.kitazato.co.jp/biopharma/VT60.html. Accessed 5 December 2013.
47. Kuwayama M (2007) Highly efficient vitrification for cryopreservation of human oocytes and embryos: the Cryotop method. Theriogenology 67: 73–80.

25

Oxidative Stress in Zebrafish (*Danio rerio*) Sperm

Mary Hagedorn[1,2]*, Megan McCarthy[3], Virginia L. Carter[1,2], Stuart A. Meyers[3]

1 Department of Reproductive Sciences, Smithsonian Conservation Biology Institute, Washington, District of Columbia, United States of America, **2** Hawaii Institute of Marine Biology, University of Hawaii, Kaneohe, Hawaii, United States of America, **3** Department of Anatomy School of Veterinary Medicine, University of California at Davis, Davis, California, United States of America

Abstract

Laboratories around the world have produced tens of thousands of mutant and transgenic zebrafish lines. As with mice, maintaining all of these valuable zebrafish genotypes is expensive, risky, and beyond the capacity of even the largest stock centers. Because reducing oxidative stress has become an important aspect of reducing the variability in mouse sperm cryopreservation, we examined whether antioxidants might improve cryopreservation of zebrafish sperm. Four experiments were conducted in this study. First, we used the xanthine-xanthine oxidase (X-XO) system to generate reactive oxygen species (ROS). The X-XO system was capable of producing a stress reaction in zebrafish sperm reducing its sperm motility in a concentration dependent manner ($P<0.05$). Second, we examined X-XO and the impact of antioxidants on sperm viability, ROS and motility. Catalase (CAT) mitigated stress and maintained viability and sperm motility ($P>0.05$), whereas superoxide dismutase (SOD) and vitamin E did not ($P<0.05$). Third, we evaluated ROS in zebrafish spermatozoa during cryopreservation and its effect on viability and motility. Methanol (8%) reduced viability and sperm motility ($P<0.05$), but the addition of CAT mitigated these effects ($P>0.05$), producing a mean 2.0 to 2.9-fold increase in post-thaw motility. Fourth, we examined the effect of additional cryoprotectants and CAT on fresh sperm motility. Cryoprotectants, 8% methanol and 10% dimethylacetamide (DMA), reduced the motility over the control value ($P<0.5$), whereas 10% dimethylformamide (DMF) with or without CAT did not ($P>0.05$). Zebrafish sperm protocols should be modified to improve the reliability of the cryopreservation process, perhaps using a different cryoprotectant. Regardless, the simple addition of CAT to present-day procedures will significantly improve this process, assuring increased and less variable fertilization success and allowing resource managers to dependably plan how many straws are needed to safely cryopreserve a genetic line.

Editor: Shuang-yong Xu, New England Biolabs, Inc., United States of America

Funding: This work was supported by a grant to MM from National Institutes of Health (NIH) # R24 RR016581 and SM from NIH # RO1 2RR016581. The funders had no role in study design, data collection and analysis, decision to publish, or preparation of the manuscript.

Competing Interests: The authors have declared that no competing interests exist.

* E-mail: hagedornm@si.edu

Introduction

In the past decade, laboratories around the world have produced tens of thousands of mutant and transgenic zebrafish lines. As with mice, maintaining all of these valuable zebrafish genotypes is expensive, risky, and beyond the capacity of even the largest stock centers [1]. Our long-term goal is to preserve genetic resources from aquatic model organisms, specifically zebrafish, that are vital to advancing biomedical research and knowledge. Although conventional cryopreservation is a proven method for long-term, safe storage of genetic material, protocols used by zebrafish researchers are not standardized and yield inconsistent results with high post-thaw variability [2–6], thereby putting the security of many genotypes in individual laboratories and stock centers at great risk.

Fish have similar homeostatic enzyme systems to those reported in many mammalian systems [7]. Specifically, reactive oxygen species (ROS) are produced during respiration and can interact with organic substances. They can also be produced by intracellular enzyme systems like NADPH oxidase and cytoplasmic xanthine [7]. When these reactive oxygen species overcome the buffering capacity of the cell, the cell then enters oxidative stress, potentially leading to damage of DNA/RNA, proteins and lipids. Several cell adaptations exist which counteract the negative effects of oxidative stress, such as a thiol reducing buffer composed of glutathione and thioredoxin and enzymes to remove reduced oxygen species, such as catalase (CAT), superoxide dismutase (SOD) and glutathione peroxidase. However, zebrafish somatic cells have electrophile responsive genes that encode proteins that inactivate oxidants, thus avoiding oxidative stress [8,9] When zebrafish are exposed to oxidative stress, this induces a genetic pathway that produces antioxidants to help ward off the negative aspects of the stress [8,9]. However, zebrafish sperm cells are terminal haploid cells with tightly coiled DNA. These cells do not have access to their natural anti-oxidant pathway simply because this terminal cell type does not transcribe DNA or translate RNA.

An increase in ROS has been linked to abnormal or damaged spermatozoa [10–14], and this may be especially true in fish. For example, duroquinone induced ROS in carp spermatozoa that caused DNA damage in the sperm and subsequently impaired reproductive success [15]. The xanthine-xanthine oxidase (X-XO) system is one of several ROS generating systems frequently used in biology to investigate the cytotoxic effects of superoxide anion and hydrogen peroxide. Generation of ROS *in vitro* by the (X-XO) system results in a reduction in sperm motility [16–19], viability [16,20], ionophore-induced acrosome reaction [18,21] and sperm-oocyte fusion [18,22]. Hydrogen peroxide, in contrast to superoxide anion, is more stable, less polar, and can readily cross the plasma membrane [23]. Consequently, hydrogen peroxide appears to be the primary ROS responsible for oxidative damage to spermatozoa *in vitro* [16–19,22,24,25], and membrane lipid peroxidation is believed to be an important mechanism of action [10,18,26]. Spermatozoa are particularly susceptible to lipid peroxidation because they contain high concentrations of unsaturated fatty acids [27] and, as terminally differentiated cells, have limited repair mechanisms [28,29]. As a consequence of lipid peroxidation, the plasma membrane loses the fluidity and integrity it requires for participation in the membrane fusion events associated with fertilization [26,30,31].

Spermatozoa and seminal plasma possess a number of enzymes and low molecular weight antioxidants that scavenge ROS in order to prevent cellular damage. Spermatozoa contain a limited volume of cytoplasm and are predominantly dependent upon the antioxidant support of seminal plasma [32–34]. Sperm preparation for cryopreservation involves the removal of seminal plasma, thereby increasing the susceptibility of spermatozoa to oxidative stress. In addition, research suggests that the antioxidant activity of the spermatozoa themselves may be decreased by cryopreservation [35–37]. Furthermore, freeze-thawing of equine spermatozoa [14] and cryopreservation of human [38] and bovine [39,40] spermatozoa was associated with an increase in reduced oxygen species generation. Oxidative damage to DNA in cryopreserved bovine sperm had a high negative correlation with pregnancy rates [41]. The addition of enzyme scavengers or antioxidants to sperm preparations *in vitro* has been successful at counteracting the effects of oxidative stress on sperm motility, viability, lipid peroxidation, sperm-oocyte fusion and DNA fragmentation [18,22,42–46].

Because reducing oxidative stress has proven to be an important aspect of reducing the variability in mouse sperm cryopreservation [47–49], we examined whether this might show the same types of improvements for zebrafish sperm. It was not known whether antioxidants that scavenge ROS in order to prevent cellular damage would be beneficial for zebrafish sperm cryopreservation. In order to examine this issue, four experiments were conducted in this study. In Experiment 1, the X-XO system was established with zebrafish sperm and levels of ROS production were determined. In Experiment 2, three antioxidant substances, SOD, CAT and vitamin E were examined to determine whether they could mitigate the exogenous stress produced by the X-XO system. These antioxidants are commonly found in fish tissue [50]. In Experiment 3, we cryopreserved zebrafish sperm [6], evaluated the level of ROS production, and used CAT in pre-freeze and post-thaw samples to determine whether these antioxidants would reduce stress and improve motility and viability. Finally, in Experiment 4, we analyzed whether antioxidants might improve pre-freeze motility of three commonly used cryoprotectants [2–6]. We hypothesized that a potent antioxidant could be used to improve post-thaw motility and reduce variability in this valuable biomedical model. In future applications of cryopreservation, zebrafish sperm protocols could be modified to improve the reliability of the cryopreservation process and allow resource managers to reliably plan how many straws are needed to safely cryopreserve a line.

Materials and Methods

Maintenance of Animals

Fish were housed in standard microcosms (Aquatic Ecosystems, Apopka, FL) that have independent, water, temperature and waste management with sensors on each rack constantly monitoring the pH, temperature and conductivity of the water. The Zebrafish International Resource Center (ZIRC) at the University of Oregon has prepared detailed user manuals that describe standard operating procedures [51]. We followed their recommended facilities operations including care and maintenance of adults, breeding and obtaining gametes and embryos, record keeping, sending and receiving fish from other laboratories, quarantine and other procedures relating to disease control, and euthanasia. Briefly, AB wild-type fish were obtained from stocks at ZIRC at approximately 4 months of age. They were maintained in recirculating dechlorinated systems at 26–28°C with an artificial light cycle (14 h light; 10 h dark). Feeding schedule consisted of twice-daily provision of live brine shrimp (*Artemia* nauplii) and "Master Mix" dry food (a combination of Nelson's Silver Cup Tropical No. 1, Spirulina Flake, Golden Pearl and Cyclopeeze, see Westerfield [51] for details). Fish were examined daily for any signs of stress or disease and any fish in poor health were removed from the experimental group. Condition factors were standardized to use males of 5 to 10 months old with a condition factor of 1.5 to 2.5. Condition factor is defined as $K = 100{,}000 \times \text{weight}/(\text{standard length})^3$ [52]. All care and welfare for the animals met NIH animal care standards. Full details of the study approval are listed with the Smithsonian CRC-IACUC (approval ID #06-19) and the University of Hawaii, Hawaii Institute of Marine Biology IACUC (protocol ID# 06-022).

Collection of Germplasm

The procedure used to obtain sperm was gentle abdominal squeezing of mature males [51]. The afternoon prior to squeezing, males were chosen and removed from mixed-sex tanks and placed into a single tank as a group. Briefly, males were immersed in a solution of tricaine methane sulfonate (MS-222) made according to Westerfield [51] until gill movements have slowed (~30 sec). To collect sperm, males were rinsed with clean aquarium water, excess water was removed by placing fish on a KimWipe, so that standard length and weight could be taken and recorded. Fish were then placed on a slit, damp sponge and oriented with their dorsal surface down. While viewed under a dissecting microscope, the anal fin area was dried and gentle pressure was exerted using forceps to squeeze both sides of the fish at a point just anterior to the pelvic fins. The sperm was collected with a 1-ml capillary tube, amount recorded and then placed into an Eppendorf tube on ice to await more sperm, if pooling, or to be expressed into chilled buffer. Fish were returned to a recovery tank for observation after squeezing. Sperm was then diluted to appropriate concentrations for each experiment into Hanks balanced salt solution (HBSS) at 300 mOsm/kg (0.137 M NaCl, 5.4 mM KCl. 1.3 mM CaCl. 1.0 mM $MgSO_4$, 0.25 mM Na_2HPO_4, 4.2 mM $NaHCO_3$ and 5.55 mM glucose, pH 7.2) to maintain the sperm and prevent activation. Sperm was held on ice until used in experiments. Generally, ejaculate volume of a single male was ~1 µl at 10^{10} cells/ml [53]. Depending on the number of treatments, usually a sample consisted of pooled sperm from 3 to 5 males, diluted ~1:1000 with HBSS, then tested across the array of treatments. If more treatments were performed, the number of males in the pooled sample was increased. Throughout the study, the sperm concentration present in all treatment vials was 5×10^6 to 1×10^7 cells/ml.

Chemicals

Superoxide Dismutase (SOD), Xanthine Oxidase (XO), α-Tocopherol (Vitamin E dissolved in 95% methanol), dimethylacetamide (DMA), dimethylformamide (DMF) and catalase (*Aspirgillus niger*, Sigma C3515, CAT), were obtained from Sigma Chemical Co. (St. Louis, MO, USA). Xanthine was purchased from Calbiochem (La Jolla, CA, USA). Dihydroethidium (DHE), 4,4-difluro-5-(4-phenyl-1,3-butadienyl)-4-bora-3a,4a-diaza-s-indacene-3-undecanoic acid (C_{11}-BODIPY® $^{581/591}$), propidium iodide and SYTOX® Green, were obtained from Invitrogen (Eugene, OR, USA). All percent solutions were made up vol/vol in HBSS, unless otherwise stated.

Cellular Analysis

We measured the number of membrane intact cells and superoxide production within samples with two types of fluorescent stain pairs (DHE and SYTOX® Green) and (BODIPY® and propidium iodide) measured on a flow cytometer (see below).

DHE indicates the presence of general superoxide production because it permeates the cell and exhibits a blue-fluorescence in the cytosol. When oxidized, however, it intercalates within the cell's DNA, staining its nucleus a bright fluorescent red. SYTOX® Green nucleic acid stain only penetrates cells with compromised membranes and does not cross membranes of living cells. After various treatments (see details below), DHE was added to each sample (2 µM for 10 min) followed by SYTOX® Green (0.05 µm for 5 min).

BODIPY® incorporates into cell membranes and responds to lipid peroxidation by going through a spectral emission shift from red to green [54]. The nucleic acid stain, Propidium Iodide, only penetrates membranes of compromised cells and subsequently binding to nuclear DNA. Samples were loaded with BODIPY® (1 µM) for 30 min, centrifuged at 2300 g for 5 min, then exposed to the various treatments (see below), then stained with Propidium Iodide (10 µM).

Immediately following staining, cells were analyzed by flow cytometry that was performed using an Accuri C6 flow cytometer (Accuri Cytometers, Inc. Ann Arbor, MI USA) and data analyzed using the Accuri software (CFlow Plus, Ver. 1.0.202.1). The C6 is equipped with a blue and a red laser, two scatter detectors, and four fluorescence detectors (FL1 533/30 nm; FL2 585/40 nm; FL3>670 nm and FL4 675/25 nm) whose range displayed data across 6.2 logs. Because of this, no pre-gating was required as the C6 has a digital data collection system that allowed it to display and analyze all possible data post-collection without any alteration of the original data file. Forward scatter and side scatter were plotted, as well as florescence detected by plotting detection on FL-1 versus FL-2. Gating and fluorescence compensation values were set after data collection.

In order to limit the evaluation of DHE and BODIPY® fluorescence to viable spermatozoa, the subpopulation of SYTOX® Green or propidium iodide-positive cells (membrane compromised cells) were removed from the evaluation of 10,000 events (cells) per sample. Sample cell numbers were determined by measuring the volume needed to achieve 10,000 events with the C6.

Along with the other treatment groups (detailed below), two controls were included in all experiments to help identify live and dead cells for the flow cytometer; 1) unstained, untreated cells; and 2) cells destroyed by 3 cycles of freeze-thawing followed by staining with DHE and SYTOX® Green or BODIPY® and propidium iodide.

Since the motility of activated zebrafish sperm lasted less than 1 min, their motility measurements must be taken quickly. A single observer determined the sperm motility with a phase contrast microscope (200 x, Olympus BX41), measuring the mean percent progressive motility. Specifically, two µl of sperm at 10^7 cells/ml were placed onto the surface of a slide, 18 µl of deionized water was added to activate the sperm, the drop was gently mixed on the slide, and the motility measured within 5 to 10 sec of mixing. The slide was moved to assess at least 3 full frames of sperm motility and estimated at <10, 25, 50, 75 or > 90% progressive motility.

Experiment 1: Establishing of the Xanthine-Xanthine Oxidase (X-XO) System and DHE Method to Evaluate ROS Production with Zebrafish Spermatozoa

Before examining ROS production in zebrafish sperm cryopreservation, we established ROS production using a known ROS generating system. The X-XO system was used to generate the reactive oxygen species, superoxide and hydrogen peroxide. An aliquot (250 µl at 1×10^7 cells/ml) of the sperm sample (N = 10) was treated with one of 3 doses of X-XO. The range of concentration or dose of X-XO used in McCarthy and Meyers [40] served as a guide for our preliminary experiments. The treatments contained: (1) sperm treated with DHE (control), (2) low X-XO (0.1 mM X and 0.01units/ml XO), (3) medium X-XO (0.2 mM X and 0.0175 units/ml XO) and (4) high X-XO (0.3 mM X and 0.025 units/ ml XO). Treated samples were incubated for 30 min then a small sample was taken for motility analysis, as described above. The remaining samples were stained with DHE and SYTOX® Green and analyzed on the flow cytometer (see above) for cell viability and oxidative stress.

Experiment 2: Mitigation of ROS by Candidate Anti-Oxidants

The addition of scavengers for reactive oxygen species, SOD, CAT and Vitamin E was examined to determine whether they could decrease ROS. After squeezing males, the concentration of spermatozoa was adjusted to 1×10^7/mL in 250 µl aliquots and pre-treatment motility evaluated. The samples were then incubated at room temperature for 30 min, according to the following treatments: (1) sperm treated (control), (2) X-XO (0.2 mM X and 0.0175 U/ml XO), (3) SOD (200 U/ml), (4) X-XO (0.2 mM X and 0.0175 U/ml XO) + SOD (100 U/ml), (5) X-XO (0.2 mM X and 0.0175 U/ml XO) + SOD (200 U/ml), (6) Vitamin E (100 µM), (7) X-XO (0.2 mM X and 0.0175 U/ml XO) + Vitamin E (100 µM), (8) X-XO (0.2 mM X and 0.0175 U/ml XO) + Vitamin E (200 µM), (9) CAT (200 U/ml), (10) X-XO (0.2 mM X and 0.0175 U/ml XO) + CAT (100 U/ml), (11s) X-XO (0.2 mM X and 0.0175 U/ml XO) + CAT (200 U/ml). After the 30 min treatment- exposure, a sub-sample was taken from each treatment for motility assessments. The remaining sample was stained with DHE for 10 minutes, followed by SYTOX® Green for 5 minutes, then examined on the flow cytometer to measure the number of DHE (ROS)- and SYTOX® Green-positive cells (viability), as described above.

Experiment 3: Evaluation of Cryopreservation Stress and the Subsequent Effect on ROS Production with Zebrafish Spermatozoa

Cells were subjected to cryopreservation stress to determine (1) if the stress resulted in the production of ROS and (2) if CAT could mitigate stress contributed by hydrogen peroxide, or increase viability or motility over control values. After squeezing males, sperm samples were diluted to either 1×10^7 cells/ml or 2×10^7 cells/ml, depending on their treatment. Samples treated with cryoprotectant were diluted 1:1, so an initial cell concentration of 2×10^7 cells/ml was used in these samples. The following sperm treatments were analyzed using DHE or BODIPY®): (1) sperm treated (control), (2) CAT (200 U/ml), (3) 8% methanol, (4) 8% methanol + CAT (200 U/ml), (5)

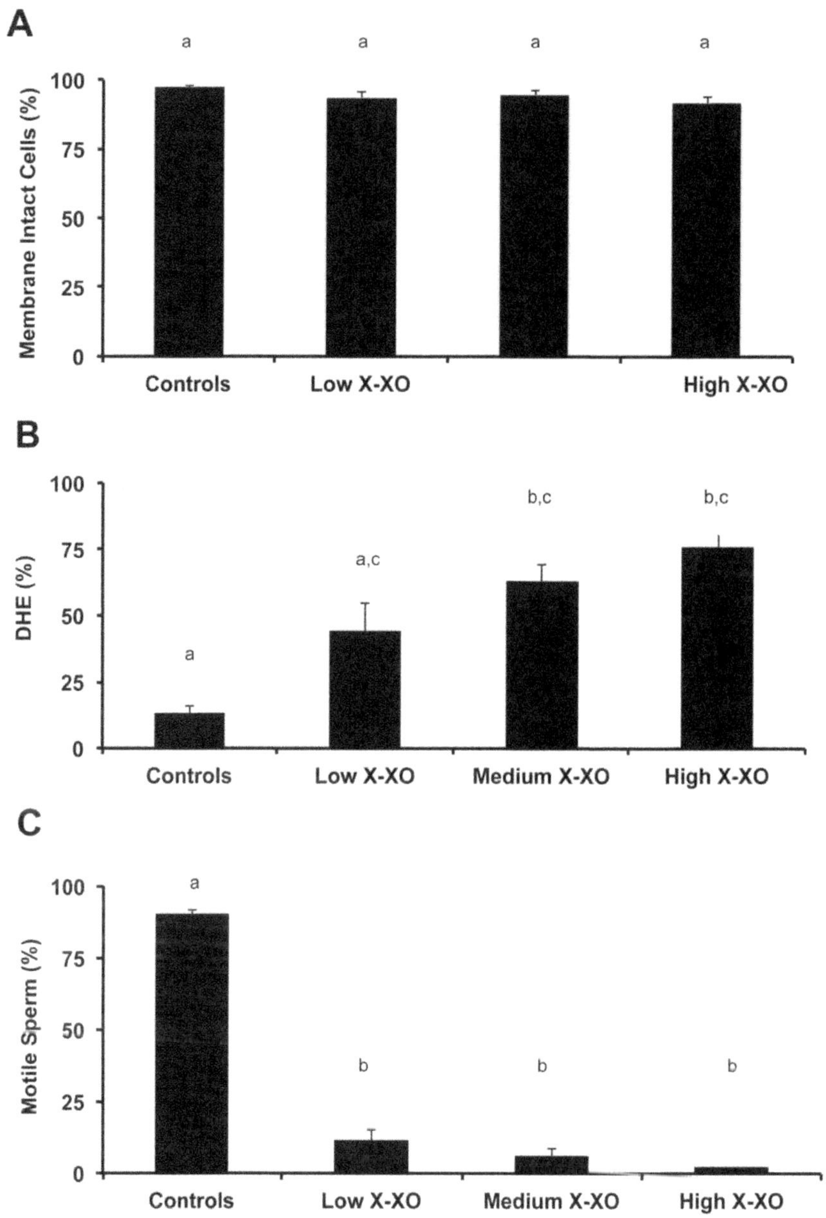

Figure 1. Evaluation of the xanthine-xanthine oxidase (X-XO) system and methods to identify DHE production (general oxidative stress) during ROS in zebrafish spermatozoa. In a concentration dependent manner, induced oxidative stress (X-XO): (**A**) did not alter the membrane integrity or viability (as assayed by propidium iodide) from the control values (P>0.05; Kruskal-Wallis test and Dunn's Multiple Comparison test); (**B**) increased stress as assayed by DHE-positive cells (P<0.05; Kruskal-Wallis test and Dunn's Multiple Comparison test); and, (**C**) reduced sperm motility as assayed by visual inspection with phase microscope (P<0.05; Kruskal-Wallis test and Dunn's Multiple Comparison test). Bars with the same letter are not significantly different (P>0.05), but bars with different letters are (P<0.05). Low X-XO exposure = 0.1 mMX + 0.01 units/ml XO; Medium X-XO exposure = 0.1 mMX + 0.01 units/ml XO; High X-XO exposure = 0.3 mM X + 0.025 units/ml XO. In all treatments, N = 10 samples and each sample consisted of pooled sperm from 3 to 5 males.

cryopreserved sperm with 8% methanol, (6) cryopreserved sperm with 8% methanol + CAT (200 U/ml) added pre-freeze. Preliminary experiments determined that CAT added prefreeze instead of post-thaw, produced the best results. Cryopreservation was performed as described by Yang and coworkers [6]. Briefly, sperm were incubated for 10 minutes on ice in an 8% methanol solution in HBSS, loaded into 0.25 ml French straws, then frozen at a rate of 10°C/min from 4°C to −80°C, using a Planer Kryo 320 - 1.7 controlled rate freezer (Planer PLC, Middlesex, UK). Sperm samples were then quenched in liquid nitrogen for at least 10 minutes before thawing (20 to 40 sec at 30°C). Motility was assessed immediately following the 10 min cryoprotectant exposure (for fresh samples) or post-thaw, for frozen samples. Samples were stained with DHE and SYTOX® Green (as described in Experiment 2) or BODIPY® and propidium iodide (as described above) and assessed on the flow cytometer for viability and ROS. Each sample consisted of fresh, pooled sperm from at least 3 males (N = 9 to 13 samples/treatment, except for one CAT treatment with an N = 5). Additionally, preliminary cryopreservation experiments were

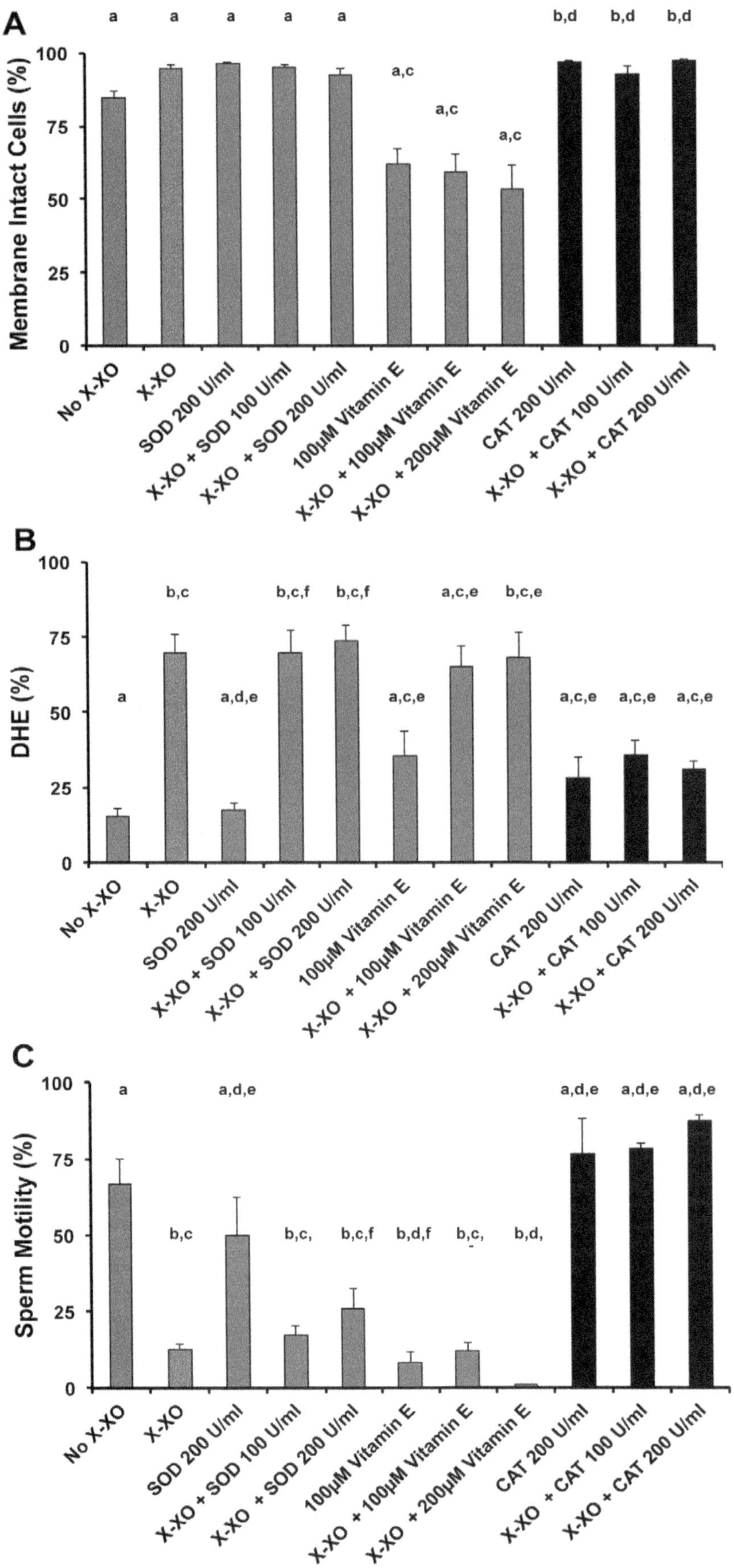
A
Membrane Intact Cells (%)
100
75
50
25
0
a
a
a
a
a
a,c
a,c
a,c
b,d
b,d
b,d
No X-XO
X-XO
SOD 200 U/ml
X-XO + SOD 100 U/ml
X-XO + SOD 200 U/ml
100μM Vitamin E
X-XO + 100μM Vitamin E
X-XO + 200μM Vitamin E
CAT 200 U/ml
X-XO + CAT 100 U/ml
X-XO + CAT 200 U/ml
B
DHE (%)
a
b,c
a,d,e
b,c,f
b,c,f
a,c,e
a,c,e
b,c,e
a,c,e
a,c,e
a,c,e
C
Sperm Motility (%)
a
b,c
a,d,e
b,c,
b,c,f
b,d,f
b,c,
b,d,
a,d,e
a,d,e
a,d,e

Figure 2. Evaluation of the xanthine-xanthine oxidase (X-XO) system and the impact of antioxidants on sperm viability, ROS and motility. In all the various treatments, the addition of CAT (black bars) performed the best. (**A**) Viability was not altered by most of the treatments as their mean values were not different than the control values (P>0.05), except CAT 200 U/ml, X-XO + CAT 100 U/ml X-XO + CAT 200 U/ml which were slightly higher than controls (P<0.05, Kruskal-Wallis and Dunn's Multiple Comparison test). (**B**) Only the combination of X-XO + CAT mitigated the oxidative stress as their values were not different than the control values (P>0.05), whereas the combinations of X-XO + SOD and X-XO + vitamin E did not (P<0.05, Kruskal-Wallis and Dunn's Multiple Comparison test). (**C**) Increased stress was correlated with decreased sperm motility that was mitigated by X-XO + CAT (P>0.05), but not by X-XO + SOD or X-XO + vitamin E (P<0.05; ANOVA and Tukey's Multiple Comparison test, F = 39.5). Bars with the same letter are not significantly different (P>0.05), but bars with different letters are different (P<0.05). The X- axis categories and sample sizes for all tests were: No X-XO = control (N = 4); X-XO (in all treatments) = 0.2 mM X and 0.0175 U/ml XO (N = 11); SOD 200 U/ml (N = 4); X-XO + SOD 100 U/ml (N = 12); X-XO + SOD 200 U/ml (N = 12); 100 μM vitamin E (N = 9); X-XO + 100 μM vitamin E (N = 9); X-X0 + 200 μM vitamin E (N = 9); CAT 200 U/ml (N = 4); X-XO + CAT 100 U/ml (N = 12); X-XO + CAT 200 U/ml (N = 12). In all trials, a medium X-XO exposure was used when X-XO was added. In all treatments, each sample consisted of pooled sperm from at least 3 males.

also performed to determine (1) if the stress resulted in the production of ROS and (2) if SOD (200 U/ml) could mitigate stress contributed by superoxide (as measured by DHE fluorescence), or increase viability or motility over control values (N = 5 samples/treatment).

Experiment 4: Evaluating Other Cryoprotectants on Motility and their Response to Antioxidants

To determine whether other cryoprotectants might be better candidates than methanol in terms of the stress they cause to the sperm, we compared 8% methanol, 8% methanol + CAT, 10% DMA, 10% DMA+CAT, 10% DMF and 10% DMF+CAT. These cryoprotectants were chosen because they have worked with zebrafish sperm to produce viable sperm post-thaw [3–6]. Each sample consisted of fresh, pooled sperm from at least 3 males. The following sperm-treatment groups were examined: (1) untreated sperm, N = 16; (2) CAT, N = 16; (3) 8% methanol, N = 16; (4) 8% methanol + CAT, N = 16; (5) 10% DMF, N = 8; (6) 10% DMF + CAT, N = 8; (7) 10 % DMA; and, (8) 10 % DMA+CAT, N = 8. The samples were held for 10 min at 0°C, then it's motility assessed on a phase microscope (200 x as described above). The concentration of the CAT used in these experiments was (200 U/ml).

Data Analysis

All data analyses in this study were performed using Graphpad Prism 5.0 (San Diego, CA) and Microsoft Excel (version 2007). All percent data were either LOG or arcsine transformed, and difference in the means was analyzed with either a one-way analysis of variance and a Tukey's Multiple Comparison test or with a Kruskal-Wallis test and a Dunn's Multiple Comparison test. A value of $P<0.05$ was considered significant, and all data was expressed as mean ± SE unless otherwise stated.

Results

Experiment 1: Establishing of the Xanthine-Xanthine Oxidase System and Methods to Evaluate ROS Production with Zebrafish Spermatozoa

The induced oxidative stress (X-XO) system was functional for zebrafish sperm. In these studies, X-XO- generated oxidative stress was evaluated using DHE as an indicator of the superoxide anion. An increase in percentage of cells expressing high levels of DHE indicated increasing superoxide anion and, hence, oxidative stress in the system. The X-XO system was capable of producing a stress reaction in zebrafish sperm in a concentration dependent manner, which correlated with decreased motility, but had no affect on cell viability (Fig. 1). For the remainder of the tests, a medium X-XO concentration (0.2 mM X and 0.0175 U/ml XO) was used.

Experiment 2: Mitigation of ROS by Candidate Anti-Oxidants

Overall, the addition of the X-XO and combinations of X-XO + SOD or X-XO + and vitamin E did not impact the cell viability (P>0.05, Kruskal-Wallis and Dunn's Multiple Comparison test), but the CAT and X-XO + CAT treatments improved viability over control values (P<0.05, Kruskal-Wallis and Dunn's Multiple Comparison test, Fig. 2A). The X-XO system increased DHE-positive cells from about 15% (control) to about 70% (X-XO) (P<0.05, Kruskal-Wallis and Dunn's Multiple Comparison test), but only the addition of CAT mitigated this stress to control levels (P>0.05, Kruskal-Wallis and Dunn's Multiple Comparison test, Fig. 2B). Additionally, only CAT maintained sperm motility at or slightly above control levels (P>0.05, Kruskal-Wallis and Dunn's Multiple Comparison test, Fig. 2C), whereas X-XO alone or in combination with SOD or vitamin E produced over a 50% reduction in motility (Kruskal-Wallis, P<0.05, Fig. 2C). Due to the fact that CAT produced superior results with the zebrafish sperm X-XO system, it was used in the subsequent experiments.

Experiment 3: Cryoprotectant Stress and Mitigation by CAT

We examine both intracellular ROS and lipid peroxidation in zebrafish spermatozoa during cryopreservation and its correlated effects on viability and motility (Figs. 3 and 4). Several sperm treatments were examined including: (1) Control = no treatment; (2) CAT (200 U/ml, concentration used in all CAT treatments); (3) 8% Methanol; (4) 8% Methanol + CAT; (5) cryopreservation with 8% Methanol; (6) cryopreservation with 8% Methanol + CAT.

Both intracellular ROS (Fig. 3) and lipid peroxidation (Fig. 4) showed similar responses to the treatments, except for a slight difference in the viability measurements. Specifically, ROS was increased 2.7(DHE)- to 3.0(BODIPY®)-fold after cryopreservation, (P<0.05), but this ROS was not mitigated by CAT (P>0.05, ANOVA). The addition of 8% methanol to fresh zebrafish sperm reduced its motility of (P<0.05, ANOVA), but adding 8% methanol + CAT maintained its motility at control levels (P>0.05, ANOVA) improving the post-thaw motility 2.5 (DHE)- to 2.9 (BODIPY®)-fold. Using DHE and SYTOX® Green to measure viability demonstrated a ~50% loss in viability after cryopreservation with 8% methanol alone, but viability was improved 13% with the addition of CAT to the cryopreservation process (P<0.05, ANOVA, Fig. 3). Using BODIPY® and propidium iodide to measure viability both the fresh and cryopreserved sperm experienced a loss of viability that was improved with the addition of CAT to the process (P<0.05, ANOVA, Fig. 4).

Our preliminary results with SOD and cryopreservation did not vary from those reported here for CAT, except there was no improvement of post-thaw sperm motility with SOD (5%±2.2%) or without SOD (6%±2.2%).

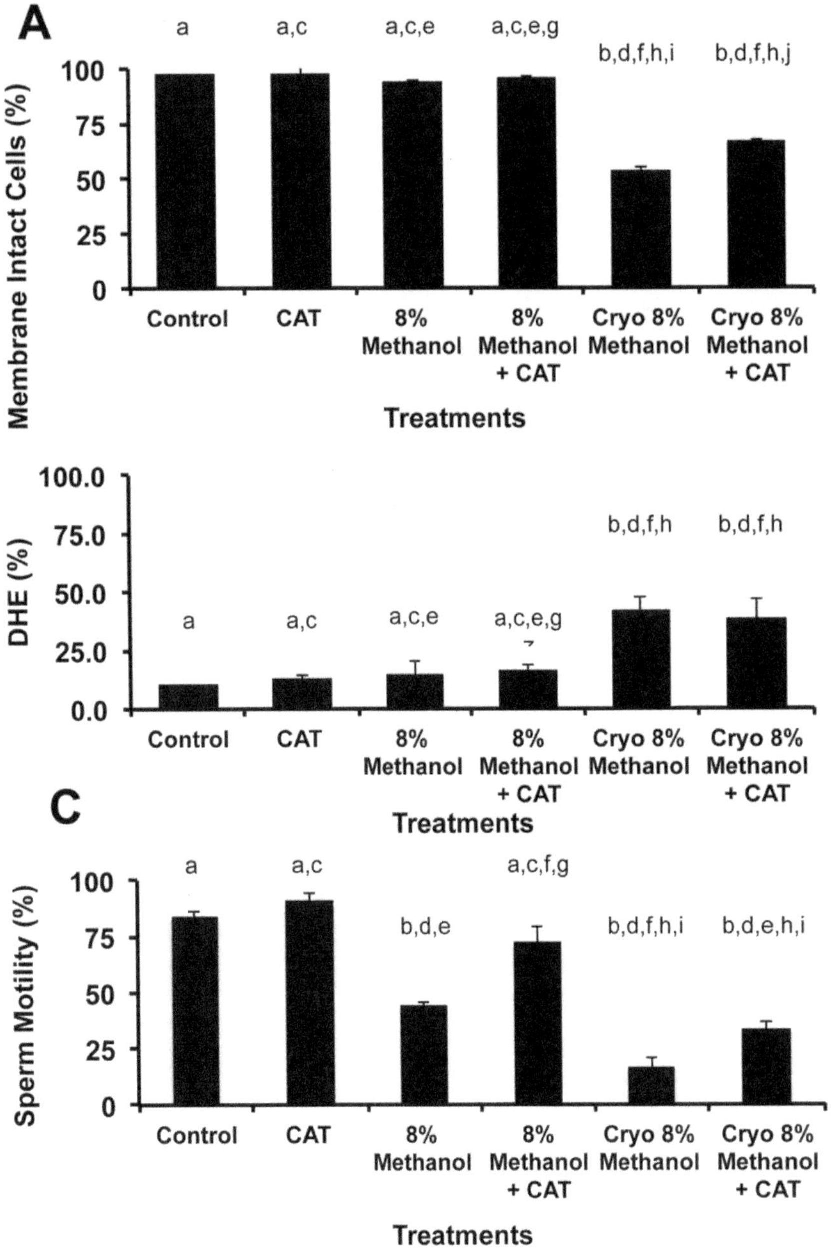

Figure 3. Evaluation of the intracellular ROS in zebrafish spermatozoa during cryopreservation and its effect on viability and motility. (**A**) Overall, cryopreservation decreased the viability of zebrafish sperm by approximately 50%, but the addition of CAT to the methanol improved its post-thaw viability 16% ($P<0.05$, ANOVA and Tukey's Multiple Comparison test, $F=60.5$). (**B**) Cryopreservation increased ROS over 2.7-fold ($P<0.05$, ANOVA, and Tukey's Multiple Comparison test $F=6.2$), but the addition of CAT did not mitigate it ($P>0.5$, ANOVA). (**C**) Motility demonstrated the greatest sensitivity in our treatments. Prior to cryopreservation, the 8% methanol treatment reduced motility by ~50%, and the addition of CAT to the 8% methanol returned the motility to control levels. In parallel, the cryopreserved sperm (8% methanol) decreased the pot-thaw motility of zebrafish sperm from 83% (control) to 17%, however the addition of CAT doubled this post-thaw motility ($P<0.05$, ANOVA and Tukey's Multiple Comparison test, $F=22.6$). Bars with the same letter are not significantly different ($P>0.05$), but bars with different letters are ($P<0.05$). X- axis categories and sample sizes were: Control= live sperm stained with DHE, N = 13; CAT (200 U/ml, concentration used in all CAT treatments), N = 9; 8% Methanol, N = 12; 8% Methanol + CAT 200, N = 12; Cryo 8% Methanol = sperm cryopreserved according to [6] with 8% methanol, N = 13; Cryo 8% Methanol + CAT, N = 13.

Experiment 4: Evaluating Other Cryoprotectants on Motility and their Response to Antioxidants

In zebrafish, pre-freeze and post-thaw motility is correlated fertilization success [2]. For successful cryopreservation, the process must begin with highest quality material possible, especially in regards to motility. Ideally, the addition of the cryoprotectant should not negatively impact the motility and viability of the sperm. However, in experiment 3, we observed that the routinely-used cryoprotectant (8% methanol) decreased motility and viability of the fresh sperm prior to the cryopreservation process. Some of this loss in motility was improved with CAT, but even with the help of CAT the motility values did not

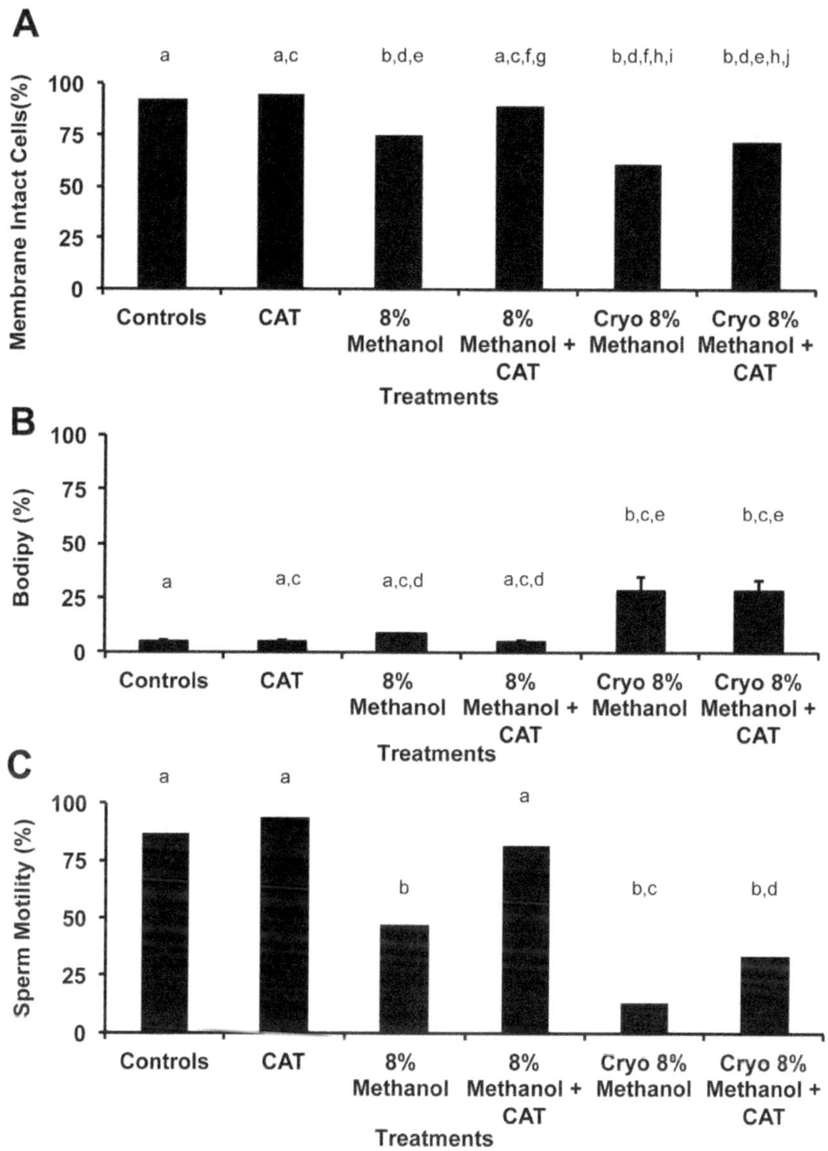

Figure 4. Evaluation of the lipid membrane ROS in zebrafish spermatozoa during cryopreservation and Its effect on viability and motility demonstrated the same pattern as in Fig. 3. (**A**) Methanol (8%) decreased the viability of zebrafish sperm in fresh and cryopreserved treatments, and the addition of CAT to the methanol improved its viability in pre- and post cryopreservation treatments ($P<0.05$, ANOVA, $F = 19.0$). (**B**) Cryopreservation increased ROS 3.0-fold ($P<0.05$), but CAT did not mitigate it ($P>0.5$, ANOVA and Tukey's Multiple Comparison test, $F = 13.4$). (**C**) The addition of 8% methanol decreased zebrafish sperm motility in fresh and cryopreserved treatments ($P<0.05$), and the addition of CAT to the methanol improved its viability in pre- and post cryopreservation treatments ($P<0.05$, ANOVA and Tukey's Multiple Comparison test, $F = 19.0$). Bars with the same letter are not significantly different ($P>0.05$), but bars with different letters are ($P<0.05$). X- axis categories and sample sizes were: Control = live sperm stained with DHE, N = 12; CAT (200 U/ml, concentration used in all CAT treatments), N = 5; 8% Methanol, N = 12; 8% Methanol + CAT 200, N = 12; Cryo 8% Methanol = sperm cryopreserved according to [6] with 8% methanol, N = 12; Cryo 8% Methanol + CAT, N = 12.

approach control values. To determine whether other cryoprotectants might impact the motility less, we examined 10% DMA and 10% DMF with and without CAT in addition to 8% methanol with and without CAT. In the fresh sperm treatments, both 10% DMF and 10% DMF + CAT maintained motilities at the control level (Fig. 5, $P>0.05$, ANOVA), whereas all the other treatment pairs did not ($P<0.05$, ANOVA, $F = 15.4$).

Discussion

Preservation of genetic resources from a variety of animal disease models is paramount for maintenance of genetic diversity and the study of disease. The sheer numbers of genetically unique zebrafish lines makes systematic preservation of these models essential to preserve this valuable resource. There are simply too many such genetic lines to warrant maintenance of live animals for the entire resource. A long-term goal for this resource has been to preserve gametes, embryos, and stem cells since numerous methods will likely be necessary to ensure preservation of each of these resources. Cryopreservation of zebrafish sperm has been one method that has been used for several decades although research groups working with this resource have reported varying success rates.

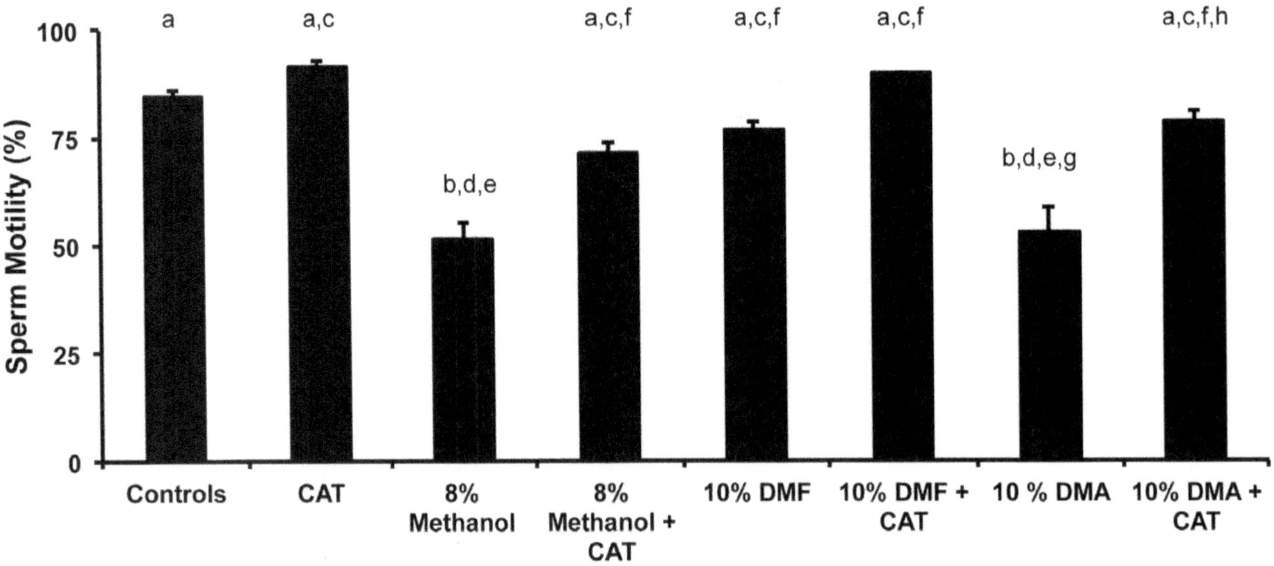

Figure 5. The effect of cryoprotectants and antioxidants on the motility of fresh zebrafish sperm. Each sample consisted of fresh, squeezed sperm pooled from 3 males that was treated with cryoprotectants alone or with cryoprotectants and CAT to examine the effect on the motility of the sample. X-axis categories and sample sizes were: Control, no treatment, N = 16; CAT (200 U/ml for all treatments), N = 16; 8% methanol, N = 16; 8% methanol + CAT, N = 16, 10% dimethylformamide (DMF), N = 8; 10% DMF + CAT, N = 8; 10% dimethylacetamide (DMA), N = 8; and, 10% DMA + CAT, N = 8. Of all the treatments, the 8% methanol and 10% DMA reduced the motility over the control value ($P<0.5$, ANOVA and Tukey's Multiple Comparison Test, $F = 15.4$), whereas the other cryoprotectants (with or without CAT) did not ($P>0.05$). When CAT was added to the cryoprotectants, all the mean values increased, but the 10% DMF + CAT produced the highest mean motility at 90%. Bars with the same letter are not significantly different ($P>0.05$), but bars with different letters are ($P<0.05$). Thus, the addition of catalase had a beneficial effect on the motility of zebrafish sperm.

Among vertebrates, successful cryopreservation of gametes has been shown to be highly variable and dependent on the species and conditions of cooling, media, and storage. Cryopreservation of sperm induces severe stresses in sperm cells owing largely to biophysical stressors such as intracellular and extracellular ice formation, osmotic stress, and oxidative stress. In our study, the positive response of zebrafish sperm to CAT, a mediator of hydrogen peroxide radicals, suggests that hydrogen peroxide could be the primary source of oxidative cell damage during cryopreservation. Although both the X-XO ROS generating system and the process of cryopreservation have been demonstrated to produce both superoxide anion and hydrogen peroxide in sperm cell systems in mammals, we found that a significant degree of mitigation of ROS-induced damage could result from the addition of the ROS scavenging enzyme, CAT, in comparison to the specific blockage of superoxide anion by SOD.

Mammalian sperm have been shown to be particularly susceptible to lipid peroxidation because they contain high concentrations of unsaturated fatty acids and because of the sperm cell's inherent limited self-repair mechanism. Relatively little is known regarding the fatty acid composition of zebrafish sperm however, it seems clear from the present study that these cells are also highly susceptible to the effects from lipid peroxidation and oxygen free radicals. In previous studies, we have demonstrated that both equine and rhesus monkey sperm are susceptible to peroxidative damage resulting from cryopreservation [19,55,56] and that both superoxide anion and hydrogen peroxide are major sources of this damage. However, in both of these species it is apparent that hydrogen peroxide is the more significant of the two regarding ROS-induced cell damage. Especially since SOD had no effect on the post-thaw motility of the cryopreserved zebrafish sperm, suggesting that superoxide was not produced by cryopreservation. In porcine sperm, it has been demonstrated that superoxide anion plays a less significant role as a source of oxidation than does hydrogen peroxide [57].

An increase in ROS has been linked to abnormal or damaged spermatozoa in other species [10,12,14,18]. It has also been reported in fish that ROS may induce DNA damage to sperm. For example, duroquinone-induced ROS caused DNA damage in carp sperm, which impaired reproductive success [15]. The X-XO system is one of several ROS generating systems frequently used to investigate the cytotoxic effects of the superoxide anion and hydrogen peroxide. Generation of ROS *in vitro* by the X-XO system results in reduced sperm motility [16–19], viability [16,20], ionophore-induced acrosome reaction [18], and sperm-oocyte fusion [18,22]. Hydrogen peroxide, in contrast to superoxide anion, is more stable and can readily cross the plasma membrane [58]. Consequently, hydrogen peroxide appears to be the primary ROS responsible for oxidative damage to spermatozoa *in vitro* in some systems [16,19,21,22,24]. Membrane lipid peroxidation also is believed to be an important mechanism of action [10,18,59]. Spermatozoa are particularly susceptible to this condition, in part, because they contain high fatty acid concentrations [27] and, as terminally differentiated cells, have limited repair mechanisms [28]. Lipid peroxidation causes loss of plasma membrane fluidity and integrity required for fusion events associated with fertilization [30,59].

One of the goals of this work was to determine whether an *in vitro* ROS system could be established in zebrafish to examine cryopreservation-related stresses. In this way, a rational approach to minimizing the damage that sperm incur during cryopreservation can be determined. The antioxidant, CAT, performed best in our studies. It maintained viability and motility, and mitigated stress in the X-XO system, returning it to control levels. In contrast, SOD and vitamin E did not improve any measures of sperm viability/membrane integrity and motility. These studies may not be an accurate assessment of vitamin E, however, because it was dissolved in methanol, which may have negatively impacted its physiological action. Elevated BODIPY®-labeling of zebrafish sperm was associated with cryopreservation and X-XO induced ROS production in this study, suggesting that hydrogen peroxide production may be the dominant ROS in the zebrafish model.

Methanol is one of the main cryoprotectants for zebrafish sperm used in laboratories around the world [51]. Unfortunately, 8% Methanol performed very poorly in our experiments reducing sperm motility, and viability prior to the cryopreservation process (as measured by BODIPY® and propidium iodide). Most importantly, motility correlates with fertilization success in zebrafish [2]. Methanol, DMA and DMF are a class of cryoprotectants that do not cause osmotic swelling of zebrafish sperm during the solute equilibration process [60] which makes them excellent candidates for understanding and improving the cryopreservation process. However, 10% DMF maintained control levels of motility and viability, whereas 8% methanol and 10% DMA did not. These experiments did not address whether 10% DMF or 10% DMF + CAT would out perform 8% methanol or 8% methanol + CAT in improving fertilization success. This is because additional studies need to accomplished to examine all three cryoprotectants and CAT in parallel using the exact same procedures (i.e., sperm concentrations, buffers, *in vitro* fertilization assays etc.) to understand their strengths and weaknesses. The cryopreservation process did increase intracellular and lipid ROS, and CAT mitigated this by improving post-thaw motility and viability. However, we also observed that the addition of CAT did not decrease the ROS production after cryopreservation. This is puzzling considering CAT reduced ROS stress in the X-XO system. The reasons for these mixed observations remain unclear. Additional studies will determine the nature of lipid composition of fish sperm membranes and their susceptibility to membrane damage.

In summary, these experiments indicated that the X-XO system could be used for determination of zebrafish sperm oxidative damage and selection of appropriate antioxidants to help mitigate ROS during zebrafish sperm cryopreservation. Further studies will be necessary to evaluate other potential ROS and Reactive Nitrogen Species, including nitric oxide and others, and their possible roles in oxidation of membranes and DNA of zebrafish sperm. In addition, our study demonstrated that sperm from this aquatic species behaves similar to the more extensively studied mammalian sperm. Although we used an accepted cryopreservation protocol [6] for the cryopreservation experiments with the 8% methanol, this and many of the other related protocols will need improvement in the future. Certainly, the addition of CAT to present procedures may be able to more than double the post-thaw motility, assuring increased and less variable fertilization success for zebrafish laboratories and stock centers around the world.

Acknowledgments

We would like to thank Dr. Bruce Draper for his kind assistance during the preliminary studies for this work and the Zebrafish International Research Center for their support and comments on this manuscript. In addition, we would like to thank the three anonymous reviewers whose comments greatly improved this paper. This manuscript has been approved for publication by the Hawaii Institute of Marine Biology as contribution #1490.

Author Contributions

Conceived and designed the experiments: MH VC MM SM. Performed the experiments: VC MH MM. Analyzed the data: VC MM MH. Contributed reagents/materials/analysis tools: SM. Wrote the paper: MH VC MM SM.

References

1. Knight J, Abbot A (2002) Full house. Nature 417: 785–786.
2. Harvey B, Kelley RN, Ashwood-Smith MJ (1982) Cryopreservation of zebra fish spermatozoa using methanol. Can J Zool 60: 1867–1870.
3. Morris JP, Berghmans S, Zahrieh D, Neuberg DS, Kanki JP (2003) Zebrafish sperm cryopreservation with N,N-dimethylacetimide. Biotechniques 35: 956–958.
4. Draper BW, McCallum CM, Stout JL, Slade AJ, Moens CB (2004) A high-throughput method for identifying N-ethyl-N-nitrosourea (ENU)-induced point mutations in zebrafish. Meth Cell Biol 77: 91–112.
5. Sood R, English MA, Jones M, Mullikin J, Wang DM, et al. (2006) Methods for reverse genetic screening in zebrafish by resequencing and TILLING. Methods 39(3): 220–7.
6. Yang H, Jones C, Varga ZM, Tiersch TR (2007) Development of simplified and standardized protocol for sperm cryopreservation in zebrafish, *Danio rerio*, with potential for high throughput. Theriogenology 68(2): 128–36.
7. Rocha MJ, Rocha E, Resende AD, Lobo-da-Cunha A (2003) Measurement of peroxisomal enzyme activities in the liver of brown trout (*Salmo trutta*), using spectrophotometric methods. BMC Biochem. 4: 2.
8. Takagi Y, Kobayashi M, Li L, Suzuki T, Nishikawa K, et al. (2004) MafT, a new member of the small Maf protein family in zebrafish. Biochem Biophys Res Commun 16;320(1): 62–9.
9. Suzuki T, Takagi Y, Osanai H, Li L, Takeuchi M, et al. (2005) Pi class glutathione S-transferase genes are regulated by Nrf 2 through an evolutionarily conserved regulatory element in zebrafish. Biochem J. 388(Pt 1): 65–73.
10. Aitken RJ, Clarkson JS, Hargreave TB, Irvine DS, Wu FC (1989) Analysis of the relationship between defective sperm function and the generation of reactive oxygen species in cases of oligozoospermia. J Androl 10: 214–220.
11. Aitken RJ, Krausz C, Buckingham D (1994) Relationships between biochemical markers for residual sperm cytoplasm, reactive oxygen species generation, and the presence of leukocytes and precursor germ cells in human sperm suspensions. Mol Reprod Dev 39: 268–279.
12. Rao B, Soufir JC, Martin M, David G (1989) Lipid peroxidation in human spermatozoa as related to midpiece abnormalities and motility. Gamete Res 24: 127–134.
13. Iwasaki A, Gagnon C (1992) Formation of reactive oxygen species in spermatozoa of infertile patients. Fertil Steril 57: 409–416.
14. Ball BA, Vo AT (2001) Osmotic tolerance of equine spermatozoa and the effects of soluble cryoprotectants on equine sperm motility, viability, and mitochondrial membrane potential. J Androl 22: 1061–9.
15. Zhou B, Liu W, Siu WH, O'Toole D, Lam PK, et al. (2006) Exposure of spermatozoa to duroquinone may impair reproduction of the common carp (Cyprinus carpio) through oxidative stress. Aquat Toxicol 77(2): 136–42.
16. de Lamirande E, Gagnon C (1992a) Reactive oxygen species and human spermatozoa. I. Effects on the motility of intact spermatozoa and on sperm axonemes. J Androl 13: 368–78.
17. de Lamirande E, Gagnon C (1992b) Reactive oxygen species and human spermatozoa. II. Depletion of adenosine triphosphate plays an important role in the inhibition of sperm motility. J Androl 13: 379–386.
18. Aitken RJ, Buckingham D, Harkiss D (1993) Use of a xanthine oxidase free radical generating system to investigate the cytotoxic effects of reactive oxygen species on human spermatozoa. J Reprod Fertil 97: 441–450.
19. Baumber J, Ball BA, Gravance CG, Medina V, Davies-Morel MC (2001) The effect of reactive oxygen species on equine sperm motility, viability, acrosomal integrity, mitochondrial membrane potential, and membrane lipid peroxidation. J Androl 21: 895–902.
20. Baiardi G, Ruiz RD, Fiol de Cuneo M, Ponce AA, Lacuara JL, et al. (1997) Differential effects of pharmacologically generated reactive oxygen species upon functional activity of epididymal mouse spermatozoa. Can J Phys Pharm 75: 173–178.
21. Griveau JF, Dumont E, Renard P, Callegari JP, Le Lannou D (1995a) Reactive oxygen species, lipid peroxidation and enzymatic defence systems in human spermatozoa. J Reprod Fertil 103: 17–26.
22. Blondin P, Coenen K, Sirard MA (1997) The impact of reactive oxygen species on bovine sperm fertilizing ability and oocyte maturation. J Androl 18: 454–460.
23. Halliwell B (1991) Reactive oxygen species in living systems: source, biochemistry and role in human disease. Am J Med 91 (suppl 3c): 14s–22s.
24. Alvarez JG, Storey BT (1989) Role of glutathione peroxidase in protecting mammalian spermatozoa from loss of motility caused by spontaneous lipid peroxidation. Gamete Res 23: 77–90.
25. Griveau JF, Dumont E, Renard P, Callegari JP, Le Lannou D (1995b) Reactive oxygen species, lipid peroxidation and enzymatic defense systems in human spermatozoa. J Reprod Fertil 103: 17–26.

26. Storey BT (1997) Biochemistry of the induction and prevention of lipoperoxidative damage in human spermatozoa. Mol Human Reprod 3: 203–213.
27. Jones R, Hamilton DW, Fawcett DW (1979) Morphology of the epithelium of the extratesticular rete testis, ductuli efferentes and ductus epididymidis of the adult male rabbit. Am J Anat 156(3): 373–400.
28. Aitken RJ, Clarkson JS (1987) Cellular basis of defective sperm function and its association with the genesis of reactive oxygen species by human spermatozoa. J Reprod Fert 81: 459–469.
29. van Loon AA, Den Boer PJ, Van der Schans GP, Mackenbach P, Grootegoed JA, et al. (1991) Immunochemical detection of DNA damage induction and repair at different cellular stages of spermatogenesis of the hamster after in vitro or in vivo exposure to ionizing radiation. Exp Cell Res 193: 303–309.
30. Ohyashiki T, Ohtsuka T, Mohri T (1988) Increase of the molecular rigidity of the protein conformation in the intestinal brush-border membranes by lipid peroxidation. Biochim Biophys Acta 939: 383–392.
31. Block ER (1991) Hydrogen peroxide alters the physical state and function of the plasma membrane of pulmonary artery endothelial cells. J Cell Phys 146: 362–369.
32. Alvarez JG, Touchstone JC, Blasco L, Storey BT (1987) Spontaneous lipid peroxidation and production of hydrogen peroxide and superoxide in human spermatozoa. Superoxide dismutase as major enzyme protectant against oxygen toxicity. J Androl 8: 338–348.
33. Lewis SEM, Donnelly ET, Sterling ESL, Kennedy MS, Thompson W, et al. (1996) Nitric oxide synthase and nitrite production in human spermatozoa: evidence that endogenous nitric oxide is beneficial to sperm motility. Mol Hum Reprod 2: 873–878.
34. Vernet P, Aitken RJ, Drevet JR (2004) Antioxidant strategies in the epididymis. Mol Cell Endocrinol 216: 31–39.
35. Lasso JL, Noiles EE, Alvarez JG, Storey BT (1994) Mechanism of superoxide dismutase loss from human sperm cells during cryopreservation. J Androl 15: 255–265.
36. Bilodeau JF, Chatterjee S, Sirard MA, Gagnon C (2000) Levels of antioxidant defenses are decreased in bovine spermatozoa after a cycle of freezing and thawing. Mol Reprod Dev 55: 282–288.
37. Gadea J, Sellés E, Marco MA, Coy P, Matás C, et al. (2004) Decrease in glutathione content in boar sperm after cryopreservation: Effect of the addition of reduced glutathione to the freezing and thawing extenders. Theriogenology 62: 690–701.
38. Wang AW, Zhang H, Ikemoto I, Anderson DJ, Loughlin KR (1997) Reactive oxygen species generation by seminal cells during cryopreservation. Urology 49: 921–925.
39. Chatterjee S, Gagnon C (2001) Production of reactive oxygen species by spermatozoa undergoing cooling, freezing, and thawing. Mol Reprod Dev 59: 451–458.
40. McCarthy M, Meyers S (2011) Antioxidant treatment in the absence of exogenous lipids and proteins protects rhesus macaque spermatozoa from cryopreservation-induced cell membrane damage. Theriogenology. 2011 Jul 1;76(1): 168–76.
41. van der Schans GP, Haring R, van Dijk-Knijnenburg HC, Bruijnzeel PL, den Daas NH (2000) An immunochemical assay to detect DNA damage in bovine sperm. J Androl 21: 250–257.
42. Aitken RJ, Clarkson JS (1988) Significance of reactive oxygen species and antioxidants in defining the efficacy of sperm preparation techniques. J Androl 9: 367–376.
43. Griveau JF, Le Lannou D (1997) Influence of oxygen tension on reactive oxygen species production and human sperm function. Int J Androl 20(4): 195–200.
44. Twigg J, Fulton N, Gomez E, Irvine DS, Aitken RJ (1998) Analysis of the impact of intracellular reactive oxygen species generation on the structural and functional integrity of human spermatozoa: lipid peroxidation, DNA fragmentation and effectiveness of antioxidants. Human Reprod 13: 1429–1436.
45. Donnelly ET, McClure N, Lewis SE (1999) The effect of ascorbate and alpha-tocopherol supplementation in vitro on DNA integrity and hydrogen peroxide-induced DNA damage in human spermatozoa. Mutagenesis 14: 505–512.
46. Donnelly ET, McClure N, Lewis SE (2000) Glutathione and hypotaurine in vitro: effects on human sperm motility, DNA integrity and production of reactive oxygen species. Mutagenesis 15: 61–68.
47. Koshimoto C, Mazur P (2002) The effect of the osmolality of sugar-containing media, the type of sugar, and the mass and molar concentration of sugar on the survival of frozen-thawed mouse sperm. Cryobiology 45: 80–90.
48. Katkov II, Nadezhda Katkova N, Critser JK, Mazur P (1998) Mouse spermatozoa in high concentrations of glycerol: chemical toxicity vs osmotic shock at normal and reduced oxygen concentrations. Cryobiology 37(4): 325–338.
49. Ostermeier GC, Wiles MV, Farley JS, Taft RA (2008) Conserving, distributing and managing genetically modified mouse lines by sperm cryopreservation. PLoS One 3(7): e2792.
50. Morris SM, Albright JT (1981) Superoxide dismutase, catalase, and glutathione peroxidase in the swim bladder of the physoclistous fish, *Opsanus tau* L. Cell Tissue Res 220: 739–752.
51. Westerfield M (2005) The zebrafish book. A guide for the laboratory use of zebrafish (*Danio rerio*). 5th ed., Eugene: Univ. of Oregon Press.
52. Williams JE (2000) The Coefficient of Condition of Fish. In: Schneider, James C, editor. Manual of fisheries survey methods II: with periodic updates. Ann Arbor: Michigan Department of Natural Resources, Fisheries Special Report 25.
53. Hagedorn M, Carter VL (2011) Zebrafish reproduction: revisiting the basics to increase sperm cryopreservation success. PLoS ONE 6(6): e21059.
54. Aitken RJ, Wingate JK, De Iuliis GN, McLaughlin EA (2007) Analysis of lipid peroxidation in human spermatozoa using BODIPY C11. Mol Hum Reprod (4): 203–11.
55. Baumber J, Ball BA, Linfor JJ, Meyers SA (2003) Reactive Oxygen Species and Cryopreservation Promote DNA Fragmentation in Equine Spermatozoa. J Androl 24(4): 621–628.
56. McCarthy M, Baumber J, Kass PH, Meyers S (2010) Osmotic Stress Induces Oxidative Cell Damage to Rhesus Macaque Spermatozoa. Biol Reprod 82(3): 644–651.
57. Awda BJ, Mackenzie-Bell M, Buhr MM (2009) Reactive oxygen species and boar sperm function. Biol Reprod 81(3): 553–561.
58. Halliwell B (1996) Mechanisms involved in the generation of free radicals. Pathol Biol (Paris). 44(1): 6–13.
59. Storey BT, Noiles EE, Thompson KA (1998) Comparison of glycerol, other polyols, trehalose, and raffinose to provide a defined cryoprotectant medium for mouse sperm cryopreservation. Cryobiology 37(1): 46–58.
60. Hagedorn M, Ricker J, McCarthy M, Meyers SA, Tiersch TR, et al. (2009) Biophysics of zebrafish (*Danio rerio*) sperm. Cryobiology 58(1): 12–9.

PERMISSIONS

All chapters in this book were first published in PLOS ONE, by The Public Library of Science; hereby published with permission under the Creative Commons Attribution License or equivalent. Every chapter published in this book has been scrutinized by our experts. Their significance has been extensively debated. The topics covered herein carry significant findings which will fuel the growth of the discipline. They may even be implemented as practical applications or may be referred to as a beginning point for another development.

The contributors of this book come from diverse backgrounds, making this book a truly international effort. This book will bring forth new frontiers with its revolutionizing research information and detailed analysis of the nascent developments around the world.

We would like to thank all the contributing authors for lending their expertise to make the book truly unique. They have played a crucial role in the development of this book. Without their invaluable contributions this book wouldn't have been possible. They have made vital efforts to compile up to date information on the varied aspects of this subject to make this book a valuable addition to the collection of many professionals and students.

This book was conceptualized with the vision of imparting up-to-date information and advanced data in this field. To ensure the same, a matchless editorial board was set up. Every individual on the board went through rigorous rounds of assessment to prove their worth. After which they invested a large part of their time researching and compiling the most relevant data for our readers.

The editorial board has been involved in producing this book since its inception. They have spent rigorous hours researching and exploring the diverse topics which have resulted in the successful publishing of this book. They have passed on their knowledge of decades through this book. To expedite this challenging task, the publisher supported the team at every step. A small team of assistant editors was also appointed to further simplify the editing procedure and attain best results for the readers.

Apart from the editorial board, the designing team has also invested a significant amount of their time in understanding the subject and creating the most relevant covers. They scrutinized every image to scout for the most suitable representation of the subject and create an appropriate cover for the book.

The publishing team has been an ardent support to the editorial, designing and production team. Their endless efforts to recruit the best for this project, has resulted in the accomplishment of this book. They are a veteran in the field of academics and their pool of knowledge is as vast as their experience in printing. Their expertise and guidance has proved useful at every step. Their uncompromising quality standards have made this book an exceptional effort. Their encouragement from time to time has been an inspiration for everyone.

The publisher and the editorial board hope that this book will prove to be a valuable piece of knowledge for researchers, students, practitioners and scholars across the globe.

LIST OF CONTRIBUTORS

Jianfeng Xiao, Zhiling Li, Yongcui Zhou, Hong Lin, Xiaoyan Wu, Man Chen and Wanfen Xiao
Reproductive Center, The First Affiliated Hospital of Shantou University Medical College, Shantou University, Shantou, Guangdong, People's Republic of China

Yanmei Liu
Reproductive Center, Women and Children Healthcare Hospital, Qinzhou, Guangxi, People's Republic of China

Suzanne Crowe
Division of Cardiac Surgery, University of Ottawa Heart Institute, Ottawa, Canada

Tanja Sofrenovic, Jenelle Marier, Erik J. Suuronen and Drew Kuraitis
Division of Cardiac Surgery, University of Ottawa Heart Institute, Ottawa, Canada
Department of Cellular and Molecular Medicine, University of Ottawa, Ottawa, Canada

Kimberly McEwan
Division of Cardiac Surgery, University of Ottawa Heart Institute, Ottawa, Canada
Department of Mechanical Engineering, University of Ottawa, Ottawa, Canada

Robbie Davies
Department of Statistics, University of Ottawa Heart Institute, Ottawa, Canada

Yusuke Makino, Toshio Kukita and Takayoshi Yamaza
Department of Molecular Cell Biology and Oral Anatomy, Graduate School of Dental Science, Kyushu University, Fukuoka, Japan

Lan Ma
Department of Molecular Cell Biology and Oral Anatomy, Graduate School of Dental Science, Kyushu University, Fukuoka, Japan
Department of Pediatric Dentistry, Graduate School of Dental Science, Kyushu University, Fukuoka, Japan

Haruyoshi Yamaza, Yoshihiro Hoshino and Kazuaki Nonaka
Department of Pediatric Dentistry, Graduate School of Dental Science, Kyushu University, Fukuoka, Japan

Kentaro Akiyama and Songtao Shi
Center for Craniofacial Molecular Biology, Herman Ostrow School of Dentistry of USC, University of Southern California, Los Angeles, California, United States of America

Guangtai Song
Department of Pedodontics, School of Stomatology, Wuhan University, Wuhan, China

Céline Borlon, Peter van den Eede and Annette Erhart
Unit of Malariology, Institute of Tropical Medicine Antwerp, Belgium

Florian Noulin
Unit of Malariology, Institute of Tropical Medicine Antwerp, Belgium
Stem cells institute, Catholic University Leuven, Belgium

Umberto D'Alessandro
Unit of Malariology, Institute of Tropical Medicine Antwerp, Belgium
Medical Research Council Unit, Fajara, The Gambia

Luc Boel
Unit of Immunology, Institute of Tropical Medicine Antwerp, Belgium

Catherine M. Verfaillie
Stem cells institute, Catholic University Leuven, Belgium

Yujie Zou and Jing Yang
Reproductive Medicine Center, Renmin Hospital of Wuhan University, Wuhan, China

Tailang Yin
Reproductive Medicine Center, Renmin Hospital of Wuhan University, Wuhan, China
Center of Reproductive Medicine, Department of Obstetrics and Gynecology Peking University Third Hospital, HaiDian District, Beijing, China

Shijing Chen and Weihua Huang
Key Laboratory of Analytical Chemistry for Biology and Medicine (Ministry of Education), College of Chemistry and Molecular Sciences, Wuhan University, Wuhan, China

Bianca Lawson, Simon Clulow, Michael J. Mahony and John Clulow
School of Environmental and Life Sciences, University of Newcastle, Callaghan, New South Wales, Australia Abstract

Ortal Mizrahy, Maya Bar-Dolev, Shlomit Guy and Ido Braslavsky
The Institute of Biochemistry, Food Science, and Nutrition, The Robert H. Smith Faculty of Agriculture, Food and Environment, The Hebrew University of Jerusalem, Rehovot, Israel

Munekazu Naito, Shuichi Hirai, Hayato Terayama, Naoyuki Hatayama and Masahiro Itoh
Department of Anatomy, Tokyo Medical University, Tokyo, Japan

Makoto Mihara
Department of Plastic and Reconstructive Surgery, Graduate School of Medicine, University of Tokyo, Tokyo, Japan

Shogo Hayashi
Medicinal Education Center, School of Medicine, Aichi Medical University, Aichi, Japan

Masayuki Matsushita
Molecular and Cellular Physiology, University of the Ryukyus, Okinawa, Japan

Éva Váradi, Barbara Végi, Krisztina Liptói and Judit Barna
Institute for Small Animal Research and Co-ordination Centre for Gene Conservation, Gödöllő , Hungary

Hang Wun Raymond Li, Vivian Chi Yan Lee, Estella Yee Lan Lau, William Shu Biu Yeung, Pak Chung Ho and Ernest Hung Yu Ng
Department of Obstetrics and Gynaecology, Centre of Assisted Reproduction and Embryology, The University of Hong Kong, Queen Mary Hospital, Hong Kong

Sandeep Gupta, Shunan Li, Md. Joynal Abedin, Kajohnsak Noppakun, Lawrence Wang and Tarundeep Kaur
Department of Medicine, University of Minnesota Medical School, Minneapolis, Minnesota, United States of America

Clifford J. Steer
Department of Medicine, University of Minnesota Medical School, Minneapolis, Minnesota, United States of America
Department of Genetics, Cell Biology and Development, University of Minnesota Medical School, Minneapolis, Minnesota, United States of America

Behzad Najafian
Department of Laboratory Medicine and Pathology, University of Minnesota Medical School, Minneapolis, Minnesota, United States of America

Cecília M. P. Rodrigues
Research Institute for Medicines and Pharmaceutical Sciences (iMed.UL), Faculty of Pharmacy, University of Lisbon, Lisbon, Portugal

Marc Suquet, Christian Mingant, Myrina Boulais, Isabelle Queau and Dominique Ratiskol
Ifremer, UMR 6539, PFOM Department, Station Expérimentale d'Argenton, Argenton, France

Catherine Labbé
INRA, UR 1037, LPGP, Rennes, France

Sophie Puyo, Benjamin Quittet and Pierrick Haffray
SYSAAF, LPGP, Rennes, France

Blandine Diss
Satmar, Barfleur, France

Lee D. Spate, Clifton N. Murphy, Randall S. Prather
Division of Animal Sciences, University of Missouri, Columbia, Missouri, United States of America

Maria Cristina Vinci, Francesca Prandi, Rosaria Santoro and Maurizio Pesce
aboratorio di Ingegneria Tissutale Cardiovascolare, Centro Cardiologico Monzino-IRCCS, Milan, Italy

Giulio Tessitore and Gianluca Polvani
Dipartimento di Scienze Cardiovascolari, Universitá degli Studi di Milano, Milan, Italy

Laura Castiglioni, Francesca Colazzo and Luigi Sironi
Dipartimento di Farmacologia, Universitá di Milano, Milan, Italy

Monica Soncini and Filippo Consolo
Dipartimento di Bioingegneria, Politecnico di Milano, Milan, Italy

Barbara Micheli
Banca dei Tessuti Cardiovascolari della Regione Lombardia, Milan, Italy

Stephanie E. Woods, Peimin Qi, Elizabeth Rosalia, Tony Chavarria, Allan Discua, John Mkandawire, James G. Fox and Alexis García
Transgenic Core Facility, Division of Comparative Medicine, Massachusetts Institute of Technology, Cambridge, Massachusetts, United States of America

Alexandre Kerkis, Cícera M. Gomes and Camila F. Oliveira
Laboratory of Genetics, Butantan Institute, Sao Paulo, SP, Brazil

Nelson F. Lizier and Irina Kerkis
Laboratory of Genetics, Butantan Institute, Sao Paulo, SP, Brazil
Departament of Morphology of Federal University of Sao Paulo (UNIFESP), Sao Paulo, SP, Brazil

Josimeri Hebling
Department of Orthodontics and Pediatric Dentistry of State University of Sao Paulo (UNESP), Araraquara, SP, Brazil

Arnold I. Caplan
Skeletal Research Center, Department of Biology of Case Western Reserve University, Cleveland, Ohio, United States of America

Farman Ali and David A. Wharton
Department of Zoology, University of Otago, Dunedin, New Zealand

Jayanthi Nadarajan and Hugh W. Pritchard
Seed Conservation Department, Royal Botanic Gardens Kew, Wellcome Trust Millennium Building, Ardingly, West Sussex, United Kingdom

Craig A. Gedye, Joshua Paterson, Alannah Smrke, Harleen Saini, Danylo Sirskyj, Christopher Go, Jenny Ho, Maria Kondratyev, Twan van den Beucken, Michael Jewett Jean Wang and Robert Rottapel
Princess Margaret Cancer Centre, University Health Network, Toronto, ON, Canada

Ali Hussain, Keira Pereira, Nazleen Lobo, Jocelyn Stewart and Mauricio Medrano
Dept. of Medical Biophysics, University of Toronto, Toronto, ON, Canada

Benjamin Neel, Laurie E. Ailles and Bradly Wouters
Princess Margaret Cancer Centre, University Health Network, Toronto, ON, Canada
Dept. of Medical Biophysics, University of Toronto, Toronto, ON, Canada

Elzbieta Hyatt, Julie Yuan, Stevan Lauriault, Peter Dirks, Cynthia J. Guidos and Jayne Danska
Hospital for Sick Children Research Institute, Toronto, ON, Canada

Manuela Mandl, Susanne Schmitz and Michael Hristov
Institute for Cardiovascular Prevention, Ludwig-Maximilians-University (LMU), Munich, Germany

Christian Weber
Institute for Cardiovascular Prevention, Ludwig-Maximilians-University (LMU), Munich, Germany
Munich Heart Alliance, Munich, Germany

Marina Buzzi andv Sabrina Manara
Cardiovascular Tissue Bank of Emilia-Romagna, Azienda Ospedaliero-Universitaria Policlinico S. Orsola-Malpighi, Bologna, Italy

Anna Guarino, Luca Dainese and Gianluca Polvani
Cardiovascular Tissue Bank of Lombardia, Centro Cardiologico Monzino, Milan, Italy

Claudio Gatto and Jana D'Amato Tóthová
Research and Development department, AL.CHI. MI.A. S.r.l., Ponte San Nicoló , Italy

José Julián Garde
SaBio, Instituto de Investigación en Recursos Cinegéticos (IREC), CSIC-UCLM-JCCM, Campus Universitario, Albacete, Spain

José Luis Ros-Santaella
SaBio, Instituto de Investigación en Recursos Cinegéticos (IREC), CSIC-UCLM-JCCM, Campus Universitario, Albacete, Spain
Department of Animal Science and Food Processing, Faculty of Tropical AgriSciences, Czech University of Life Sciences, Prague, Czech Republic

Álvaro Efrén Domínguez-Rebolledo
Instituto Nacional de Investigaciones Forestales, Agrícolasy Pecuarias (INIFAP), Mocochá, Yucatán, México

Oleksandr Gryshkov, Denys Pogozhykh, Nicola Hofmann and Birgit Glasmacher
Institute for Multiphase Processes, Leibniz Universitaet Hannover, Hannover, Germany

Olena Pogozhykh and Thomas Mueller
Institute for Transfusion Medicine, Hannover Medical School, Hannover,
Germany

Derek G. Pyne, Jun Liu and Yu Sun
Department of Mechanical and Industrial Engineering, University of Toronto, Toronto, Ontario, Canada

Mohamed Abdelgawad
Department of Mechanical Engineering, Assiut University, Assiut, Egypt

Mary Hagedorn and Virginia L. Carter
Department of Reproductive Sciences, Smithsonian Conservation Biology Institute, Washington, District of Columbia, United States of America
Hawaii Institute of Marine Biology, University of Hawaii, Kaneohe, Hawaii, United States of America

Megan McCarthy and Stuart A. Meyers
Department of Anatomy School of Veterinary Medicine, University of California at Davis, Davis, California, United States of America

Index